James W. S. Longhurst (Ed.)

Acid Deposition

Origins, Impacts and Abatement Strategies

With 130 Figures

Springer-Verlag
Berlin Heidelberg New York
London Paris Tokyo
Hong Kong Barcelona
Budapest

Dr. James W.S. Longhurst
Acid Rain Information Centre
Department of Environmental and Geographical Studies
Manchester Polytechnic
Chester Street
Manchester Ml 5GD, England

On this topic, J.W.S. Longhurst also edited the volume "Acid Deposition: Sources, Effects and Controls", published jointly by the British Library and Technical Communications in 1989.

ISBN-13:978-3-642-76475-2 e-ISBN-13:978-3-642-76473-8
DOI: 10.1007/978-3-642-76473-8

Softcover reprint of the hardcover 1st edition 1991

31/3145-543210 – Printed on acid free paper

Preface

The subject of acid deposition remains one of the most urgent of our contemporary environmental problems. Research programmes are continually redefining our understanding of cause and effect and hence the continuing need for a timely and authoritative series addressing these issues. This volume seeks to review and define our contemporary understanding of acid deposition by reference to new international data and as a consequence assist the definition of our future research requirements and policy developments. International contributions to the volume are drawn from the Federal Republic of Germany, the U.S.A., Canada, Brazil, Switzerland, Austria, Israel, France and the United Kingdom. Some of these nations have experienced acid deposition on a regional scale for considerable periods of time; for others the phenomenon is an emerging problem.

This collection of papers has been compiled by invitation to eminent members of the acid deposition research community and by selection from a carefully targeted call for papers.

It is primarily designed to meet the needs of researchers, lecturers and postgraduate students in environmental disciplines and for environmental policy makers. It is of interest to professionals in related disciplines and essential as a reference text for libraries.

The volume is divided into four broad themes : **Emissions, chemistry and deposition; Ecosystem effects** (freshwater, soils and forest systems); **Effects on structural materials; Mitigation, control and management.** Each of these sections provides an overview of contemporary understanding, presents new experimental or field evidence and provides guidance for our future research agenda.

October 1990

James Longhurst
Director, Acid Rain Information Centre
Manchester Polytechnic

CONTENTS

3. EFFECTS ON STRUCTURAL MATERIALS

4. MITIGATION, CONTROL AND MANAGEMENT

Introduction

Acid deposition is a global phenomenon with its effects most clearly experienced in the industrial nations of the northern hemisphere. However as papers in this volume illustrate, the effects are now occuring in developing nations as well. The precursors to acid deposition are emitted to the atmosphere from natural and anthropogenic sources, of the latter transportation and power generation are the most important sources. Transport and transformation reactions in the atmosphere distribute the emissions over large areas.

Initially, effects to freshwater systems attributed, at least in part, to acid deposition were reported from Sweden, Norway and Canada. Now, however, effects are reported for both freshwater and terrestrial components of ecosystems in a large number of nations. Acid deposition may be considered an additional stress factor for terrestrial systems. Materials of technical, economic and cultural importance are also at risk from acid deposition. The immediate importance of these problems tends to indicate that the phenomenon of acid deposition is of recent origin, however, the subject has a relatively long history. As early as 1852 R. A. Smith collected and analysed rainwater in north west England, neologised the term acid rain and described its effects upon terrestrial ecosystems and materials. At this time, sulphur was the major pollutant and the effects of acid deposition were most clearly experienced at the meso scale.

The emergence of acid deposition upon the global political agenda is far more recent, and followed the work of the Swede, S. Oden in the latter part of the 1960's. He identified a relationship between deposition of sulphur and fresh water acidification on a regional scale where sensitive geology and climatology were coincident with pollutant source areas. This work became the basis of the Swedish government's case study to the United Nations Conference on the Human Environment held in Stockholm in 1972. From this date the intensive study of the processes and causes of emission, conversion, deposition and effects of acidifying pollutants and their control has been driven by a political impetus. At first this impetus came from only a few nations but as time progressed and the real scale, and interlinkage of the problem with other environmental variables emerged, more and more nations entered into national and international research programmes.

Through such bodies as the United Nations Economic Commission for Europe and the Commission of the European Community international agreements to limit the emission of acidifying air pollutants, to measure their distribution, conversion and deposition, and to assess the scale and impact of deposition upon sensitive systems have been made. As this work has developed many of the original conceptions regarding acid deposition have been reassessed in the light of field evidence. The importance of sulphur as an acidifying pollutant has been demonstrated but the interlinked role of photochemical oxidants, nitrogen oxides, reactive hydrocarbons and ammonia in explaining acid deposition has emerged over this period. Concomitantly the importance of cultural factors has been demonstrated in ameliorating or exacerbating effects as has the timescale over which long range transport of air pollutants has occurred.

Emissions, chemistry and deposition

Two papers in this section reflect the global nature of our concern with acid deposition Y. Mamane and J. Gottlieb (Israel) discuss the influences upon the chemistry of precipitation in Israel. For Europeans and North Americans this provides an interesting comparison with the situation in Europe and North America where technological, ecological and geological circumstances are quite different. They report an increase in the acidity of precipitation in Israel which may be due to anthropogenic activities. Further comparison of national circumstances is provided by the review of acid precipitation research in Brazil by I. Ovalle and E. Filho (Brazil). As less developed nations industrialise there will be a risk that emissions of acidifying pollutants will also increase. Developed countries must make their experience freely available to ensure preventative actions are taken.

Whilst our understanding of the national scale has increased in recent years a significant number of questions remain unanswered. The importance and rate of marine derived sulphur deposition is one such area reviewed by C. Hewitt and B. Davison (U. K.). The role and importance of acidic aerosols in atmospheric chemistry and their contribution to biological effects is an emerging area requiring further work. F. Lipfert (U.S.A.) discusses the current position from an American perspective and identifies a number of research needs particularly with respect to human health.

Ecosystem effects : soils, freshwater, and forest systems

The mechanisms by which aluminium is released to soil solutions and drainage waters are reviewed by B. Bache and G. Townsend (U.K.). Their work on flow rates and pathways in contrasting soils demonstrates that both concentration and speciation of aluminium differ from predictions based upon equilbrium conditions and they propose a generalised model to account for their observations. The timescale and magnitude of acidification of surface waters in the U.K. is reviewed by A. Tickle (U.K.) who describes palaeoecological studies indicating the post 1850 timescale of acidification. Tickle concludes that currect emission reduction plans are insufficient to halt acidification in many areas of the U.K. The reduction in agricultural liming in the catchments of the Rivers Esk and Duddon (English Lake District) may be a major factor in the development of acid episodes and consequent fish kills. The impact on water quality of controlled application of lime to the Esk catchment is described by M. Diamond et al (U.K.).

The current status of forest decline in Central Europe is reviewed by H. Essman (FRG) and W. Zimmermann (Switzerland). The effects of acid deposition on high altitude forests in the U.S.A. are reported by M. Adams and C. Eager (U.S.A.) who evaluate the hypotheses for spruce - fir forest decline. Theoretical, experimental and field studies of the effect of acid deposition upon forest health are evaluated by J. Innes (U.K.) with particular reference to the U.K.

The morphology of beech crowns in relation to forest damage is reviewed by A. Roloff (FRG). An externally induced stress factor in a forest system is the leaching of minerals from the canopy to mediate acid deposition. Mineral leaching has a series of indirect consequences for a tree which are difficult to separate from the effects of acid deposition. These are discussed by S. Leonardi (FRG) who provides comparative data and experimental evidence to evaluate the physiological importance of leaching processes. Nutrient cycling responses of hardwood forests in the Turkey Lakes watershed of Ontario are described by N. Foster and P. Hazlett (Canada) in relation to reductions in sulphate deposition. Nutritional and physiological disturbances in firs growing in the Pyrenees are discussed by F. Fromard et al. (France). Analysis of the mineral content of leaves indicates potassium deficiency in trees growing on schistose materials and deficiency of manganese and iron in trees on limestone. Biochemical analysis indicates reduced carbon dioxide fixation and modification of seasonal patterns of starch and soluble glucides. Metabolic modifications ascribable to the biochemical and physiological effects of atmospheric pollution on *Picea abies* are discussed by A. Santerre and V. Villaneuva (France). Both these studies on fir trees form part of the extensive French DEFORPA research programme on forest decline.

Effects on structural materials

One effect of acidic air pollutants recognised by R. A. Smith as early as the middle of the nineteenth century is the decay of structural materials. In this volume B. Smith et al (U.K.) review the background and local contributions to acid deposition in Northern Ireland and their relative impact on building stone decay. Located to the northwest of mainland Europe, Northern Ireland provides an ideal environment to examine natural and anthropogenic causes of stone decay. The surface chemistry involved in the weathering of limestone is reviewed by R. Bradley (U.K.) who presents new data illustrating the importance of atmospheric sulphur species, particularly through surface sulphate formation, in the deterioration of limestone.

Mitigation, control and management

As the scientific understanding of cause and effect has improved so attention has focused upon both the technologies available to control emissions of acidifying pollutants from both fixed plant and mobile sources and the most effective means to introduce abatement strategies. The control of oxides of nitrogen emissions from fossil fuel combustion plants is reviewed by A. Hjalmarsson (U.K.) who outlines the main approaches available, indicates their applicability, state of development and commercial application whilst D. C. Gibbs (U.K.) provides a comparative study of the sulphur dioxide emission control programmes for large combustion plant in the U.K. and the U.S.A.

The development and assessment of strategies to combat acid deposition in Europe are reviewed by R. Shaw (Austria) whilst M. Chadwick and J. Kuylenstierna (U.K.) review the concept of a critical load based abatement strategy. Such a strategy based upon ecologically defined limits may be one of the most effective policy responses to acid deposition.

The papers published in this volume provide a significant contribution to our knowledge of the sources and effects of this multi - faceted environmental problem.

James Longhurst,
Acid Rain Information Centre, Manchester Polytechnic, U.K.

ARIC is supported by the Association of Greater Manchester Authorities and the Department of the Environment.

1

EMISSIONS, CHEMISTRY AND DEPOSITION

CHEMISTRY OF PRECIPITATION IN HAIFA, ISRAEL, 1981-1989

Y. Mamane, and J. Gottlieb
Environmental and Water Resources Engineering,
Technion–Israel Institute of Technology,
Haifa 32000, Israel

Abstract

Rain event samples have been collected in Haifa, Israel, since the winter of 1981-1982. The sampling and analysis program, based on WMO recommendations for background networks, includes the following parameters: precipitation amount, pH, $SO_4^{=}$, NO_3^-, Cl^-, NH_4^+, Na^+, K^+, Ca^{++}, Mg^{++}, conductivity and alkalinity. The sampling was performed manually, and the analysis was based on wet chemistry for ions and atomic absorption for metals. Analysis of 185 rain samples collected during eight hydrological years, 1981-1989, showed that the average pH was 5.3 (± 0.3). The variability in pH was very high with 30% of the rain events below 5.6 and 27% above pH of 7.0. Natural sources, sea salt and soil carbonates, are the main contributors to rain chemistry. However, cases with low pH, that suggest the impact of manmade emissions, were also observed.

Key Words: precipitation chemistry, alkalinity, sea salt, calcium carbonates, acidic precipitation, Mediterranean Sea.

Introduction

Rain near arid regions, such as in the Midwest and Southwest United States, tends to show net alkalinity due to the washout of alkaline soil components (Felly and Liljestrand, 1983; Doty and Semonin, 1984). Studies in India reported pH levels of 6.1 to 8.4, higher than the so-called 5.6 reference level, during the monsoon summer precipitation (Subramanian and Saxena, 1980; Khemani et al., 1985). In Israel, earlier studies reported average pH of 6.5 (± 0.8) for the winters of 1979 to 1982 (Mamane, 1987; Mamane et al., 1987). Other locations along the Mediterranean Sea also reported high variations in rain chemistry attributed to acidity and alkalinity transported from manmade and natural sources (Loye-Pilot et al., 1986; Glavas, 1988; Ezcurra et al., 1988 and Dikaiakos et al., 1990). Large variations in rain pH were also observed in other regions affected by the desert belt such as Japan and Ivory Coast in Africa (Hara et al., 1989 and Lacaux et al., 1987, respectively).

In this paper rain chemistry data collected during 1981-1989 in Haifa are presented, with the purpose of identifying the major sources that determine the chemistry of precipitation in the East Mediterranean.

Experimental

Precipitation samples were collected daily at the Technion site on the Mount Carmel in Haifa near the Mediterranean Coast. A simple collector made of a funnel and a polyethylene bottle was mounted in a cabinet. The bottle and the funnel were washed every day to avoid settling of dust that accompanies rain in Israel. The samples were analyzed on - site immediately after collection, using procedures recommended by the World Meteorological Organization (WMO) , 1978. The following chemical and instrumental methods were employed:

Precipitation - Volumetric
pH - with glass electrode
$SO_4^{=}$ - Barium precipitation
NO_3^- - Hydrazine reduction
Cl^- - Diphenyl carbazone method
NH_4^+ - Blue complex with hypochlorite
Ca^{++} and Mg^{++} - Volumetric titration
Na^+ and K^+ - Atomic absorption
Alkalinity - Titration
Conductivity - Conductivity cell

Each rain event, collected on a daily basis (excluding weekends when rain event may consist of two-day sample) from morning to morning, was analyzed for the above twelve parameters. The hydrogen ion concentration was calculated from pH values. Average of these variables took into account precipitation volume (volume - weighted).

Results and Discussion

Table 1 summarized the chemical analysis of 185 samples collected from November 17, 1981 to March 16, 1989. Volume - weighted average concentrations of the major ions are expressed in microequivalent per liter (μeq/l) as well as standard deviation, minimum and maximum. Table 1 leads to the following points :

Table 1. Summary of rain chemistry data for precipitation collected in Haifa, Israel, during eight hydrological years (1981-1989)

Variable	Weighted[a] Average	Standard Deviation	Minimum	Maximum
pH	5.3	0.3	4.2	9.5
Rain (mm)	21.7	16.4	2.0	110.0
Alkalinity[b]	131.4	141.5	5.0	739.0
Cl^-	312.1	208.7	59.0	1190.0
$SO_4^{=}$	182.6	135.5	41.0	983.0
NO_3^-	14.1	13.6	1.0	90.0
NH_4^+	18.2	16.5	1.0	93.0
K^+	18.5	33.1	1.0	300.0
Mg^{++}	113.4	88.0	8.0	477.0
Na^+	294.8	200.4	20.0	1090.0
Ca^{++}	208.2	174.3	8.0	983.0

a - Rain volume weighted average of 185 samples

b - Alkalinity and ions concentration are given in μeq/l

(a) The relative variation of most ions, expressed as the ratio of standard deviation to average value is remarkably high.
(b) Sea salt contribution (Na^+, Cl^-, Mg^{++}, K^+, and $SO_4^=$) is very large.
(c) Soil contribution is also very high (Ca^{++}, Mg^{++}, $CO_3^=$ as expressed by alkalinity, and $SO_4^=$).

Precipitation pH in Haifa has a volume - weighted average of 5.3, with a low pH of 4.2 obtained on November 3, 1986 to a high pH of 9.2 sampled on February 8, 1984. Glavas (1988) and recently Dikaiakos et al. (1990) also reported large variability of pH in Patras and Athens, Greece, ranging from about 4.1 to 8.8. Similar results were reported for northern Spain by Ezcurra et al. (1988). Previous rain chemistry data (1979-1982) in Israel pointed at similar pH range from 4.1 to 8.7; although the average pH was significantly higher: 6.5 (Mamane, 1987). In past studies the rain was collected on an event basis (4 to 7 days) and not daily. Often that caused deposition of carbonates, associated with cold fronts and dust storms, onto the collection bottle. Daily collection of rain, as has been done in this study, minimized that effect.

Figure 1 is a plot of the rain pH for the 185 rain daily samples collected from 1981 to 1989. The rain pH data can be grouped into three sections: the acidic range (pH < 5.6) - 30% of the cases, the moderate basic section (5.6 < pH < 7.0) - 40%, and the alkaline region (pH > 7.0) - 30%. Note that during the first winter (1981-1982), the pH range (not the average) is

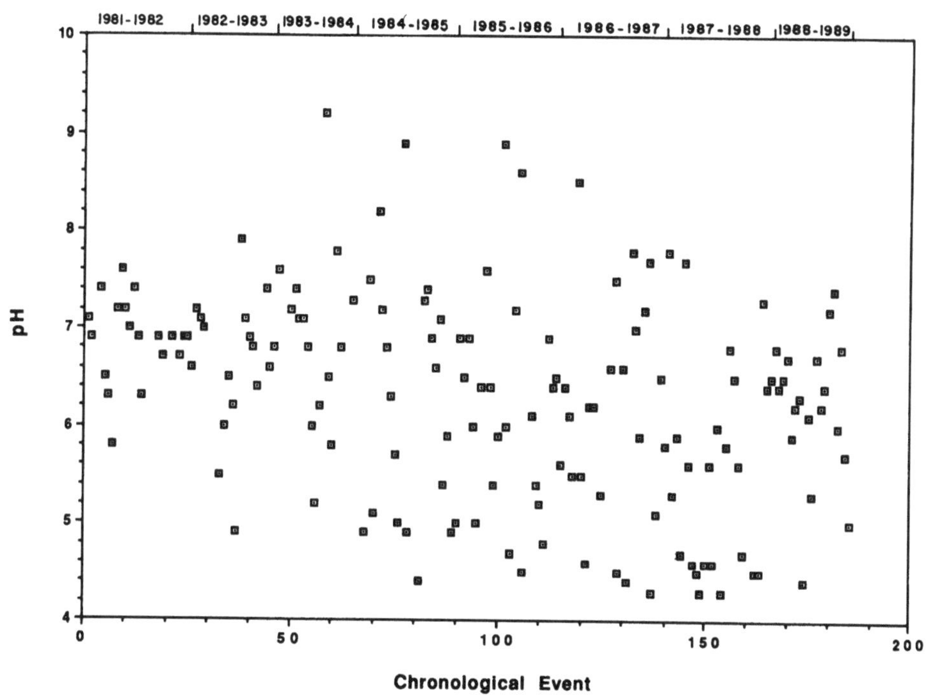

Figure 1. A plot of pH values for rain samples collected from 1981 to 1989.

significantly narrower than in later years. This may be associated with a smaller number of rain events in that year, more dry spells between rains, and therefore more airborne soil minerals (mostly carbonates) that are capable of buffering any acidity in the air. Figure 2 is an histogram of the grouped pH data showing large variation in H^+, over five orders of magnitude.

Table 2 is the correlation matrix of the main ions. The following points are made:

(a) Sea salt components are intercorrelated (0.98 for Na^+/Cl^- and 0.47 for Mg^{++}/Cl^-.
(b) Calcium ion is not only correlated with alkalinity (0.80) but also with sulfates (0.68).
(c) Sulfates are negatively correlated with the hydrogen ion (-0.28).
(d) Ammonium and nitrate ions do not show any correlation with the other rain chemical components.

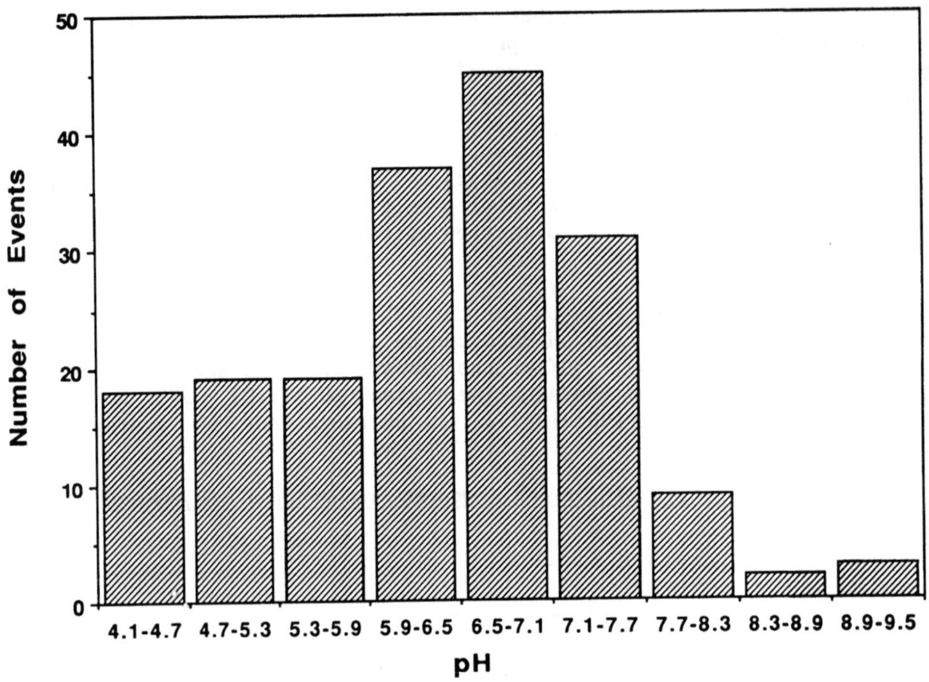

Figure 2. Histogram of pH data pointing at a large pH range from 4.2 to 9.2.

Table 2. Correlation matrix for rain chemistry data

Variable	Alk	Cl^-	$SO_4^=$	NO_3^-	NH_4^+	K^+	Mg^{++}	Na^+	Ca^{++}	H^+
Alk	1.00									
Cl^-	0.36	1.00								
$SO_4^=$	0.41	0.32	1.00							
NO_3^-	0.06	-0.20	0.03	1.00						
NH_4^+	0.20	-0.04	0.15	0.09	1.00					
K^+	0.14	0.17	0.26	-0.08	0.07	1.00				
Mg^{++}	0.44	0.47	0.63	0.06	-0.05	0.11	1.00			
Na^+	0.38	0.98	0.35	-0.22	0.00	0.15	0.44	1.00		
Ca^{++}	0.80	0.33	0.68	0.13	0.13	0.06	0.44	0.32	1.00	
H^+	-0.37	0.00	-0.28	-0.14	-0.25	-0.16	-0.25	0.00	-0.31	1.00

Figure 3 is a plot of Cl^- versus Na^+ for all daily rain events obtained in this study. As expected, a linear dependence is visible; it is indicative of the contribution of sea salt to rain composition. The Cl^-/Na^+ ratio in Haifa is not different from the ratio of 1.16 (concentrations are in eq/l) reported for other coastal sites (Lebowitz and de Pena, 1985). However, the values obtained in Israel are some of the highest ever

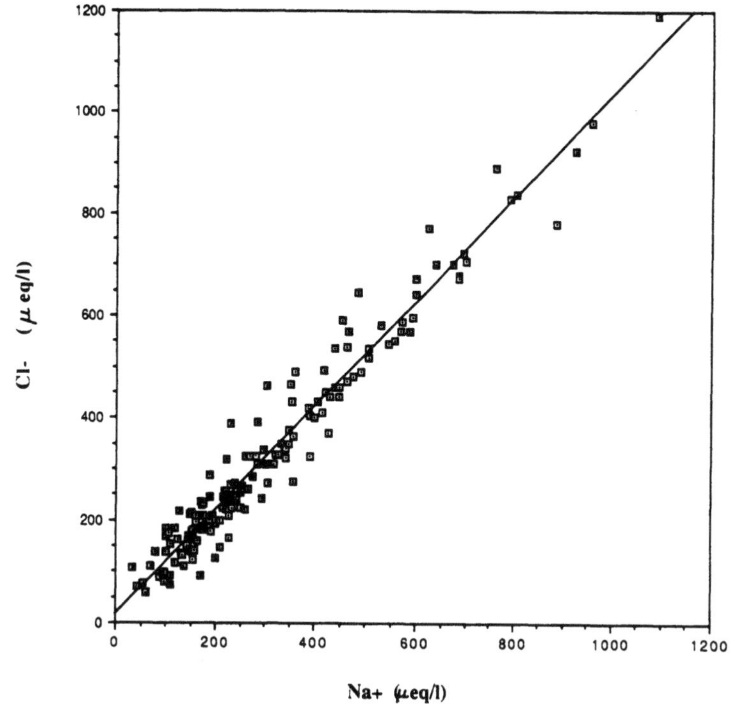

Figure 3. A scatter diagram of Cl^- versus Na^+ in rainwater.

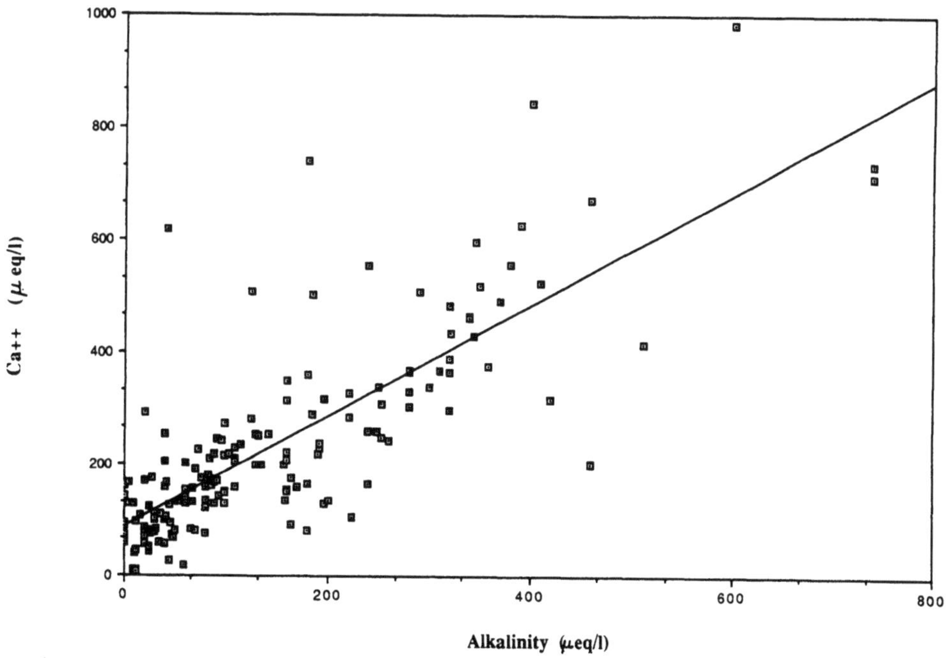

Figure 4. Calcium versus alkalinity in rainwater. The correlation between the two is fairly high, 0.80.

reported. Figure 3 suggests also that no excess of Cl^- (from industrial emissions) are observed in Haifa.

Calcium concentration shown in Table 1, 208.2 μeq/l, is one of the highest values reported for rainwater. Ca^{++} correlated fairly well with alkalinity as shown in Figure 4. For the rain pH measured in Haifa and assuming that sulfate, nitrate and chloride are the main ions, alkalinity is a measure of bicarbonate concentration. As explained in other studies (Dayan, 1986; Mamane, 1987; Mamane et al., 1987), alkaline minerals are linked with precipitation in East Mediterranean. These minerals contain a large fraction of calcites that are washed out by falling rain. Ca^{++} was also correlated with the sulfate ion as seen in Figure 5, a plot of excess sulfate versus Ca^{++}. Possibly the sulfates in rain are related to soil minerals, and not to sulfuric acid or ammonium sulfates.

In Northwest and Central Europe and Northeast United States , where acidity is governed by oxides of sulfur and nitrogen, the main neutralizing agent is ammonia (Schuurkes et al., 1988). In Israel, Ca^{++} is the most dominant neutralizing agent. Similar results were obtained by Moreira et al. (1989) for the coast of Brazil.

Nitrates in rainwater are probably of anthropogenic origin, since they do not show any correlation with sources of natural origin such as sea salt and soil minerals.

A strong correlation of 0.992 was obtained between the sum of anions and the sum of cations. The ratio between the two was almost equal to one. The agreement between anions and cations obtained by different measurements is a necessary condition for good quality data set.

The ionic ratios for sea water (concentrations are in μeq/l) for Cl^-/Na^+, $SO_4^=/Na^+$, Mg^{++}/Na^+, Ca^{++}/Na^+ and K^+/Na^+ are 1.17, 0.12, 0.22, 0.044 and 0.021 , respectively. In rain, the presence of soil minerals in air and in clouds could modify those ratios. If Na^+ is chosen as a tracer, then the sea salt contribution to rain chemistry amounts to 19.4% for $SO_4^=$, 57.2% for Mg^{++}, 6.2 for Ca^{++}, and 33.5% for K^+. Na^+ is assumed to be solely contributed by marine sources. The contributions of sea spray, clay minerals, and calcite may be estimated from concentrations of the respective elements in rainwater. Mg^{++} is found in sea water and in clay minerals, and to some extent in calcite. Thus Mg^{++} in clay is estimated to be given by:

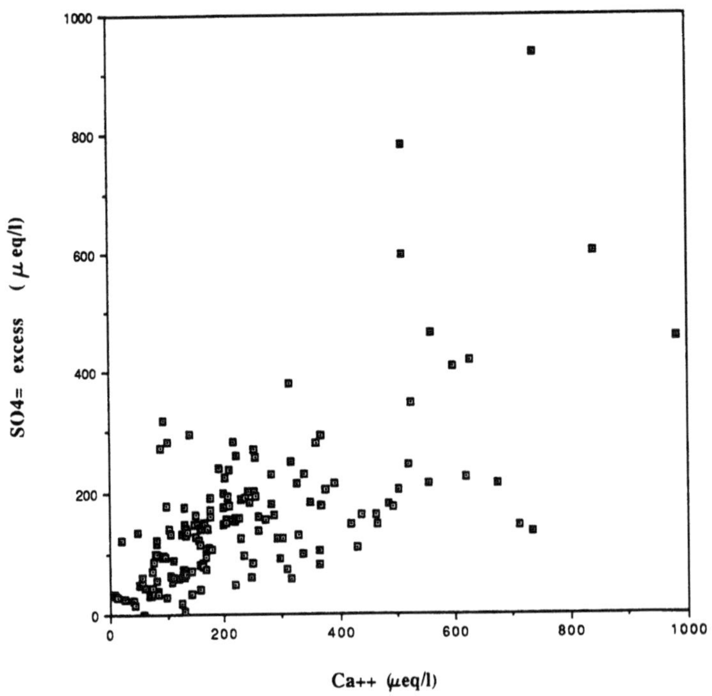

Figure 5. A plot of excess sulfates versus calcium. Ca^{++} correlates with excess sulfate, although the correlation is quite weak.

$$[Mg]_{sea\ water} = (Mg/Na)_{sea\ water} * [Na]_{rain} = 64.8\ \mu eq/l$$

$$[Mg]_{clay} = [Mg]_{rain} - [Mg]_{sea\ water} = 48.6\ \mu eq/l$$

In Northern Spain and other Mediterranean regions where rain is not always dominated by soil minerals, Mg^{++} originates from sea salt. Ezcurra et al. (1988) obtained a correlation of 0.81 for the Mg^{++}/Cl^- pair in comparison to 0.47 in our case. Calcium of marine origin was estimated to be only 6.2% or 12.9 µeq/l. On the basis of crustal abundances of elements (Ichikuni, 1978), Ca^{++} is twice the amounts of Mg^{++} in clay minerals, or 97.2 µeq/l. Thus

$$[Ca]_{calcite} = [Ca]_{rain} - [Ca]_{sea\ water} - [Ca]_{clay} = 98.1\ \mu eq/l$$

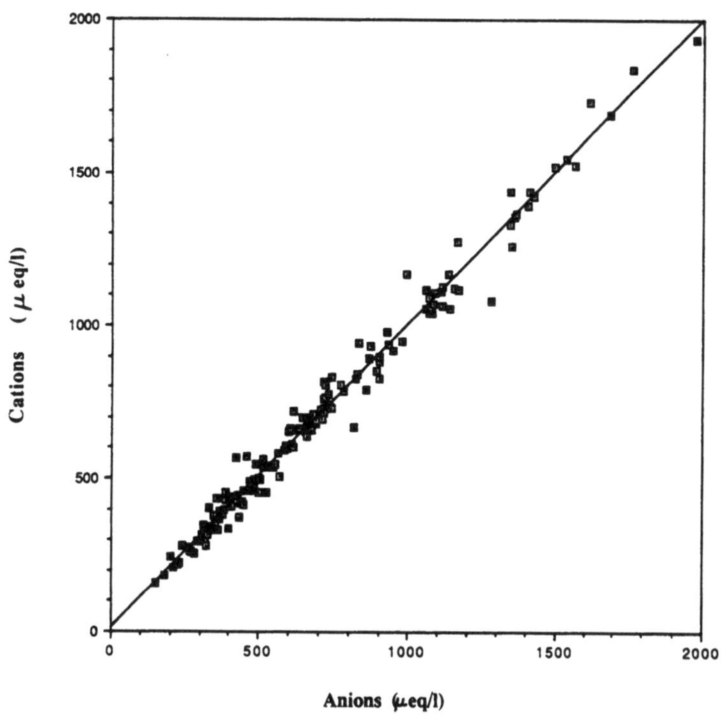

Figure 6. A scatter diagram of the sum of cations versus the sum of anions. The ratio of the sums is almost 1.00.

Alkalinity, as bicarbonate, is attached to 98.1 µeq/l of Ca (as calcite). The rest, 33.3 µeq/l is attached to Ca^{++} in clay (other possibilities may exist). The sulfate ion contributed by sea water was calculated to be 35.4 µeq/l. The rest of the sulfate, known as excess sulfate, 147.2 µeq/l, is divided between the following:

$$[SO_4^=]_{excess} = [SO_4^=]_{soil\ minerals} + [SO_4^=]_{pollution} = 147.2\ \mu eq/l$$

$$[SO_4^=]_{poll} = 147.2 \mu eq/l_{exc} - 48.6 \mu eq/l_{with\ Mg} - 63.9 \mu eq/l_{with\ Ca} = 34.7\ \mu eq/l$$

In the last equation excess K^+ was excluded since its concentration was too small when compared to the other components. Thus on the average sulfate from pollution is estimated to be about 19.0% of the total sulfate measured in rain, the rest is divided between sea salt (19.4%) and soil components (61.6%). Soil minerals in Israel and Saharan dust storms impacting the East Mediterranean (Ganor and Mamane, 1982) are rich in calcite, dolomite, and some sulfates. Lacaux et al. (1988) also reported that Saharan air mass, containing basic terrigeneous particles ($CaCO_3$, $CaSO_4$ for example), impacted the content of rainwater in the Tropical

Forest of the Ivory Coast. In Southwestern semi-arid regions of China similar results were obtained: sulfates are associated mainly with Ca^{++} and not with H^+ (Zhao et al., 1988).

Summary and Conclusions

The purpose of this study was to summarize rain chemistry data in Haifa, Israel. Rain samples were collected daily at the Technion site on Mount Carmel, near the Mediterranean Coast, during eight hydrological years, and were analyzed on site. Data indicated that sea salt (Na^+, Cl^-, Mg^{++}, K^+, and $SO_4^=$) and soil minerals (Ca^{++}, Mg^{++}, HCO_3^- and $SO_4^=$) still dominate the composition of the precipitation. However, an average pH of 5.3 suggested that manmade contributions (H^+, $SO_4^=$, Cl^-, NO_3^-, and metals) may become significant. Elevated values of sea salt constituents, of Ca^{++} and of sulfates are mainly due to the stormy winds and dust storms sweeping the East Mediterranean during active low pressure systems in the winter.

Acknowledgements

The technical assistance of Mrs. E. Melamed is greatly appreciated.

References

Dayan, U. (1986), "Climatology of back-trajectories from Israel based on synoptic analysis", **J. Climate Appl. Meteorol.**, 25, 591-595.

Dikaiakos, J. G., C. G. Tsitouris, P. A. Siskos, D. A. Melissos, and P. Nastos (1990), "Rainwater composition in Athens, Greece ", **Atmos. Environ.**, 24B, 171-176.

Doty, K. G. and R. G. Semonin (1984), " A case study of the effects of local and distant sources of alkaline material of a precipitation in the Midwest and Plains", Paper 84-20.3 presented at the 77th APCA Annual Meeting , San Francisco, June 24-29, 1984.

Ezcurra, A., H. Casado, J. P. Lacaux, and C. Garcia (1988), "Relationships between meteorological situations and acid rain in Spanish Basque country " **Atmos. Environ.**, 22, 2779-2786.

Felly, J. A. and H. M. Liljestrand (1983), "Source contributions to acid precipitation in Texas", **Atmos. Environ.**, 17, 807-814.

Ganor, E. and Y. Mamane (1982), "Transport of Saharan dust across the Eastern Mediterranean", **Atmos. Environ.**, 16, 581-587.

Glavas, S. (1988), "A wet only precipitation study in a Mediterranean site, Patras, Greece", **Atmos. Environ.**, 22, 1505-1507.

Hara, H., K. Taguchi, K. Sekiguchi, M. Tamaki, T. Okita, T. Katou, Y. Kitamura, T. Komeiji, E. Ito, M. Oohara, K. Yoshimura, and Y. Yamawara (1989), "Acid precipitation chemistry over Japan". International Conference on Global and Regional Environmental Atmospheric Chemistry, May 3-10, 1989. Beijing, China.

Ichikuni, M. (1978), "Calcite as a source of excess calcium in rainwater", **J. Geophys. Res.**, 83, 6249-6252.

Khemani, L. T., G. A. Monin, M. S. Naik, P. S. Prakasarao, R. Kumar, and B. H. V. Ramana Murty (1985), "Impact of akaline particulates on pH of rain water in India", **Water, Air, Soil Pollution**, 25, 365-376.

Lacaux, J. P., J. Servant, and J. G. R. Baudet (1987), "Acid rain in the tropical forests of the Ivory Coast", **Atmos. Environ.**,21, 2643-2642.

Lebowitz, L. G. and R. G. de Pena (1985), "Chloride and sodium contents in Northeastern United States precipitation", **J. Geophys. Res.** 90, 8149-8154.

Loye - Pilot, M. D., J. Morelli, J. M. Martin, J. M. Gras and B. Strauss (1986), "Impact of Saharan dust on the rain acidity in the Mediterranean atmosphere". In Proc. of the Fourth European Symp. Physico-Chemical Behaviour of Atmospheric Pollutants, pp. 489-499. Stresa, Italy.

Mamane, Y. (1987), "Chemistry of precipitation in Israel", **The Science of the Total Environment**, 61, 1-13.

Mamane, Y., U. Dayan, and J. M. Miller (1987), "Contribution of alkaline and acidic sources to precipitation in Israel", **The Science of the Total Environment**, 61, 15-22.

Moreira Nordemann, L. M., C. Ferreira, L. A. Magalhaes, W. Z. Mello, E. Silva Filho, M. M. Santiago, C. M. N. Panitz, and C. F. Souza (1989) , "Rain water chemistry in the coast of Brazil". International Conference on Global and Regional Environmental Atmospheric Chemistry, May 3-10, 1989, Beijing, China.

Schuurkes, J. A. A. R., M. M. J. Maenen, and J. G. M. Roelofs (1988), "Chemical characteristics of precipitation in NH_3 affected areas", **Atmos. Environ.**,22, 1689-1698.

Subramanian, V. and K. K. Saxena (1980), "Chemistry of Monsoon rainwater at Delhi", **Tellus**, 32, 558-661.

WMO (1978), "International Operations Handbook for Measurement of Background Atmospheric Pollution", 110pp. **WMO - No. 491**, World Meteorological Organization, Geneva, Switzerland.

Zhao, D., J. Xiong, Y. Xu, and W. H. Chan (1988), "Acid rain in Southwestern China", **Atmos. Environ.**, 22, 349-358.

ACID AEROSOLS

Frederick W. Lipfert
Environmental Consultant, Newport, N.Y., U.S.A.

Abstract

This paper reviews the emerging issue of acid aerosols, primarily from an American perspective. Emission sources, ambient levels, measurement methods, and health effects are discussed, as well as the background of the issue in the United States. Ambient acidity levels are seen to depend on both oxidants and sulfur oxides. Identified research needs include better characterization of population exposures and reconciliation of the exposures used for health effects research with realistic ambient concentrations, exposure times, and aerosol composition.

INTRODUCTION AND BACKGROUND

As discussed throughout this volume, acidity is one of the properties of airborne substances that raises environmental concerns. Deposition of acids in sufficient quantities can damage ecosystems and corrode building materials. Breathing acidic substances can lead to respiratory irritation if the body's natural defenses are overwhelmed; sulfuric acid was one of the agents suspected of contributing to mortality during the severe air pollution episodes of the 1930s to 1960s. Following a request by the Clean Air Scientific Advisory Committee (CASAC) of the U.S. Environmental Protection Agency (EPA), relating to concerns about possible adverse health effects, that agency is now considering adding acid aerosols to the present list of "criteria" air pollutants (CASAC, 1988). Such an action could lead to a variety of regulatory measures potentially further reducing emissions of sulfur and nitrogen oxides, as has been the case with the current list of criteria pollutants in the United States.

Definitions

An aerosol is defined as a suspension of solid or liquid particles in a gaseous medium. The acidity of an aerosol normally refers to the particles, not to the suspension medium, although in the real world it is not uncommon to find acid gases and acid particles together. Commonly found acid gases include SO_2, HNO_3, HCl, and a variety of organic acids such as formic (HCOOH) and acetic (CH_3COOH). Acid particles include sulfuric acid, ammonium bisulfate (NH_4HSO_4) and certain organic compounds. (Most of the common organic acids are in the gaseous phase.) In the Eastern United States, most of the acid particles are sulfuric acid and its salts of ammonia, collectively referred to as "sulfates" (Lipfert, 1988).

Aqueous solutions are important in the chemistry of atmospheric acids, which may include gases, solids or liquids. The strength of an acid refers to the degree of dissociation, i.e., the ease with which hydrogen ions are liberated in solution. "Strong acids," which include all the inorganic acids listed above, dissociate completely. Thus at a given pH, solutions of all strong acids are equivalent. The organic acids are considered weak acids; their production of H^+ depends on concentrations and the other ions present in solution. Ammonium bisulfate (NH_4HSO_4) is a strong acid under normal environmental conditions; ammonium sulfate ($[NH_4]_2SO_4$) acts as a very weak acid but is considered "neutral" for most practical purposes.

Characteristics

Acid aerosols are classified further according to the physical nature of the particles, which in turn relates to the way they are formed in the atmosphere. For example, sulfuric acid mist can be produced from manufacturing operations and normally exists as particles of 5 um diameter or larger. "Primary" sulfates can also be emitted from combustion of sulfur-bearing fuels, including those sources equipped with scrubbers to remove (gaseous) SO_2. "Secondary" aerosols are formed in the atmosphere from gas-phase precursors, involving some of the same chemical reactions that can acidify precipitation. These particles begin as very small condensation nuclei and grow over time, due to both agglomerating with other particles and by absorbing water vapor. Particle size thus tends to be a function of relative humidity (Figure 1). The characteristically submicron size of acid aerosols (at humidities less than about 85%) reduces their rates of atmospheric deposition, thus increasing atmospheric residence times and transport distances. Particles of this size also penetrate deep into the lung. Thus, in contrast to many other topics in this volume, for acid aerosols air concentrations rather than deposition rates are the preferred metric and the direct effects of inhalation are the chief concern.

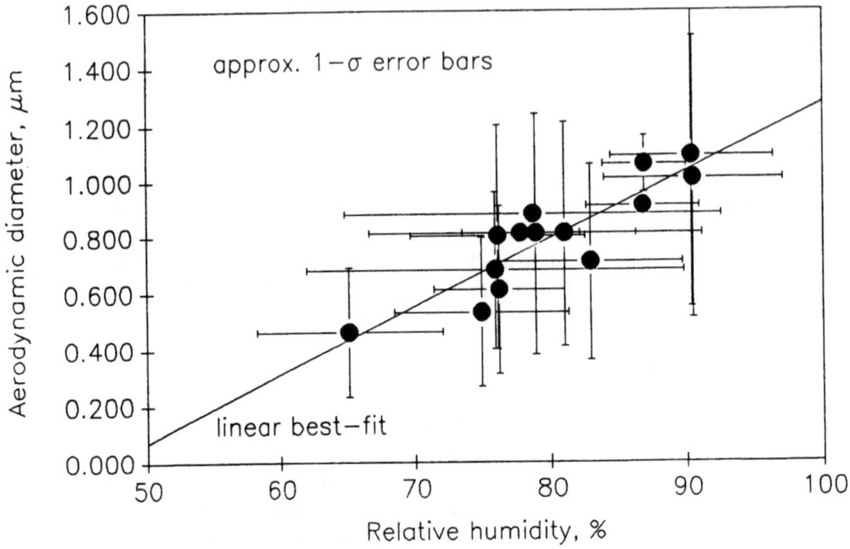

Figure 1. Aerosol aerodynamic diameter vs. average relative humidity (data from Koutrakis et al., 1989).

Although sulfate aerosols may play an important role in global climate change by virtue of their effects on clouds and albedo (Schwartz, 1988), these topics are not a part of this review.

Fog offers still another opportunity to form or modify atmospheric processes. Since the common sulfate particles are quite hygroscopic, they are readily scavenged by the relatively large (ca. 5-50 um*) fog water droplets, as is nitric acid vapor. In addition, if gaseous SO_2 is absorbed into the droplets, it may be oxidized to form H_2SO_4 by any of several chemical reactions involving either oxidants or catalysts within the droplet. These processes constitute one of the natural "sinks" for SO_2. The acidity of fog is usually expressed in pH rather than mass units. Fog water may reflect either the composition of the atmosphere before the fog formed or it may modify that composition through chemical reactions that would have been much slower in the absence of fog. After fog droplets evaporate and the fog clears, sulfate aerosol particles may be left behind, which is also the case with clouds. These precipitated particles tend to be larger (about 0.7 um) than the aerosol particles formed by condensation of gas phase precursors (John et al., 1989)

<u>Units of Measure or Effect</u>

Air pollution is traditionally measured in either mass or volumetric units, per volume of air sampled. Such concentrations are determined using specific "reference" methods of sample preparation and chemical analysis; such a reference method does not yet exist in the U.S. for acid aerosols. Multiplying concentration by the rate of breathing (typically about 15 cubic meters per day or 10 liters per minute) gives the amount of pollutant inspired. This is straightforward for a

----

* In this manuscript, the Greek prefix mu (10^{-6}) is denoted by the letter u. For particles, sizes in microns are given as um, mass median aerodynamic diameter (MMAD).

well-defined stable substance such as ozone or carbon monoxide. However, for chemically or physically reacting substances, it is more difficult to predict the effective dose actually received.

For example, strong acids are characterized chemically by their pH or by the concentration of hydrogen ions (H^+). For fog, concentrations are referred to the liquid content of the droplet; for other aerosols, to the volume of air sampled. The two may be related by the liquid water content of the aerosol (LWC), which ranges from about 0.01 to 1 g H_2O per cubic meter of air for fogs and about four orders of magnitude less for clear air aerosols (depending on the relative humidity). For this reason, the same mass concentrations of acid fog particles and acid aerosol particles represent greatly different ionic strengths and pH's. For example, a fog with a liquid water content of 1 g/m^3 and pH of 3.7 corresponds to an air concentration of 10 ug/m^3 as sulfuric acid. That concentration in clear air aerosol could correspond to pH values less than 1.

Inspired aerosols may be changed chemically and physically during breathing. The respiratory tract is characterized by high humidity (ca. 98%) and varying amounts of endogenous NH_3, which can neutralize some of the inspired acids. More ammonia is found in the oral portion of the respiratory tract, probably because of food particles. Thus, the amount of aerosol neutralization depends on whether the nose or the mouth is the primary route of entry into the tract. In addition to being neutralized by ammonia, hygroscopic particles can absorb moisture, which increases their average diameter and reduces pH (but not the mass of H^+).

A further consideration in the characterization of acid aerosols is the mix of compounds present. Newly emerging information on biological responses (see below) shows that the net H^+ concentration is not always sufficient to predict the response. This may relate to the physical behavior of the various compounds as they pass through the respiratory tract. Given the absence of specific measurement methods for the common sulfate salts, the overall molar ratio, either $H^+/SO_4^=$ or $NH_4^+/SO_4^=$, is a convenient descriptor for the degree of aerosol neutralization.

The particle size is very important in determining the specific region of the respiratory tract in which particles deposit. Highly soluble gases (such as SO_2 or HNO_3) are removed by the moist surfaces of the upper respiratory tract, where they may react with some of the NH_3. Large particles (> 1-2 um) are more likely to deposit higher in the airways; particles smaller than this may penetrate deeper into the lung and some of them will be exhaled. Research is in progress to model these chemical and physical processes to develop a better understanding of the relationships between various measures of ambient concentrations and the actual dosage delivered to various parts of the respiratory system.

Background of the Acid Aerosols Issue in the United States

It has been recognized for over a century that there are other atmospheric sulfur compounds besides SO_2 that may cause adverse effects. The criteria documents produced by the U.S. Federal Government in 1967, 1969, and 1982 discussed the effects of SO_2 and of H_2SO_4; the 1982 document (U.S. E.P.A., 1982) discussed the general effects of particulate matter in combination with sulfur oxides. However, in the 1969 document (U.S. HEW, 1969), which was the basis for the present U.S. national ambient air quality standards (NAAQS), the epidemiological discussion and suggested threshold levels were limited to SO_2, as are the present U.S. NAAQS. (Several U.S. states now have ambient standards for sulfates.) The emphasis on SO_2 at that time may have reflected the limitations of ambient sampling technology or perhaps the hope that if ambient SO_2 were reduced, proportional reductions in sulfates would follow. Recent experience reveals the fallacy in this assumption, since the relationship between ambient H^+ and SO_2 varies by season and location, as discussed below.

The State implementation plans promulgated in the U.S. in the early 1970s to meet the NAAQS for SO_2 resulted in reductions in the allowable sulfur contents of fuels used, especially for smaller, distributed sources. Such limitations resulted in reductions in ambient levels of all sulfur compounds near the sources, i.e., in cities. An alternative strategy to meet the NAAQS for SO_2 for sources that could not easily switch fuels involved increasing stack heights, which greatly reduces the local surface air concentrations but does not reduce the total atmospheric sulfur burden. As a result, over the past 20 years, U.S. sulfate air concentrations have not improved as much as urban SO_2 concentrations and may have actually increased in some remote areas. U.S. total SO_x emissions peaked about 1970 and remain at about the levels of the late 1960s.

In the early 1970s, EPA began the Community Health and Environmental Surveillance System (CHESS) program, which included a variety of epidemiological studies and the associated ambient air quality monitoring with a focus on sulfur oxides (U.S. EPA, 1974). Particulate sulfate was one of the species monitored and was reported to be associated with adverse health effects in many of the CHESS studies. However, problems were later shown with EPA's sulfate measurement methods and some of their conclusions were not confirmed by independent analyses of the data (U.S. EPA, 1980; Roth et al., 1977). The net result was a disparagement of the CHESS findings and a subsequent reduction of EPA research efforts in community health studies.

In the late 1970s, the Electric Power Research Institute (EPRI) completed the Sulfate Regional Experiment (SURE), which was designed to create a regional aerometric data base of information on sulfur oxides and related air pollutants in the northeastern states (Mueller and Hidy, 1983). The SURE showed a maximum in regional sulfate concentrations within 100-300 km of major SO_2 source areas and that the temporal variability in sulfate concentrations was most strongly influenced by meteorology.

By 1980, environmental concerns regarding sulfur oxides had broadened to include acidified precipitation and the potential for damage to natural ecosystems. The U.S. National Acid Precipitation Assessment Program (NAPAP) began a multi-agency 10-year program of research on nearly all aspects of sulfur and nitrogen oxides emissions, transport, transformations, deposition and effects, with the exception of direct (inhalation) health effects. In 1984, a conference on the "health effects of acid precipitation" was held by the National Institute of Environmental Health Sciences (NIEHS) and by 1985, information on acid aerosols was becoming more available, for example as presented at the U.S.-Dutch conference on aerosols at Williamsburg, VA (Lee et al., 1986). By this time the emphasis had clearly shifted from sulfates per se to *acid* sulfates, with concomitant and more difficult requirements for ambient monitoring.

In 1987, NAPAP added inhalation health effects to their "final" assessment, which is scheduled for completion by the end of 1990; a descriptive account of the "state of the science" was released in 1989 (Graham et al., 1989). In 1988, CASAC requested EPA to take steps toward listing acid aerosols as a criteria pollutant; these steps included a 1988 symposium on measurements (Tropp, 1989) and plans for an intercomparison of measurement methods in 1990. NIEHS sponsored several new epidemiological studies focusing on acid aerosols, beginning in 1988.

SOURCES OF ACID AEROSOLS

Acid aerosols are the end product of a chain of processes, as depicted schematically in Figure 2. A number of pathways are possible, and not all elements will be present in all cases. The important elements include:

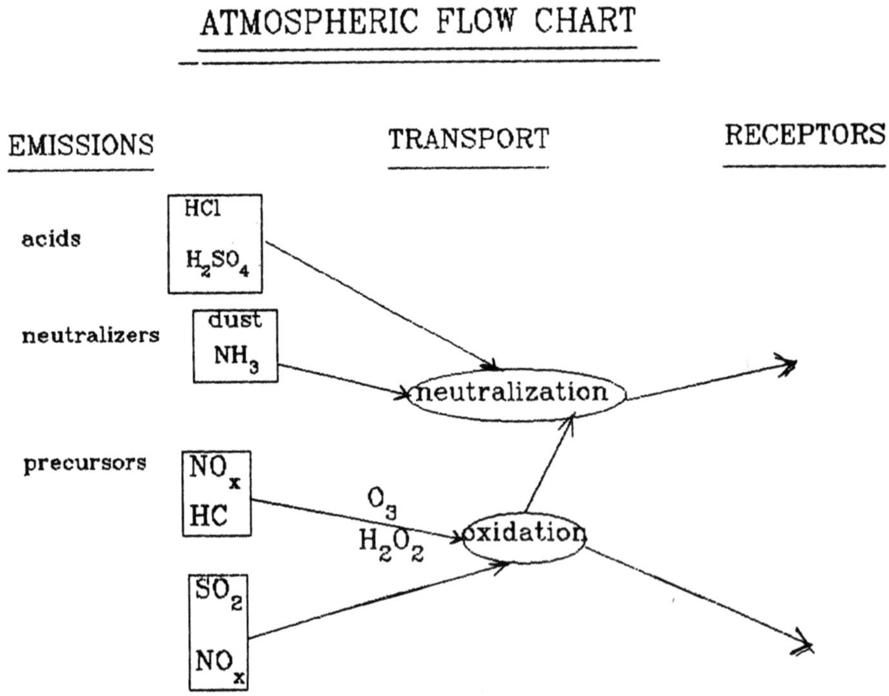

Figure 2. Atmospheric process flow chart.

emissions

- primary acids (H_2SO_4, HCl, HNO_3)
- acid precursors (SO_2, NO_x)
- oxidizing agents and catalysts (hydrocarbons, soot, metals)
- neutralizing agents (ammonia, dust)

transformations

- formation of oxidants
- gas phase reactions
- aqueous phase reactions
- surface reactions
- evaporation of fog and clouds

removal processes

- dry deposition
- wet deposition
- neutralization

Note that non-acidic sulfate aerosols are produced from sea spray and crustal materials; these compounds tend to have larger particle sizes than the typically acidic particles. With the exception of primary emissions of acids, the transformation processes are just as important as the source terms; both are needed to produce acid aerosols. Reactions in the aqueous phase (fog) are much faster than in the gas phase and can produce acids on a local scale. The slower gas-phase reactions tend to produce acids on a larger scale, often regional in extent. Organic acids are usually the products of photochemical smog reactions.

The removal processes act to alleviate the air concentrations that would otherwise exist:

- precipitation effectively scavenges both acid gases and particles.
- dry deposition tends to be more rapid for the acid gases and for fog than for the typically small diameter acid particles.
- neutralization may take place to varying degrees in either the atmosphere or within the human airways.

The acid aerosol atmospheric process chain (emission-transport-oxidation-neutralization-deposition) is inherently more complicated than the simple emission-transport-deposition path for a primary pollutant such as SO_2 or CO. Although in theory, the short-term ground-level impact from primary acid emissions could approach the maximum values measured during regional episodes (Lipfert, 1988), this has not been confirmed by measurements. The high correlations between acidity and SO_2 or smoke in London (Ito and Thurston, 1989) suggest that some portion of the acid may have been of primary origin (HCl, H_2SO_4).

The particle size ranges can depend strongly on relative humidity and may include very small submicron-sizes (<0.4 um MMAD) typical of freshly-formed aerosol, intermediate sizes (0.4 - 1 um) typical of evaporated cloud droplets, and fog droplets (up to about 50 um). The lower range of fog droplet sizes has not been well explored; this may be an important source of uncertainty since stratus cloud sampling has shown that the smaller fog droplets can be the more acidic, because larger ones tend to be associated with larger (alkaline) aerosol particles (Munger et al., 1989). Most of the fogwater is contained in the larger droplets; Waller (1963) found 50% of the total acid content of a London aerosol sample (at 85% relative humidity) in the size range below 0.5 um and only 16% above 4 um. The very small aerosol particles will be more readily neutralized in the airways (Larson, 1989).

Since many of the aqueous reaction rates are (positively) dependent on pH, the degree of neutralization can also be important with regard to transformation processes. As a droplet becomes acidified, the solubility of SO_2 decreases and some reaction rates slow down, so that the transformation process tends to be self-limiting. For example, the highest sulfate concentrations in California fogs are usually accompanied by correspondingly high ammonium levels, resulting in near-neutral pH values (Waldman and Hoffmann, 1987). The low pH cases there are usually characterized by high concentrations of nitrates, hence the need to consider HNO_3. Similarly, for clear air aerosols, the highest sulfate levels do not always correspond to the most acidic compositions since the time required for extensive SO_2 gas-phase oxidation also allows more contact with neutralizing species such as ammonia.

METHODS OF MEASUREMENT FOR ACID AEROSOLS

Although a reference method has yet to be defined in the U.S. for acid aerosols, acid aerosol measurement methods have evolved over the years into a complex technology. As is the case with most air pollution measurements, the interpretation of concentration levels must be keyed to the measurement techniques used. The methodology now in most common use involves the following steps:

- selecting the desired range of particle sizes.
- capturing a representative sample.
- protecting the collected sample from inadvertent neutralization or other chemical reactions or physical losses during transport to the laboratory or while awaiting analysis.
- preparing an aqueous solution of the collected particles
- determining the chemical composition of this solution.

Much of the recent research effort has emphasized the third step. The difficulties encountered in each of these steps depends on the objectives of the monitoring. For example, sampling to characterize the long-term average properties of the atmosphere *per se* is much easier than developing integrated atmospheric exposure data for free-living populations. The latter task requires information on microenvironments for indoors (home, work, transit) as well as outdoors, by time of day and season. In addition, data may be required for a range of particle sizes, since different sizes tend to deposit in different regions of the respiratory system and thus may result in different types of health effects. While generalized atmospheric data may suffice for determining compliance with some future NAAQS, the more detailed data are essential for the development of such a standard with scientific rigor.

Current Technologies

In most current research programs, acid aerosol monitoring is typically based on 24-hour filter samples of particles smaller than about 2-3 um (Waldman et al., 1989; Brauer et al., 1989; Koutrakis et al., 1989). Since most of the acid particles are smaller than 1 um, this size cut allows the potential for mixing of small acid particles with larger alkaline particles, which could result in an understatement of the true acidity of the smaller particles. Using shorter sampling times and a smaller particle size range will reduce the mass of collected material, which can create analytical difficulties. In addition, shorter sampling times will result in more samples for laboratory analysis. It thus appears that the optimum sampling times have not yet been defined.

The samples are protected against inadvertent neutralization by ambient ammonia in two ways. First, the sampling apparatus strips the ammonia from the gas stream to prevent a subsequent gas flow from neutralizing previously-collected particles on the filter. This process has the disadvantage of altering the phase equilibrium of the gas stream and might not be required for shorter sampling times. After removal from the sampler, the filters are further protected against neutralization and loss of acidity by use of a citric acid atmosphere during transit and in the laboratory (Koutrakis et al., 1989).

Extraction of the collected particles from the filter into solution is now a well-developed process with no particular problems. Determination of the concentrations of the major ions in solution is also straightforward; ion chromatography is the preferred technology. However, direct determination of the acidity of the solution has several options. The simplest method measures pH, the free acidity of the solution. This will be equal to the strong acid content if no buffering agents are present in the sample. Determination of the total acidity requires titration to a specified end point (i.e., total titratable acidity). For example, titration to pH=3 yields strong acidity; to ph=7, strong and weak acidity; titration to ph=10 will include very weak acids such as ammonium sulfate (Tanner, 1989). Considerable differences were found between these two methods in early European experiments (Junge and Scheich, 1971).

This rather complicated technology is now in routine use by several U.S. research groups but yields no direct information on the chemical species present. To the extent that the effects of say, H_2SO_4 differ from those of NH_4HSO_4 at the same pH, for example, this is an important unfulfilled need. Methods of selectively extracting H_2SO_4 from filters are not in common use because of uncertainties regarding their efficiency. Methods of determining H_2SO_4 without using filters were first developed in the 1970s and have been refined since then. The basic method uses the flame-photometric SO_2 analyzer, modified to accept small particles (Cobourn et al., 1978). Thermal volatilization is used to separate H_2SO_4 from the ammonium salts, which, unfortunately, cannot be further speciated through this technique. If the ionic composition is known and the sulfuric acid content has been determined independently, the split between ammonium bisulfate and ammonium sulfate may be inferred.

Another technology for speciation of ammonium salts is based on differentiating their infrared spectral signatures (Johnson et al., 1983). This method offers fast response and non-destructive analysis but thus far its application has been limited to the laboratory which developed it.

Estimates Based on Charge Balance

An approximate measure of aerosol acidity levels may be obtained from ion charge balances if reliable measures of sulfate, nitrate, and ammonium ion concentrations are available (Lipfert, 1988). This method has often been used to check the validity of pH determinations in precipitation samples. While sulfate ion concentrations are usually reliable, there can be problems with both ammonium and nitrate. Ordinary filter sampling will likely capture some gaseous nitric acid on the filter, resulting in an overstatement of particulate nitrate. This was probably the case with the SURE data (Mueller and Hidy, 1983), but particulate nitrate levels are sufficiently low in the eastern U.S. that this error does not significantly affect the ion balance. Ammonium concentrations may be subjected to either positive or negative artifacts. Ambient ammonia may add to the ammonium present in the air sample if allowed to contact acidic particles, as discussed above. If ammonium sulfate contacts basic particles (which are usually in the larger size fractions), ammonia gas may be liberated, resulting in a negative artifact. This is more likely to occur in non-size selective sampling such as high-volume sampling, and is a likely explanation for the inordinately low NH_4^+ levels typically found in the older EPA data bases (Lipfert, 1988).

HISTORICAL LEVELS OF ACID AEROSOLS

Early studies (pre-1970) of aerosol acidity are scarce, which makes it difficult to compare current levels with those of the highly polluted urban atmospheres of the past. In London, observations of H_2SO_4 were reported as early as 1936 (Coste and Courtier, 1936) and routine observations began there in the mid-1950's (Commins and Waller, 1967). Although these data are reported in equivalent mass units of H_2SO_4 and often labeled "H_2SO_4" as if this compound had been specifically observed, the titration methodology used would have not been able to separate H_2SO_4 from NH_4HSO_4. Annual average levels were reported as about 5 ug/m^3; the sampling methods did not preclude artifacts from either SO_2 or NH_3 passing through the collection filter, so that acidity levels must be regarded as approximate. The maximum 1-hr concentration reported was 678 ug/m^3 as H_2SO_4, which occurred during the 1962 fog episode in conjunction with an SO_2 level of about 2 ppm. Some of the acidity/SO_2 ratios from high pollution episodes in London are plotted in Figure 3. The ratios are not greatly sensitive to the overall pollution (i.e., SO_2) level and do not differ greatly from current values in the U.S. (as represented by the "SURE" data, discussed below). Current aerosol acidity values in London have not been reported.

Thomas (1962) measured SO_2 and aerosol acidity at two sites in Los Angeles for ten weeks during the winter of 1961, using the hydrogen peroxide method. H^+ ranged up to about 15% of total sulfur (which is a higher ratio than in London) and was a strong function of the oxidant level (Figure 4); H^+ was not systematically dependent on SO_2 in this data set. The range of H^+/S ratio values was higher in Los Angeles than in London; although no London data on oxidants exist for this early period, oxidants were likely higher in Los Angeles.

Corn and DeMaio (1965) analyzed samples taken in central Pittsburgh on high humidity winter days, using the method of Commins (1963). They reported "12 to 16 ug/m^3 of acid," presumably as H_2SO_4. However, they also reported a strong inverse dependence of total sulfate concentration on sample collection time, which is a key indication of artifacts due to SO_2 oxidizing on the filters. They did not report the sample time for the acid determinations and thus it is difficult to assess the validity of these values. By way of comparison, SO_2 in Pittsburgh was in the range 80-300 ug/m^3 during 1961-63; these values would yield H^+/S ratios similar to those found in London and Los Angeles.

In summary, these early determinations of aerosol acidity suggest that levels were in the range up to about 15% of total sulfur and strongly dependent on ambient oxidants.

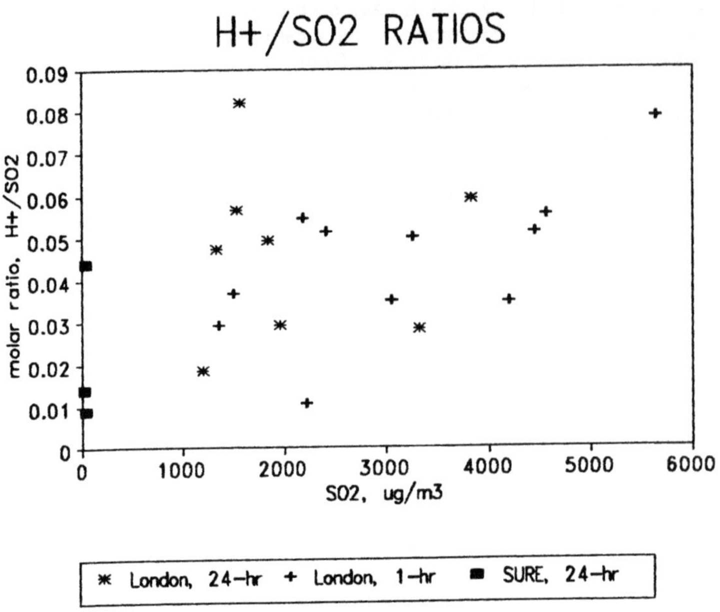

Figure 3. Ratios of H^+ to SO_2 (London data from Commins and Waller, 1967; SURE data from Lipfert, 1988).

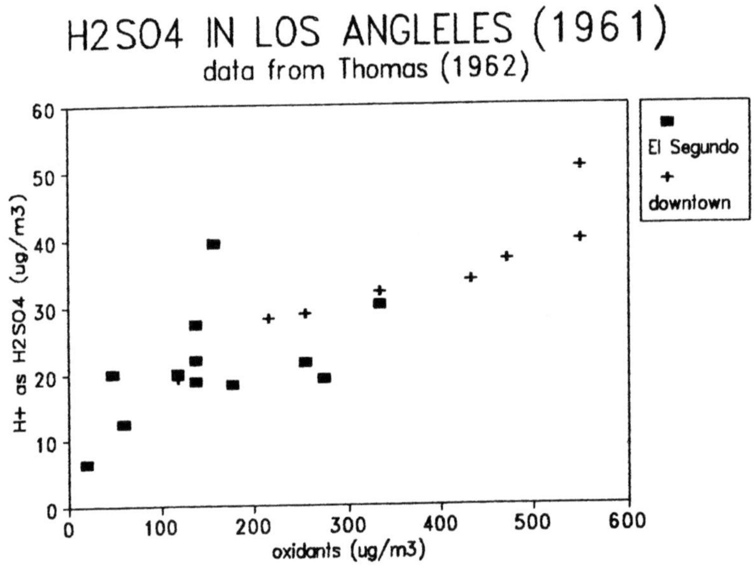

Figure 4. Aerosol acidity measurements from Los Angeles, 1961 (data from Thomas, 1962).

CURRENT AMBIENT LEVELS OF ACID AEROSOLS IN THE U.S. AND CANADA

No routine monitoring data bases have been developed for aerosol acidity, but a number of research campaigns have accumulated useful data (Kelly, 1985,1987; Pierson, et al., 1989; Waldman et al., 1990). Data on H_2SO_4 and strong acidity were monitored for about one year at the six cities being studied by the Harvard School of Public Health (Spengler et al., 1989; Spengler, 1989). The EPRI SURE data base was adapted for the purpose of estimating approximate acid aerosol levels by using charge balance to estimate the apparent sulfate acidity at the 54 SURE monitoring sites in the Eastern U.S. (Lipfert, 1988). Fog chemistry has been determined at a number of sites in California and elsewhere (Munger et al., 1983). Gaseous acids and particulate species were measured recently in the South Coast (CA) Air Basin (Solomon et al., 1988).

Temporal Patterns

As with most air quality data, the concentrations of acid aerosols depend on time of day, season, and on sampling and averaging times. However, since acid aerosol concentrations display the relatively slow moving patterns typical of secondary pollutants, the temporal patterns are less "spiky" than, for example, SO_2, and typically consist of periods of low levels or zeroes punctuated by "episodic" periods of a few days' duration at most, as shown in Figure 5 for Allegheny Mountain, PA (data from Pierson et al., 1989) and in Figure 6 for Whiteface Mountain, NY, for example (there are also periods of missing data in the Whiteface Mountain data set).

Long-term Acid Aerosol Data

Typical long-term average ion concentrations are shown in Figures 7 - 9. In these graphs, each observation is represented by two stacked bar graphs; the negative ions ($SO_4^{=}$ and NO_3^{-}) are shown in the first bar and the positive ions (NH_4^+ and H^+) in the second. The sums of positives and negatives must be equal to satisfy electroneutrality; the differences shown represent either missing ions or measurement errors, except for the SURE data in Figure 7 and Figure 8, for which H^+ was estimated from ion balance and was not measured directly.

Figure 7 compares several eastern locations; the "SURE" (Mueller and Hidy, 1983; Lipfert, 1988) and Whiteface (WFC) Mountain (Kelly, 1985,1987) data are long-term averages (> 1 yr) from rural locations (1979 and 1984-6, respectively); the New York City (Tanner et al., 1979) and Toronto (Waldman et al., 1990) data are for a few weeks in the summer (1976 and 1986,

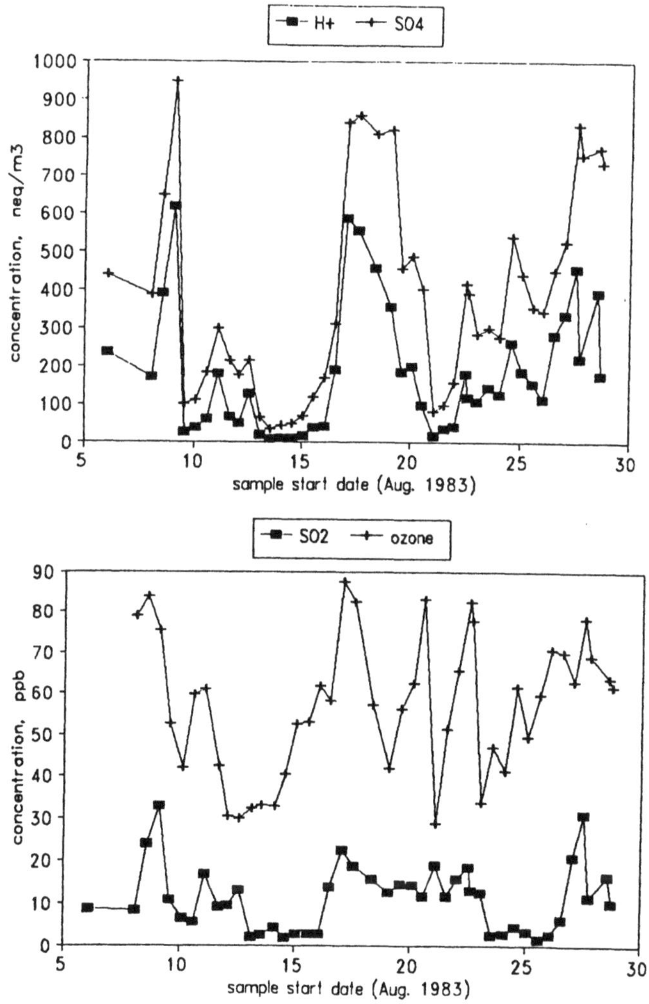

Figure 5. Aerosol and gas concentration data from Allegheny Mtn., PA (data from Pierson et al., 1989).

respectively), and the "episds" data (Lipfert, 1988) are for heavily polluted days at the same Class I SURE sites in July and October (1978). The figure shows that sulfate makes up the overwhelming bulk of the negative ions at these locations, that ammonium is the most abundant positive ion, and that there is a substantial proportion of unmeasured positive ions (besides H^+) at the urban sites. The lack of proportionality between H^+ and $SO_4^=$ is also apparent from this plot; during the "episodes" at the SURE sites, sulfate was higher but acidity remained about the same. Acidity was higher at Whiteface Mountain than at Toronto, in spite of much lower sulfate levels.

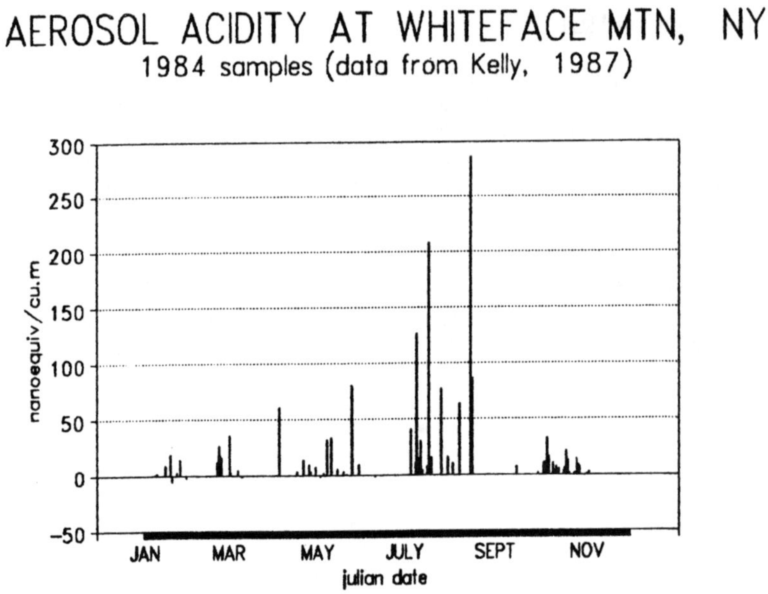

Figure 6. Aerosol acidity measurements from Whiteface Mtn., NY (1984). (data from Kelly, 1987).

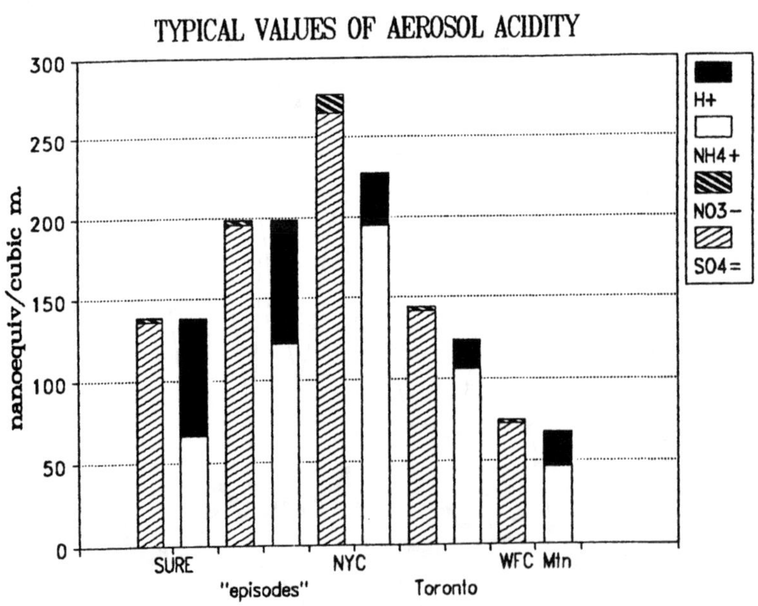

Figure 7. Typical average values of aerosol components from various locations (see text for site identifications and data sources).

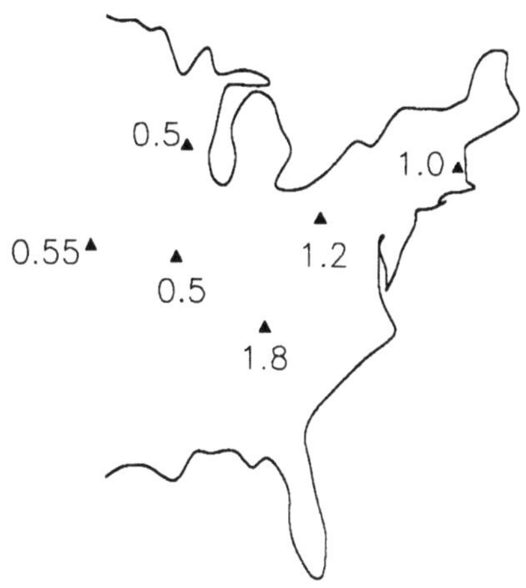

Figure 8. Long-term average aerosol acidity (ug/m^3 as H_2SO_4) at the Harvard Six-Cities Sites (data from Spengler, 1989).

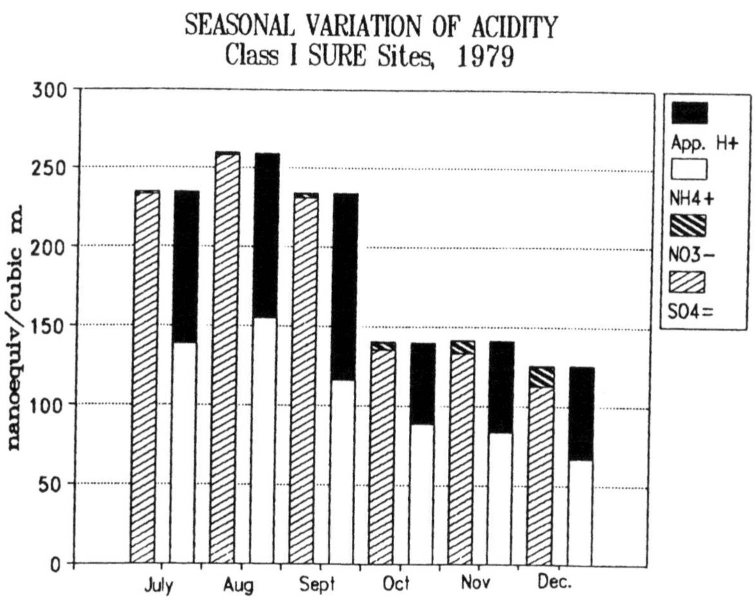

Figure 9. Monthly averages of aerosol components averaged over 9 rural sites (data from Mueller and Watson, 1982; Lipfert, 1988). H^+ values are based on ion balance.

Seasonal variability is shown from the 1979 ERAQS data (Mueller and Watson, 1982) averaged over nine SURE Class I sites in Figure 8; ammonium drops faster than sulfate from summer to winter, resulting in a more acidic composition but lower mass concentrations. In rural areas, the composition is approximately that of ammonium bisulfate, but the aerosol is more nearly neutralized in cities.

Additional urban aerosol acidity data for the Eastern U.S. were obtained from the Harvard School of Public Health (Spengler, 1989), based on their epidemiological research and using the latest acid aerosol measurement technology. Long term average acidity values, plotted in Figure 9, ranged from about 10-37 neq/m^3 (0.5-2 ug/m^3 as H_2SO_4). The highest 95th %ile reported was about 100 neq/m^3. Only a small portion of these aerosols was estimated to be in the actual form of H_2SO_4; the balance was apparently ammonium bisulfate. The Harvard group has also collected aerosol acidity based on personal and indoor sampling; indoor levels averaged about 25% of outdoor levels, presumably because of additional neutralization indoors (Brauer et al., 1989).

Data on Fog Chemistry

Average fog composition data are compared in Figure 10, on the basis of air concentrations (fog water concentrations multiplied by liquid water content). This places the values in the same range as for clear air aerosols in the Eastern U.S., except for the South Coast Air Basin (SCAB) data. These data are based on the averages of several sampling campaigns conducted by the Cal-Tech group (Munger et al., 1983; Waldman and Hoffmann, 1987). At the coastal sites (north of SCAB), ammonium is less abundant than the other alkaline species, nitrate levels are low, and acidity is moderate (pH=4.05). In the Bakersfield area, where SO_2 levels are the highest in California, sulfate is more important but since ammonium is quite abundant, in part because of local agricultural sources, average net acidity is also moderate (ph=3.9). At this site, additional negative ions (17 neq/m^3), which could be organic acids, are required to complete the charge balance. In SCAB, nitrate is the major factor and ammonium levels are higher than near Bakersfield but not high enough to neutralize the fog. The average fog pH is 3.1. Note that fog sulfate levels in SCAB are higher than in Bakersfield, which illustrates the importance of the oxidizing agents, in addition to SO_2 levels. The fog pH values reflect the liquid water contents as well as the H^+ air concentrations shown in the figure (aerosol equivalent pH values are about 1 pH unit higher).

Fog water at locations in the Eastern U.S. (data from Whiteface Mountain [Arons et al., 1988] are shown in Figure 10 for comparison) and in Europe tends to contain less nitrate than in California, but a much higher proportion of nitrate is seen in fog than is typically found in aerosol at the same locations, because of the absorption of nitric acid (vapor).

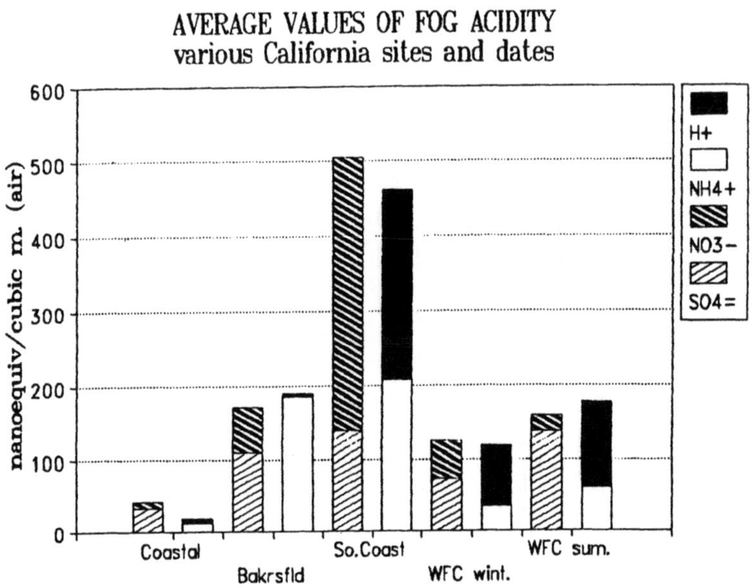

Figure 10. Average fogwater composition at various sites, on the basis of the air sampled.

Effects of Ozone

Figure 11 plots aerosol acidity versus ozone at various sites, with a linear regression line for each site. Most of these data fall into a common envelope defined by H^+ = 30 ug/m^3 (as H_2SO_4) at an ozone level of 200 ug/m^3 (24-hr averages), suggesting that the ozone level at a given site has an influence on the aerosol acidity (this envelope is reasonably consistent with the early Los Angeles data shown in Figure 4). Note that ozone is reduced in many urban areas because of titration by nitric oxide. The aerosol acidity levels at Allegheny Mountain and Laurel Hill, PA (Pierson et al., 1989, stand out as significantly higher than the other sites, perhaps because of the relative proximity of these sites to large sources of SO_2.

Comparison of Acidic Species

Data collected during the South Coast Air Basin Air Quality Study (SCAQS) allow a comparison

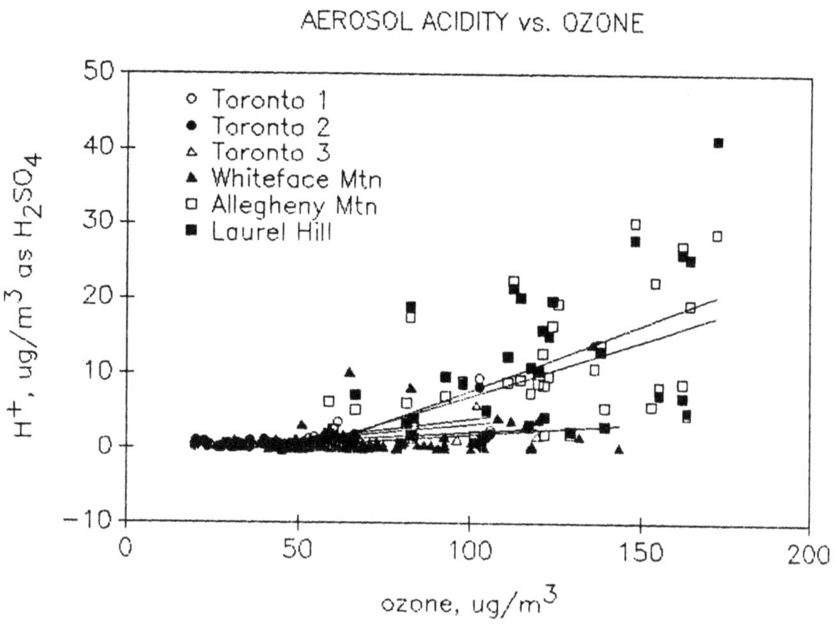

Figure 11. Aerosol acidity as a function of ozone concentration level, for various urban and rural sites.

of various gaseous species, based on the average of eight sites for the year 1986 (Solomon et al., 1988):

Species	Annual Av'g	Range in Max 24-hr Av'gs.
gas	(data in neq/m^3)	
HNO_3	75.	100-300
HCl	39.	90-170
HCOOH	175.	200-550
CH_3COOH	278.	500-850
NH_3 (alkaline)	346.*	100-4300*

*including one "outlier" site with an average of 1600 neq/m^3

These data indicate the relative importance of the gaseous species, either in their own right or as potential absorbers of breath ammonia. Although direct measurements of particle acidity were not made at this time, estimated annual averages based on ion balance were in the range 10-60 neq/m^3, i.e., mostly lower than the gaseous species. HCl, which may result either from direct emissions or from the the interaction of HNO_3 and sea salt, may be an overlooked species; Harrison and Allen (1990) found gaseous strong acidity to be of the same order as particle strong acidity in a small area study in Eastern England.

SUSPECTED HEALTH EFFECTS

Bases for Concern

Firm evidence that air pollution can seriously affect health came from the air pollution episodes of the Meuse Valley (1930), Donora, PA (1948), and London (1952), in which excess mortality was clearly demonstrated. Air pollution measurements were sparse during these episodes but the presence of sulfuric acid was strongly suspected in all of them. Direct emissions sources of H_2SO_4 were present in the Meuse Valley and Donora, and it was later shown that the foggy conditions in London, together with soot and metal catalysts, could have converted some portion of the SO_2 to H_2SO_4 (Meetham, 1981).

This long held suspicion notwithstanding, definitive (positive) associations of acid aerosol association with adverse human health effects have yet to be demonstrated in any epidemiological studies completed to date, in part because of the lack of appropriate exposure measurements (Lioy and Waldman, 1989; Lipfert et al., 1989). Several past epidemiological studies (U.S. EPA, 1974) have postulated an association between sulfates and health responses; however, the evidence for an association between sulfate concentrations and human health effects is inconsistent (U.S. EPA, 1980; Roth et al., 1977). Nevertheless, epidemiological studies that find an association between health and (total) sulfate levels have been cited as supporting the hypothesis that hydrogen ion is the active agent (U.S. EPA, 1988; Lippmann, 1989). Epidemiological studies are emphasized in a review report by Morris and Lipfert (1989).

The current basis for concern about acid aerosol health effects thus rests largely on the results of animal toxicological and human clinical studies. [For comprehensive reviews of the health studies, see the EPA Issues Paper (U.S. EPA, 1988) and the review articles by Shy (1989), Graham (1989), and Follinsbee (1989).] The animal studies have found several endpoints responsive to acid aerosol exposure, ranging from changes in lung clearance rates (Schlesinger, 1989) to changes in lung morphology (Gearhart and Schlesinger, 1989) and rates of cell turnover (Kleinman et al., 1988). Human clinical studies have emphasized transient lung function and clearance effects, in response to sulfuric acid at concentrations greater than current ambient levels. Early studies found responses only after exposure to very high levels of H_2SO_4, often in excess of 1000 ug/m^3. More recent work has reported responses by asthmatic subjects at lower levels (68-100 ug/m^3) (Koenig et al., 1989; Utell et al., 1989). Exposure times for the reported laboratory studies vary from single doses of less than an hour to repeated doses of a few hours per day. The magnitude of the inspired dose also depends on the breathing rate, which can be elevated by exercise.

Pending Health Effects Issues

As shown above, monitoring data indicate that sulfuric acid is usually not the major acid constituent in the ambient environment; moreover, current measurement programs are unable to provide definitive information on the actual sulfate compounds present in an aerosol sample. It has generally been assumed that H^+ is the agent of principal concern and that biological responses to other acidic species may be estimated by using the H^+ contribution as a measure of relative potency, based on responses to H_2SO_4. Some experimental evidence is consistent with this hypothesis (Schlesinger, 1989); however, there have also been contradictions: Schlesinger (1989), for example, found that NH_4HSO_4 exposures to rabbits had significantly less effect on alveolar clearance than did equivalent concentrations of H_2SO_4. These findings were confirmed by comparisons of phagocytic activity of alveolar macrophages in rabbits (Schlesinger et al., 1990). Thus

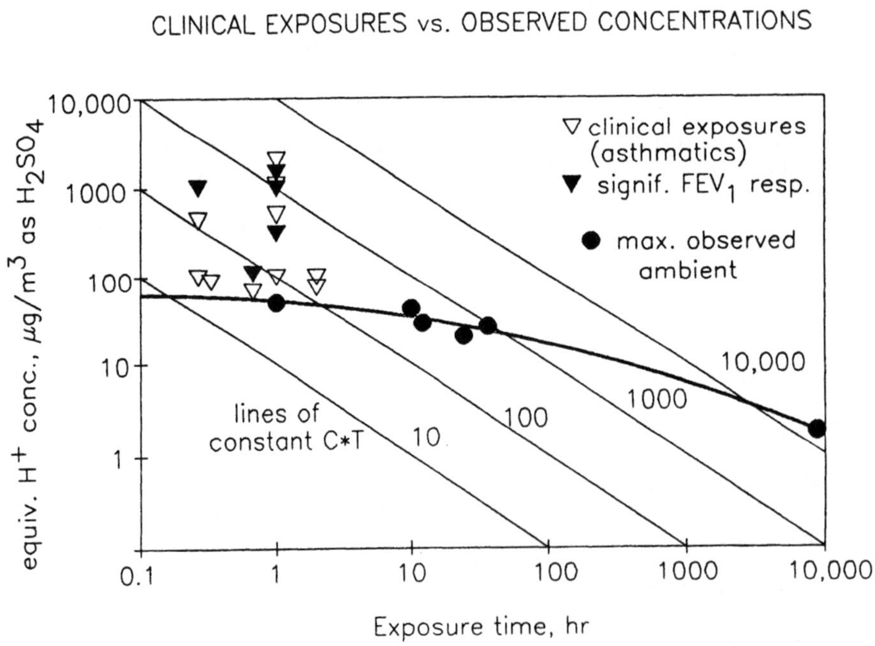

Figure 12. Comparison of ambient sulfate concentrations and composition with exposure levels used in human clinical experiments (Wyzga and Lipfert, 1990). Ambient data are from the SURE (Mueller and Hidy, 1983; Lipfert, 1988); clinical exposures were summarized by EPA (1988).

the question of predicting human responses to identical H^+ loads delivered by means of differing aerosol compositions remains an important subject for research. The extant body of experimental acid aerosol health effects research deals almost entirely with sulfuric acid exposures.

The disparity between experimental exposures and ambient levels is illustrated in Figure 12, which plots aerosol composition (ratio of NH_4^+ to $SO_4^=$, or the degree of neutralization, where a value of 2.0 represents complete neutralization and 0.0 represents sulfuric acid) against sulfate concentration level. The ambient data are from the SURE data base (Lipfert, 1988) and the clinical exposure data are taken from EPA, 1988. The plot shows little overlap between the two exposure regimes, since clinical exposures emphasize H_2SO_4 at high concentrations and the ambient data are highly neutralized, at much lower concentrations.

The results of animal studies have been used to support the hypothesis that the total integrated exposure (concentration x time [CxT]) may be the most appropriate acid aerosol dose measure (Lioy and Waldman, 1989; Spengler et al., 1989; Schlesinger, 1989; Gearhart and Schlesinger, 1989). For lung clearance, there are limited supporting data (Schlesinger, 1989); however, these data are also consistent with an effect due to variations in concentration alone, since concentration and integrated exposure were highly correlated in these experiments. Figure 13 contrasts recent U.S. ambient data with concentrations and exposure times for clinical exposures of asthmatic subjects. Note that acidity concentrations do not continue to rise with shorter averaging times, as might be expected with a primary pollutant. The clinical exposures are characterized by high concentrations and short times; the two regimes do not overlap. The solid triangles in the figure represent conditions under which a statistically significant (reversible) lung function change was observed; the lowest of these (Koenig et al., 1989) is somewhat problematic since subsequent experiments have failed to replicate this outcome (Koenig, 1989; Linn et al., 1990). By following the diagonal lines of constant CxT in the figure, it appears that exposure times of 12 hours or more will be needed to inspire a total acidity dose similar to those reliably inducing lung function changes.

The validity of the CxT hypothesis needs to be investigated more systematically (Graham, 1989), especially at lower concentrations and longer exposure times, conditions for which thresholds due to neutralization by endogenous ammonia may be important. Note that elevated breathing rates induced by exercise are unlikely to be sustained long enough to substantially influence doses greater than a few hours.

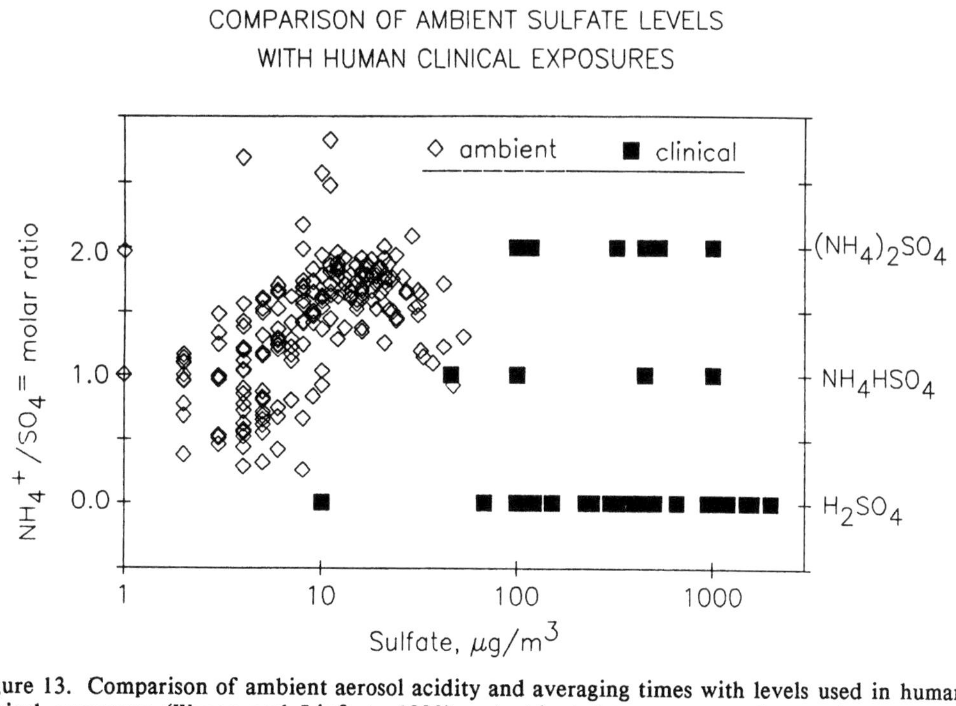

Figure 13. Comparison of ambient aerosol acidity and averaging times with levels used in human clinical exposures (Wyzga and Lipfert, 1990). Ambient data represent the maximum values reported in the United States at any location, for that averaging time. Clinical exposures represent levels used for testing asthmatics (EPA, 1988).

Most clinical experiments to date have only considered concentration levels substantially higher than ambient. It is not clear how to extrapolate responses observed under these conditions to more realistic (lower) ambient conditions. Such an extrapolation may be particularly confounded by endogenous (breath) ammonia, since experiments have determined that artificial depletion of oral NH_3 (by dental brushing or ingesting acids) can enhance the lung function responses obtained at a given sulfuric acid dose level:

- Utell et al.(1989) showed about 150% increase in the responses of exercising asthmatics when they gargled with lemon concentrate before exposure to 350 ug/m^3 H_2SO_4.

- Koenig (1989) showed that lung function response to about 70 ug/m^3 H_2SO_4 was statistically significant only when the subjects drank lemonade before exposure.

The ability to predict responses to current ambient acid levels requires further understanding of the role of endogenous ammonia and the extent to which it may offer some protection from breathing acids. Also, exercise will complicate the neutralization process, by changing the breathing mode and the residence times of particles.

Although the emphasis to date has been on acid sulfate particles, it is possible that gaseous acids (HNO_3, organic acids) can be important. Koenig et al.(1988) reported significant lung function decrements in adolescent asthmatics exposed to 0.05 ppm HNO_3. Kleinman et al.(1989) have reported a significant synergistic biological response associated with joint exposures to ozone and NO_2, which in combination will yield HNO_3 and HNO_2. Also, Graham (1989) reports acid aerosol responses in the upper airways, which is the locus of responses expected from acid gases. Finally, acid gases may react with endogenous ammonia that would otherwise be available to neutralize acid particles.

Mixtures may be important since acid aerosols do not occur in the ambient in isolation and peak levels may coincide with those of other pollutants. Animal studies have suggested that synergism can occur between high doses of acid aerosols and other pollutants, such as ozone (Kleinman et al., 1988; Last, 1989).

The effects of particle size are not well defined. The epidemiology seems to implicate larger particles (fog and mist) which also have been shown to exacerbate symptomatic responses at high concentrations (Linn et al., 1989), but asthmatics seem to respond more to submicron aerosols (Hackney et al., 1989).

Koenig et al.(1989) have reported transient changes in lung function among exercising adolescent asthmatics exposed to 68 ug/m3 H_2SO_4; however, replication of this experiment with different groups of exercising adolescent asthmatics failed to confirm this result (Koenig, 1989; Linn et al.,1990). Further research is needed to identify whether especially susceptible subpopulations exist.

The medical significance of the different symptomatic and biological responses observed in health studies is not always clear and may differ according to health status. For example, the health significance of small transient decrements in lung function is not clear and small transient increases and decreases in lung clearance rates are difficult to interpret. Research to achieve a better understanding of biological mechanisms of action and their relationships to clearly adverse responses in humans would be one way to clarify significance.

CONCLUDING REMARKS

Summary of Ambient Data

In general, the levels of U.S. population exposure to acid aerosols are quite modest in comparison with the levels at which health effects have been demonstrated. The most likely acid sulfate species appears to be NH_4HSO_4; particle sizes depend on the age of the aerosol but are in the submicron range. Elevated concentrations of acidity are likely to persist at a given location for a few days, at most. Gaseous acidity levels are likely to equal or exceed particulate acidity levels in most locations.

Research Needs

Although data indicate that acidic aerosols occur in the Eastern U.S. on a regular basis, much remains to be done to characterize the extent of population exposures and their potential health significance. The following questions appear to be relevant to this task:

1. Which aerosol species should be measured?
2. What is the optimum averaging time for sampling?
3. What particle size ranges should be segregated?
4. Which co-pollutants are important?

Information on detailed particle size resolution and speciation for periods less than 24 hours, sampling in major Eastern population centers, and indoor sampling are particularly needed in the U.S.

The major question on health effects is:

> What are the human dose-response relationships for relevant biological endpoints, at conditions similar to current actual population exposures?

Since current U.S. population exposures remain to be adequately defined, there may be important interactions among the monitoring and health questions.

ACKNOWLEDGMENTS

This review has benefited from the cooperation and assistance of many individuals and organizations. Financial support for previous review articles, on which this work is partially based, was provided by the Electric Power Research Institute and the U.S. Department of Energy. The collaboration of Drs. R.E. Wyzga and S.C. Morris on previous review efforts is specifically acknowledged. Data were provided by Drs. J.M. Waldman, R.M. Gorse, and J.D. Spengler. Any errors or omissions in this review are my responsibility alone.

REFERENCES

Arons, T., et al. (1988), Wintertime Trace Gas and Cloudwater Chemistry Studies, Whiteface Mountain, 1984-1986, Report EP 82-10, Atmospheric Sciences Research Center, State University of New Yorkl at Albany, Albany, NY.

Brauer, M., P. Loutrakis, J.M. Wolfson, and J.D. Spengler (1989), Evaluation of the gas collection of an annular denuder system under simulated atmospheric conditions, *Atm. Env.* 23:1981-1986.

Brauer, M., P. Koutrakis, and J.D. Spengler (1989), paper presented at the 1989 EPA/AWMA International Symposium on Measurement of Toxic and Related Air Pollutants, Raleigh, NC, May 1989. Also see Paper 89-161.1 presented at the Annual Meeting of the Air and Waste Management Association, Anaheim, CA, June 1989.

CASAC (1988), Subcommittee on Acid Aerosols, Report on Acid Aerosol Research Needs, EPA-SAB/CASAC-89-002, U.S. Environmental Protection Agency, Washington, DC. Oct. 19, 1988.

Cobourn, W.G., R.B. Husar, and J.D. Husar (1978), Continuous *in-situ* monitoring of ambient particulate sulfur using flame photometry and thermal analysis, *Atm. Env.* 12:89-98.

Commins, B.T. and R.E. Waller (1967), Observations from a Ten-Year-Study of Pollution at a Site in the City of London, *Atm. Env.* 1:49-68.

Corn, M. and L. DeMaio (1965), Particulate Sulfates in Pittsburgh Air, *J.APCA* 15:26-30.

Coste, J.H. and G.B. Courtier (1936), Sulphuric Acid as a Disperse Phase in Town Air, *Trans. Faraday Soc.* 32:1198-1202.

Folinsbee, L.J. (1989), Review, discussion, and summary: acute pulmonary responses in humans, *Environ. Health Persp.* 79:195-200.

Gearhart, J.M., and R.B. Schlesinger (1989), Sulfuric acid-induced changes in the physiology and structure of the tracheobronchial airways, *Environ. Health Persp.* 79:127-136.

Graham, J.A. (1989), Review, discussion, and summary: Toxicology, *Environ. Health Persp.* 79:191-194.

Graham, J.A. et al. (1989), Direct Health Effects of Air Pollutants Associated with Acidic Precursor Emissions, NAPAP State of Science Technology Report 22, National Acid Precipitation Assessment Program, Washington, DC, Dec. 1989.

Hackney, J.D., W.S. Linn, and E.L. Avol (1989), Acid fog: Effects on respiratory function and symptoms in healthy and asthmatic volunteers, *Environ. Health Persp.* 79:159-162.

Harrison, R.M., and A.G. Allen (1990), Measurements of Atmospheric HNO_3, HCl, and Associated Species on a Small Network in Eastern England, *Atm. Env.* 24A:369-376.

Ito, K., and G.D. Thurston (1989), Characterization and reconstruction of historical London, England, acidic aerosol concentrations, *Environ. Health Persp.* 79:35-42.

John, W., S.M. Wall, J.L. Ondo, and W. Winklmayr (1989), Acidic Aerosol Size Distributions During SCAQS, report to California Air Resources Board CA/DOH/AIHL/SP-51, California Air Resources Board, Sacramento, CA.

Johnson, S.A., R. Kumar, and P.T. Cunningham (1983), Airborne detection of acidic sulfate aerosol using an ATR impactor, *Aerosol Sci. Technol.* 2:401-405.

Junge, C. and G. Scheich (1971), Determination of the Acid Content of Particles, *Atm. Env* 5:165-175. See *Atm. Env.* 3:423-441 (1969) for the original German version and tables and figures.

Kelly, T.J. (1985,1987), Trace Gas and Aerosol Measurements at Whiteface Mountain, NY. Brookhaven National Laboratory Reports BNL 37110 (Sept. 1985) and BNL 39464 (Jan. 1987).

Kleinman, M.T., T.R. McClure, W.J. Mautz, R.F. Phalen, and T.T. Crocker, (1988), *Ann. Occup. Hygiene*, 32, Suppl.1: 239-245.

Kleinman, M.T., R.F. Phalen, W.J. Mautz, R.C. Mannix, T.R. McClure, and T.T. Crocker (1989), Health effects of acid aerosols formed by atmospheric mixtures, *Environ. Health Persp.* 79:137-145.

Koenig, J., An Assessment of Pulmonary Function Changes and Oral Ammonia Levels after Exposure of Adolescent Asthmatic Subjects to Sulfuric or Nitric Acid. Paper 89-92.4, presented at the 82nd Annual Meeting of the Air and Waste Management Association, Anaheim, CA, June 1989.

Koenig, J.Q., D.S. Covert, and W.E. Pierson (1989), Effects of inhalation of acidic compounds on pulmonary function in allergic adolescent subjects, *Environ. Health Persp.* 79:173-178.

Koenig, J.Q., D.S. Covert, W.E. Pierson, and M.S. McManus (1988), The Effects of Inhaled Nitric Acid on Pulmonary Function in Adolescent Asthmatics (abstract), *Am. Rev. Respir. Dis.* 137:A169.

Koutrakis, P., Wolfson, J.M., Spengler, J.D. (1989), Equilibrium Size of Atmospheric Aerosol Sulfates as a Function of the Relative Humidity, *J. Geophys. Res.* 94(D5): 6442-48.

Koutrakis, P., et al. (1989), Design of a personal annular denuder sampler to measure atmospheric aerosols and gases, *Atm. Env.* 23:2767-2774.

Larson, T.V., (1989) The influence of chemical and physical forms of ambient air acids on airway doses, *Environ. Health Persp.* 79:7-14.

Last, J.A. (1989), Effects of inhaled acids on lung biochemistry, *Environ. Health Persp.* 79:115-119.

Lee, S.D., T. Schneider, L.D. Grant, and P.K. Verkerk, eds. (1986) Aerosols: Research, Risk Assessment, and Control Strategies. Lewis Publishers, Inc., Chelsea, MI.

Linn, W.S., E.L. Avol, K.R. Anderson, E.A. Shamoo, R-C. Peng, and J.D. Hackney (1989), Effect of Droplet Size on Respiratory Responses to Inhaled Sulfuric Acid in Normal and Asthmatic Volunteers, *Am. Rev. Respir. Dis.* 140:161-166.

Linn, W.S., E.L. Avol, E.A. Shamoo, K.R. Anderson, R-C. Peng, and J.D. Hackney (1990), Respiratory response of Young Asthmatics to Sulfuric Acid Aerosol (abstract), *Am. Rev. Respir. Dis.* 141:A74.

Lioy, P.J. and J.M. Waldman (1989), Acidic sulfate aerosols: Characterization and exposure, *Environ. Health Persp.*, 79:15-34.

Lipfert, F.W. (1988), Exposure to Acidic Sulfates in the Atmosphere, EPRI EA-6150, Electric Power Research Institute, Palo Alto, CA.

Lipfert, F.W., Morris, S.C., Wyzga, R.E. (1989), Acid aerosols: The next criteria pollutant. *Env. Sci. Tech.* 23:1316-1322.

Lippmann, M. (1989), Progress, prospects, and research needs on the health effects of acid aerosols, *Envir. Health Perspect.* 79:203-205.

Meetham, A.R. (1981), Atmospheric Pollution, Its History, Origins, and Prevention, 4th ed., Pergamon Press, Oxford.

Mueller, P.K., and J.G. Watson (1982), "Eastern Regional Air Quality Measurements," EPRI EA-1914, Electric Power Research Institute, Palo Alto, CA, 1982.

Mueller, P.K., and G.M. Hidy (1983), "The Sulfate Regional Experiment: Report of Findings," EPRI EA-1901, Electric Power Research Institute, Palo Alto, CA 1983.

Morris, S.C. and F.W. Lipfert (1989), Health Effects of Acid Aerosols and Their Precursors, Brookhaven National Laboratory Report to the U.S. Department of Energy, Brookhaven National Laboratory, Upton, NY. May, 1989.

Munger, J.W., D.J. Jacob, J.M. Waldman, and M.R. Hoffmann (1983), Fogwater Chemistry in an Urban Atmosphere, *J. Geophys. Res.* 88D:5109-5121 (1983).

Munger, J.W., J. Collett, Jr., B. Daube, Jr., and M.R. Hoffmann (1989), Chemical Composition of Coastal Stratus Clouds: Dependence on Droplet Size and Distance from the Coast, *Atm. Env.* 23:2305-2320 (1989).

Pierson, W.R et al. (1989), Atmospheric Acidity Measurements on Allegheny Mountain and the Origins of Ambient Acidity in the Northeastern United States, *Atm. Env.* 23:431-459.

Roth, H.D., J.R. Viren, and A.V. Colucci (1977), "Evaluation of CHESS: New York Asthma Data 1970-71," EPRI EA-460, Electric Power Research Institute, Palo Alto, CA, 1977.

Schlesinger, R.B. (1979), Factors affecting the response of lung clearance systems to acid aerosols: Role of exposure concentration, exposure time, and relative acidity, *Environ. Health Persp.* 79:121-126.

Schlesinger, R.B. (1989), Factors affecting the response of lung clearance systems to acid aerosols: Role of exposure concentration, exposure time, and relative acidity, *Environ. Health Persp.* 79:121-127.

Schlesinger, R.B., L.C. Chen, I. Finkelstein, and J.Z. Zelikoff (1990), Comparative Potency of Inhaled Acidic Sulfates: Speciation and the Role of Hydrogen Ion, submitted to Env.Res.

Schwartz, S.E. (1988), Are global cloud albedo and climate controlled by marine phytoplankton? *Nature* 336:441-445.

Shy, C.M. (1989), Review, discussion, and summary of epidemiological studies, *Environ. Health Persp.* 79:187-190.

Solomon, P.A., et al. (1988), Acquisition of Acid vapor and Aerosol Concentration Data for Use in Dry Deposition Studies in the South Coast Air Basin, EQL Report 25, Environmental Quality Laboratory, California Institute of Technology, Pasadena, CA.

Spengler, J.D. (1989), personal communication, Nov. 22, 1989.

Spengler, J.D., et al. (1989), Exposures to acidic aerosols, *Environ. Health Persp.* 79:43-52.

Tanner, R. L. (1989), The measurement of strong acid in atmospheric samples, in Methods of Air Sampling and Analysis, 3rd ed., J.P. Lodge, ed., Lewis Publishers, Chelsea, MI, 1989, pp. 703-714.

Tanner, R.L. et al. (1979), Chemical Composition of Sulfate as a Function of Particle Size in New York Summer Aerosol, *Ann. NY Acad. Sci.* 322:99-113.

Thomas, M.D. (1962), Sulfur Dioxide, Sulfuric Acid Aerosol, and Visibility in Los Angeles, *Int. J. Air Wat. Poll.* 6:443-454.

Tropp, R.J. (1989) Acid Aerosol Measurement Workshop, EPA/600/9-89/056, U.S. Environmental Protection Agency, Research Triangle Park, NC.

U.S. Department of Health, Education, and Welfare (1969), "Air Quality Criteria for Sulfur Oxides," National Air Pollution Control Administration Publication No. AP-50, Washington, DC. 1969.

U.S. Environmental Protection Agency (1974), Health Consequences of Sulfur Oxides: A Report from CHESS, 1970-71. EPA-650/1-74-004, May 1974.

U.S. Environmental Protection Agency (1980), Addendum to "Health Consequences of Sulfur Oxides: A Report from CHESS, 1970-71," May 1974. EPA-600/1-80-021. April 1980.

U.S. Environmental Protection Agency (1982), Air Quality for Particulate Matter and Sulfur Oxides, EPA-600/8-82-029a, Dec. 1982.

U.S. Environmental Protection Agency (1988), Acid Aerosols Issue Paper, EPA-600-8-88-005a, Washington, DC, 1988.

Utell, M.J., J.A. Mariglio, P.E. Morrow, F.R. Gibb, and D.M. Speers (1989), Effects of Inhaled Acid Aerosols on Respiratory Function: The Role of Endogenous Ammonia, *J. Aerosol Med.* 2:141-147.

Waldman, J.M., P.J. Lioy, G. D. Thurston, and M. Lippmann (1990), Spatial and temporal patterns in summertime sulfate aerosol acidity and neutralization within a metropolitan area, *Atm. Env.* 24B:115-126.

Waldman, J.M., and M.R. Hoffmann (1987), Depositional aspects of pollutant behavior in fog and intercepted clouds. In Sources and Fates of Aquatic Pollutants, ed. by R.A. Hites and S.J. Eisenreich. Advances in Chemistry Series 216, pp. 80-129, American Chemical Society, Washington, DC.

Waller, R.E. (1963), Acid Droplets in Town Air, *Int. J. Air Water Poll.* 7:773-778.

Wyzga, R.E., and F.W. Lipfert (1990), The Need to Reconcile Experimental and Ambient Exposures to Acid Aerosols, presented at the NAPAP 1990 International Conference on "Acidic Deposition: State of Science and Technology," Hilton Head, SC, Feb. 1990.

OCEANIC SOURCES OF SULPHUR AND THEIR CONTRIBUTION TO THE ATMOSPHERIC SULPHUR BUDGET: A REVIEW

C. Nicholas Hewitt and Brian Davidson
Institute of Environmental and Biological Sciences
Lancaster University, Lancaster, U.K.

Abstract

This paper reviews current knowledge concerning the biogenic emissions of sulphur from the oceans and their contribution to the atmospheric sulphur budget. In particular, the temporal and spatial distributions of such emissions are considered, as are the magnitude of their fluxes. The importance of the reduced sulphur species to the deposition of acidity from the atmosphere, relative to anthropogenic sources, is also discussed.

1. Introduction

All models of the natural sulphur cycle require a volatile or gaseous sulphur compound to allow transfer of sulphur from the sea to air. As hydrogen sulphide (H_2S) was originally thought to be the most abundant primary natural sulphur compound it was suggested for this role. However attempts failed to detect it in the atmosphere at the concentrations predicted. Challenger (1951) first observed dimethyl sulphide (DMS) production by marine algae and subsequent measurements of its concentrations in sea water lead to the suggestion that it may be the volatile sulphur compound responsible for the sea to air transfer of the element (Lovelock et al, 1972). DMS is now known to be the most abundant reduced sulphur species in sea water and to be a major contributor to the atmospheric sulphur budget.

Here we review the measurements made to date of the most important reduced sulphur compounds in the atmosphere and attempt to quantify their contribution to the atmospheric sulphur budget and the deposition of acidity.

2. Primary Sulphur Compounds

Dimethyl sulphide

DMS is one of the breakdown products of dimethyl sulphoniopropionate (DMSP), a compound involved in regulating cellular osmotic pressure in algae (Dickson et al,1980). Dacey and Wakeham (1986) observed that ingestion of phytoplankton by zooplankton releases DMS into the water column and this "harvesting" of phytoplankton during bloom periods releases substantial quantities of DMS into the atmosphere, with values increasing 40 to 60 fold over those during non-bloom periods (Turner and Liss 1985).

During cruises in the Atlantic Ocean and elsewhere Barnard et al(1982) and Andreae and Barnard (1983a,1984) observed a wide range of DMS concentrations in the water column (17-700 $ng(S)L^{-1}$, the average being around 90 $ng(S)\ L^{-1}$) with the highest values being found in the more productive waters of the continental shelf. The vertical distribution of DMS through the water column showed a subsurface maxinmum around 20m and a gradual decrease in concentration in the euphotic zone there after. Early work (Andreae and Barnard, 1984) suggested a correlation between DMS and chlorophyll (a). Subsequent studies have failed to confirm this. Although DMS is produced by a number of phytoplankton its production varies with species and stage of growth (Turner et al,1989). Other factors, such as salinity, temperature etc., may also be of importance in the amount of DMS produced by phytoplankton, though their significance has yet to be determined. Bates et al's (1987) comprehensive study of DMS concentrations in Pacific waters off the coast of north and south America found seasonal variations most noticeable in the 5-50° latitude belt with concentrations ranging from 20 $ng(S)L^{-1}$ in winter to 60-70 $ng(S)L^{-1}$ in summer. Values around estuaries and in coastal waters were generally higher as were values close to the equator. These values were in general agreement with those obtained by Andreae at similar sites. Fewer data are available for Arctic and Antarctic waters but those of Berresheim (1987) show similar values, $60ng(S)L^{-1}$ of DMS in samples from Antarctic and sub-Antarctic waters.

Detection of atmospheric DMS has proved more difficut, as the co-trapping of atmospheric oxidants with the sample will lead to decomposition of the sample prior to analysis. This problem casts doubt on the validity of some early measurements of DMS in the atmosphere. Various measurements have been made in the Pacific Ocean and those considered to be most reliable are summarized here. During sampling cruises in the equatorial Pacific and Sargasso Sea, Andreae et al (1985) observed mean DMS concentrations of 160 $ng(S)m^{-3}$ and 230 $ng(S)m^{-3}$ respectively while values from Cape Grim, Tasmania were in the range 34-381 $ng(S)m^{-3}$ with a mean of 167$ng(S)m^{-3}$. Nguyen et al (1983) found mean concentrations of 20 $ng(S)m^{-3}$ in the Pacific.

A diurnal cycle in the concentration of DMS was first observed in the Pacific Ocean with a daytime minimum of 120 $ng(S)m^{-3}$, attributed to DMS oxidation by photolytically produced hydroxyl radicals (OH), and a night time maximum of 200 $ng(S)m^{-3}$ (Andreae and Raemdonck, 1983(b)). Saltzman and Cooper (1988) observed a similar diurnal cycle in the trade winds of the Caribbean during a north-south transect, with daytime values of 56 $ng(S)m^{-3}$ and a nighttime maximum of 83 $ng(S)m^{-3}$. In clean Antarctic air with a mean atmospheric concentration of 120$ng(S)m^{-3}$ a similar cycle has been found (Berresheim, 1987). Values in coastal regions were found to be very variable during this study, possibly due to algae growth on icebergs. Measurements in North Atlantic air east to the Azores showed a high mean concentration, 180$ng(S)m^{-3}$, decreasing to 60$ng(S)m^{-3}$ in continentally influenced air. Mid Atlantic atmospheric DMS concentrations of 32-38 $ng(S)m^{-3}$ were found during spring between latitudes 49N - 34S by Burgermeister and Georgii (1990), while values from the more productive English Channel and North Sea were considerable higher, 185 and 561 $ng(S)m^{-3}$.

Continentally influenced air masses are found to have lower DMS concentrations and lack a clear diurnal cycle. Andreae et al(1985) suggested the overall lower levels of DMS were due to increased concentration of atmospheric oxidants with the lack of a diurnal cycle being due to nighttime oxidation of the DMS by nitrate radicals (NO_3).

Both Andreae et al(1983c) and Saltzman and Cooper (1988) observed oxidant scrubber failure during sampling in polluted air and the accuracy of early DMS values in such air still remains questionable. Using a new KI oxidant scrubber Saltzman and Cooper (1989) conducted a cruise in the western Atlantic Ocean and found a strong diurnal cycle to DMS with a diurnal variation of 3.6, the greatest observed to date.

So far all the DMS concentrations looked at have been shipboard measurements. The first vertical profile of DMS in marine air was conducted by Ferek et al (1986) who found a steady decline in DMS concentration from 100 $ng(S)m^{-3}$ at ground level to 60 $ng(S)m^{-3}$ at the cloud top level at 1 km under stable meteorological conditions off the coast of Barbados. Above this altitude there was a more rapid decline to a few $ng(S)m^{-3}$ at 2 - 3 km. Under more disturbed convective conditions the DMS concentrations at >2 km altitude were an order of magnitude higher than previously found. Luria et al (1986) made DMS measurements over the Gulf of Mexico and observed a similar decline in concentrations with increase in altitude. During this study marine boundary layer values of 39 $ng(S)m^{-3}$ were observed, the concentrations in continentally influenced boundary layer being lower at 10 $ng(S)m^{-3}$. In both cases free tropospheric values for DMS were below 4 $ng(S)m^{-3}$. A similar profile has been observed over the north east Pacific Ocean but with lower boundary layer concentrations attributed to low seawater DMS concentrations during the sampling period (Andreae et al, 1988). Similar results were found by Ockelmann et al (1986) during flights over the Azores where low concentrations were attributed to low marine production of DMS.

The kinetics and mechanism of DMS oxidation by OH and NO_3 radicals have been studied by a number of groups. Hynes et al (1986) investigated the initial step of the OH-DMS reaction and found a strong temperature dependence with addition favoured over abstraction at lower temperatures. At 288 K abstraction is favoured 55:45 over addition:

$$OH + CH_3SCH_3 \xrightarrow{2O_2} CH_3O_2 + CH_3SO_3H \quad \text{addition}$$

$$OH + CH_3SCH_3 \xrightarrow{O_2} CH_3SCH_2O_2 + H_2O \quad \text{abstraction}$$

This work contradicts previous studies which favoured

addition reactions (Atkinson et al, 1978; Kurylo, 1978). However recent work by Barnes et al(1988) agrees with the findings of Hynes and suggests dimethyl sulphoxide (DMSO) and HO_2 to be the products of OH addition to DMS. The abstraction of hydrogen from DMS either by OH or NO_3 leads to the formation of the peroxy radical $CH_3SCH_2O_2$ which eventually leads to the production of sulphur dioxide (SO_2).

The increasing body of data available on atmospheric DMS concentrations and a greater understanding of its reaction mechanisms and kinetics have led to a refining of models used to predict concentrations and behaviour. An early such model by Graedel (1979) considered OH radicals as the only loss mechanism for DMS. Andreae et al (1985) expanded on this by considering the diurnal cycle of DMS and its nighttime oxidation by NO_3. Chameides and Davis (1980) suggested that methyl iodide released from the sea into the atmosphere may produce the IO radical which could react rapidly with DMS to produce dimethyl sulphoxide (DMSO). This rapid reaction would provide another sink for DMS and help reconcile the difference between predicted and observed air concentrations. If oxidation by IO is a major reaction pathway for DMS, DMSO and dimethyl sulphone ($DMSO_2$) would be produced, but to date they have not be observed in the troposphere at the concentrations expected. However great uncertainty exists in the estimates of atmospheric concentrations of IO and the rate of possible DMSO depletion reactions. As methyl iodide is most abundant in coastal areas and high IO concentrations are likely to be localised its contribution to the chemistry of DMS may be small (Fletcher, 1989).

An important parameter in all models of the sulphur cycle is the sea to air flux of DMS. This has not been obtained by direct measurement but is arrived at by calculation:

$$\text{Flux} = k\ ([DMS_{sw}] - x\ P_{[DMS]g})$$

where x = DMS solubility in sea water and k = exchange constant (m/day), also known as the piston velocity (V_p).

Since sea water is supersaturated with DMS relative to the atmosphere, the term $x\ P_{[DMS]g}$ is negligible compared with $[DMS_{sw}]$ and the expression simplifies to

$$\text{Flux} = k\ [DMS_{sw}]$$

The regional and global flux rates of DMS from the oceans to the atmosphere may therefore be estimated by using estimates of the average sea water concentration of DMS and the piston velocity. This latter parameter is poorly known (see Liss, 1983 for a review) but, for example, Liss and Slater (1974) calculated a value of 4.6 m/day using $^{14}CO_2$ and transfer across a thin film while Broecker and Peng (1974) estimated a value of 2.8 m/day based on comparison with a radon deficit method. Global atmospheric fluxes of 39 x 10^{12} g S/yr and 16 x 10^{12} g S/yr have been estimated by Andreae and Raemdonck (1983(b)) and Bates et al (1987) respectively. The lower of these values is due mainly to the use of lower sea water concentration data obtained from non-tropical and coastal areas. These models still do not take into full account the seasonal nature of emissions, for example the dramatic increase in DMS concentrations observed in some northern seas during spring (Turner and Liss 1985). As DMS concentration data become available for more areas of the world's oceans the estimates of a global DMS flux to the atmosphere will become more certain.

Various attempts have been made to model the behaviour of DMS in the troposphere using box models with a DMS flux into the box and diurnal variations in OH concentration (Saltzmann and Cooper, 1988). However it has proved difficult to reconcile observed and calculated DMS concentrations, suggesting that either higher concentrations of oxidants are present in the "clean" troposphere or lower piston velocities have to be used to give lower calculated flux rates of DMS into the atmosphere.

Hydrogen sulphide (H_2S):

Prior to the suggestion of DMS being the predominant reduced sulphur gas in the atmosphere (Lovelock et al, 1972) this role had been assigned to H_2S with its distinctive odour recognised from tidal flats. However attempts to measure H_2S in the atmosphere in the expected quantities failed. The production of H_2S occurs by non-specific reduction of organic sulphur, or by sulphate reduction by anaerobic bacteria. Such conditions are found in swamps and tidal flats where the absence of oxygen can lead to sulphate reduction.

The majority of marine H_2S concentrations reported use the analytical method of Natusch et al (1972) which utilizes the formation of relatively stable silver sulphide and which does not suffer from the same potential oxidation problems as do the DMS methods. However a slight interference from COS has been reported (Cooper and Saltzman, 1987) which may have amplified the diurnal cycle in H_2S concentrations reported in early measurements. The measurements of Herrmann and Jaeschke (1984) in clean air over the Atlantic Ocean range from below detection at 14ng(S)m^{-3} to 145 ng(S)m^{-3} with the higher values only being observed in coastal waters, and with a mean value of 38 ng(S)m^{-3}. A diurnal cycle with a daytime minimum reflecting oxidation by OH was observed. During a cruise in the Gulf of Mexico and Caribbean Saltzman and Cooper (1988) observed concentrations up to 340 ng(S)m^{-3} with a mean of 11 ng(S)m^{-3}. Lower values were observed in continentally affected air.

Recent measurements of total sulphide in sea water show concentrations of about 1nM (Cutter and Krahforst, 1988) but how much of this is as free sulphide and hence available for exchange into the atmosphere is not clear. Using simultaneous DMS and H_2S measurements Saltzman and Cooper (1988, 1989) suggest that H_2S emissions may contribute about 10% of the non-sea salt sulphate in marine air but caution against the extrapolation of these data to the world ocean. Unfortunately there are too few data available at the present time to give a representative picture of temporal and spatial variations in H_2S concentrations and emission rates.

Carbon disulphide (CS_2)

Fewer data are available for CS_2. Sea and air measurements taken in the North Atlantic in spring and the Sargasso Sea in September indicated seawater to be supersaturated with CS_2 in relation to marine air (Kim and Andreae, 1987). Measurements in coastal waters of the North Atlantic from New York to Miami showed higher values in September: 3ng(S) L^{-1} as opposed to 2 ng(S) L^{-1} in April. Open ocean concentrations were lower than those in coastal water. Atmospheric CS_2 levels taken at the same time showed little temporal variation; means of 37 ng(S)m^{-3} in April and 31 ng(S)m^{-3} in September were found.

Atmospheric CS_2 concentrations at varying altitudes over the Azores were found to range from about 29 ng(S)m^{-3} at low altitudes to below the detection limit of 6 ng(S)m^{-3} in the free troposphere (Ockelmann et al, 1986). The free tropospheric concentrations observed by Carroll (1985) at 6 km were 329 ng(S)m^{-3} and 66 ng(S)m^{-3} at 7 km. These values are higher than those found by Kim and Andreae (1987) in the marine boundary layer and are attributed by Carroll to strong updraft from the intense cumulonimbus activity at the time of sampling. The suggestion of a marine source of free tropospheric CS_2 is not supported by the lower values found by Kim and Andreae (1987) who found a strong correlation between aerosol soot particles and atmospheric CS_2 concentrations, so suggesting a continental influence.

The flux rate of CS_2 to the troposphere from the oceans has been estimated as about 0.22 Tg (S) yr^{-1} (Kim and Andreae, 1987; Ockelmann et al, 1986) and the reaction of CS_2 with OH is believed to produce SO_2 and COS (Baulch et al, 1984; Barnes et al, 1983).

Carbonyl Sulphide (COS):

COS is the most stable of the atmospheric sulphur gases with an atmospheric lifetime estimated in years (Hewitt and Davison, 1988). It is therefore well mixed throughout the atmosphere and is believed to sustain a stratospheric sulphate layer (Carroll, 1985). COS is one of the products of OH oxidation of CS_2, as well as being found in sea water (Rasmussen et al, 1982). Aircraft measurements by Ockelmann et al (1986) in maritime air showed small variations between 643-786 ng(S)m^{-3} with an increase with altitude during strong convection. Higher concentrations were observed in continental air. Other measurements include those of Carroll (1985) who found 740 ng(S)m^{-3} over the western North Atlantic, Van Valin et al (1987) who found similar concentrations over the Gulf of Mexico and Khalil and Rasmussen (1984) who found concentrations of 670 ng(S)m^{-3}.

These latter data were extrapolated to give an estimated flux to the atmosphere of 1 Tg(S) yr^{-1} as COS.

Various other reduced sulphur gases, including methyl mercaptan and dimethyl disulphide, are released into the atmosphere but in most cases there are very limited data available on their concentrations and distribution. What is available seems to indicate the contribution of such species is negligibly small to the global sulphur budget.

3. Secondary Sulphur Compounds

DMS is generally acknowledged to be the major reduced sulphur gas emitted from the oceans and its temporal and spatial concentration fluctuations and atmospheric chemistry have therefore been investigated more thoroughly than those of other reduced sulphur gases. The oxidation pathway of DMS once transmitted to the atmosphere is still uncertain and the object of much research. However it is thought that MSA and SO_2 (leading to SO_4^{2-}) are two of the major products of oxidation. The detailed mechanisms of their production and their concentrations are still uncertain although measurements of the concentrations of MSA, SO_2 and DMSO (another possible oxidation product) have been undertaken in the last few years to give more information on the oxidation pathway of atmospheric DMS. The atmospheric concentrations and distribution of these secondary sulphur compounds must therefore be considered in understanding the pathways of the global sulphur cycle.

Sulphur dioxide (SO_2):

Sulphur dioxide in the atmosphere has two sources: it is one of the products of the oxidation of a number of biogenically produced reduced sulphur compounds and is also released into the atmosphere in large quantities by a variety of industrial processes. Hence true natural background concentrations, free from anthropogenic influences, are hard to obtain with values from the most pristine environment of the Antarctic probably being the only ones that may be classed as true background levels.

A vertical profile of SO_2 concentrations obtained over the Azores showed levels of 430 ng(S)m^{-3} at 500 m dropping to 60 ng(S)m^{-3} at 2000 m and increasing again to 130 ng(S)m^{-3} at around 3600 m (Ockelmann et al, 1986). There was believed to be some local anthropogenic input during these flights.

Similar values were observed in maritime air over the coastal Gulf of Mexico by Van Valin et al (1987) with some substantially higher values up to 4 $\mu g(S)m^{-3}$ being found within the boundary layer in continental air. A profile of SO_2 above the NE Pacific by Andreae et al (1988) found a decrease in SO_2 concentrations with increased altitude and this was attributed to the ocean being a sink for SO_2. Concentrations above 3 km were in the range 14-42ng$(S)m^{-3}$ compared with 5-8 ng$(S)m^{-3}$ within the boundary layer. These values are substantially lower than those previously reported.

One of the the most comprehensive data sets for maritime SO_2 is that of Nguyen et al (1983) who took shipboard measurements in the Indian, Atlantic and sub-Antarctic oceans and observed an average value of 50 ng$(S)m^{-3}$. Areas of greater productivity, eg, continental shelf areas and the Peruvian upwelling,gave higher values around 100-150 ng$(S)m^{-3}$ while lower values around 15 ng$(S)m^{-3}$ were observed in less productive seas such as the tropical Indian Ocean. Values at Dome C on the Antarctic continent were below the detection limit of 10 ng$(S)m^{-3}$.

Non sea salt sulphate (NSSS):

Non sea salt or excess sulphate (NSSS) is the extra sulphate in the aerosol above that originating from sea salt. It may be calculated from the relationship between the total SO_4^{2-} and the Na^+ or Cl^- concentration, these elements being considered to originate solely from sea salt. The excess SO_4^{2-} in remote marine regions, where SO_4^{2-} arising from anthropogenic SO_2 is assumed to be negligible, results from the oxidation of reduced sulphur gases produced locally. However, the ubiquitous presence of anthropogenic SO_2 makes true background values of NSSS difficult to estimate. As the small NSSS values are calculated from the measurement of two larger numbers there is uncertainty in the final estimate and this may explain the negligible or even negative NSSS values sometimes arrived at.

Saltzman et al (1986) measured NSSS throughout the year at a number of remote island sites in the Pacific Ocean and found a range of values between 100 - 300 ng$(S)m^{-3}$. Simultaneous measurements of methane sulphonic acid where made and concentrations in the range 3 - 60 ng(S)m^{-3} were found. A seasonality in NSSS concentrations was observed, with maximum concentrations occurring during spring and early summer. Other estimates of NSSS have been made in the vicinity of Bermuda: 630 ng(S)m^{-3} in the boundary layer and 180 ng(S)m^{-3} in the free troposphere (Luria et al, 1989). Recent Antarctic measurements showed very low levels of 10 ng(S)m^{-3} during March/April and these low values are attributed to a high MSA yield from DMS oxidation, due to the low ambient temperatures (Berresheim, 1987 and Pszenny, 1989).

Dimethyl sulphoxide (DMSO):

Dimethyl sulphoxide is one of the products of oxygen addition to DMS and according to laboratory studies, is the sole product of the DMS and IO reaction. DMSO in sea water samples were first detected by reducing the DMSO to DMS, with subsequent analysis by GC/FPD. Concentrations around 400 ng(S) L^{-1} were found in surface waters, decreasing with depth (Andreae, 1980). However DMSO has a low volatility and so significant transfer to the atmosphere does not occur.

Both DMSO and dimethyl sulphone ($DMSO_2$) have proved difficult to determine in air samples. Harvey and Lang (1986) published the first reported values of atmospheric DMSO and $DMSO_2$ by trapping on an adsorbent polymer, solvent extraction and analysis by GC. Watts et al(1987) found aerosol DMSO and $DMSO_2$ concentrations in samples collected near Plymouth, England in the range 0.2 - 0.4 ng(S)m^{-3} with springtime maxima and winter minima. This small data set are the only published to date. The low concentrations have led to the conclusion that either DMSO has a very short residence time in the atmosphere or that oxidation of DMS by IO cannot be a major sink for DMS.

Methane sulphonic acid (MSA):

MSA is a product of the oxidation of DMS by OH. It was first identified in the aerosol phase in continental air over Germany at concentrations between 6 - 99 ng(S)m^{-3} by a method involving esterification with identification by GC/FPD (Panter and Penzhorn, 1980). It has since been determined in maritime air by ion chromotography (Saltzman et al, 1983) and its occurrence in the sub 0.5 µm particle size range suggests that it is formed by a gas - particle conversion (Davison and Hewitt, unpublished data). A seasonality in MSA concentrations has been observed in Pacific air (Saltzman et al, 1986) and near Plymouth, UK with a spring/early summer maxima of 30 $ng(S)m^{-3}$ and a winter minimum of 4 $ng(S)m^{-3}$ (Watts et al, 1987).

Samples taken from a platform in the North Sea during May gave concentrations of 60 $ng(S)m^{-3}$ (Burgermeister and Georgii, 1990). During a period of stable easterly winds a diurnal cycle with a daytime maximum of 18 $ng(S)m^{-3}$ and a nighttime minimum of 10 $ng(S)m^{-3}$ was observed and this was attributed to MSA production by the photochemical oxidation of DMS by OH. These concentrations agree well with the model predictions of Fletcher (1989) which indicated MSA concentrations between 16-22 hrs to be twice those during 4-10 hrs in air off the Scandinavian coast. A similar diurnality has been observed in N.W. Scotland (Davison and Hewitt, unpublished data).

4. The contribution of biogenic sulphur compounds to acidity

The dissolution of CO_2 in cloud and rain droplets may lower the pH of rainwater to about 5.6. However, in some remote areas the pH of rainwater is typically around 5 (Galloway et al, 1982) with values ranging between 4.6 and 5.6. These levels of acidity have been observed even in remote, unpolluted, areas of the Southern Hemisphere and hence are attributed to natural sources of acidity, rather than to anthropogenic emissions. The most important contributor of acidity in unpolluted marine precipitation is H_2SO_4, the other acids HNO_3, HCOOH and CH_3SO_3H contributing substantially less (Galloway et al, 1982; Saltzman et al, 1986) and the oxidation of biogenic reduced sulphur species from the oceans to sulphate may therefore play a major role in determining the acidity of pristine precipitation. This model has been tested by Vong et al (1988) by the simultaneous sampling of submicrometer aerosol particles and rainwater in maritime air masses. They found that the sulphuric acid component of these particles was sufficiently large to account for the observed rainwater sulphate and acidity and that the NSSS controlled rainwater pH in these air masses.

An assessment of the role of biogenic sulphur from the Gulf of Mexico to the sulphur budget of North America has been attempted by measuring sulphate and DMS concentrations in offshore and onshore air during aircraft flights (Luria et al, 1986). DMS concentrations were found to vary significantly, being less than $4 ng(S) m^{-3}$ above the boundary layer, and averaging 10 ± 4 $ng(S) m^{-3}$ and 39 ± 43 $ng(S) m^{-3}$ in the offshore and onshore flowing boundary layer respectively. Non sea salt sulphate was found to account for the majority of the total aerosol loading in all samples. By assuming that DMS is removed from the atmosphere only by reaction with OH and NO_3, the flux rate of DMS into the boundary layer from the sea surface (R(DMS)) was calculated to be 2×10^9 molecules $cm^{-2} s^{-1}$ or 0.1 Tg (S) y^{-1} from the Gulf of Mexico.

The actual amount of sulphur transported over the North American coast depends upon the lifetimes of the various sulphur compounds in the air and the volume of air crossing the coast. Whilst the DMS lifetime is easily obtained from the gas kinetics equations, estimating the lifetime of sulphate aerosol is more problematical. Psuedo-first order rate constants for the wet and dry deposition processes (k_w and k_d) may be calculated, knowing the removal rate of sulphate, giving an atmospheric lifetime for SO_4^{2-} of 90 hours. However, it is also possible to calculate the SO_4^{2-} lifetime from the chemistry of DMS:

$$d[SO_4^{2-}]/dt = (k_1[OH] + k_2[NO_3])\ [DMS] - (k_w + k_d)[SO_4^{2-}]$$

and this gives a value of 530 hours.

Using both extreme values of the lifetime of SO_4^{2-} Luria et al (1986) calculated limit values for the amount of sulphur crossing the Gulf Coast. These were 0.004 and 0.25 $Tg(S) y^{-1}$, representing < 0.25 - 1.5% of the total sulphur emission of the US. Using the method of Charlson and Rodhe (1982) they also calculated the effect of the biogenic sulphur on cloud water acidification at three different cloud water concentrations (0.1, 0.5 and 2.5 g m^{-3}). The resultant estimated pHs lay in the range 2.9 - 5.1, suggesting that natural acidification of rainwater may be occurring.

A more complex Lagrangian chemical model has been developed to study the contribution of biogenic emissions from the

North Sea to sulphate over Scandinavia (Fletcher, 1989). This is based on a 102 step reaction scheme encompassing 47 species and uses a DMS emission rate of 3.5×10^{-13} - 1.1×10^{-11} $kg(S)m^{-2}s^{-1}$ (or 40 $Tg(S)y^{-1}$ when extrapolated to the world's oceans). Provided MSA is the major product of the DMS-OH reaction the model predicts that it could account for 30-50% of the total sulphur acids in Scandinavian air during phytoplankton bloom periods and may make a significant contribution to NSSS concentrations.

The emission of biogenic sulphur compounds vary substantially during the year, being at a minimum in the winter, when anthropogenic SO_2 output is at a maximum. Using spring and summer DMS values Turner et al (1989) calculated a flux of 32 $\mu mol(S)m^{-2}$ d^{-1} from seas around Europe. The anthropogenic emission of SO_2 in Europe was 638 Gmol (S) y^{-1} in 1983, or 128 μmol (S) m^{-2} d^{-1}, which implies that the biogenic emissions are equivalent to ~25% of man's on an area basis. Decreasing SO_2 emissions over recent years may make this proportion still higher.

References

Andreae MO,1980, Determination of trace quantities of dimethyl sulhphoxide in aqueous solution. *Anal. Chem.* 52, 150-153.

Andreae MO and Barnard WR,1983(a), Determination of trace quantities of dimethyl sulphide in aqueous Solution. *Anal. Chem.* 55, 608-612.

Andreae MO and Raemdonck H,1983(b), Dimethyl sulphide in the surface ocean and the marine atmoshere: A global view. *Science* 221,744-747.

Andreae MO, Barnard WR and Ammons JM,1983(c), The biological production of dimethyl sulphide in the ocean and its role in the global atmospheric sulphur budget. *Environ. Biochem., Ecol. Bull.* 35, 167-177.

Andreae MO and Barnard WR,1984, The marine chemistry of dimethyl sulphide. *Marine Chem.*14, 267-279.

Andreae MO, FerekRJ, Bermond F, Byrd KP, Engstrom RT, Hardin S, Houmere PD, Marrec H, Raemdonck H and Chatfield RB, 1985, Dimethyl sulphide in the marine atmosphere. *J. Geophys. Res.* 90, D7, 12891-12900.

Andreae MO, Berresheim H, Andreae TW, Kritz MA, Bates TS and Merrill JT,1988, Vertical distribution of dimethyl sulphide, sulphur dioxide, aerosol ions and radon over the north east Pacific ocean. *Atmos. Chem.* 6,150-173.

Atkinson R, Perry RA and Pitts JN, 1978, Rate constant for the reaction of OH radical with COS, CS_2 and CH_3SCH_3 over the temperature range 299-430 K. *Chem. Phys. Lett.* 54,14 –18.

Barnard WR, Andreae MO, Watkins WE, Bingemer H and Georgii HW, 1982, The flux of dimethyl sulphide from the ocean to the atmosphere. *J. Geophys. Res.* 87, C11, 8787-8793.

Bates TS, Cline JD, Gammon RH and Kelly-Hansen SR, 1987, Regional and seasonal variations in the flux of oceanic dimethyl sulphide to the atmosphere. *J. Geophys. Res.* 93,C3,2930-2938.

Barnes I,Becker KH,Fink EH, Reimer A, Zabel F and Niki H, 1983, Rate constant and products of reaction CS_2+ OH in the presence of O_2, *Int. J. Chem. Kinetics,* 15, 631-645.

Barnes I, Bastian V and Becker KH, 1988, Kinetics and mechanisms of the reaction of OH radicals with dimethyl sulphide. *Int. J. Chem. Kinet.* 20,415-431.

Baulch DL,Cox RA,Hampson RF, Kerr JA, Troe J and Watson RT, 1984, Evaluated kinetic and photochemical data for atmospheric chemistry : II, Codata task group on gas-phase chemical kinetics, *J. Chem. Phys. Ref. Data,*13,1259-1280.

Berresheim H,1987, Biogenic sulphur emissions from the subantarctic and Antarctic oceans. *J. Geophys. Res.* 92, D11,13,245-13262.

Broecker WS and Peng TH,1974, Gas exchange rates between air and sea. *Tellus* 26,21-35.

Burgermeister S and Georgii HW,1990, Distribution of methane sulphonate, NSS-sulphate and dimethyl sulphide over the Atlantic and north sea. *Atmos. Environ.*

Carroll MA, 1985, Measurements of OCS and CS_2 in the free troposphere. *J. Geophys. Res.* 90, D6, 10483-10486.

Challenger R,1951, Biological methylation, *Advan. Enzymol.*12,429-12,491.

Chameides WL and Davis DD, 1980, Iodine: Its possible role in tropospheric photochemistry. *J. of Geophys. Res.* 85,C11,7383-7398.

Charlson RJ and Rodhe H, 1982, Factors controlling the acidity of natural rainwater, *Nature*,295,683-685.

Cooper DJ and Saltzman ES, 1987, Uptake of carbonyl sulphide by silver nitrate impregnated filters: Implications for the measurement of low level atmospheric H_2S, *Geophys. Res Lett.,* 14, 206-209.

Cutter GA and Krahforst CF, 1988, Sulphide in surface waters of the western Atlantic ocean, *Geophys. Res. Lett.*,15,1393-1397.

Dacey JWH and Wakeham SG,1986, Oceanic dimethyl sulphide: Production during zooplankton grazing on phytoplankton. *Science*, 223,1314-1316.

Dickson DM, Wyn Jones RG and Davenport J, 1980, Steady state osmotic adaption in ulva lactuca. *Planta,* 150, 158-165.

Ferek RJ, Chatfield RB and Andreae MO, 1986, Vertical distribution of dimethyl sulphide in the marine atmosphere. *Nature*. 320, 514-516.

Fletcher I,1989, North sea dimthyl sulphide emissions as a source of background sulphate over Scandinavia: A model, in *Biogenic Sulphur In The Environment*, Eds ES Saltzman and WJ Cooper, ACS Symp. 393, 489-501.

Galloway JN, Likens GE, Keene WC, and Miller JM, 1982, The composition of precipitation in remote areas of the world, *J. Geophys. Res.* 87,8871-8886.

Graedel TE,1979, Reduced sulphur emission from the open oceans. *Geophys. Res. Lett.,* 6,4,329-331.

Harvey GR and Lang RF, 1986, Dimethyl sulphoxide and dimethyl sulphone in the marine atmosphere. *Geophys. Res. Lett.* 13, 1, 45-51.

Herrmann J and Jaeschke W,1984, Measurements of H_2S ansd SO_2 over the Atlantic ocean. *J. Atmos. Chem.*,1,111-123.

Hewitt CN and Davison BM, 1988, The lifetime of organosulphur compounds in the troposphere. *App. Organomet. Chem.* 2,407-415.

Hynes AJ, Wine PH and Semmes DH, 1986, Kinetics and mechanism of OH reactions with organic sulphides. *J. Phys. Chem.* 90,4148-4156.

Khalil MAK and Rasmussen RA, 1984, Global sourses, lifetimes and mass balance of carbonyl Sulphide (OCS) and carbon disulphide CS_2 in the earth's atmosphere. *Atmos. Environ.* 18,1805-1813.

Kim KH and Andreae MO, 1987, Carbon disulphide in seawater and the marine atmosphere over the north Atlantic. *J. Geophys. Res.* 92,D12,14733 -14738.

Kurylo MJ,1978, Flash photolysis resonance fluorescence investigation of the reaction of OH radicals with dimethyl sulphide. *Chem. Phys. Lett.* 58, 233-237.

Liss PS and Slater PG,1974, Flux of gases across the air-sea interface. Nature,247,181-184.

Liss PS, in *Air-Sea Exchange Of Gases And Particles*;1983, Eds Liss PS and Slinn WGN, 241-298.

Lovelock JE, Maggs RY and Rasmussen RA, 1972, Atmospheric dimethyl sulphide and the natural sulphur cycle. *Nature.* 237,452-453.

Luria M, Van Valin CC, Wellman DL and Pueschel RF, 1986, Contribution of Gulf area natural sulphur to the N. American sulphur budget. *Environ. Sci. Technol.* 20,91-95.

Luria M, Van Valin CC, Galloway JN, Keenes WC, Wellmann DL, Sievering H and Boatman JF, 1989, The relationship between dimethyl sulphide and particulate sulphate in the mid-Atlantic ocean atmosphere. *Atmos. Environ.* 23,1,139-147.

Natusch DFS, Klonis HB, Axelrod HD, Teck RJ and Lodge JP,1972, Sensitive method for measurement of atmospheric hydrogen sulphide. *Anal. Chem.* 44,12, 2067-2070.

Nguyen BC, Bonsang B and Gaudry A,1983, The role of the ocean in the global amospheric sulphur cycle. *J. Geophys. Res.* 88,C15,10903-10914.

Ockelmann G, Burgermeister S, Ciompa R and Georgii HW, 1986, Aircraft measurements of various sulphur compounds in a marine and continental environment. in *4th European Symp. of Atmos. Pollutants*, Reidel Press, 596-603.

Panter R and Penzhorn RD,1980, Alkyl sulphonic acids in the atmosphere. *Atmos. Environ.,* 14,149-151.

Pszenny AAP, Castelle AJ, Galloway JN and Duce RA,1987, A study of the sulphur cycle in the Antarctic marine boundary layer, *J. Geophys. Res.*,94,D7,9819-9830.

Rasmussen RA, Hoyt SD and Khalil MAK, 1982, Atmospheric carbonyl sulphide (OCS): Techniques for measurement in air and water. *Chemosphere,* 11,9,869-875.

Saltzman ES, Savoie DL, Zika RG and Prospero JM, 1983, Methane sulphonic acid in the marine atmosphere. *J. Geophys. Res.* 88, C15, 10897-10902.

Saltzman ES, Savoie DL, Prospero JM and Zika RG, 1986, Non-sea salt sulphate in Pacific air: Regional and seasonal variations. *J.Atmos. Chem.*, 4, 227-240.

Saltzman ES and Cooper DJ,1988, Shipboard measurements of atmospheric dimethyl sulphide and hydrogen sulphide in the Caribbean and Gulf of Mexico. *J. Atmos. Chem.* 14,7,191-209.

Saltzman ES and Cooper DJ, 1989, Dimethyl sulphide and hydrogen sulphide in marine air, in *Biogenic Sulphur In The Environment*, Eds. ES Saltzman and WJ Cooper, ACS Symp. 393,330-351.

Turner SM and Liss PS, 1985, Measurements of various sulphur gases in a coastal marine environment. *J. Atmos, Chem.* 2,223-232.

Turner SM, Malin G and Liss PS, 1989, Dimethyl sulphide and (dimethylsulphonio) propionate in European coastal and shelf waters in *Biogenic Sulphur In The Environment*, Eds ES Saltzman and WJ Cooper, ACS Symp. 393, 183-200.

Van Valin CC,Berresheim H, Andreae MO and Luria M, 1987, Dimethyl sulphide over the western Atlantic ocean. *Geophys. Res. Lett.* 14,7,715-718.

Vong RJ, Hansson H-C, Covert DS, and Charlson RJ, 1988, Acid rain: Simultaneous observations of a natural marine background and its acidic sulphate aerosol precursor. *Geophys. Res. Lett.,* 15,4,338-341.

Watts SF, Watson A and Brimblecombe P, 1987, Meaasurements of the aerosol concentrations of methane sulphonic acid, dimethyl sulphoxide, and dimethyl sulphone in the marine atmosphere of the British Isles. *Atmos. Environ.* 21,12, 2667-2672.

ACID PRECIPITATION RESEARCH IN BRAZIL: A SHORT REVIEW

Alvaro Ramon Ceolho Ovalle and Emmanoel Vieira da Silva Filho
Department of Geochemistry, Fluminense Federal University,
Rio de Janeiro, Brazil

INTRODUCTION

The acid precipitation phenomenon has been extensively studied in various regions of the northern hemisphere, where anthropogenic influences have disrupted natural biogeochemical cycles. The impact of this acidity ranges from acidification of lakes, rivers, groundwater and soils, to changes in agricultural and forest crop productivity (Mason & Seip, 1985).

The acidity of rainfall is associated with carbon, sulphur and nitrogen oxidation, principally from industrial and fossil fuel combustion sources. These anthropogenic emissions may induce a disequilibrium in the atmosphere redox balance, a system more susceptible to these modifications because of its lower dimensions when compared to the hydrosphere and lithosphere (Stum et al., 1987).

In Brazil industrialisation and urbanisation processes have accelerated in the last 30 years, with installation of industries and urban centre growth concentrated along the Brazilian Atlantic Coast (Fig. 1). However, this development has occured without considering the full environmental impact. At this time gaseous, particulate and liquid effluents were freely discharged in soils, rivers and the atmosphere. In view of their more dynamic nature, materials injected into the atmosphere induce long distance impacts, making atmospheric pollution a chronic problem, principally at urban centres and industrial zones concentrated in Sao Paulo and Rio de Janeiro States (Fig. 1).

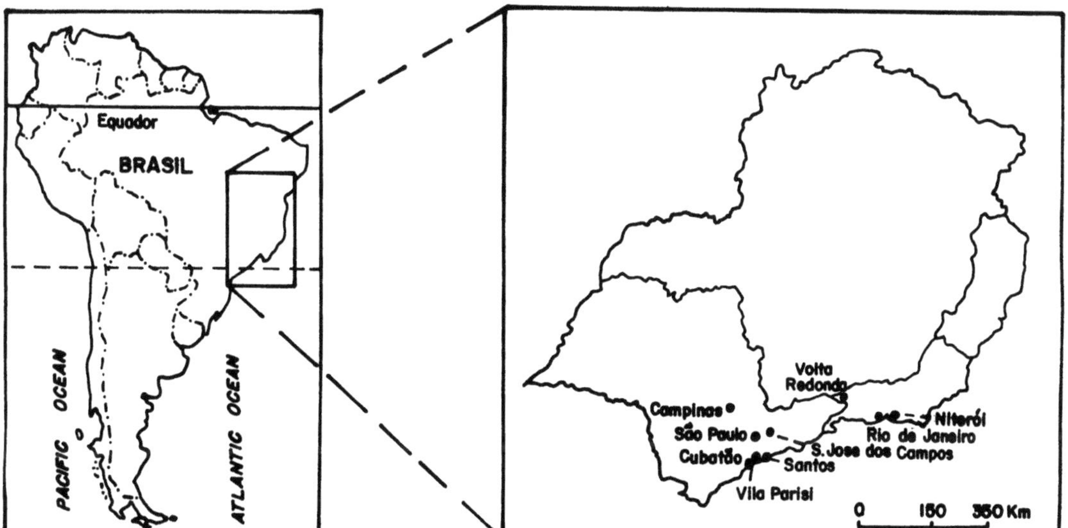

FIGURE 1 - Map of the Southeastern region of Brazil, showing principal areas of precipitation chemistry research.

Air quality has become a serious issue in some areas, such as Sao Paulo and Cubatao cities, where this problem assumed critical proportions, with serious damage to water resources, vegetation and population (CETESB, 1981; Kucinski, 1982; Monteleone Neto, 1982; and Queiroz Neto et al.,1984).The serious health implication of air pollution in these areas has meant that gaseous and particulate matter has been researched and controlled more fully than has acidic precipitation.

The first record of acidic precipitation in Brazil appeared in the literature at the end of the 1970's as part of nutrient cycling studies in forested areas, but only in the 1980's does the literature begin to contain specific papers concerning precipitation chemistry in urban centres and industrial zones.

The objective of this short review is to critically discuss the data available concerning acid precipitation in Brazil, classifying the various levels of the phenomena ranging from the detection of the problem to free acidity budget.

Methodology Used in Studies of Precipitation Chemistry

A critical aspect in analysing precipitation chemistry in Brazil is the absence of a standard method, which usually involves collection apparatus and periodic sampling. Collectors used vary from automatic wet - only samplers (Stallard and Edmond, 1981) to acrilic (Ferreira and Moreira - Nordemann, 1985) and plastic funnels (Silvo Filho, 1985). The latter are permanently opened and the results reported as bulk precipitation. As dry deposition plays an important role in the atmospheric inputs in Brazilian Atlantic Coast ecosystems (Ovalle et al., 1987), results obtained from bulk and wet - only collectors must be critically compared, as dry deposition may have great spatial variation, related to the nature of material injected into the atmosphere and local meteorological conditions.

The data discussed here have a sampling interval varying from single events (Mello & Motta, 1988) to weekly integrated samples (Silva Filho, 1985), and the duration of the study ranging from a few days (Stallard & Edmond, 1981) to some years (Ovalle et al., 1987). Variability of chemical composition of rain water either between events or between years indicates that comparison of data from different locations must be done carefully.

The Detection of the Problem

As precipitation chemistry is intimately related to atmospheric composition, air quality becomes a starting point to understand the acidity of rain water. An evaluation of atmospheric SO_2 content in Brazilian natural, industrial and urban areas is presented in Table 1.

High values for industrial and urban areas when compared to natural regions and rural sites in the United Kingdom, attest the critical state of the atmospheric problem in Brazil. These data indicate the relative position of South America in terms of global man - made emissions of sulphur dioxide.

The Brazilian scientific community involved with atmospheric pollution was alerted to acidity of rain water when mineral cycling studies in forested ecosystems on the Brazilian Atlantic Coast reported pH values ranging from 3.8 to 5.5 (Meguro et al., 1979; Silva Filho et al., 1984a,b; Silva Filho, 1985). These studies were carried out in forested areas between the metropolitan areas of Sao Paulo and Rio de Janeiro States. The frequency that precipitation pH lower than 4.0 was detected leaves no doubt about the magnitude of the problem, principally because these lower values are frequently associated to low volume events (< 10 mm/day), resulting in strongly aggressive solutions (Silva Filho et al., 1987). These data led to a pilot study of acidity related to fog in Rio de Janeiro city, a frequent phenomenon associatedwith coastal montane forested ecosystems during winter. Pedlowski (1986) showed that fog pH ranges between 3.5 and 7.5, with weighted mean of 4.9. Despite the tentative nature of the research, it is clear that there is an acid component present in the fog in the Rio de Janeiro metropolitan region. A study of precipitation chemistry in Cubatao, a highly industrialised zone with significant environmental problems, also reported the problem (Moeira - Nordmann et al., 1983).

Table 2 summarises the pH range in precipitation chemistry in some Brazilian areas. With the exception of the Amazon Forest, the other areas are localised in a zone of some 500km along the

Table 1 : Annual Mean Sulphur Dioxide in different Brazilian regions ; (after Moreira - Nordmann , 1987) and Rural sites in United Kingdom (United Kingdom Review Group on Acid Rain, 1987)

Location	ug SO_2 / m^3	Source
Sao Paulo	44.3	CETESB, 1985
Rio de Janeiro and Volta Re donda	60 - 80	FEEMA, 1984
Rural areas of Sao Paulo State	3.0	CETESB, 1985
Vila Parisi (1984*)	36	CETESB, 1985
Cubatao (1984*)	50	CETESB, 1985
Natural regions	0.15	Lawson & Winchester, 1978
Rural sites in United Kingdom	1.3 - 16.5	United Kingdom 1987

*** = mean value**

Table 2 : pH ranges in precipitation of some Brazilian areas

Location	pH range	Source
Rio de Janeiro	3.8 - 5.4	Silva Filho, 1985
Sao Paulo	3.8 - 6.8	Alonso et al., 1985
Sao Paulo	3.8 - 4.6	Meguro et al., 1979
Cubatao	3.7 - 4.7	Moreira - Nordeman et al., 1983
Niteroi	4.3 - 5.3	Mello & Motta, 1988
Amazon Forest	4.7 - 5.7	Stallard & Edmond, 1981

Brazilian Atlantic Coast (Fig.1), characterising acid precipitation as a well defined regional problem, associated to urban centres and highly industrialised zones.

In the Amazon region acid precipitation with pH values close to 4.0 have been reported (Ungemach, 1969; Brinkmann & Santos, 1973; apud Stallard & Edmond, 1981), but the authors assumed these values as natural in origin, as there is no plausible pollution source to which one could attribute this acidity. Large - scale deforestation in the Brazilian Amazon frequently associated with large scale burning is a potential source of atmospheric CO_2 and other gases (Malingreau and Tucker, 1988),that must be taken into account when assessing precipitation chemistry in the Amazon region.

Precipitation Chemistry and Sources of Elements

Table 3 summarises precipitation chemistry data available in the literature for various areas in Brazil, including natural regions (Amazon), urban centres (Rio de Janeiro, Niteroi, Sao Paulo, San Jose dos Campos, Salvador and Santos), and high industrialised areas (Cubatao and Vila Parisi). The Amazon, as would be expected for a natural region with minimum marine influence, presents low values for all chemicals and could be used as a background in discussion.

In a meq basis SO_4 and NO_3 together have a variable contribution to total anions in precipitation: Amazon - 34%; Santos - 47%; Cubatao - 69%; Sao Paulo - 87%; Vila Parisi - 93%. In the Amazon with no plausible pollution source, chloride is the principal anion, whereas in the other locations with high anthropogenic contribution, SO_4 and NO_3 tend to be dominant. Vila Parisi is a critical case, where SO_4 alone corresponds to 88% of total anions in precipitation.

In a general view we can separate a coastal group with high ionic content precipitation including Salvador, Rio de Janeiro, Niteroi, Cubatao, Vila Parisi and Santos, and an interior group with low ionic content precipitation comprising Sao Paulo, Sao Jose dos Campos and the Amazon.

Table 3 : Chemical composition of precipitation in some Brazilian regions; values in umol/1 mean/standard deviation); Amazon data are mean value for the whole basin, except snow samples.

Local	Na	K	Ca	Mg	Cl	NH_4	NO_3	SO_4	pH
Amazon	13/13	1.0/0.7	1.0/1.1	1.2/1.5	15/15	0.8/1.3	2.1/1.7	5.3/2.8	5.1/0.2
Sao Paulo	16.5/41.3	5.9/10.2	23/20	12/18	20/33	46/31	35/22	21/35	4.9/0.7
San José	3.5/---	2.6/---	5.7/---	1.7/---	7.3/---	23/---	-------	-------	4.4/---
Salvador	89/67	6.1/4.6	23/18	10/7.4	111/89	3.9/3.3	-------	-------	5.7/0.3
Rio de Janeiro	130/117	15/8.7	16/8.5	21/23	158/121	-------	-------	-------	4.6/0.4
Niterói	196/278	22/18	-------	21/28	140/200	19/12	6.1/5.0	22/27	4.7/0.3
Santos	221/326	18/19	9.7/10	45/54	296/423	26/41	25/27	61/91	6.4/---
Cubatao	144/148	113/128	15/15	45/45	166/166	72/89	40/26	85/93	4.2/---
Vila Parisi	139/113	143/205	893/1929	193/317	330/420	200/222	84/124	563/599	6.2/0.4

In the coastal group Salvador and Santos present normal pH values, with the latter showing SO_4 and NO_3 contents which could justify a higher acidity in precipitation. Data concerning sulphur and nitrogen compounds in Rio de Janeiro city are not available, but proximity with Niteroi and similar composition with respect to other species, suggest that Niteroi data could be used as an indicator for Rio de Janeiro metropolitan region. SO_4 content in Niteroi precipitation indicates that SO_2 emission from combustion of fossil fuel is the principal precursor of acidity in this area, with NO_3 having a minor contribution. On the other hand Salvador data reflects an insignificant anthropogenic contribution.

Cubatao and Vila Parisi are characterised by an intense industrial activity, with abnormal rain water ionic content for various species. Vila Parisi precipitation presents SO_4 and NO_3 contents that could justify a lower pH than those measured, but unfavourable topographic conditions for pollutant dispersion and the variety of materials injected in the atmosphere with further interaction with rain water generate a solution so complex that pH values tend to neutrality. On one occasion after 20 rain free days, long pH values up to 9.0 were measured (Moreira - Nordmann et al., 1986).

The interior group contains the Amazon data that represent a natural condition, whereas Sao Paulo and San Jose dos Campos present acid precipitation with high values for SO_4, NO_3 and NH_4, principally at Sao Paulo, related to the diverse nature of its sources. The metropolitan regions have a strong contribution from combustion of fossil fuel, and also from a great number of industrial areas with significant emission of atmospheric pollution. SO_4 values for these two locations are underestimated in view of analytical methods used (Moreira - Nordmann et al., 1985).

The sources of elements in precipitation can be grouped into general categories of marine and non - marine (including anthropogenic). Generally a reference element is used to quantify the marine contribution to precipitation chemistry. Data presented in Table 3 have a strong anthropogenic component that alters Cl content in precipitation collected in Cubatao and Vila Parisi, (Miller et al., 1985; apud Moreira - Nordmann et al., 1986). Sodium is also affected, but to a lesser extent than chlorine. Thus we used sodium as the reference element for evaluation of the proportion of marine source in precipitation chemistry.

In Table 4 we can observe that all ratios are above sea salt values, except calcium and chloride in the Amazon Basin, suggesting a significant contribution of a terrestrial source in precipitation chemistry. Various ratios present abnormal values, in particular those related to Vila Parisi for all chemicals, Sao Paulo for K, Ca, Mg and SO_4, San Jose dos Campos for K, Ca, Mg and Cl, Cubatao for K, Mg and SO_4 and the Amazon for SO_4. Except for the Amazon where the SO_4 ratio could be related to a great extent to natural emissions from soils and vegetation, the others reflect a strong anthropogenic contribution. The sulphur ratio shows extremely high values for Vila Parisi, Sao Paulo and Cubatao.

Table 4 : Mole Ratio between various elements and Na, compared to sea salt ; data from Table 3 used in calculations.

	Na	K	Ca	Mg	Cl	SO_4
Sea salt	1.000	0.021	0.119	0.022	1.173	0.060
Amazon	1.000	0.076	0.076	0.092	1.154	0.408
Sao Paulo	1.000	0.357	1.393	0.727	1.212	1.273
San Jose	1.000	0.743	1.628	0.486	2.086	-------
Salvador	1.000	0.073	0.277	0.120	1.337	-------
Rio de Janeiro	1.000	0.115	0.123	0.161	1.215	-------
Niteroi	1.000	0.112	-------	0.107	0.714	0.112
Santos	1.000	0.081	0.044	0.203	1.333	0.275
Cubatao	1.000	0.785	0.104	0.313	1.153	0.590
Vila Parisi	1.000	1.029	6.424	1.388	2.374	4.050

Long Distance Tranport and Acidity Deposition

Despite the importance of the long distance transport of acidity in the regionalisation of acid precipitation phenomena, little effort has been made to quantify the magnitude of this process in Brazil.

A single and superficial study of this aspect was carried out in Rio de Janeiro city by Silva Filho et al.(1987). They observed an association between events of acidity deposition in coastal forested ecosystems, and the passage of cold fronts, that originate in the south of the continent. During their pathway to the north, the fronts pass over the industrialised regions of Sao Paulo State before reaching Rio de Janeiro. Residence time in the atmosphere of between 2 to 5 days for nitrogen and sulphur oxides (Irwin & Williams, 1988; Babich et al., 1980; Likens & Borman, 1974) indicates that, besides a local contribution, long distance transport of acidity is probably occuring in Brazil's Atlantic Coast.

Silva Filho (1985) in a one year study of atmospheric inputs to a premontane forested ecosystem, showed that free acidity deposition varied weekly from 0.02 to 2.9 keq/ha/year with a mean value of 0.58 keq/ha/year. The deposition was characterised by occasional extremely high values, for example, the four weeks of higher precipitation volume were responsible for 13% of total annual flux. Comparing two collectors 1,500m distant and with 600m of elevation difference, he also noted that acidity deposition was spacialy uniform, despite fractioning observed for other elements such as chlorine, sodium and calcium.

Brown et al (1985) collected 180 precipitation samples at Tijuca National Park and reported a range of + 0.4 to + 12.5 % suggesting that despite a local component rain water suffered long distance transport before precipitation, acting as a mass transport agent for pollutants injected into the atmosphere.

Free Acidity Budget and Possibles Effects

The effect of acid precipitation on tropical ecosystems, particularly in Brazil, has not been evaluated. An outline of the possible environmental consequences of this phenomenum in Brazil, can be made by grouping data concerning Tijuca National Park in Rio de Janeiro city. By using data from Ovalle et al., 1987 , Silva Filho, 1985 and Ovalle, 1985, we can compose a free acidity budget based on pH measurements of rain and stream water in a 350 ha forested catchment.

The weekly fluxes are characterised by long periods of low values punctuated by occasional extreme values associated with precipitation pulses higher than 80mm/day.

During an 80mm storm, the values found were (eq/ha/year): atmospheric input = 11.7 and stream output = 0.55. Assuming this event as representative of the pulses > 80mm/day, and its annual frequency of occurrence along the studied period, they are responsible for 16 - 36% of annual free acidity input and > 95% of output measured.

Figure 2 shows weekly pH variations in rainwater, throughfall and stream water at Tijuca National Park. It is clear that during its course to stream channel, rainwater acidity decreases,

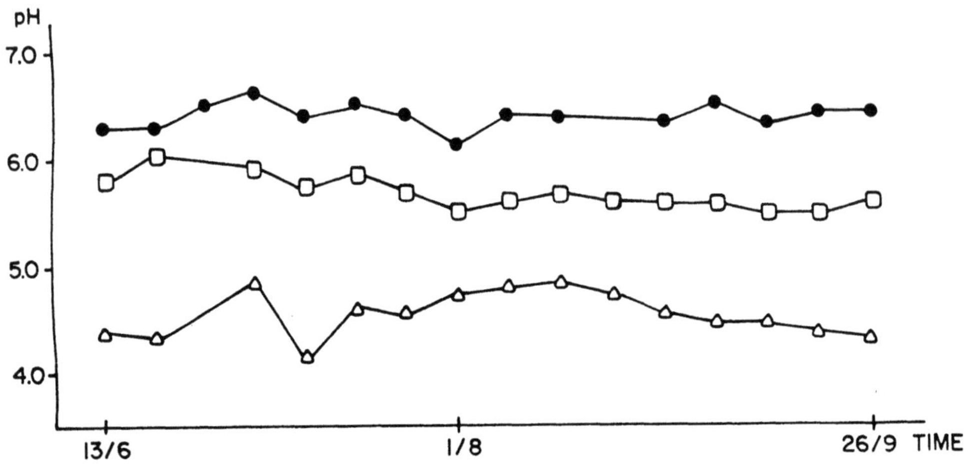

FIGURE 2 – pH weekly variation of rainwater (△), throughfall (□) and stream water (●) Tijuca National Park, Rio de Janeiro between June and September of 1984.

suggesting that this forested ecosystem acts as a natural filter. Silva Filho & Ovalle (1984) show that interaction with vegetation cover provides a pH change from 4.7 to 5.6, associated with a strong increase in water ionic content. They suggest that despite a simple washing of dry deposition from leaf surfaces, anionic exchange between rainwater and foliage could be responsible for acidity neutralisation, as calcium content in throughfall shows an increase of an order of magnitude. As forested ecosystems are associated with nutrient depleted soils, high losses of essential elements such as calcium could not be expected.

The acid nature of the soils (pH close to 4.0), the slow rate of weathering reactions (Ovalle et al., 1984), and mean pH of stream water (6.1), suggest that the neutralisation of acid precipitation at tree canopy level is an important mechanism in the filter action in these forested ecosystems.

Meguro et al. (1979) also report a similar behaviour between rainwater and throughfall in a secondary forested area at Sao Paulo city metropolitan area, suggesting that Brazilian coastal forested ecosystems may act as a natural filter for acid precipitation. The cost of this mechanism for forest nutrient cycling and consequently productivity has not been evaluated.

CONCLUSION

At present acid precipitation research in Brazil is in an initial stage, with data concerning acid precipitation being by - products of institutional research programmes. With the identification of the problem, appropriate research programmes are being formulated.

The major problem at this stage is related to the techniques of collection, preservation and analysis of rainwater samples that must be standardised in order to permit a regional approach to precipitation chemistry research in Brazil, and avoiding dispersion of efforts incomparable data, and optimising the often limited financial support.

In this context a programme was initiated in 1988 involving 6 different research centres distributed along the Brazilian coast, using the methodology suggested by Galloway and Likens (1978), and now used in a worldwide programme of precipitation chemistry. The starting point is to analyse the extent of the problem, and to create an interchange between emergency research groups and experienced ones.

In terms of environmental impacts the absence of data is almost total. The absence of long - term monitoring programmes investigating the dynamics of the diverse ecosystems existing in Brazil does not permit a realistic evaluation of the impacts. As acid precipitation effects are long - term, and although the effects would be small for some ecosystems they must be seen as additional added stress to ecosystems. The effects in most cases are unrecognised and often confounded by other sources of pollution (Likens & Borman, 1974). Other critical environmental problems such as the rate of forest clear - cutting turn studies of acid precipitation - vegetation interaction into a paleo - ecological exercise. Industrial pollution and cultural euthrophication of surface waters, make impacts of acid rain on aquatic resources a minor part of the complex environmental problems in Brazil.

REFERENCES

Alonso, C.D. ; Romano, J. e Massaro, S. 1985. Chuvas acidas em Sao Paulo - Medicoes de pH e titulacao em aguas pluviais em um ponto urbano. Suplemento Ciencia e Cultura 37 (7) : 604.

Babich, H. ; Davis, D.L. and Storzsky, G. 1980. Acid precipitation: Causes and Consequences. Environment, 22 (4) : 6 - 13.

Brinkmann, W.L.F. and Santos, A. 1973. Natural Waters in Amazonia. VI. soluble calcium properties. Acta Amazonica 3 (2) : 33 - 40.

Brown, I.F. ; Souza, M.A.T. e Vitoria, R.L. 1985. Variacoes em 180 nas aguas de chuva e do rio na bacia do Alto Rio Cachoeira, PNT, RJ. Suplemento de Ciencia e Cultura 37 (7) : 582.

CETESB. 1981. Degradacao da cobertura vegetal da Serra do Mar em Cubatao - Avaliacao preliminar. CEYESB/DEAR - ACR 140p.

Ferreira, C. e Moreira - Nordmann, L.M. 1985. Ocorrencia de ions nas precipitacoes pluviais de Salvador, BA. Revista Brasileira de Geofisica, 3 : 9 - 13.

Galloway, J.N. and Likens, G.E. 1978. The Collection of Precipitation for Chemical Anaysis. Tellus (1978), 30, 71 - 82.

Irwin, J.G. and Williams, M.L. 1988. Acid rain : chemistry and transport. Environmental Pollution, 50 : 29 - 59.

Kucinski, B. 1982. Cubatao, uma tragedia ecologica. Ciencia Hoje, 1 (1) : 11 - 23.

Likens, G.E. and Borman, F.H. 1974. Acid rain : a serious environmental problem. Science, 184 : 1176 - 1179.

Malimgreau, J.P. and Tucker, C.J. 1988. Large - scale deforestation in the southeastern Amazon Basin of Brazil. Ambio, 17 (1) : 49 - 55.

Mason, J. and Seip, H.M. 1985. The current state of knowledge on acidification of surface waters and guidelines for further research. Ambio, 14 (1) : 45 - 51.

Mello, W.Z. e Motta, J.S.T. 1988. Estudo preliminar da composicao quimica de fracoes de aguas pluviais em Niteroi - RJ. Acta Limnologica Brasiliensia, 2 : 897 - 909.

Meguro, M. ; Vinueza, G.N. e Delitti, W.B.C. 1979. Ciclagem de nutrientes na mata mesofila secundaria, Sao Paulo. II. O papel da precipitacao na importacao de potassio e fosforo. Boletim de Botanica, Universidade de Sao Paulo, 7 : 61 - 67.

Miller, E.A. ; Cooper, J.A. ; Frazier, C.A. e Pritchett, L.C. 1985. Cubatao aerosol source apportionment study - final report, Vol I. CETESB, Sao Paulo.

Monteleone Neto, R. 1982. Em Cubatao, uma nova chave para o entencimento de Anomalias. Ciencia Hoje, 1 (1) : 12.

Moreira - Nordeman, L.M. ; Bertoli, J.L.R. ; Cunha, R.C.A. e Palombo, C.R. 1983. Analise quimica preliminar das aguas de chuva de Cubatao - Impactos ambientais. Anais do V Simposio Brasileiro de Hidrologia de Recursos Hidricos, Vol 3 p. 339 - 350, Blumenau - SC, Novembro de 1983.

Moreira - Nordeman, L.M. ; Forti, M.C. ; Andrade, F. e Orsini, C.M. 1985. Composicao ionica das chuvas da cidade de Sao Paulo. VI Simposio Brasileiro de Hidrologia e Recursos Hidricos, Sao Paulo, Anais, 11-14.

Moreira - Nordeman, L.M. ; Danelon, D.M. ; Forti, M.C. ; Santos, C.M.E. ; Sardela, D.D. ; Lopes, J.C. ; Filho, B.M. e Abbas, M.M. 1986. Caracterizacao quimica das aguas de chuva de Cubatao. INPE - 3965 - PRE/515, 68p.

Moreira - Nordeman, L.M. 1987. A geoquimica e o meio ambiente. Geochimmica Brasiliensis, 1 (1) : 89 - 107.

Ovalle, A.R.C. ; Silva Filho, E.V. e Brown, I.F. 1984. Intemperismo e composicao dos rios no Parque Nacional da Tijuca, Rio de Janeiro. XXXIII Congresso Brasileiro de Geologia, Anais, 4717 - 4728.

Ovalle, A.R.C. 1985. Estudo Geoquimico de aguas fluviais da bacia do Alto Rio Cachoeira, Parque Nacional da Tijuca, Rio de Janeiro. Master Thesis, Federal Fluminense University, Rio de Janeiro, 85p.

Ovalle, A.R.C. ; Silva Filho, E.V. e Brown I.F. 1987. Element fluxes in a tropical pre - montane wet forest, Rio de Janeiro, Brazil. International Workshop on Geochemistry and Monitoring in Representative Basins, Extended Abstracts, Moldan, B. and Paces, T. Eds. Tchecoeslovaquia, 16 - 18.

Pedlowski, M.A. 1986. Estudo preliminar da contribuicao da neblina na ciclagem de nutrientes e na deposicao de acidez no Parque Nacional da Tijuca. Graduation Monograph, Depto de Geografia, Federal Rio de Janeiro University.

Queiroz Neto, J.P. ; Monteleone Neto, R. e Marques, R.A. 1984. Cubatao, 1984. Ciencia Hoje, 3 (13) : 80 - 86.

Silva Filho, E.V. ; Brown, I.F. e Ovalle, A.R.C. 1984a. Ocorrencia de chuva acida e deposicao de acidez livre no Parque Nacional da Tijuca, Rio de Janeiro. Suplemento Ciencia e Cultura, 36 : 651.

Silva Filho, E.V. ; Ovalle, A.R.C. e Brown, I.F. 1984b. Estudo biogeoquimico de entradas atmosfericas de Na, K, Ca e Mg na bacia do Alto Rio Cachoeira, Parque Nacional da Tijuca, Rio de Janeiro. XXXIII Congresso Brasileiro de Geologia, Anais, 4729 - 4745.

Silva Filho, E.V. e Ovalle, A.R.C. 1984. O papel de vegetacao na neutralizacao da chuva acida no Parque Nacional da Tijuca, Rio de Janeiro. IV Seminario Regional de Ecologia, Anais, 353 - 373.

Silva Filho, E.V. 1985. Estudos de chuva acida e entradas atmosfericas de Na, K, Ca, Mg e Cl na bacia do Alto Rio Cachoeira, Parque Nacional da Tijuca, Rio de Janeiro. Master Thesis, Federal Fluminense University, Rio de Janeiro. 85p.

Silva Filho, E.V. , Ovalle, A.R.C. e Brown, I.F. 1987. Precipitacao acida no Parque Nacional da Tijuca, Rio de Janeiro. Ciencia e Cultura, 39 (4) : 419 - 422.

Stallard, R.F. and Edmond, J.M. 1981. Geochemistry of the Amazon. 1.Precipitation chemistry and the marine contribution to the dissolved load at the time of peak discharge. Journal of Geophysical Research, 86 (c10) : 9844 - 9858.

Stum, W. ; Sigg, L. and Schnoor, J.L. 1987. Aquatic chemistry of acid deposition. Environmental Science and Technology, 21 (1) : 8 - 13.

Ungemach, H. 1969. Chemical rain water studies in the Amazon region. II Simposio y Foro de Biologia Tropical Amazonica, Anais, Idroba, J.M.Ed., 354 - 359.

United Kingdom Review Group on Acid Rain 1987 Acid Deposition in the United Kingdom 1981 - 1985, a second report. Published by Warren Spring Laboratory, Department of Trade and Industry, Gunnels Wood Road, Stevenage, Herts. SG1 2BX.

2
ECOSYSTEM EFFECTS

THE APPLICATION OF CAUSE-EFFECT CRITERIA TO THE RELATIONSHIP BETWEEN AIR POLLUTION AND FOREST DECLINE IN EUROPE

John L. Innes
Forestry Commission, Alice Holt Lodge,
Wrecclesham, Farnham, Surrey, U.K.

ABSTRACT

Many past studies relating forest decline to air pollution have failed to establish cause-and-effect, yet air pollution is widely regarded as the cause of forest decline. Recent work has enabled some hypotheses to be eliminated and the available evidence is beginning to suggest that the direct effects of gaseous air pollution may have been over-estimated. The application of rigorous cause-effect criteria brings into question some of the common assumptions about the role of air pollution in forest decline in Europe. Most studies now consider the chemical status of the soil to be important and that acidic deposition, as well as natural factors, may influence this.

Introduction

Forest decline, defined here as the progressive deterioration in the condition of a forest, is a widespread phenomenon. In some cases, there are clear pathological or entomological causes. However, in others, no ready explanation is apparent and air pollution has been suggested as the major cause of the deterioration of the trees. In an earlier review of the relationship between air pollution and forest decline, Innes (1989) drew attention to a number of problems facing any evaluation of the effects of acidic deposition and other forms of air pollution on forest health. It was argued that while the interaction between air pollution and forest decline was widely taken for granted, the evidence linking the two was weak and, in many cases, circumstantial. While it was clear that air pollution was the cause of some forms of forest decline in some areas, its role in others was much less obvious. Over the last two years, there have been considerable developments in the understanding of the factors leading to forest decline. These mostly stem from a critical evaluation of past work and detailed observations of trees in forest stands. As a result, it is now possible to make some inferences about cause-and-effect which previously were impossible because of the lack of evidence.

An American report (Committee on Biologic Markers of Air-Pollution Damage in Trees (CBMAPDT), 1989) lists a number of criteria for establishing cause-effect relationships. These provide a useful basis for an examination of the European literature related to forest decline. The five criteria are:

1. Strong correlation.
2. Plausibility of mechanism.
3. Responsiveness or experimental replication.
4. Temporality.
5. Weight of evidence.

In the following review, the evidence for the role of air pollution (including both gaseous pollution and acidic deposition) in forest decline is examined in relation to the above criteria. Particular emphasis has been given to data from Britain but, since forest damage is clearly an international problem, appropriate evidence has also been drawn from European and North American studies.

STRONG CORRELATION

According to this criterion, there should be a "consistent relationship between the measured effect and the suspected cause(s)" (CBMAPDT 1989). The criterion is based on the principle of Koch (1876) and Mosteller and Tukey (1977) that there should be a consistency of association between the cause and the effect.

The criterion pre-supposes two things; that there is a measurable effect and a measurable cause. Neither of these is always met. Innes (1989) drew attention to the problem of measuring forest decline; to date, measurements have largely been confined to estimates of the crown density of trees. These estimates are a poor measure of tree vitality (Bauch *et al.* 1985, Westman and Lesinski 1985, Innes and Cook 1989, Mahrer 1989, Schmid-Haas 1989, Rehfuess 1989) but can be sufficiently objective to enable comparisons through time and space to be made (Innes and Boswell 1988, 1989, 1990). Measurements of tree growth, which are a much more useful index, have rarely been made on a regional basis (examples include Strand 1980, Schweingruber *et al.* 1983, Hornbeck *et al.* 1986, Becker 1987, Barnard and Scott 1988, Johnson *et al.* 1988, Peterson *et al.* 1989).

There are numerous different types of pollution and the responses of trees to specific ones are extremely variable, depending on a wide range of other environmental factors. For example, Cronan *et al.* (1989) state that the sensitivity of trees to aluminium in the soil solution varies with solution *pH*, chemical speciation of the aluminium, calcium concentration of the growing medium, temperature, overall ionic strength of the growing medium, the form of any inorganic nitrogen in the soil solution, mycorrhizal interactions, soil moisture, plant nutrient status, initial vigour and species and genetic stock of the plant. This severely restricts any attempt to define a generality of association.

Given these limitations, it is hardly surprising that attempts to demonstrate a correlation between the extent of forest decline and air pollution have been largely unsuccessful. Detailed analyses of data from Britain (Innes and Boswell 1988, 1989) have not indicated any correlation between poor condition and air pollution; in fact the reverse is the case. If anything, the correlations suggest that forest condition is best in areas with the highest levels of pollution. Given the possibility of a nitrogen-fertilisation effect, the results are consistent with much of what is known about the response of trees to low levels of nitrogenous air pollution.

Elsewhere, there is little, if any, evidence of a correlation between air pollution and tree health or growth. Where such studies have been undertaken, the correlations generally indicate that air pollution is less important than other environmental factors, particularly those related to stand characteristics and soil conditions (Brooks 1989,Hauhs 1989, Neumann 1989).

Nevertheless, some significant correlations between air pollution and forest decline have been identified. One of the most convincing of these is the relationship between altitude, forest damage and ozone. A number of studies have identified a relationship between altitude and forest damage, with damage peaking at altitudes of between 900m and 1200m and at about 1600m (Decker and Gürth 1985, Ammer *et al.* 1988). This corresponds with the altitude where ozone concentrations are highest for the longest periods of time (Paffrath and Peters 1988). However, it is also correlated with a number of other altitude-dependent variables which need to be considered before inferring any cause-effect relationship between ozone and forest damage.

Innes and Boswell (1989b) also identified a relationship between the sulphur content of needles of Sitka spruce, Norway spruce and Scots pine and atmospheric sulphur dioxide concentrations, although there was no relationship between the needle sulphur concentrations and the usual parameters of tree health. Similarly, Lange *et al.* (1989) found no effect of needle sulphur content on photosynthetic capacity. Relationships between sulphur dioxide and foliage sulphur contents have also been identified both in Britain (Farrar *et al.* 1977) and elsewhere (Materna 1982).

There are several possible reasons why the relationship between the spatial extent of forest damage and the distribution of air pollution might be obscured. The actual dose of specific pollutants received by trees growing in forests is very difficult to assess and figures are generally taken from area-averaged models. This can sometimes result in a considerable underestimate of the pollution load (Fowler *et al.* 1989). Another problem is that pollution may only be important at sites already stressed by other factors. This means that a specific pollutant load may have different effects in different circumstances; an effect that will invalidate any standard regression analysis.

Currently, there is a need to develop alternative techniques for the analysis of spatial data on forest health. The existing techniques do not take into account the complex nature of any possible relationships between air pollution and forest health and new methods are urgently required. The use of geostatistics has been explored by Innes and Boswell (1989) and offers considerable potential; other techniques undoubtedly exist but have yet to be tested.

PLAUSIBILITY OF MECHANISM

The second criterion listed by the CBMAPDT (1989) is that any observed association must have a reasonable biological explanation and it should not contradict other known mechanisms. In the absence of any well-established regional associations between forest health and air pollution in Europe, this criterion is difficult to apply.

However, the association between the sulphur content of needles and sulphur dioxide is well-established. The elevated sulphur content may be due to several mechanisms. Firstly, sulphate aerosols can be deposited on the needle surface where they remain unless the needle waxes are chemically removed. In the study by Innes and Boswell (1989b), the needles were washed in distilled water, but no attempt was made to remove either the epicuticular wax or particles embedded in it. Secondly, sulphur can be absorbed by the needles, with transfer through the stomata being an important uptake mechanism (Pfanz *et al.* 1987). Thirdly, sulphur can reach the needles via root uptake. All three mechanisms have been documented and experimental work has confirmed the validity of the mechanisms.

RESPONSIVENESS OR EXPERIMENTAL REPLICATION

The third criterion states that it should be possible to duplicate any observed effect under controlled conditions. In addition, it should be possible to either stop or prevent the effect by removing the causal agent (CBMAPDT 1989). This is one of the basic criteria used in epidemiological studies and it is crucial to any cause-and-effect inference.

To meet this criterion, it is necessary to describe adequately the effect. With all the research that has been done on forest decline, it might be assumed that the effect was well-known, but this is not the case. Generally, discoloration and loss of foliage are the two main effects that have been observed in European forests, although a number of other symptoms of ill-health may also be present (Innes and Boswell 1990). A variety of different forms of foliage loss occur; there are also many different forms of discoloration. In addition, there are inter- and intra-specific differences in the symptoms that have been observed. For example, the Forschungsbeirat Waldschäden (1986) listed five types of damage present on Norway spruce in West Germany:

1. Needle yellowing and needle loss at higher elevations in the "Mittelgebirge".
2. Thinning of crowns at medium to high altitudes in the "Mittelgebirge", with no yellowing being present.
3. Reddening of needles and needle loss in older stands in southern Germany.
4. Yellowing and needle loss at higher altitudes in the calcareous Alps of southern Bavaria.
5. Thinning of crowns in coastal areas in the north of Germany, with no yellowing.

There are undoubtedly more types of damage to Norway spruce, and Lesinski and Westman (1987) describe a whole range of different forms of defoliation. However, so far, there has been no attempt to assess the distribution of these in relation to air pollution.

Similar classifications need to be developed for all species being investigated, and it seems likely that such classifications should be regionally based. Several of the Norway spruce decline types listed by the Forschungsbeirat Waldschäden are not found on other species, and provenance trials have indicated that even the same species planted in a uniform environment can vary considerably, depending on the origin of the individual trees. It seems obvious that experiments should be aimed at reproducing observed symptoms, and they should not claim to have demonstrated a cause-effect relationship by creating an effect on a species other than the one on which the effect has been observed in the field.

In Germany, there have been many attempts to simulate the different types of symptoms seen in Norway spruce. Most work has concentrated on the first type of damage, that involving both yellowing and needle loss. The most distinctive characteristic of this type of damage is the needle yellowing. Yellowing develops from the tips of the needles and progresses back towards the base (Rehfuess 1987). It occurs predominantly on the upper surfaces of needles and, in the majority of cases, is restricted to older needles and develops shortly after the current-year needles have flushed. Sutinen (1987) has reported that cytological changes initially occur in the needle vascular bundle, followed by a breakdown of the chloroplasts in cells beneath the upper epidermis. This in turn is followed by a gradual loss of chlorophyll down through the mesophyll.

If a cause-effect mechanism is to be suggested, it must follow the above pattern. The presence of yellowing on the upper surfaces of needles led some to suggest that ozone might be the cause of damage. However, when Norway spruce was fumigated with ozone (usually at concentrations well above those experienced in the field), upper-surface yellowing was not observed. Instead, a fine chlorotic mottle developed on all year classes (Guderian *et al.* 1985, Prinz *et al.* 1985, Skeffington and Roberts (1985), Kandler *et al.* 1987, Senser *et al.* 1987). Important differences were observed in the biochemistry of mottled needles and yellow needles from symptomatic trees (Senser *et al.* 1987). When ozone-damaged needles are examined microscopically, it is evident that damage starts in the sub-stomatal cavity and moves towards the inner mesophyll (Fink 1988), the reverse of what is seen in the field. In addition, there have now been a number of investigations using exclusion chambers in the field which have failed to demonstrate an effect of gaseous air pollution (Schulze *et al.* 1989a, Lange *et al.* 1989, Koch 1989, Koch and Lautenschlager 1989).

Needle yellowing *has* however been reproduced in laboratory studies. Lange *et al.* (1987) removed the buds of young trees prior to budburst; the older needles remained green whereas those on control trees, which had no buds removed, turned yellow. The yellow needles seen in Type 1 spruce decline are associated with magnesium deficiency; yellow needles have magnesium contents of less than 300 $\mu g\ g^{-1}$ (Zech and

Popp 1983, Zöttl and Mies 1983, Kaupenjohann *et al.* 1989). Fertilization with magnesium of symptomatic plots normally results in recovery (Isermann 1985, Hüttl and Wisniewski 1987, Zöttl and Hüttl 1986, 1989). Laboratory studies of yellow needles induced by magnesium deficiency indicate that the histological and cytological changes are the same as observed in the field (Fink 1983). Further evidence that the yellowing is not directly associated with gaseous pollution comes from grafting experiments. Hüttl and Mehne (1988) have demonstrated that when shoots bearing yellow needles are grafted onto healthy trees, the yellowing disappears.

It is clear that yellowing is associated with severe magnesium deficiency. However, it less evident what has caused the magnesium deficiency. It may be due to increased leaching of base cations from the needles, but the available evidence does not support this hypothesis (Roberts *et al.* 1989). Reduced uptake arising from damage to the fine roots caused by soil acidification has been proposed (Ulrich *et al.* 1980), but the evidence for this is also limited (Roberts *et al.* 1989). The most likely explanation appears to be a reduction in the availability of exchangeable magnesium, possibly caused by increased soil leaching associated with acidic deposition (Matzner *et al.* 1982, Hildebrand 1986, Rost-Siebert and Jahn 1988, Hauhs 1989 Ulrich *et al.* 1989), combined with nutritional imbalances created by increased levels of nitrogen deposition (Zöttl *et al.* 1989, Schulze *et al.* 1989a).

In Britain, type 1 Norway spruce decline has not been seen. A survey of the magnesium contents of foliage indicated that levels were well above the deficiency threshold (Binns *et al.* 1986) and there was no indication of an unbalanced nitrogen:magnesium ratio (Innes and Boswell 1990). Magnesium deficiency would not be expected given the large amounts of this cation that are deposited annually in rainfall (Roberts *et al.* 1989). Consequently, the value of experimental attempts to induce magnesium deficiency in forest trees commonly planted in Britain must be questioned.

A much more frequent symptom is the general thinness of tree crowns. This is widespread in Britain and is also characteristic of all countries where assessments of forest condition are undertaken. The symptom can develop for a variety of reasons, although it is widely believed that stress from air pollution is the main factor. This is clearly untenable, although it seems to have become firmly established in the literature. In some cases, crown thinness may not even be a problem: in Britain, the speed of growth of some conifers, as indicated by the branching density, is a significant factor leading to apparent crown thinness (Innes and Boswell 1990). There are numerous examples of crown thinness developing as a result of factors other than air pollution; a good example in Britain is the decline and death of elms over much of the country during the last 20 years.

The variety of factors causing crown thinness is important as it effectively means that experimental work cannot confirm cause-effect relationships in the field. This could only be done if the development of the crown thinness was described in sufficient detail to enable most of the confounding processes to be eliminated. It might then be possible to identify a form of development that was unique to the effects of a particular pollutant. Some progress has been made in this area, and the detailed annual assessments that are now made of trees in Britain (Innes 1990) may enable a better separation of the different types of crown thinness.

TEMPORALITY

The fourth criterion listed by CBMAPDT (1989) is that a cause must precede an effect or should be present at the appropriate time.

The widespread development of forest damage in Europe is reported to have occurred in the late 1970s and early 1980s, during a period when the awareness of the air pollution problem was also increasing. Air pollution has actually been increasing for over 100 years and claims that the forest damage developed in line with air pollution concentrations are therefore dubious. However, there is an increasing amount of information which indicates that the decline problem started earlier. There are numerous old pictures illustrating thin-crowned trees (Cramer 1984, Kandler 1989) and tree-ring studies are increasingly indicating that growth declines set in well before the onset of visual symptoms (e.g. Athari 1983, Evers *et al.* 1986, Becker 1987, Eckstein and Saß 1989).

Good time series are required to identify a temporal correlation between the development of damage and rising concentrations of an air pollutant. In practice, series exist for neither forest health nor air pollution. Reliable assessments of the extent and distribution of forest damage are available only from the mid-1980s in most countries. Given the long lag intervals that occur after a tree has been stressed by an event such as drought (Becker 1989), a short series of 5-10 years is unlikely to enable the "noise" to be removed. For example, in Britain, severe defoliation of Sitka spruce by the green spruce aphid (*Elatobium abietinum* Walker (Hemiptera)) occurred in 1989 (Innes and Boswell 1990a). In many cases, only one or two years' needles remained. As Sitka spruce normally retains its needles for a minimum of eight years, and frequently as much 12-15 years, the trees affected in 1989 will take several years to recover fully.

An example of the rejection of a hypothesis on the basis of lack of temporality is cited in Roberts *et al.* (1989). Ulrich *et al.* (1980) claimed that the fine root biomass of trees in the Solling area of northeast Germany had decreased in parallel with an increase in the concentration of aluminium in the soil solution. However, a re-analysis of the data by Rehfuess (1981) indicated that the fine roots had declined prior to the rise in aluminium concentrations and the decline could be related to moisture stress.

It is frequently argued that the only phenomenon that correlates temporally with forest decline is air pollution. This is not the case. As already argued, forest declines have occurred in the past and will no doubt occur in the future. The widespread reports of decline in the early 1980s can be partly attributed to increased awareness although long-term monitoring of plots clearly indicates that a decline did occur in 1983-1984. The decline occurred in a variety of areas, often with very different pollution regimes. Furthermore, since 1985, forests in West Germany have been improving, whereas air quality has not shown a similar trend. The most likely explanation for the simultaneous onset of a number of different declines is climate. Many trees suffered badly in the 1976 drought, but there appear to be no records of the response of forest condition on a regional scale. By 1984, annual surveys had been established, and the effects of the droughts could be determined.

The onset of yellowing after 1983/1984 can be explained by drought. The dry summers would have caused an increase in the mineralisation and nitrification of soil nitrogen (Rehfuess 1989, Zöttl *et al.* 1989), resulting in increased nitrogen availability. This would have disrupted the nitrogen:magnesium balance of the trees, inducing yellowing at sites where magnesium availability was low. This has been confirmed by studies of the nitrogen:magnesium ratios in needles (Hüttl 1988, Zöttl *et al.* 1989, Schulze *et al.* 1989b).

WEIGHT OF EVIDENCE

The final criterion is that there must be general agreement between the previous four criteria. The first four criteria do not in themselves provide sufficient evidence for cause-and-effect. A firm conclusion can be reached only when they are taken together (CBMAPDT 1989). Currently, one of the strongest arguments for the case against air pollution is the reported concensus amongst scientists that air pollution must be involved in forest decline. This conclusion is based on the absence of any other factor that could explain the decline in forests and not on sound experimental evidence (Cowling 1989). However, this view is being increasingly challenged and it is now widely accepted that forest decline can occur without invoking air pollution (Loehle 1988, Mueller-Dombois 1987, 1988).

The role of air pollution is still intensely debated, although agreement has been reached on some issues. For example, it is clear that the yellowing of needles in type 1 Norway spruce decline is the result of magnesium deficiency. The majority of scientists agree that the deficiency is the result of insufficient uptake from the soil rather than increased loss from needles. However, the cause of the insufficient uptake is uncertain; it may be due to a reduction in the ability of tree roots to take up available magnesium or to a reduction in the available magnesium. The weight of evidence seems to favour the latter explanation, although at some sites, the former, or a combination of the two, seems feasible.

The evidence for a direct role of normal ambient gaseous air pollution is equivocal. In some cases, such as in the San Bernadino Mountains of California, there is little doubt that ozone has damaged trees (Miller and McBride 1989). All four criteria have been satisfied in reaching this deduction. Elsewhere, the evidence is much less certain. Locally, high concentrations of certain pollutants can cause severe damage and mortality of trees; this has been known for many years. However, the evidence for regional-scale effects is scanty and frequently conflicting.

CONCLUSIONS

Many of the difficulties currently facing any assessment of the effects of air pollution on forests stem from the lack of coordination between research groups. This has resulted in an illogical programme of research, with laboratory scientists frequently trying to explain phenomena that have never been observed outside the laboratory. Insufficient care has been taken over the documentation of the symptoms of individual decline types and of the spatial and temporal development of each decline type. This has resulted in a tendency to reject certain hypotheses on the basis of a lack of generality of association, yet the evidence for the rejection may be faulty as different types of decline are used. For example, soil acidification cannot be rejected as a cause of the yellowing in Norway spruce in the German Mittelgebirge on the basis that trees growing in calcareous soils in the Alps show yellowing; the latter is a specific decline type which is clearly different from type 1 decline. Conversely, the soil acidification hypothesis cannot, and indeed should not, be applied to all decline types.

This review has concentrated on field evidence for cause-effect relationships. Laboratory studies have shown that air pollution can cause subtle changes to the metabolism of trees that are not necessarily apparent as visual symptoms (e.g. Fink 1988). Whether these would appear if the experiments were continued for a sufficiently long period is unknown. Another doubt concerns the adaptation of trees to ambient levels of pollution. The extent to which a tree adjusts to pollution over a period of decades is unknown, and no models appear to have been developed to approach this question. With the increasing public awareness of pollution issues, it is likely that emissions of many forms of pollution will fall in the future, although some will continue to rise. How trees will respond to these changes in their chemical environment is unknown.

BIBLIOGRAPHY

Ammer, U., Burgis, M., Koch, B. and Martin, K. (1988). Untersuchungen über den Zusammenhang zwischen Schädigungsgrad und Meereshöhe im Rahmen des Schwerpunktsprogramms zur Erforschung der Wechselwirkungen von Klima und Waldschäden. *Forstwissenschaftliches Centralblatt* 107, 145-151.

Athari, S. (1983). Zuwachsvergleich von Fichten mit unterschiedlich starken Schadsymptomen. *Allgemeine Forst Zeitschrift* **38**, 653-655.

Barnard, J.E. and Scott, C.T. (1988). *Changes in tree growth rates in Vermont*. United States Department of Agriculture, Forest Service, South-east Forests Experimental Station Research Note SE-350.

Bauch, J., Rademacher, P., Berneike, W., Knoth, J. and Michaelis, W. (1985). Breite und elementgehalt der Jahrringe in Fichten aus Waldschadensgebieten. In, *Waldschäden. Einflussfaktoren und ihre Bewertung. Kolloquium Goslar, 18 bis 30 Juni 1985*, ed. H. Stratmann, 943-959. VDI-Berichte 560.

Becker, M. (1987) Bilan de santé actuel et rétrospectif du sapin (*Abies alba* Mill.) dans les Vosges. Étude écologique et dendrochronologique. *Annales des Sciences Forestières* **44**, 379-401.

Becker, M. (1989). The role of climate on present and past vitality of silver fir forests in the Vosges Mountains of northeastern France. *Canadian Journal of Forest Research* **19**, 1110-1117.

Binns, W.O., Redfern, D.B., Boswell, R.C. and Betts, A.J.A. (1986). *Forest health and air pollution: 1985 survey*. Forestry Commission Research and Development Paper 147. Forestry Commission, Edinburgh.

Brooks, R.T. (1989). An analysis of regional forest growth and atmospheric deposition patterns, Pennsylvania (USA). In, *Air Pollution and Forest Decline*, ed. J.B. Bucher and I. Bucher-Wallin, 283-288. Eidgenössische Anstalt für das Forstliche Versuchswesen, Birmensdorf.

Committee on Biologic Markers of Air-Pollution Damage in Trees (1989). *Biologic Markers of Air-Pollution Stress and Damage in Forests*. National Academy Press, Washington.

Cowling, E.B. (1989). Recent changes in chemical climate and related effects on forests in North America and Europe. *Ambio* **18**, 167-171.

Cramer, H.H. (1984). On the predisposition to disorders of Middle European forests. *Pflanzenschutz-Nachrichten Bayer* **37**, 97-207.

Cronan, C.S., April, R., Bartlett, R.J., Bloom, P.R., Driscoll, C.T., Gherini, S.A., Henderson, G.S., Joslin, J.D., Kelly, J.M., Newton, R.M., Parnell, R.A., Patterson, H.H., Raynal, D.J., Schaedle, M., Schofield, C.L., Sucoff, E.I., Tepper, H.B. and Thonton,F.C. (1989). Aluminium toxicity in forests exposed to acidic deposition: the ALBIOS results. *Water, Air and Soil Pollution* **48**, 181-192.

Decker, B. and Gürth, P. (1985). Das Waldsterben im Forstbezirk Müllheim/Südlicher Schwarzwald. *Allgemeine Forst- und Jagdzeitung* **156**, 233-240.

Eckstein, D. and Saß, U. (1989). Dendroecological assessment of decline and recovery of fir and spruce in the Bavarian forest. In, *Forest Decline and Air Pollution*, ed. J.B. Bucher and I. Bucher-Wallin, 255-260. Eidgenössische Anstalt für das forstliche Versuchswesen, Birmensdorf. 7

Evers, F.H., Hildebrand, E.E., Kenk, G. and Kremer, W.L. (1986). Boden-, ernährungs- und ertragskundliche Untersuchungen in einem stark geschädigten Fichtenbestand des Buntsandstein Schwarzwaldes. *Mitteilungen des Vereins für Forstliche Standorstkunde und Forstpflanzenzüchtung* **32**, 72-80.

Farrar, J.F., Relton, J. and Rutter, A.J. (1977). Sulphur dioxides and the scarcity of *Pinus sylvestris* in the industrial Pennines. *Environmental Pollution* **14**, 63-68.

Fink, S. (1983). Histologische und histochemische Untersuchungen an Nadeln erkrankter Tannen und Fichten im Südschwarzwald. *Allgemeine Forst Zeitschrift* **38**, 660-663.

Fink, S. (1988). Histological and cytological changes caused by air pollutants and other abiotic factors. In, *Air Pollution and Plant Metabolism*, ed. S. Schulte-Hostede, N.M. Darrall, L.W. Blank and A.R. Wellburn, 36-54. Elsevier Applied Sciences, Oxford.

Forschungsbeirat Waldschäden (1986). *2. Bericht*. Karlsruhe.

Fowler, D., Cape, J.N. and Unsworth, M.H. (1989). Deposition of atmospheric pollutants on forests. *Philosophical Transactions of the Royal Society, Series B* **324**, 247-265.

Guderian, R. Küppers, K. and Six, R. (1985). Wirkungen von Ozon, Schwefeldioxid und Stickstoffdioxid auf Fichte und Pappel bei unterschiedlicher Versorgung mit Magnesium und Kalzium sowie auf die Blattflechte *Hypogymnia physodes*. In, *Walschäden. Einflussfaktoren und ihre Bewertung. Kolloquium Goslar, 18 bis 23 Juni, 1985*, ed. H. Stratmann, 657-701. VDI-Berichte 560.

Hauhs, M. (1989). Lange Bramke: an ecosystem study of a forested catchment. In, *Acidic precipitation. Volume 1: case studies*, ed. D.C. Adriano and M. Havas, 275-305. Springer-Verlag, New York.

Hildebrand, E.E. (1986). Zustand und Entwicklung der Austauschereigenschaften von Mineralböden aus Standorten mit erkrankten Waldbeständen. *Forstwissenschaftliches Centralblatt* **105**, 60-76.

Hornbeck, J.W., Smith, R.B. and Federer, C.A. (1986). Growth decline in red spruce and balsam fir relative to natural processes. *Water, Air, and Soil Pollution* **31**, 425-430.

Hüttl, R.F. and Mehne, B.M. (1988). New-type of forest decline, nutrient deficiencies and the Virus-hypothesis. In, *Air Pollution and Ecosystems*, ed. P. Mathy, 870875. D. Rediel, Dordrecht.

Hüttl, R.F. and Wisniewski, J. (1987). Fertilisation as a tool to mitigate forest decline associated with nutrient deficiencies. *Water, Air, and Soil Pollution* **32**, 265-276.

Innes, J.L. (1989). Acid rain and trees. In, *Acid deposition. Sources, effects and controls*, ed. J.W.S. Longhurst, 229-242. British Library, London.

Innes, J.L. (1990). Monitoring of forest condition in the United Kingdom. Results from six years of investigation. *Proceedings of the International Congress on Forest decline Research: state of knowledge and perspectives*, Friedrichshafen, 2-6 October 1989. In press.

Innes, J.L. and Boswell, R.C. (1987). *Forest health surveys 1987. Part 1: results*. Forestry Commission Bulletin 74. HMSO, London.

Innes, J.L. and Boswell, R.C. (1988). *Forest health surveys 1987. Part 2: analysis and interpretation*. Forestry Commission Bulletin 79. HMSO, London.

Innes, J.L. and Boswell, R.C. (1989a) *Monitoring of forest condition in the United Kingdom - 1988*. Forestry Commission Bulletin 88. HMSO, London.

Innes, J.L. and Boswell, R.C. (1989b). Sulphur contents of conifer needles in Great Britain. *GeoJournal* **19**, 63-66.

Innes, J.L. and Boswell, R.C. (1990). *Monitoring of forest condition in the United Kingdom - 1989*. Forestry Commission Bulletin, in press.

Innes, J.L. and Cook, E.D. (1989). Tree-ring analysis as an aid to evaluating the effects of pollution on tree growth. *Canadian Journal of Forest Research* **19**, 1174-1189.

Isermann, K. (1985). Diagnose und Therapie der 'neuartigen Waldschäden' aus der Sicht der Waldnährung. In, *Waldschäden. Einflussfaktoren und ihre Bewertung. Kolloquium Goslar, 18 bis 20 Juni 1985*, ed. H. Stratmann, 897-920. VDI-Berichte 560.

Johnson, A.H., COOK, E.R. and SICCAMA, T.G. (1988). Climate and red spruce growth and decline in the northern Appalachians. *Proceedings of the National Academy of Sciences, U.S.A.*, **85**, 5369-5373.

Kandler, O. (1989). Epidemiological evaluation of the course of 'Waldsterben' from 1983 to 1987. In, *Forest decline and air pollution*, ed. J.B. Bucher and I. Bucher-Wallin, 297-302. Eidgenössische Anstalt für das forstliche Versuchswesen, Birmensdorf.

Kandler, O, Miller, W. and Ostner, R. (1987). Dynamik der "akuten Vergilbung" der Fichte. Epidemiologische und physiologische Befunde. *Allgemeine Forst Zeitschrift* **42**, 715-723. et al. 1987 5

Kaupenjohann, M., Zech, W., Hantschel, R., Horn, R. and Schneider, B.U. (1989). Mineral nutrition of forest trees: a regional survey. In, *Forest decline and air pollution. A study of spruce (Picea abies) on acid soils*, ed. E.-D. Schulze, O.L. Lange and R. Oren, 282-296. Springer-Verlag, Berlin.

Koch, R. (1876). Die aetiologie der Milzbrand-Krankheit, begundet auf die Entwicklungschiachte des Baiollus Antracis. *Beiträge zur Biologie der Pflanzen* **2**, 277.

Koch, W. (1989). Der Reinluft/Standortsluft - Vergleich an Fichte. *Forstwissenschaftliche Centralblatt* **108**, 73-82.

Koch, W. and Lautenschlager, K. (1989). Vergleichende Gaswechselmessungen unter Reinluft und Standortsluft an Fichte zur Quantitativen Ehrmittlung von Primärschäden durch Gasförmige Luftverunreinigungen. In, *Forest decline and air pollution*, ed. J>B> Bucher and I. Bucher-Wallin, 119-124. Eidgenössische Anstalt für das forstliche Versuchswesen, Birmensdorf.

Lange, O.L., Weikert, R.M., Wedler, M., Gebel, J. and Heber, U. (1989). Photosynthese und Nährstoffversorgung von Fichten aus einem Waldschadensgebiet auf basenarmen Untergrund. *Allgemeine Forst Zeitschrift* **44**, 55-64.

Lange, O.L., Zellner, H., Gebel, J., Schramel, P., Kostner, B. and Czygan, F.C. (1987). Photosynthetic capacity, chloroplast pigments, and mineral content of the previous year's spruce needles with and without the new flush: analysis of the forest decline phenomenon of needle bleaching. *Oecologia* **73**, 351-357.

Lesinski, J.A. and Westman, L. (1987). Crown injury types in Norway spruce and their applicability for forest inventory. In, *Acid rain: scientific and technical advances*, ed. R. Perry, R.M. Harrison, J.N.B. Bell and J.N. Lester, 657-662. Selper, London.

Loehle, C. (1988). Forest decline: endogenous dynamics, tree defenses, and the elimination of spurious correlation. *Vegetatio* **77**, 65-78.

Mahrer, F. (1989). Problems in the determination and interpretation of needle and leaf loss. In, *Air Pollution and forest decline*, ed. J.B. Bucher and I. Bucher-Wallin, 229-231. Eidgenössische Anstalt für das forstliche Versuchswesen, Birmensdorf.

Materna, J. (1982). Concentration of sulphur dioxide in the air and sulphur content in Norway spruce seedling (*Picea abies* Karst.). *Communicationes Instituti Forestalis Cechosloveniae* **12**, 137-146.

Matzner, E., Khanna, P.K., Meiwes, K.S., Linhei, M., Prenzel, J. and Ulrich, B. (1982). Elementflüsse in Waldökosystemen in Solling. Datendokumentation. *Göttinger Bodenkundliche Berichte* **71**, 1-267.

Miller, P.R. and McBride, J.R. (1989). Trends of ozone damage to conifer forests in the western United States, particularly southern California. In, *Air pollution and forest decline*, ed. J.B. Bucher and I. Bucher-Wallin, 61-68. Eidgenössische Anstalt für das forstliche Versuchswesen, Birmensdorf.

Mosteller, F. and Tukey, J.W. (1977). *Data analysis and regression: a second course in statistics*. Addison-Wesley, Reading (Massachussetts).

Mueller-Dombois, D. (1987). Natural dieback in forests. *BioScience* **37**, 575-583.

Mueller-Dombois, D. (1988). Towards a unifying theory for stand-level dieback. *GeoJournal* **17**, 249-251.

Neumann, M. (1989). Einfluss von Standortsfaktoren auf den Kronenzustand. In, *Air pollution and forest decline*, ed. J.B. Bucher and I. Bucher-Wallin, 209-214. Eidgenössische Anstalt für das forstliche Versuchswesen, Birmensdorf.

Paffrath, D. and Peters, W. (1988). Betrachtung der Ozonvertikalverteilung in Zusammenhang mit den neuartigen Waldschäden. *Forstwissenschaftliche Centralblatt* **107**, 152-159.

Peterson, D.L., Arbaugh, M.J. and Robinson, L.J. (1989). The effects of ozone stress on tree growth and vigor in the Sierra Nevada of California, USA. In, *Air pollution and forest decline*, ed. J.B. Bucher and I. Bucher-Wallin, 289-296. Eidgenössische Anstalt für das forstliche Versuchswesen, Birmensdorf.

Pfanz, H., Martinoia, E., Lange, O.-L. and Heber, U. (1987). Flux of SO_2 into leaf cells and cellular acidification by SO_2. *Plant Physiology* **85**, 928-933.

Prinz, B., Krause, G.H.M. and Jung, K.D. (1985). Untersuchungen der LIS Essen zur Problematik der Walschäden. In, *Walschäden - Theme und*

Praxis auf der Suche nach Antworten, ed. V. Kortzfleisch, 143-194. Oldenburg, München-Wien.
Rehfuess, K.E. (1981). Über die Wirkungen der sauren Niederschläge in Waldökosystemen. *Forstwissenschaftliche Centralblatt* **100**, 363-381.
Rehfuess, K.E. (1987). Perceptions of forest diseases in central Europe. *Forestry* **60**, 1-11.
Rehfuess, K.E. (1989). Acidic deposition - extent and impact on forest soils, nutrition, growth and disease phenomena in central Europe: a review. *Water, Air, and Soil Pollution* **48**, 1-20. 1987 5
Roberts, T.M., Skeffington, R.A. and Blank, L.W. (1989). Causes of Type 1 spruce decline in Europe. *Forestry* **62**, 179-222.
Rost-Siebert, K. and Jahn, G. (1988). Veränderungen der Waldbodenvegetation während der letzten Jahrzehnte - Eignung zur Bioindikation von Immissionswirkungen? *Forst und Holz* **43**, 75-81.
Schmid-Haas, P. (1989). Do the observed needle losses reduce increments. In, *Air pollution and forest decline*, ed. J.B. Bucher and I. Bucher-Wallin, 271-275. Eidgenössische Anstalt für das forstliche Versuchswesen, Birmensdorf.
Schulze, E.-D., Oren, R. and Lange, O.L. (1989a). Processes leading to forest decline: a synthesis. In, *Forest decline and air pollution. A study of spruce (Picea abies) on acid soils*, ed. E.-D. Schulze, O.L. Lange and R. Oren, 459-468. Springer-Verlag, Berlin.
Schulze, E.-D., Oren, R. and Lange, O.L. (1989b). Nutrient relations of trees in healthy and declining Norway spruce stands. In, *Forest decline and air pollution. A study of spruce (Picea abies) on acid soils*, ed. E.-D. Schulze, O.L. Lange and R. Oren, 392-417. Springer-Verlag, Berlin.
Schweingruber, F.H., Kontic, R. and Winkler-Seifert, A. (1983). *Eine jahrringanalytische Studie zum Nadelbaumsterben in der Schweiz.* Eidgenössische Anstalt für das forstliche Versuchswesen, Berichte 253. Birmensdorf.
Senser, M., Hopker, K., Peuker, A. and Glashagen, B. (1987). Wirkungen extremer Ozonkonzentrationen auf Koniferen. *Allgemeine Forst Zeitschrift* **42**, 709-714.
Skeffington, R.A. and Roberts, T.M. (1985). The effects of ozone and acid mist on Scots pine saplings. *Oecologia* **65**, 201-206.
Strand, L. (1980). *Acid precipitation and regional tree ring analyses.* SNSF Project, Report IR 73/80, Oslo.
Sutinen, S. (1987). Cytology of Norway spruce needles -II. *European Journal of Forest Pathology* **17**, 74-85.
Ulrich, B., Mayer, R. and Khanna, P.K. (1980). Chemical changes due to acid precipitation in a loess-derived soil in Central Europe. *Soil Science* **30**, 193-199.
Ulrich, B., Meyer, H., Jänich, K. and Büttner, G. (1989). Basenverluste in den Böden von Hainsimsen-Buchenwäldern in Südniederschsen zwischen 1954 und 1986. *Forst und Holz* **44**, 251-253.
Westman, L. and Lesinski, J.A. (1985). Thinning out of the tree crown - what is hidden in that integrated measure of forest damage? In, *Inventorying and monitoring of endangered forests*, ed. P. Schmid-Haas, 223-228. Eidgenössische Anstalt für das forstliche Versuchswesen, Birmensdorf.
Zech, W. and Popp, E. (1983). Magnesiummangel, einer der Gründe für das Fichten- und Tannensterben in NO-Bayern. *Forstwissenschaftliche Centralblatt* **102**, 50-55.
Zöttl, H.W. and Hüttl, R.F. (1986). Nutrient supply and forest decline in south-west Germany. *Water, Air, and Soil Pollution* **31**, 449-462.
Zöttl, H.W. and Hüttl, R.F. (1989). Nutrient deficiencies and forest decline. In, *Air pollution and forest decline*, ed. J.B. Bucher and I. Bucher-Wallin, 189-193. Eidgenössische Anstalt für das forstliche Versuchswesen, Birmensdorf.
Zöttl, H.W., Hüttl, R.F., Fink, S., Tomlinson, C.H. and Wisniewski, J. (1989). Nutritional disturbances and histological changes in declining forests. *Water, Air, and Soil Pollution*, **48**, 87-109.
Zöttl, H.W. and Mies, E. (1983). Der Fichtenkrankung in den Hochlagen des Südschwarzwaldes. *Allgemeine Forst- und Jagdzeitung* **154**, 110-114.

EFFECTS OF ACIDIC DEPOSITION ON HIGH-ELEVATION SPRUCE-FIR FORESTS IN THE UNITED STATES

Mary Beth Adams & Christopher Eager
USDA Forest Service, Northeastern Forest Experiment Station
Radnor Corporate Centre, Radnor, U.S.A.

Abstract

Large numbers of dead red spruce trees in high elevation spruce-fir forests of the eastern United States have caused much concern, and have been linked with elevated levels of air pollutants. These high elevation forests receive a significant amount of sulfate and nitrate from cloudwater deposition. The Spruce-Fir Research Cooperative, an integrated multi-institutional program, is investigating a number of hypothesized effects on the spruce-fir forests: (1) soil-mediated effects, (2) altered physiological processes, (3) increased foliar injury, and (4) increased susceptibility to winter injury. The results of Spruce-Fir Research Cooperative research are discussed in this paper.

Introduction

Recently, much attention has been focused on effects of air pollutants, in particular "acid rain", on forest ecosystems. In the United States, the significant visible decline, growth reduction, and mortality of high elevation spruce-fir forests has been hypothethically linked to regional air pollution. The Spruce-Fir Research Cooperative (SFRC) was formed in 1985 as part of the Forest Response Program of the National Acid Precipitation Assessment Program, with the mission of explaining possible air pollution-related changes in the structure, function, and composition of the spruce-fir ecosystem. Two particular questions are addressed by researchers in the SFRC: (1) What are the effects of air pollutants and acidic deposition on spruce-fir forests in the eastern United States? and (2) Are red spruce declining at high elevations? This paper will present a synopsis of SFRC findings to date, drawing predominantly from research which has already received peer review. Research continues through 1990, thus some conclusions may change as new information is revealed.

The Resource

Forests dominated by red spruce (Picea rubens Sarg.) and associated fir (balsam fir (Abies balsamea L. Mill) in the north, or Fraser fir (Abies fraseri (Pursh) Poir.) in the south) are found at the higher elevations of the northern and southern Appalachians, and at lower elevations in the northeastern U.S. and southeastern Canada (Figure 1). Although this forest type comprises a small portion of the forested area of the eastern United States, the spruce-fir forests are highly valued for recreation and aesthetics and as habitat for unique and endangered species. There are

This research was supported by funds provided by the Northeastern Forest Experiment Station, Spruce-fir Research Cooperative within the joint US Environmental Protection Agency-USDA Forest Service Forest Response Program. The Forest Response Program is part of the National Acid Precipitation Assessment Program. This paper has not been subject to EPA or Forest Service policy review and should not be construed to represent the policies of either Agency.

about 158,000 ha of high elevation spruce-fir forests in the Appalachians with 26,600 occurring in the southern Appalachians. There are over 4 million ha of low elevation spruce-fir forests in northern Vermont, northern New Hampshire and Maine, but these forests are not the focus of this paper.

Red spruce is a shade-tolerant species, often living longer than 300 years. In the northern Appalachians, red spruce and balsam fir forests are found at elevations between 750 and 1200 m, with hardwood forests below 750 m and pure balsam fir forests and alpine vegetation above 1200 m. In the southern Appalachians red spruce and Fraser fir forests begin at about 1500 m and extend to the highest summits (2000 m).

Current Status and Rates of Change

Several factors suggest an unreversed decline in high elevation red spruce in the northern Appalachians during the past 20 years. Decline symptoms are varied and include reduced growth, needle discoloration and loss, crown thinning, apical bud dieback, altered branch structure, and ultimately branch and tree mortality (Woodman 1987, Evans 1986). Between 1964 and 1983, basal area and density of red spruce decreased by 50% to 70% in the Green Mountains of Vermont (Vogelman et al. 1985, Siccama et al. 1982) and at Whiteface Mountain (Scott et al. 1984) (See Figure 2 for locations). Other associated species (i.e. balsam fir) did not experience such dramatic reductions in basal area and stand density. In all stands red spruce accounted for more than 75% of stand-level reductions in both basal area and density.

Johnson and Siccama (1983) reported the percent standing dead red spruce above 900 m elevation at Whiteface Mountain, Mount Mansfield (VT), and Mount Washington (NH) ranged from 20% to 60% in 1982. When resurveyed in 1987, the mean percent dead canopy spruce across all sites had increased from 26% to 37% (Silver et al. in press). In both 1982 and 1987, a trend toward a higher percentage of dead spruce

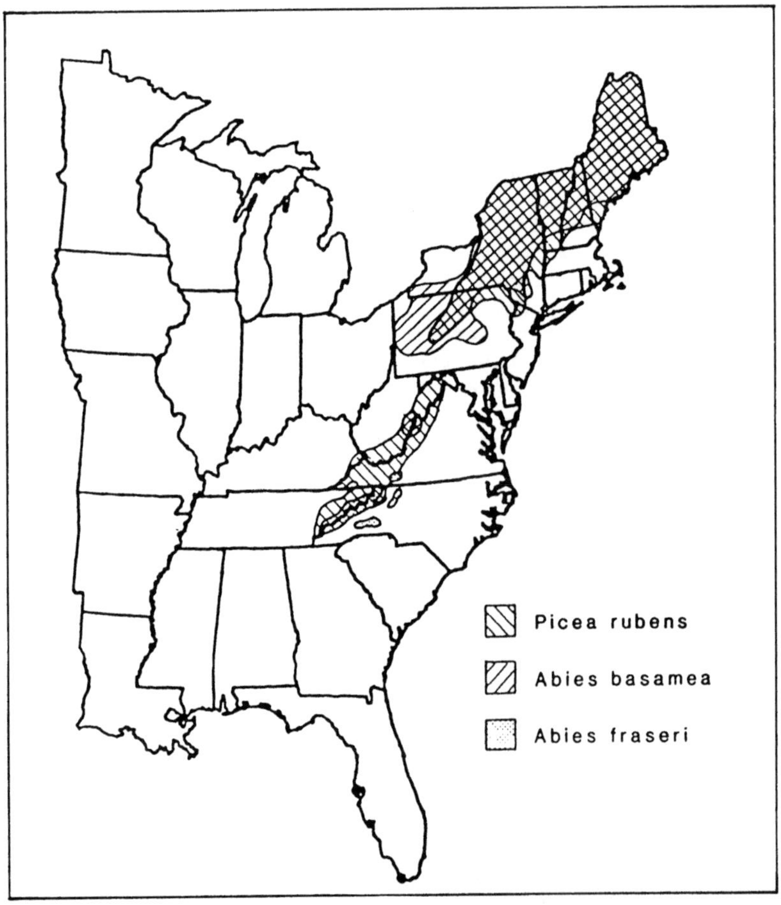

Figure 1. Extent of spruce-fir forests in the eastern United States.

with increasing elevation was observed, with greater than 50% standing dead red spruce above 1000 m on Whiteface Mountain and Mount Mansfield. Similar results were reported from a regional survey of red spruce condition in the Adirondacks and Green Mountains (Friedland 1989) and for the east face of Mt. Moosilauke, NH (Peart et al. 1988).

A detailed study of Whiteface Mountain during 1986 and 1987 revealed that between 50% and 70% of the red spruce were dead at elevations above 1000 meters, with the higher percentages occurring at the higher elevations (Battles et al. 1988). More dead red spruce were recorded in old-growth stands than in stands that were disturbed by logging about 100 years ago. However, mortality occurred across all size/age classes in these old-growth stands indicating that age alone was not a factor in the death of red spruce.

Further evidence of a decline comes from tree ring records from surviving high elevation red spruce in the northern Appalachians. A region-wide abrupt, unreversed reduction in annual growth increments occurred beginning in the early 1960's (Johnson, A. et al. 1988, Cook et al. 1987, McLaughlin et al. 1987). This growth decrease occurred in stands differing in age and disturbance history, and was not explained by existing concepts of growth behavior in red spruce. The anomalous nature of this growth decline is evident through detailed evaluation of the historical relationship of growth to climate and is unprecedented in the previous 160 years (Johnson, A. et al. 1988, Cook et al. 1987).

For the southern Appalachians, similar historical data do not exist with which to determine recent changes in forest structure. However, evaluations of the current status of red spruce were conducted in 1985 and 1986 at three of the seven mountains in the southern Appalachians having spruce-fir forests (Nicholasand Zedaker 1989). The percent standing dead red spruce (dbh >5 cm) ranged from 12% to 20%, well within the range expected under normal forest dynamics, and was not related to elevation or aspect. Large numbers of Fraser fir have died during the past 30 years in the south-

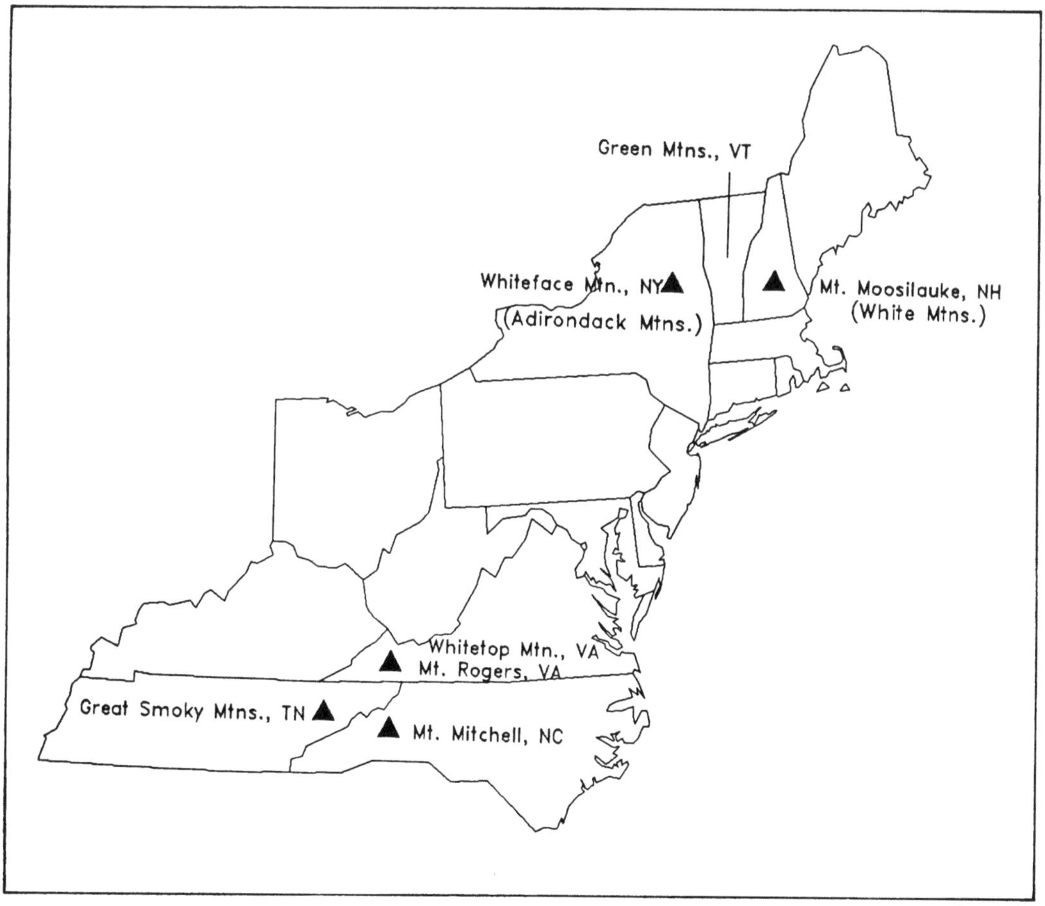

Figure 2. Locations of Spruce-Fir Research Cooperative Research Sites.

ern Appalachians due to an infestation of the balsam woolly adelgid (Adelges piceae) (Bruck 1989, Eagar 1984, Amman and Spears 1965). Only Mount Rogers in Virginia has not experienced high levels of mortality, despite the presence of the adelgid since at least 1962 (Eagar 1984).

Crown condition, as indicated by premature needle loss, has deteriorated at two of the three southern Appalachian sites (Zedaker et al. 1988). In the Great Smoky Mountains the percent red spruce classified as healthy (less than 10% needle loss) decreased from 85-90% in 1985 to about 60% in 1988. Red spruce in the Black Mountains of North Carolina showed a similar trend, but leveled off between 1987 and 1988 with about 70% of the trees classified as healthy. Crown condition at Mount Rogers has remained constant, with about 87% of the red spruce classified as healthy. In all cases, crown condition was not related to elevation or aspect.

Evidence of recent, anomalous growth reductions in the southern Appalachians is more equivocal than in the northern Appalachians. In a study using tree cores from 11 sites in the Great Smoky Mountains, McLaughlin et al.. (1987) found recent growth reductions only at the higher elevations (>1800 m). Deviations during the decline period (1960-1984) from the pre-decline growth-climate model (1900-1940) were less substantive and occurred five to ten years later than those of red spruce growing in the northern Appalachians. These findings agree with other dendrochronological studies of red spruce in the middle and southern Appalachians (Cook 1988, Ord and Duerr 1988, Van Deusen 1988, Adams et al. 1985).

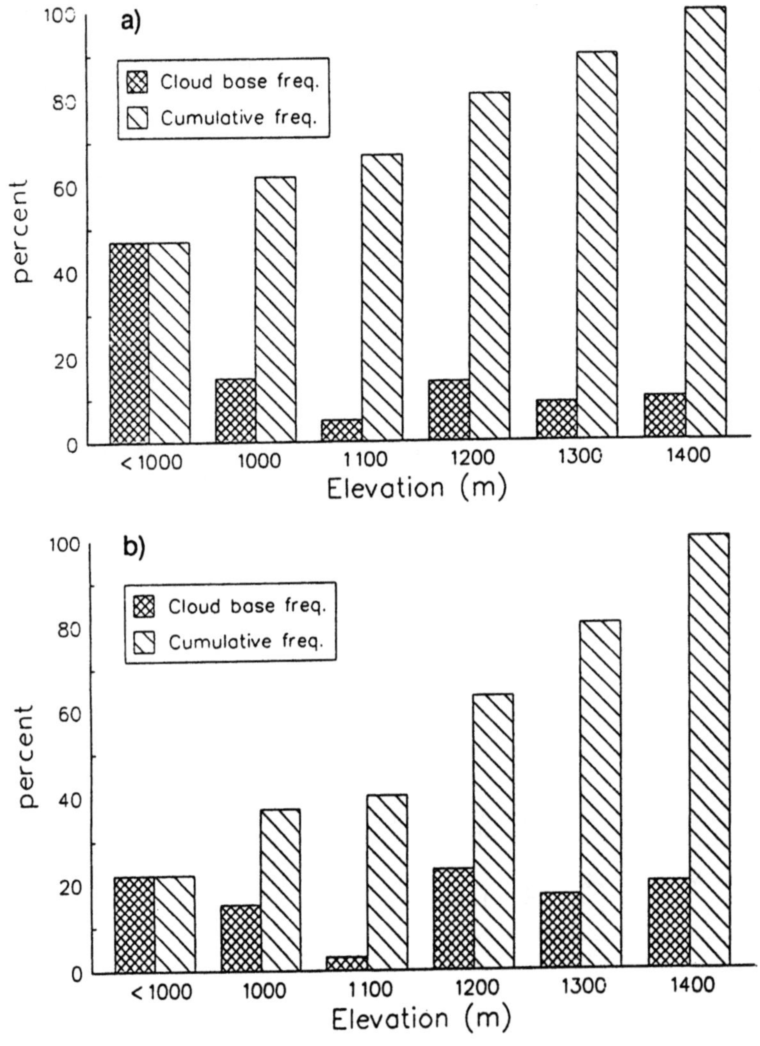

Figure 3. Cloud frequency at (a)Mt. Moosilauke, NH, and (b) Whiteface Mtn., NY. Data are for May-October 1987. (Adapted from Mohnen et al. 1989).

Atmospheric Deposition

The high elevations of the Appalachian Mountains are unique in eastern North America with respect to the deposition of acidic substances, due to frequent immersion in clouds and the subsequent deposition of cloud water to the forest canopy. The Mountain Cloud Chemistry Project (MCCP) characterized the chemical environment of these high elevation sites (Mohnen 1988). Regional cloud climatology studies indicated typical cloud base heights from 800 to 1200 m in the Appalachians, and that peaks above 1400 m were in clouds approximately 28 to 48 % of all hours during the growing season (Figure 3) (Mohnen et al. 1990). For one site with year-around data, the peak was in cloud even more frequently during the non-growing season. Data on the chemistry of the water in these clouds showed that cloud water has a lower average pH (3.5) than precipitation (4.2) and has higher concentrations of sulfate and nitrate. Warm season (May through October) cloud water deposition of sulfate and nitrate at high elevation sites in the Appalachians was calculated to be more than twice that deposited by precipitation (Table 1). Below-cloud eastern forests were exposed to both lower mean and fewer extreme concentrations than sites frequently in cloud, with the highest concentrations of wet-deposited ions found near cloud-base.

Exposure to ozone is also greater at high elevation sites than at nearby rural low elevation sites (Table 2). Maximum one hour average concentrations are not vastly different; however, high elevation sites do not experience the diurnal cycle in ozone concentrations typical of low elevation sites. Instead, the concentrations remain high throughout the night and early morning. Median ozone concentrations ranged from 45 ppb to 60 ppb, but hourly concentrations in excess of 100 ppb were recorded. Ozone exposure data also suggest a significant north-south gradient and pronounced year-to-year differences. Concentrations of NO_2 and SO_2 were low, usually below the detection limit of the monitoring equipment; average warm season mean SO_2 concentrations were less than 2.1 ppb for all MCCP sites.

Table 1. Estimated 1987-88 growing season (April 15 - October 15) mean sulfur and nitrogen deposition at Mountain Cloud Chemistry Program sites (kg S or N/ha-mo). (Source: Mohnen et al. 1990).

Site	Wet		Cloud		Dry	
	S	N	S	N	S	N
Howland	0.43	0.15	no clouds		0.17	0.08
Moosilauke	0.84	0.29	0.69	0.35		
Whiteface-2	0.58	0.21	0.65	0.17		0.31
Shenandoah	0.99	0.28	insufficient data		1.80	0.23
Whitetop	1.01	0.28	3.58	1.23		
Mt. Mitchell	1.31	0.39	2.65	0.87		

Cause and Effect Hypotheses

Many hypotheses have been advanced to explain the dramatic decline of red spruce in the northern Appalachians, and the severely stressed spruce-fir ecosystems in the southern Appalachians. An epidemiological view of this decline suggests a relationship of visibly declining red spruce with greater deposition of acidic substances and elevated levels of ozone. This is based on a spatial relationship related to elevation, with cloud water deposition and higher ozone levels accounting for increased pollutant loading at higher elevations. The temporal correlations are less clear; however, the increase in the emissions of nitrogen oxide since the middle 1950's combined with the constant emissions of sulfur dioxide since about 1930 indicate an increased atmospheric burden of acidic substances during the past 30 years (Gschwandtner et al. 1986). Based on this correlative evidence and the lack of other obvious biological causes for the regionwide decline, a number of hypotheses for mechanisms of air pollution effects on red spruce have been developed. These include:

Table 2. Total season ozone exposure at MCCP sites for three successive growing seasons (April 15-October 15, daylight hours 7 AM -6 PM). Dose is expressed in ppm-hr as the sum of seasonal doses ≥ 0.07ppm.

Site	1986	1987	1988	Elevation (m)
Howland Forest, ME	N.D.[1]	0.82	4.16	250
Mt. Moosilauke, NH	N.D.	7.81	12.51	1000
Whiteface Mtn,-1, NY	2.29	9.68	20.85	1483
Whiteface Mtn.-3, NY	N.D.	9.41	16.5	1026
Whiteface Mtn.-4, NY	N.D.	3.47	N.D.	604
Huntington, NY	5.09	5.74	11.36	500
Shenandoah-1, VA	N.D.	9.49	23.27	1015
Shenandoah-2, VA	N.D.	9.01	39.44	716
Shenandoah-3, VA	N.D.	6.07	20.88	524
Big Meadow, VA	5.56	28.50	31.89	1071
Dickey Ridge, VA	3.21	31.07	40.25	631
Sawmill Run, VA	11.26	26.80	30.16	453
Whitetop Mtn. TN	N.D.	38.54	37.68	1689
Giles Co. TN	16.38	16.73	28.91	244
Marion, VA	4.11	9.27	26.92	710
Mt. Mitchell-1, NC	8.34	5.14	45.17	1950
Mt. Mitchell-2, NC	N.D.	6.68	19.49	1750

1 ND = not determined

1. Sulfur and/or nitrogen derived pollutants affect red spruce through soil mediated processes involving: a) direct toxicity to roots by mobilized metals in the soil, b) inhibition of nutrient uptake by roots due to the mobilization of metals, and c) leaching of nutrients from the rooting zone by acidic deposition resulting in altered tree water and nutrient balance.

2. Sulfur and/or nitrogen derived pollutants alone or in combination with oxidants affect red spruce through alteration of physiological processes and carbon dynamics.

3. Sulfur and/or nitrogen derived pollutants alone or in combination with oxidants affect red spruce through the mechanism of increased leaching of foliar nutrients.

4. Sulfur and/or nitrogen compounds alone or in combination with oxidants affect winter hardiness of red spruce.

A number of natural environmental stresses may be associated with the recent decline of red spruce, thus it is important to consider the role of interactive stresses which may be important within the spruce-fir ecosystem. First, the physical environment is harsh. High winds and very low winter temperatures are of primary concern. The second factor is related to soil nutrient status. Many of these forests are located on shallow soils or those with low available nutrients. Also, insect and disease related stresses are recognized as a potential contributor to declines of forest species. The role of insects and pathogens was assessed for both the northern and southern Appalachians. Aside from the impact of the balsam woolly adelgid on Fraser fir in the southern Appalachians, no insects or pathogens were found to contribute significantly to the observed decline (Bruck 1989, Smith 1988).

Soil-mediated Effects

Much of the concern about effects of acidic deposition on soils has centered on the likelihood of soil acidification, brought about predominantly by leaching of base cations from the soil by mobile anions (sulfate and nitrate). Increased deposition of strong acid anions could also lead to increased aluminum availability in the soil solution, with direct, toxic effects to roots, or an indirect effect on Ca and Mg uptake (Shortle and Smith 1988).

Results of experiments with red spruce seedlings provide some support for a potential involvement of aluminum in reduced root and shoot growth and altered nutrient status (Raynal et al. 1990, Joslin and Wolfe 1988, Thornton et al. 1987, Hutchinson et al. 1986). Fine root production was more sensitive to elevated aluminum than above ground biomass in these studies, with reductions in red spruce root biomass and root elongation generally observed at aluminum solution concentrations as low as 200 uM/l. Peak soil solution total aluminum concentrations collected under minimal tension ranging from 1-280 uM have been recorded in the southern Appalachians (Raynal et al. 1990, Johnson, D. et al. 1988). However, this low tension solution mostly bypasses the roots and may underestimate aluminum concentrations to which roots are exposed between rain events when soils are drier. Rapidly reactive aluminum levels exceeding 140 uM were seldom found in the forest floor on Whiteface Mountain and levels exceeding 200 uM would be very rare in either the forest floor or mineral soil

Table 3. Exchangeable aluminum, calcium, and magnesium in soils from high elevation spruce-fir sites.

Horizon	Depth	Al	Ca	Mg
	cm	meq/100g		
		Great Smoky Mountains - Becking Site[1]		
Oie		4.66	2.08	0.71
Oa		14.88	2.79	1.41
A	0-20	14.43	0.29	0.40
Bw1	20-33	7.10	0.10	0.14
Bw2	33-64	1.59	0.02	0.02
		Great Smoky Mountains - Tower Site[1]		
Oie		6.21	8.45	2.04
Oa		13.99	2.04	0.82
A11	0-3	9.54	0.21	0.20
A12	3-21	8.21	0.12	0.16
Bw1	21-36	4.87	0.07	0.10
Bw2	36-61	2.43	0.03	0.05
		Whiteface Mountain[1]		
Oa		4.50	10.40	1.60
Bhs		8.50	0.40	0.10
Bs		3.20	0.20	0.01
		Mt. Moosilauke - 1000m[2]		
Oie		7.2	6.1	1.2
Oa		16	3.0	0.6
	0-10	9.49	0.35	0.19
	10-20	7.33	0.23	0.15
	20+	4.1	0.13	0.04
		Mt. Moosilauke - 1200 m[2]		
Oie		4.1	7.9	1.5
Oa		8.2	4.2	1.0
mineral soil total		8.2	0.22	0.17

1 Source: Johnson et al. 1988.
2 Source: Huntington et al. 1988.

of similar sites (T. Huntington, personal communication). Effects on root elongation may not be totally aluminum-concentration-dependent, however, but may in part be dependent on competing ion effects and the relative contribution of aluminum ions to total solution ionic activity (Kelly et al. 1990). It is also important to consider the organic matter content of the rooting zone. Most of the fine roots observed in a study in the southern Appalachians (Kelly and Mays 1989) were found in the surface organic horizon, where organic matter could provide important complexation of aluminum, thus minimizing the effects of aluminum.

Elevated aluminum may also alter cation uptake by the roots. Reductions in root and foliar calcium and magnesium concentrations occur at lower soil solution aluminum concentrations than those causing direct injury (Raynal et al. 1990, Thornton et al. 1987). Studies of mature red spruce growing across a gradient of soil solution aluminum concentrations found an inverse relationship between soil and root aluminum levels and foliar calcium and magnesium levels, suggesting high soil aluminum may inhibit uptake of these nutrients (Joslin et al. 1988). Bondietti et al.(1990) reported an unprecedented increase in the aluminum:calcium ratio in the wood of mature red spruce growing in the Great Smoky Mountains, which coincided with reductions in annual radial increment beginning 15-40 years ago. These changes were greater in trees from high elevation sites which receive higher amounts of acidic cloud water deposition, have high soil solution aluminum concentrations, and have soil solution aluminum:calcium ratios greater than one (Johnson, D. et al. 1988). Shortle and Smith (1988) reported that aluminum:calcium ratios greater than one in fine roots are associated with reduced red spruce vigor in the northeastern U.S. In addition, increases in root aluminum:calcium ratios and reductions in sapwood area and tree vigor were most pronounced in soils which had aluminum:calcium ratios greater than one. Low tree vigor is believed to be related to an aluminum-induced calcium deficiency resulting in reduced sapwood production.

Altered Physiological Processes

To examine physiological effects, members of the SFRC conducted controlled exposure experiments with red spruce seedlings, and measured physiological parameters of sapling and larger red spruce trees in the field. It was hypothesized that photosynthesis would be sensitive to increased acidic deposition and ozone expsoure. Results to date are equivocal. Kohut et al. (in press) reported increased photosynthesis of two-year-old red spruce seedlings in response to increasing acidity. A similar increase was observed by Seiler and Paganelli (1987). Ozone fumigation of seedlings resulted in either no effect on photosynthesis (Kohut et al. in press, Taylor et al. 1986), or increased photosynthesis (Cumming et al. 1988). Detailed physiological studies are ongoing. In a study utilizing grafted material to test tissue age differences, total chlorophyll levels and rates of net photosynthesis were reduced at the highest ozone level (ambient + 150 ppb) in both juvenile and mature tissue, although results suggest juvenile tissue may be more sensitive (J. Rebbeck, personal communication).

Foliar injury in response to elevated levels of acidic deposition or ozone is of potential significance to many physiological processes. For example, it is hypothesized that increased foliar injury results in decreased net photosynthesis. Acid mist treatments produced visible foliar injury at acidities similar to the minimum values reported for ambient cloud water (pH 2.7-3.0) (Jacobson et al. 1990, Leith et al. 1989). Browning of needles was increased by sulfate acids compared to nitrate acids, and current year needles were more susceptible than older needles (Jacobson et al. 1989). Studies are continuing to determine the implications of this damage for red spruce physiology.

Comparisons of physiological parameters have also been made on larger trees growing in the montane environment. Results from the Smoky Mountains indicate substantially slower growth of sapling and mature trees at high deposition sites, despite higher rainfall and less competition at these sites (McLaughlin et al. 1989). Reduced growth of saplings was attributed to increased rates of dark respiration, reduced photosynthesis, reduced foliar chlorophyll, low levels of calcium, magnesium, and phosphorous in foliage, and high levels of soil and foliar aluminum. Slight increases in water stress have been reported in response to acidic deposition (Eamus et al. 1989), however, water stress was generally comparable between the two sites and was not considered to contribute to the observed differences (Andersen and McLaughlin 1990).

These studies and others currently ongoing (R. Amundson, personal communication) suggest that red spruce growing in sites receiving high levels of acid deposition and which are exposed to greater amounts of ozone may have a reduced capacity to assimilate carbon. However, these results must be interpreted with caution as other important environmental variables which influence the growth and physiology of red spruce also differ among the sites contrasted in these studies.

Foliar Leaching

Evidence of leaching of nutrients from foliage comes from both controlled exposures and field studies. Short term exposures of red spruce seedlings to simulated acid mist (pH 2.6-4.2) resulted in decreased calcium, magnesium, and potassium in foliage (Jacobson et al. 1989). Intermittent wetting cycles resulted in increased foliar sulfur and nitrogen levels, whereas continuous exposures produced reduced levels of calcium, magnesium, and potassium. Measurements at Whitetop Mountain documented changes in cloud water chemistry as it passed through the red spruce canopy (Joslin et al. 1988). Acidity was neutralized by exchanges of hydrogen ions for mainly calcium and magnesium ions which were subsequently leached from the needles. Calcium and magnesium leaching accelerated as the ambient cloud water acidity increased, leading the authors to suggest that losses of cations caused by acidic cloud water may contribute to nutrient deficiencies.

Foliar leaching has not yet been documented to induce nutrient deficiencies within mature red spruce trees. However, the low base status of many high elevation spruce-fir soils, the high aluminum levels in soil solutions, and the potential antagonism of nutrient uptake by aluminum suggest foliar leaching could contribute significantly to reduced nutrient status and increased plant stress. Low levels of calcium, magnesium, and phosphorous found in mature and sapling red spruce (McLaughlin et al. 1989) are consistent with the hypothesis of nutrient deficiency resulting from exposure to acidic cloud water.

Winter Injury

Analyses of the relationship of diameter growth of red spruce in the Northeast with climatic parameters revealed that growth is adversely affected by warmer than average late summer or colder than average early winter temperatures (Johnson, A. et al. 1988, McLaughlin et al. 1987, Cook et al. 1987). Detailed analyses of tree rings and temperature showed that winter in the late 1950's and early 1960's were severe, based on climatic parameters believed to contribute to winter injury. However, severe climatic conditions of similar duration were recorded around 1910, 1935, and 1945, and did not appear to trigger widespread spruce mortality or winter injury (Johnson, A. et al. 1988). Red spruce growth could be predicted prior to 1960 by a model based on temperatures from July/August and November/December. However, this model did not accurately predict growth for the period 1960 to 1981, indicating that something other than temperature was influencing red spruce growth during this period of recent decline. In addition, field observations in the northern Appalachians have linked late winter/early spring discoloration and subsequent loss of the previous season's needles to the development of crown thinning associated with declining red spruce (Friedland et al. 1988). As a result of this evidence, focus has been directed at the potential for air pollutants to alter cold hardiness of red spruce.

Red spruce is less cold-tolerant than balsam fir (DeHayes et al. in press), and has been more frequently and severely injured during winter than balsam fir during the past 40 years (Johnson et al. 1986). Rates of cold acclimation and deacclimation in red spruce are comparable to fir, but red spruce does not achieve midwinter hardiness levels recorded for other North American boreal species. Current year needles of red spruce reach midwinter hardiness levels that are barely sufficient to survive midwinter temperatures commonly encountered (around -40°C) (Sheppard et al. 1989). Winter injury to red spruce is found primarily on current year needles and these needles lag behind year old needles in development of cold tolerance (DeHayes et al. in press). Red spruce provenances growing in a common garden were found to vary in susceptibility to winter injury, but variation patterns were not consistent from year to year and were not associated with geographic features of provenances. Red spruce seedlings are prone to midwinter and early spring dehardening in response to consecutive days of above freezing temperatures, and require more than five days to regain pre-thaw hardiness levels. This suggests that red spruce trees may be prone to freezing injury throughout the winter following several days of above-freezing temperatures.

The actual mechanism(s) for winter injury is not yet clearly elucidated, but two hypothesized mechanisms are receiving study: freezing injury ("frost injury") and desiccation ("winter drying"). Herrick and Friedland (1989) measured consistently lower relative water content (RWC) throughout winter in needles which exhibited winter injury symptoms in the late winter. Trees at higher elevations had, on average, a lower RWC. Other recent work suggests red spruce is very poorly protected against mid-winter (December and January) freezing (DeHayes et al. in press, Sheppard et al. 1989). Because red spruce deharden very rapidly, it may be particularly prone to freezing after a prolonged thaw. Further research to elucidate the mechanism of winter injury to red spruce is ongoing.

An initial hypothesis concerning winter injury of red spruce suggested increased atmospheric deposition of nitrogen caused reductions in cold tolerance, due to de-

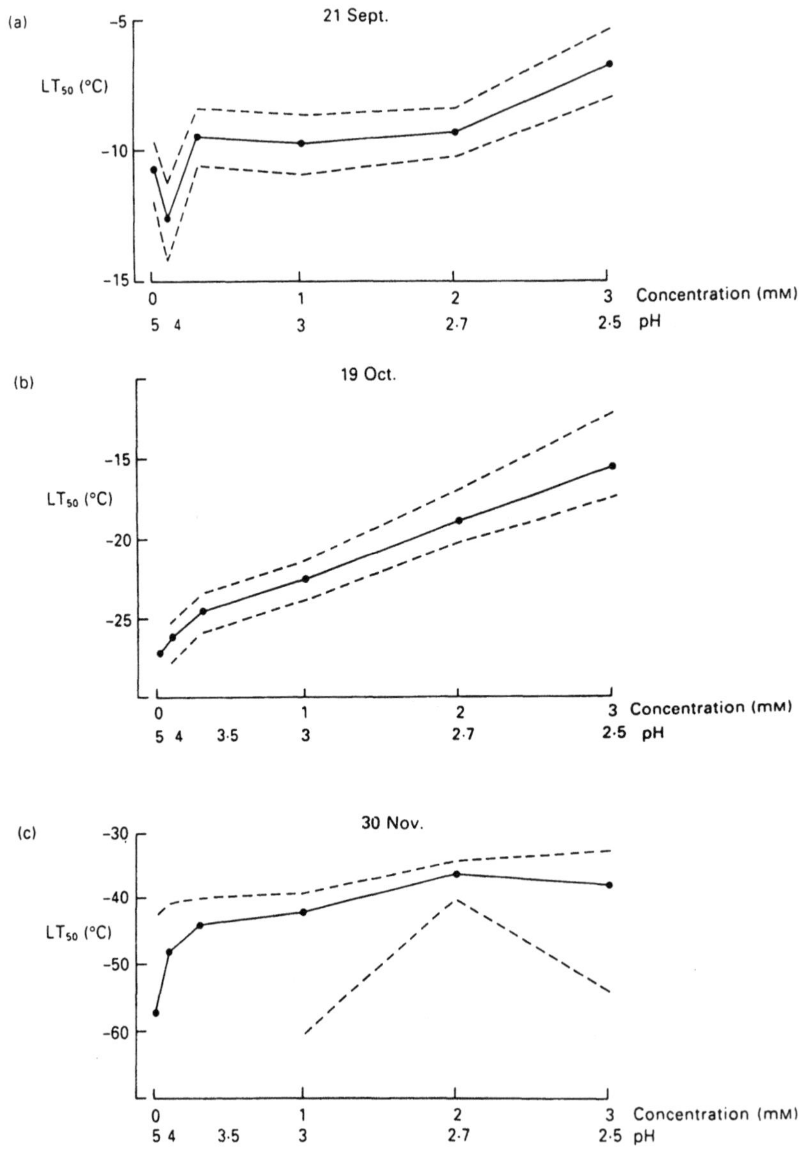

Figure 4. Effects of mist treatments on the freezing temperatures needed to kill 50% of shoots (LT_{50}) harvested from red spruce seedlings on (a) 21 September, (b) 19 October, (c) 30 November, 1987. Dotted lines represent 95% confidence intervals about the mean. (Source: Fowler et al. 1989).

layed winter hardening in the autumn. Nitrogen fertilization to the roots decreased cold tolerance of seedlings during the early stages of dormancy, but the opposite effects was observed in later stages of dormancy (Jacobson et al. 1989). DeHayes et al. (1989) reported increased cold tolerance with increasing nitrogen uptake, up to a fertilization rate of 1500 kg N ha^{-1} yr^{-1}. In general, seedlings receiving nitrogen in mid or late summer were as hardy or hardier than seedlings fertilized in early summer, regardless of the fertilizer rate. Recent results suggest increased sulfate levels are responsible for decreased cold hardiness. Red spruce seedlings receiving acid mist treatments containing nitrate but no sulfate were more cold tolerant than treatments with a mixture of nitrate and sulfate or only sulfate (Fowler et al. 1989). When the ratio of sulfate to nitrate was varied, high acidity with high sulfate resulted in delayed autumn frost hardening compared to high acidity and high nitrate or controls (deionized water) (Fowler et al. 1989). Hardening was not prevented by acidic misting, but throughout the autumn and early winter, treated seedlings were substantially less hardy than untreated ones (Figure 4).

Ozone may also affect winter hardiness. Ozone fumigation resulted in significantly higher levels of winter injury in some studies (Cumming et al. 1988, Fincher et al. 1990), while other studies reported slightly more cold tolerance in seedlings exposed to ambient or elevated ozone levels (DeHayes et al. 1989).

DISCUSSION

Individually, none of the studies discussed above provides sufficient evidence to support the hypothesis that acidic deposition is responsible for the recent, dramatic decline in high elevation red spruce. Collectively, they provide strong circumstantial evidence. Mostellar and Tukey (1977) suggest determination of causality involves an evaluation of: a) consistency in time and space between a specific effect and a causal factor, b) responsiveness in the form of a reproducible relationship between the causal factor and the effect, and c) a known mechanism or series of processes linking the effect to the causal factor. Limited causality can be inferred if two of the three criteria are met; the case is substantially strengthened when all three are met. These criteria provide an appropriate method to evaluate the data which address the hypotheses of a cause and effect relationship between atmospheric deposition and red spruce decline.

Consistency: Most soils of high elevation spruce-fir forests in the southern Appalachians, and some northern Appalachian soils, are acidic and have very low base saturations. High soil solution sulfate and nitrate concentrations are associated with elevated atmospheric deposition (Joslin et al. 1988, Johnson, D. et al. 1988, Lindberg et al. 1988). High soil solution concentrations of aluminum are associated with an increase in the aluminum:calcium ratio in the wood of red spruce, and these increases are coincident with a decrease in radial growth (Bondietti et al. 1989). High aluminum:calcium ratios in fine roots of mature red spruce are associated with reduced tree vigor (Shortle and Smith 1988). Sapling and mature red spruce growing at high elevation sites which receive high levels of acidic deposition from cloud water have a reduced capacity to assimilate carbon and have poorer foliar nutrient status compared to red spruce growing at lower elevations which do not receive cloud water deposition (McLaughlin et al. 1989, Friedland et al. 1984). Consistency in the relationship of winter injury of red spruce to spatial and temporal trends in deposition has not been thoroughly investigated, but there appears to be a higher frequency in winter injury during the past 30 years (Johnson, A. et al. 1988, Johnson et al. 1986), and the growth of red spruce has decreased over approximately the same time period (Johnson, A. et al. 1988, McLaughlin et al. 1987, Cook et al. 1987).

Responsiveness: The concentration of soil solution aluminum is related to sulfate and nitrate inputs from atmospheric deposition (Johnson, D. et al. 1988, Joslin and Wolfe 1988, Reuss and Johnson 1986). Experiments with seedlings have documented reductions in plant biomass, root elongation, and levels of calcium, magnesium, and phosphorous in roots and shoots as a function of increased aluminum concentration (Joslin and Wolfe 1988, Thornton et al. 1987, Taylor et al. 1986, Hutchinson et al. 1986,). The ratio of aluminum to calcium in wood is higher at high deposition sites than at low deposition sites (Bondietti et al. 1989). Acidic mist treatments produced visible foliar injury in proportion to treatment acidity and sulfate concentration relative to nitrate. Calcium and magnesium were leached from red spruce foliage as a function of cloud water acidity (Jacobson et al. 1989). Autumn frost hardening was delayed in red spruce seedlings as a function of acidity and sulfate concentration (Fowler et al. 1989)

Mechanisms: The processes by which aluminum is mobilized in acidic soils with low base saturations by additions of acidic anions are based on accepted concepts of ion exchange processes (Reuss and Johnson 1986), and have been validated by results from the Integrated Forest Study (D.W. Johnson, personal communication). The mechanisms of aluminum-caused dysfunction of physiological or biochemical processes of red spruce are not known. Research has not yet clearly identified the mechanisms by which acid deposition and ozone reduce the capacity of red spruce to assimilate carbon. Likewise, mechanistic research on acidic deposition-caused foliar leaching, foliar injury, and reduced cold tolerance is continuing.

Uncertainties: Our understanding of the effect of aluminum on red spruce is based on seedling studies under artificial environments. The relationship between the response of seedlings and mature trees remains unclear. Several sites in the southern Appalachians have high levels of nitrate being generated from within the ecosystem. The relative contribution (and cause) of this internal nitrate to alterations in the biogeochemistry of aluminum needs further research. Our understanding of the effects of acidic deposition on cold hardiness of red spruce is also based on seedling studies; effects on mature trees are not known. Uncertainty is also associated with the methods of determining cold tolerance. However, when treated as an assay of relative changes and not as an indicator of absolute temperature sensitivity, the results from three different studies in different laboratories have a high level of correspondence. Lack of information on the internal processes which result in winter injury also contributes uncertainty.

Assessment: A number of potentially interactive stresses and resultant impacts on physiological processes exist at high elevation sites. High pollutant loading, primarily due to cloud water deposition, appears to play a role in effects related to soil-mediated processes and increased susceptibility to winter injury. The interactions between aluminum, nutrient uptake, reduced carbon assimilation, reduced growth, and reduced fitness (e.g., susceptibility to winter injury) are complex. Additional impacts may be associated with leaching of nutrients from foliage and foliar injury (visible and non-visible).

Hypotheses related to soil-mediated factors and winter injury have stronger support than those related to carbon dynamics and foliar leaching; research to test these two hypotheses is ongoing and will address some of the uncertainties.

LITERATURE CITED

Adams, H. S., S. L. Stephenson, T. J. Blasing and D. N. Duvick. 1985. Growth trend declines of spruce and fir in mid-Appalachian subalpine forests. Environ. Expt. Bot. 25:315-325.

Amman, G. D., and C. F. Spears. 1965. Balsam woolly aphid in the Southern Appalachians. J. For. 63:18-20.

Andersen, C. P., and S. B. McLaughlin. 1990. Seasonal changes in shoot water relations of *Picea rubens* at two high elevation sites in the Smoky Mountains. Tree Phys. (in press).

Battles, J. J., A. H. Johnson, T. G. Siccama. 1988. Relationships between red spruce decline and forest characteristics at Whiteface Mountain, New York. In Hertel, G. D., (ed.), Effects of atmospheric pollutants on the spruce-fir forests of the Eastern United States and Federal Republic of Germany. USDA Forest Service Gen. Tech. Rep. NE-120. Northeastern Forest Experiment Station. Radnor, PA. pp. 163-172.

Bondietti, E. A., C. F. Baes, and S. B. McLaughlin. 1990. Radial trends in cation ratios in tree rings as indicators of the impact of atmospheric deposition on forests. Can. J. for. Res. (in press).

Bruck, R. I. 1989. Survey of diseases and insects of Fraser fir and red spruce in the southern Appalachian Mountains. Eur. J. For. Path. 11:389-398.

Cook, E. R., 1988. A tree ring analysis of red spruce in the Southern Appalachian Mountains. In P. C. Van Deusen (ed.), Analyses of Great Smoky Mountain Red Spruce Tree Ring Data, USDA Forest Service Gen. Tech. Rep. SO-69, New Orleans, LA. p. 6-20.

Cook, E. R., A. H. Johnson and T. J. Blasing. 1987. Forest decline: modeling the effect of climate in tree rings. Tree Phys. 3:27-40.

Cumming, J. R., R. G. Alscher, J. Chabot, L. H. Weinstein. 1988. Effect of ozone on the physiology of red spruce seedlings. In Hertel, G. D., (ed.) Effects of atmospheric pollutants on the spruce-fir forest of the Eastern United States and Federal Republic of Germany. USDA Forest Service Gen. Tech. Rep. NE-120. Northeastern Forest Experiment Station, Radnor, PA pp. 355-364.

DeHayes, D. G., C. E. Waite, M. A. Ingle and M. W. Williams. Winter injury susceptibilty and cold tolerance of current and year-old needles of red spruce trees from several provenances. For. Sci. (in press)

DeHayes, D. H., M. A. Ingle, and C. E. Waite. 1989. Nitrogen fertilization enhances cold tolerance of red spruce seedlings. Can. J. For. Res. 19(8): 1039-1043.

Eagar, C. 1984. Review of the biology and ecology of the balsam woolly aphid in Southern Appalachian spruce-fir forests. In P. S. White (ed.), The Southern Appalachian Spruce-Fir Ecosystem: Its Biology and Threats. USDI National Park Service, Research/Resource Mgt. Rep. SER-71. p. 36-50.

Eamus, D., I. D. Leith, and D. Fowler. 1989. Water relations of red spruce seedlings treated with acid mist. Tree Phys. 5:387-397.

Evans, L.S. 1986. Proposed mechanisms of initial injury-causing apical dieback in red spruce at high elevation in eastern North America. Can. J. For. Res. 16:1113-1116.

Fincher, J., J. R. Cumming, R. G. Alscher, G. Rubin, L. Weinstein. 1990. Long-term ozone exposure affects winter hardiness of red spruce (Picea rubens Sarg.) seedlings. New Phytol. (in press).

Fowler, D. J., N. Cape, J. D. Deans, I. D. Leith, M. B. Murray, R. I. Smith, L. J. Sheppard, and M. H. Unsworth. 1989. Effects of acid mist on the frost hardiness of red spruce seedlings. New Phytol. 113:321-335.

Friedland, A.J., G. C. Hawley, and R. A. Gregory. 1988. Red spruce (Picea rubens Sarg.) foliar chemistry in northern Vermont and New York, USA. Plant Soil 105:189-195.

Friedland, A. J. 1989. Recent changes in the montane spruce-fir forests of the northeastern U.S. Env. Monitoring and Assess. 12:237-244.

Gschwandtner, G. K., K. Eldridge, C. Mann and D. Mobley. 1986. Historic emissions of sulfur and nitrogen oxides in the United States from 1900 to 1980. J. Air Poll. Control Assoc. 36:139-149 (1986).

Herrick, G. T., and A. J. Friedland. Winter desiccation and injury of subalpine red spruce. Oecologia (in press).

Hutchinson, T.C., L. Bozie, and G. Munoz-Vega. 1986. Response of five species of conifer seedlings to aluminum stress. Water Air Soil Poll. 31:283-294.

Jacobson, J. S., L. I. Heller, K. E. Yamada, J. F. Osmeloski, T. Bethard, and J. P. Lassoie. 1990. Foliar injury and growth response of red spruce to sulfate and nitrate acidic mist. Can J. For. Res. (in press).

Johnson, A. H., E. R. Cook and T. G. Siccama. 1988. Climate and red spruce growth and decline in the northern Appalachians. Proc. Natl. Acad. Sci. USA 85:5369-5373.

Johnson, A. H., A. J. Friedland and J. G. Dushoff. 1986. Recent and historical red spruce mortality: Evidence of climatic influence. Water Air Soil Poll. 30:319-330.

Johnson, A. H., and T. G. Siccama. 1983. Acid deposition and forest decline. Environ. Sci. Tech. 17:294A-305A.

Johnson, D. W., S. E. Lindberg, and H. Van Miegroet. 1988. Nutrient Cycling Patterns in Beech and Red Spruce Forests Near Clingman's Dome, Great Smoky Mountains National Park. In Proceedings, Fifth Annual Gatlinburg Acid Rain Conference, October 31-November 1, 1988. TVA. p. 65.

Joslin, J. D., J. M. Kelly, M. H. Wolfe and L. E. Rustad. 1988. Elemental patterns in roots and foliage of mature spruce across a gradient of soil aluminum. Water Air Soil Poll. 40:375-390.

Joslin, J. D., and M. H. Wolfe. 1988. Response of red spruce seedlings to changes in soil aluminum availability in six amended forest soil horizons. Can. J. For. Res. 18:1614-1623.

Kelly, J. M., and P. A. Mays. 1989. Root zone physical and chemical characteristics in southeastern spruce-fir stands. Soil Sci. Soc. Am. J. 53(4):1248-1255.

Kelly, J. M., M. Schaedlel, F. C. Thornton, and J. D. Joslin. 1990. Sensitivity of tree seedlings to aluminum: II. Red oak, sugar maple, and European beech. J. Env. Qual. 19. (in press).

Kohut, R. J., J. A. Laurence, R. G. Amundson, R. M. Raba, and J. J. Melkonian. Effects of ozone and acidic precipitation on the growth and photosynthesis of red spruce seedlings. (in press).

Leith, I.D., M. B. Murray, L. J. Sheppard, J. N. Cape, J. D. Deans, R. I. Smith and D. Fowler. 1989. Visible foliar injury of red spruce seedlings subjected to simulated acid mist. New. Phytol. 113:313-320.

Lindberg, S., D. Silsbee, D. A. Schaefer, J. G. Owens and W. Petty. 1988. A comparison of atmospheric exposure conditions at high and low elevation forests in the Southern Appalachian Mountain Range. In: M. H. Unsworth and D. Fowler (eds), Acid Deposition at High Elevation Sites. Kluwer Academic Publishers, New York, p. 321-344.

McLaughlin, S. B., C. P. Andersen, N. T. Edwards, and W. K. Roy. 1990. Seasonal patterns of photosynthesis and respiration of red spruce saplings from two elevations in declining southern Appalachian stands. Can. J. For. Res. (in press.)

McLaughlin, S. B., D. J. Downing, T. J. Blasing, E. R. Cook and H. S. Adams. 1987. An analysis of climate and competition as contributors to decline of red spruce in high elevation Appalachian forests of the Eastern U.S.. Oecologia 72:487-501.

Mohnen. V. A. 1988. The Mountain Cloud Chemistry Program. In Hertel, G. D., (ed.). Effects of atmospheric pollutants on the spruce-fir forests of the Eastern United States and Federal Republic of Germany. USDA Forest Service NE-120. Northeastern Forest Experiment Station, Radnor, PA. pp. 381-388.

Mohnen, V. A., V. Aneja, B. Bailey, E. Cowling, S. M. Goltz, J. Healey, J. Hornig, J. A. Kadlecek, J. Meagher, S. Mueller, J. T. Sigmon. 1990. An Assessment of Atmospheric Exposure and Deposition to High Elevation Forests in the Eastern United States. Mountain Cloud Chemistry Program Final Rep. to US-Environmental Protection Agency., Research Triangle Park, NC.

Mostellar, F., and J. W. Tukey. 1977. Data Analysis and Regression. Addison-Wesley Publishing Co., Reading, MA, 588 pp.

Nicholas, N.S., and S. M. Zedaker. 1990. Ice damage in spruce-fir forests of the Black Mountains, North Carolina. Can. J. For. Res. 19:1487-1491.

Ord, J. K., and J. A. Duerr. 1988. Utilizing time series models and spatial analysis of forecast residuals for tree ring analysis of red spruce. In: P. C. Van Deusen (ed.), Analyses of Great Smoky Mountain Red Spruce Tree Ring Data, USDA Forest Service, Gen. Tech. Rep. SO-69, New Orleans, LA. p.21-39.

Peart, D. R., L. E. Conkey, W. H. Smith, F. B. Knight, M. Kiefer. 1988. Condition of the spruce-fir forest at Mt. Moosilauke, New Hampshire. In Hertel, G. D., (ed.) Effects of atmospheric pollutants on the spruce-fir forests of the Eastern United States and Federal Republic of Germany. USDA Forest Service Gen. Tech. Rep. NE-120. Northeastern Forest Experiment Station. Radnor, PA. pp. 173-182.

Raynal, D. J., J. D. Joslin, F. C. Thornton, M. Schaedle and G. S. Henderson. 1990. Sensitivity of tree seedlings to Al: III. Red spruce and loblolly pine. J. Env. Qual. 19. (in press).

Reuss, J. O. and D. W. Johnson. 1986. Acid deposition and the acidification of soils and waters. Springer-Verlag, New York, 119 pp.

Seiler, J.R., and D. J. Paganelli. 1987. Photosynthesis and growth response of red spruce and loblolly pine to soil-applied lead and simulated acid rain. For. Sci. 33:668-675.

Scott, J. T., T. G. Siccama, A. H. Johnson, and A. R. Breisch. 1984. Decline of red spruce in the Adirondacks, New York. Bull. Torrey Bot. Club 111:438-444.

Sheppard, L. J., R. I. Smith, and M. G. R. Cannell. 1989. Frost hardiness of Picea rubens growing in spruce decline regions of the Appalachians. Tree Phys. 5:25-37.

Shortle, W. C., and K. T. Smith. 1988. Aluminum-induced calcium deficiency syndrome in declining red spruce. Science 240:239-240.

Siccama, T. G., M. Bliss, and H. W. Vogelmann. 1982. Decline of red spruce in the Green Mountains of Vermont. Bull. Torrey Bot. Club 109:162-168.

Silver, W. L., T. G. Siccama, A. H. Johnson. Changes in red spruce populations in montane forest of the Appalachians,1982-1987. Can. J. For. Res. (in press).

Smith, W. H. 1988. Pathological Survey of the Spruce-fir Forest on Mt. Moosilauke. Final Report to Spruce-Fir Research Cooperative, U.S. Department of Agriculture, Forest Service, Northeastern Forest Experiment Station, Radnor, PA.

Taylor, G. E., R. J. Norby, S. B. McLaughlin, A.H. Johnson and R. S. Turner. 1986. Carbon dioxide assimilation and growth of red spruce (Picea rubens) seedlings in response to ozone, precipitation chemistry, and soil type. Oecologia 70:163-171.

Thornton, F.C., M. Schaedle, and D. J. Raynal. 1987. Effects of aluminum on red spruce seedlings in soil solution culture. Environ. Exper. Bot. 27:489-495.

Van Deusen, P.C., 1988. Red spruce tree ring analysis using a Kalman filter. In: P. C. Van Deusen (ed.), Analyses of Great Smoky Mountain red spruce tree ring data, USDA Forest Service Gen. Tech. Rep. SO-69, New Orleans, LA. pp. 57-67.

Vogelmann, H. W., G. J. Badger, M. Bliss, and R. M. Kein. 1985. Forest decline on Camel's Hump, VT. Bull. Torrey Bot. Club 112:274-287.

Woodman, J. N. 1987. Pollution-induced injury in North American Forests: facts and suspicions. Tree Phys. 3:1-15.

Zedaker, S. M., N. S. Nicholas, C. Eagar., P. S. White, T. E. Burk. 1988. Stand characteristics associated with potential decline of spruce-fir forests in the southern Appalachians. In Hertel, G. D., (ed.) Effects of atmospheric pollutants on the spruce-fir forests of the Eastern United States and Federal Republic of Germany. USDA Forest Service Gen. Tech. Rep. NE-120. Northeastern Forest Experiment Staion. Radnor, PA. pp. 123-132.

THE RELEASE OF ALUMINIUM INTO SOIL SOLUTIONS AND DRAINAGE WATERS

B. W. Bache
Department of Geography, Downing Place, Cambridge, U.K.

Abstract

One of the main effects of the acidification of the environment is the enhanced solubility of aluminium, an important toxin for biological systems. Data for Al solubility in soil solution and surface waters are discussed in terms of the processes that may be occurring. While for many systems experimental Al concentrations are consistent with the operation of solubility or cation-exchange controls, the high variability of soluble Al concentrations found in streams can only be explained by a delicate interplay between chemical and hydrological processes that are specific to the site in question.

1. INTRODUCTION

It has been shown elsewhere in these reviews that the natural trend to acidification of the land surface that occurs in all humid climatic areas of the world is accelerated by the impact of acid deposition from the atmosphere since the industrial revolution.

Acidic environments have always been characterised by restricted biological populations, but only relatively recently was it realised that the main cause of this is aluminium toxicity. The extensive literature that has developed on the biological effects of Al has been reviewed by Foy (1984) for crop plants and by Howells *et al* (1990) for fisheries. It is surprising that the most abundant metal in the earth's crust is highly toxic to organisms, but fortunately in most environments it is completely insoluble and is therefore innocuous. Only in acidic environments is Al sufficiently soluble to be harmful, and increased mobilization of Al in soils, surface waters and groundwaters is therefore an important consequence of acid deposition. A useful detailed review on Al in the environment is given by Driscoll and Schecher (1988), and on the chemistry involved in its mobilization by Bache (1986).

This review covers three relevant areas. First, it briefly outlines the main sources from which Al is derived, and the factors influencing its solubility. Secondly, a number of data sets on soluble Al in soils and waters are presented to illustrate the effect of acidification on Al concentrations in different systems, and the difficulties in explaining them by simple quantitative theory. Finally the interplay of chemical and hydrological processes that are critical in producing the observed results for a particular situation are discussed.

2. SOURCES AND COMBINATIONS OF ALUMINIUM

2.1. Mineral phases

Aluminium is an amphoteric element occurring in Group III of the periodic table of the elements. Its chemistry in the natural environment is dominated by strong coordination to six oxygen atoms in octahedral arrangement, whether in a mineral crystal lattice, or in solution when the oxygen atoms arise from water molecules. The majority of the Al in the earth's crust occurs in primary minerals in combination with silica in the feldspars and micas, and also in the aluminosilicate clay minerals such as kaolinite and the smectites. Where suitable weathering conditions occur, much of the silica is dissolved away and Al remains, mainly in the oxide/hydroxide form. This may coexist with other weathering products, or in exceptional cases may be concentrated into ore deposits such as bauxite.

The common typical Al hydroxide mineral is gibbsite, having the chemical formula $Al(OH)_3$. However, poorly crystalline and amorphous forms are more common in soils and while they may approach to this general formula, they frequently contain inclusions of other elements such as iron, manganese and particularly silicon, the Si rich ones being given the generic name of "allophane". After an intensive study of soluble Al in groundwaters, Paces (1978) concluded that a metastable cryptocrystalline aluminosilicate whose composition is pH dependent appears to control Al solubility over the pH range 4-8. Farmer *et al* (1983) summarised information on the existence of the tubular mineral imogolite in acid soils, having an Si:Al mole ratio of 0.5, compared with 1.0 for kaolinite; more importantly, a dialysable sol of similar composition, called "proto-imogolite" can be produced in the laboratory, and may be responsible for Al mobilization in acid soils. Birchall and Chappell (1988) produce evidence to show that soluble complexes having Si:Al ratio 0.25-0.35 are stable at pH>6, but unstable at pH<5, but it is not known what mineral form, if any, these may relate to. A detailed study of these aluminosilicate gels and sols was made by Mattson in the 1930s (see for example Mattson and Gustaffson, 1937), but their importance is still often overlooked, presumably because they are difficult to study.

These generally poorly-crystalline oxide/hydroxide forms of Al can be extracted from soils with a variety of reagents, such as acid oxalate, citrate-dithionite which dissolves by reduction the iron compounds that often occlude Al, and pyrophosphate. Pyrophosphate preferentially dissolves amorphous Al compounds without attacking the more-crystalline oxides, and it appears to be particularly good at complexing organically-bound Al.

2.2 Aluminium in solution

The master variable controlling Al solubility simple in mineral systems is pH, as illustrated by the following reaction:

$$Al(OH)_3 + 3H^+ \rightleftharpoons Al^{3+} + 3H_2O$$

For this equilibrium, the reaction constant can be formulated as:

$K_s = \{Al\}/\{H^+\}^3$, where { } indicates ionic activities, and has a value given by pK_s = 3pH-pAl = 8.1 for gibbsite or 9.6 for the amorphous hydroxide, at 25°C. However, this reaction is strongly temperature dependent, and pK_s for the amorphous form is quoted as being 11.6-11.8 at 2°C by Lydersen *et al* (1990), so that the equilibrium solubility increases as the temperature decreases. The maximum Al concentraion that can be expected to occur at a given pH value in an open system at equilibrium with the hydroxide can thus be readily calculated. However, this maximum solubility is rarely attained, particularly at low pH values, for a variety of reasons, but principally because of the formation of less-soluble mineral phases with silicate, and with sulphate in areas impacted by acid deposition (Bache, 1986).

The solution chemistry of Al is complicated. In acid solution it occurs as the hexahydrated ion $[Al(H_2O)_6]^{3+}$. (The water molecules are usually omitted from formulae). If the solution pH is raised by adding alkali, a number of hydrolytic products are formed, which may be mononuclear or polynuclear with respect to Al, depending on the Al concentration and the rate of addition of alkali, and therefore differ considerably in size and in the rates of their subsequent reactions. If the solution contains other components that coordinate strongly with Al, soluble complex ions may form with soluble ligands, such as fluoride, sulphate and small molecular weight polycarboxylic acids. Insoluble precipitates form with silica, sulphate, phosphate and large organic molecules. Many of these components are found in natural systems. Appropriate formation and solubility constants are available for some of these reaction products (Bache, 1986; Driscoll and Secher, 1988) which in principle should allow the species present in solution and their concentrations at equilibrium to be calculated from the gross composition of the solution. However, many of the reactions are temperature dependent and frequently equilibrium data are not available as a function of temperature. Some of the solid-phase reactions are also slow, so that equilibrium conditions are frequently not relevant to field situations.

An experimental fractionation procedure is essential for the determination of the speciation of Al in solution. The scheme devised by Driscoll (1984) is frequently used and is outlined in Figure 1. Within the total Al (Al_t) determined after acid digestion, a rapidly-reacting component is determined colorimetrically after solvent extraction, and this is assumed to be mononuclear, Al_m. A separation using a cation-exchange resin assumes that Al_o, the organically-complexed Al fraction of Al_m, is uncharged while the inorganic charged mononuclear species (Al_i) are adsorbed by the resin. This cation-exchange technique for separating organic from inorganic fractions has been critically examined by Backes and Tipping (1987); the assignment of species to the operationally-determined fractions is somewhat arbitrary, but the method is useful in providing some distinction between soluble forms differing widely in reactivity and biological importance. The inorganic mononuclear, Al_i, (shown to be the most toxic fraction, Howells *et al* 1990) can be further differentiated into its component species by the ALCHEMI programme (Schecher and Driscoll,1987) after measuring the total fluoride and sulphate concentrations, and the solution pH. A useful pretreatment of the experimental solution uses hollow fiber ultrafiltration to separate colloidal material having a molecular weight >10^4 daltons (Salbu *et al*, 1985).

Experimental procedures	← TOTAL SOLUBLE ALUMINIUM, ACID DIGESTED, Al_t →		
	rapid reaction = total monomeric, Al_m		
	passes through cation exchange column	adsorbed by cation exchange column	
Fraction designation	non-labile monomeric, Al_o	labile monomeric Al_i	acid soluble
Suggested composition	small organic complexes	Al^{3+}, $AlOH^{2+}$ charged complexes with F^-, SO_4^{2-}	polynuclear inorganic and larger organic complexes

Figure 1. Soluble aluminium fractionation scheme, after Driscoll (1984).

2.3 . Exchangeable aluminium

In addition to solid phase Al in primary and secondary minerals, and the forms found in solution, another important source of Al is the "exchangeable" fraction. This consists mainly of Al^{3+} and $AlOH^{2+}$ ions adsorbed at the negatively charged surfaces of clay and humus particles in soils. The size of this Al source is very much *smaller* than the structurally-bound Al in primary and secondary minerals, but in acid podzolic soils characteristic of northern Europe and America it is very much *larger* than the very small amounts of Al found in solution. Its importance lies in the fact that, being ionic, it can be rapidly exchanged into solution if and when other cations are leached through soils, such as may occur following a sea salt event or by an acid deposition episode. Exchangeable Al occurs more in mineral subsoil horizons of acid soils than in their organic topsoil horizons, so that it can also be exchanged by calcium ions displaced from organic topsoil horizons by acid inputs (Bache, 1984). A simple calcium-aluminium exchange reaction can be formulated thus:

$$SoilAl_2 + 3Ca(NO_3)_2 \rightleftharpoons SoilCa_3 + 2Al(NO_3)_3$$

The *amount* of exchangeable Al in a soil depends on the cation-exchange capacity of the soil material, and the degree of saturation of the cation-exchange complex with Al relative to the other exchangeable cations (principally Ca, but also Mg, K and Na). This degree of saturation is strongly pH dependent, approaching 1.0 (or 100% Al^{3+}) at pH 3.8, but being zero at pH 6. The Al *concentration* that this exchangeable fraction can maintain in solution depends in turn on three factors: the degree of Al saturation of the exchanger; the relative binding strengths of Al compared to the other exchangeable cations such as Ca and Na (reflected in the selectivity coefficients for exchange); and the total salt content of the solution in equilibrium with the exchanger. The importance and operation of cation exchange reactions in mobilising aluminium from soils into surface waters is explained in further detail by Reuss (1983) and Bache (1986).

The exchangeable Al fraction in soils is not as clearly defined experimentally as are the exchangeable fractions of the alkali and alkaline-earth cations, Ca^{2+}, Mg^{2+}, K^+ and Na^+ (Lee *et al*, 1985). This is because it is in fairly-rapid equilibrium with two other Al sinks, i.e Al ions trapped between the alumino-silicate layers of some clay minerals (the so-called interlayer Al) and the amorphous Al hydroxide/silicate material mentioned earlier.

2.4. Organically-bound aluminium

Soil organic matter is a strong sink for Al when it is present in soils in considerable quantities, rendering Al highly insoluble (Hargrove and Thomas,1981; Conyers, 1990). Early work on the stability of soluble Al-organic complexes was reported by Young and Bache (1985). Tipping and Hurley (1988) have made a more detailed study using their model of Complexation by Humic Acids in Organic Soils (CHAOS). This accounts for Al complexation by solid phase humic substances as bidentate binding to various functional groups, where the formation coefficient depends on the ionic strength of the solution and the charge at the organic surface. This is able to account for Al_i concentrations in equilibrium with organic soils over the pH range 3-5 as a result of binding of Al to weak acid groups on the humus matrix. This shows that Al_i is very low at pH>4 because most of the Al is in insoluble organic complexes, but predicted Al solubility up to 200μM at pH 3.5 seems to be at variance with other evidence that in acid peats soluble Al is less than 10μM (Bache, 1974, Hargrove and Thomas, 1981)

3. FORMS AND CONCENTRATIONS OF ALUMINIUM IN SOILS AND WATERS

3.1. Soils

Some contrasting agricultural soils with pH in the range 3.5-5.5 were equilibrated for an hour with dilute (10^{-2}) calcium chloride solutions, and gave Al concentrations from 1000-3 μM, (Bache, 1974). [The symbol μM = micromoles/litre; 1.0mg/l of Al = 37μM]. This work established a number of principles that determine soluble Al concentrations in soil solutions: (i) Al solubility is strongly dependent on pH for any one soil, but a wide range of pH-pAl relationships was shown for soils of different composition. (ii) Soils with high organic matter contents maintain much lower Al solubility than soils low in organic matter at a similar pH. (iii) The systems appeared to be under-saturated with respect to gibbsite at low pH, and supersaturated at high pH. (iv) Al concentrations vary with salt concentration, which both indicates that a cation-exchange mechanism is controlling solubility, at least in the short term, and also shows the need to measure or control other cations in the system to properly interpret results. Selectivity coefficients for Ca-Al exchange were measured and shown to vary widely for different soils. This work, however, lacked a rigorous Al fractionation procedure, and did not determine soluble Si. Later work (Bache and Sharp, 1976) included some fractionation and showed that in many soils, even at pH as low as

3.8-4.0, more than half the soluble Al is in slowly-reactive forms; it seems unlikely that all of this could be polynuclear inorganic at this pH and some high molecular weight organics could have been involved.

A similar type of study (i.e. equilibrating soils with dilute calcium chloride solutions) on some acid Australian soils (Conyers, 1990) has shown a roughly inverse relationship between soluble silica and soluble Al in the pH range 4 - 5. This supports the suggestion of Paces (1978), referred to earlier, that metastable cryptocrystalline aluminosilicates may control Al solubility. These complexes are poorly defined and their solubility is uncertain, but it seems more likely that such materials, or the amorphous allophane, controls on Al solubility rather than gibbsite, whether crystalline or amorphous. Such systems are likely to approach equilibrium with the exchange phase as well as with the solution, except during times of rapid change.

Detailed fractionation studies of water extracted from acid soils by suction lysimeters in part of the Hubbard Brook Experimental Forest were reported by Driscoll et al (1985). Unfortunately they quote only Al_m, but of this fraction concentrations around 30µM were found in various horizons of a podzolic soil in a high elevation site at pH 3.6-4.7, while around 10µM was found in horizons at lower sites at pH 4.5-5.1. In every case the majority of this Al_m was in organic form. The total salt concentrations were low, <100µM.

Van Praag and Weissen (1985) reported Al_t concentrations between 400 and 30 µM in centrifuged soil solutions under spruce and beech forests at pH 3.2-4.5 in the Belgian Ardennes. They found "no apparant relationship between total soluble Al and either pH or exchangeable Al". Between 10% and 50% of this was Al^{3+}, except for gleyed horizons where the majority was Al^{3+}, the proportion increasing with depth as organic matter decreased,

A much more detailed data set on soil waters extracted *in situ* from two Welsh sites (Plynlimon and Beddgelert, Neal *et al*, 1989) found that in in most cases Al_i was in excess of 85% of the total soluble Al, except for organic surface horizons. Mean pH values (means of a number of samples at the same site and horizon) varied from 4.0 to 4.8, and mean Al concentrations from 60 to 1µM. Figure 2, taken from their work, shows that the highest Al concentrations coincided with the lowest H^+ concentrations (highest pH), contrary to the simple theory. They point out that the increase in H^+ concentration with salt content in the soil waters is consistent with cation exchange reactions, but then surprisingly consider only H^+ as the possible cation exchanging with Al^{3+}, rather than alkali

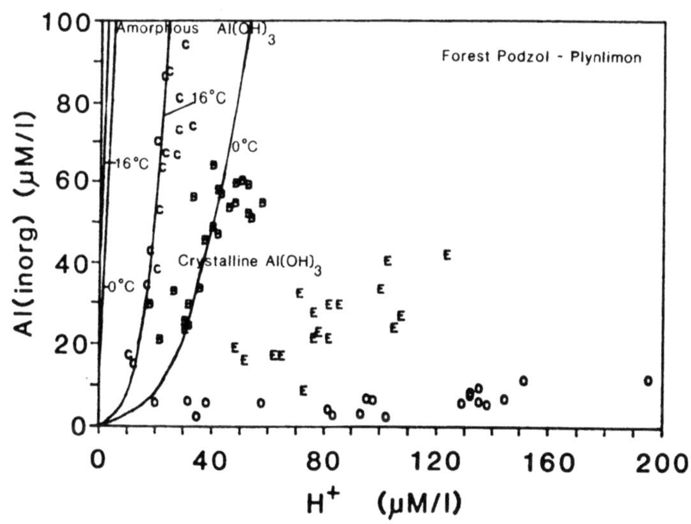

Figure 2. Aluminium-hydrogen ion relationships in soil waters of a forest podzol, Plynlimon (from Neal *et al*, 1989).

and alkaline-earth cations. Figure 3 illustrates very well the balance between H^+ and Al^{3+} as a function of soil layer in a forest podzol. [These should not be considered as complementary exchangeable cations because H^+ behaves very differently from a metal cation, *coordinating* to a negatively-charged surface, whether organic ($R.COO^-$--R.COOH) or inorganic (SiO^---SiOH), rather than behaving as an *electrostatically-adsorbed*, and therefore exchangeable, cation.]

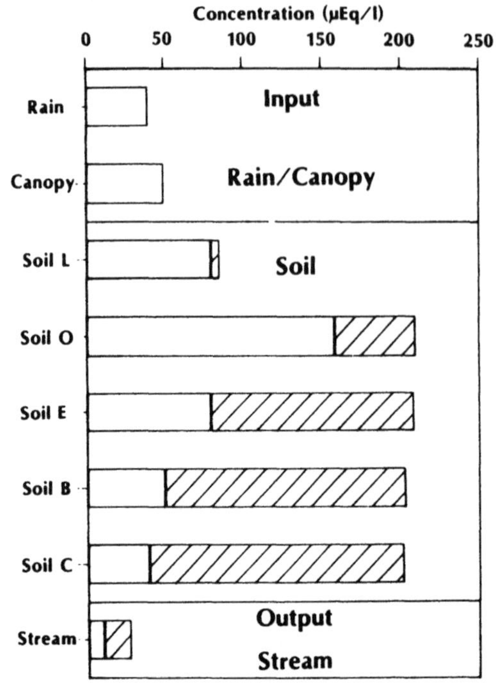

Figure 3. Aluminium and hydrogen ions in soil waters of contrasting horizons of a forest podzol, Plynlimon (from Neal *et al*, 1989).

Another intensively-studied catchment that has been impacted by acid deposition (up to $2g/m^2$ of S per year, equivalent to $125mmol/m^2$ of acidity) is at Birkenes, Norway (Seip *et al*, 1989). Recent data from tension lysimeters showed that spatial variability in soil solution chemistry was considerable, but temporal variability both between and within seasons was much less. Some of their data for Al_o and Al_i as a function of season is shown in Figure 4. Lysimeter 17, in a shallow organic soil over bedrock, gave the greatest variability in H^+ (not shown) and Al_o was the dominant form of Al in solution. Lysimeter 18 showed particularly high Al_i concentrations, coinciding with high nitrate concentrations and thus supporting the role of cation-exchange reactions in controlling Al solubility in the short term. The Al_i/pH plot over the whole of the data set (not shown here) looks somewhat similar to that at Plynlimon shown in Figure 2.

At a pristine catchment, suffering little acid deposition, at Hoylandet, Norway, Ferrier *et al* (1989) monitored snowmelt water flowing through the horizons of a humus-iron podzol soil. Figure 5 shows Al_m and Si concentrations in water passed through a 0.45µm filter. In the surface organic horizons, at a flow rate of 0.4l/h, pH = 4.5 after being initially higher, and the Al concentrations are 0-5 µM. In the B/C horizons, at lower flow rate, pH = 5 and Al = 20µM. There is no obvious relationship here Al concentrations and either silica concentrations or pH.

A recent attempt to evaluate the relationship between Al^{3+} and H^+ activites, and the relative sizes of the Al pool in the various horizons, in a podzol profile at the Birkenes catchment, has been reported by Andersen *et al* (1990). They adjusted samples of four horizons to a range of pH values at a constant ionic strength, and their data are reproduced in Figure 6. For the horizons containing organic matter (O and Bhs), the CHAOS model gives a reasonable

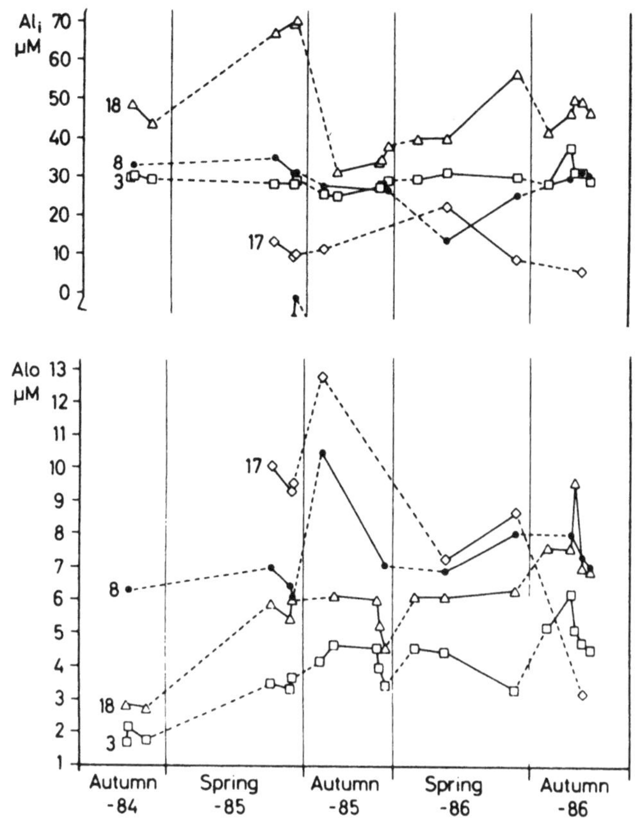

Figure 4. Concentrations of Al_i and Al_o in the waters of four lysimeters as a function of time (from Seip *et al*, 1989).

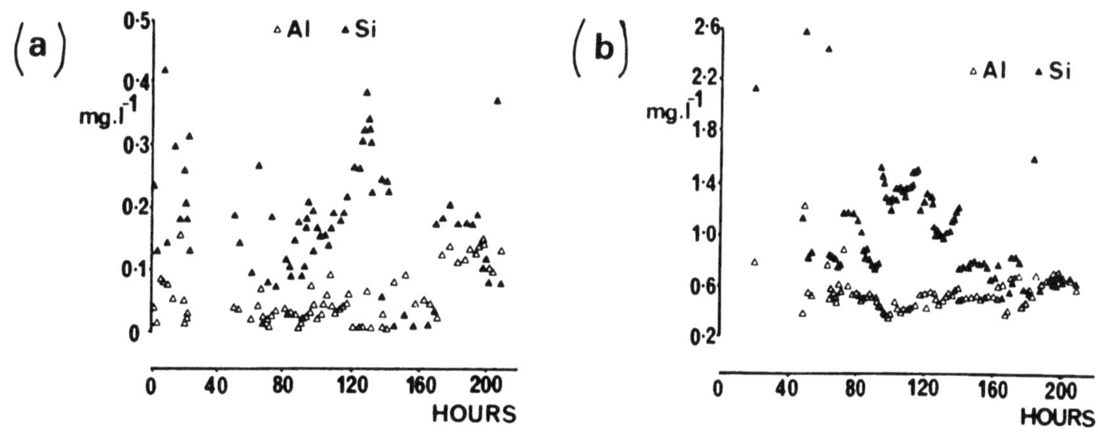

Figure 5. Aluminium and silica concentrations in leachates as a function of time in (a) the organic horizon and (b) the BC horizon of a humus-iron podzol, Hoylandet, Norway (from Ferrier *et al*, 1989).

agreement to the data, but it appears that the essential model parameters were estimated from the data. As expected from the experience described above, none of the data are near the gibbsite solubility line, and the authors suggest that either unidentified mineral phases, or an organic complexation model, may control Al solubility.

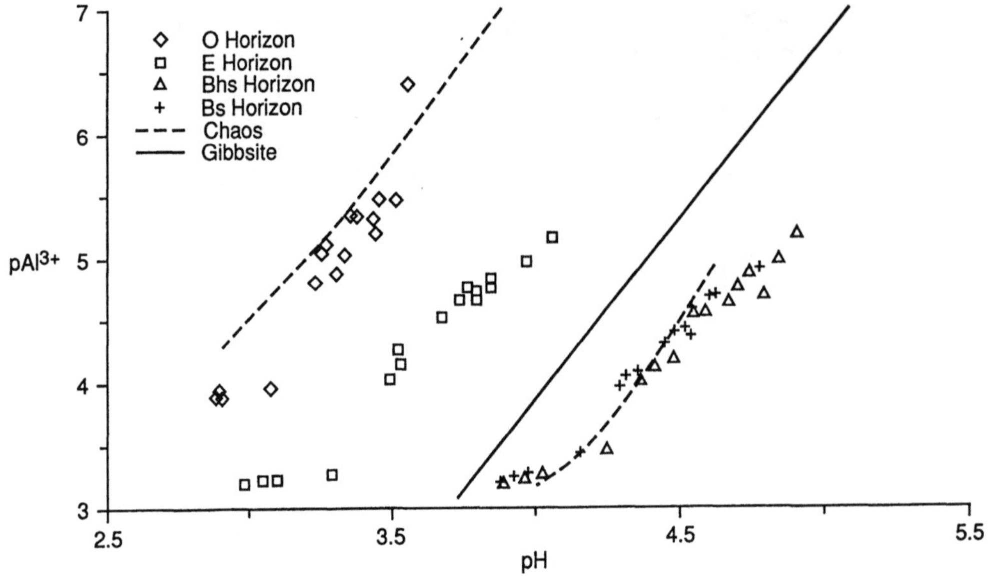

Figure 6. Aluminium-pH relationships of laboratory treated samples of the horizons of a podzol profile at Birkenes, in relation to the predictions of the CHAOS model (from Andersen *et al*, 1990).

3.2. Effects of acid depostion on soil aluminium

The last section gave examples of Al concentrations in soil solutions and how these are affected by soil composition and pH. Because pH generally controls soluble Al, other things being equal, it has been tacitly assumed that acid deposition must therefore increase Al solubility, but there is not a lot of direct evidence of this.

There are now four well-documented accounts showing considerable increases in soil acidity (decrease in soil pH) during this century. Those at Rothamsted Experimental Station in south-east England, covering 1876-1983 on a "clay-with flints" soil (Johnston *et al*, 1986), and in the Solling forest in Germany on a loess-derived soil, for the period 1966-1979 (Ulrich *et al*, 1980) are on sites that have received considerable dry deposition of sulphur in addition to acid precipitation. For those in forest soils in Sweden for 1926 to 1985 (Tamm and Hallbacken, 1988) and in north-east Scotland, 1949 to 1987 (Billet *et al*, 1990), the main acid input comes in the rainfall. In all of these cases it was considered that atmospheric deposition caused some of this increased acidity, particularly so where there has been a high input of dry deposited sulphur. In only two of these cases was some fraction of soil Al measured. In the Scottish example, "extractable Al" (approximating exchangeable Al) was found to have increased over the time period studied, mainly to a small extent but considerably so in one or two cases (Billett *et al*, 1990). In the German example, over a shorter time period, a considerable increase (10-20%) was however found in the Al concentration in the equilibrium soil solution, and more importantly in the exchangeable fraction that controls it (Ulrich *et al* 1980).

3.3. Surface waters

There are now enormous data sets on Al concentrations in lakes and streams, from those areas where acid deposition has important effects on hydrogeochemistry, that is eastern Canada and northeast USA, north and west of Britain, and Scandanavia. A number of these are summarised by Schecher and Driscoll (1990). Rarely do concentrations of total Al exceed about 10μM (roughly 300ug/l), except where very acid snowmelt is involved, in contrast to the much higher concentrations often found in soil solutions. Al fractionation and speciation studies show that organically-complexed Al seems to be a greater proportion of the total in the US and Canada (e.g. Driscoll *et al*, 1985), than in Europe (e.g. Lee, 1985), although this is probably a fortuitous reflection of the nature of the soils and hydrologic pathways in the catchments that happen to have been studied, rather than a genuine geographical difference.

A useful simple example to illustrate both typical amounts and the processes occurring in streams is that of Bull and Hall (1986)., some of whose data are reproduced in Table 1. At periods of low flow on the rivers Esk and Duddon in Cumbria, England, water pH varied from 5.3-7.1 and Al_t concentrations were relatively constant at around 1µM. However, during periods of moderate flow, there was a gradient of pH from about 5 at the headwaters to about 6.5 at the mouths of the rivers, with corresponding Al_t concentrations dropping from about 10µM to between 1 and 5µM. Where the Al_t concentrations were higher, in the acid waters of the upper reaches, the inorganic monomeric form accounted for the larger part of the total.

TABLE 1. pH and Al concentrations (uM) at sites from the headwaters to the mouth of the Rivers Esk and Duddon (Cumbria) during low and moderate flow conditions. (From Bull and Hall, 1986)

		Low Flow		Moderate Flow	
		pH	Al_t	pH	Al_t
R. Esk:	headwaters	5.3	2.4	4.8	13.7
	↓	5.8	0.2	5.1	7.0
		6.0	0.7	5.2	5.6
		6.8	0.7	5.2	9.6
		6.7	0.4	5.5	7.8
		7.1	0.7	5.7	5.9
	mouth	6.6	0.5	5.8	5.6
R. Duddon:	headwaters	6.2	0.2	5.0	7.4
	↓	5.7	0.9	4.9	15.2
		6.0	0.4	4.3	16.3
		6.5	1.6	6.3	-
		6.9	1.1	6.6	1.1
	mouth	7.3	0.9	6.8	0.7

Attention has concentrated on the more acid upland streams, whose response to rainfall events gives large and rapid changes in both flow rate and water chemistry. Continuous monitoring or frequent automatic sampling has become essential to adequately reflect the rapid changes that may occur in water chemistry (Edwards *et al*, 1984) and so understand the conditions that might be experienced by organisms living in the streams. This has produced some surprising results, particularly for Al forms and concentrations in the streams.

Both H^+ and Al_i concentrations in streams flowing from moorland and forested catchments at Plynlimon in North Wales increased strongly with flow rate, in contrast to other elements (Neal *et al*, 1989). This is contrary to what simple reasoning might suggest, i.e. that a high flow rate would dilute solute concentrations in water, which is the norm for bicarbonate concentrations, for example. The Al_i concentrations are shown here in Figure 7a, and H^+ concentrations (not shown) were highly correlated with them in approximately 1:1 ratio.

The chemistry of the main stream draining the smaller ($0.41km^2$) Birkenes catchment in Norway showed a different relationship with flow (Seip *et al*, 1989), with Al reaching a faily constant concentration with flow, and a very different H^+:Al ratio (Figure 7b). Detailed analysis of H^+ and Al fractions shows important variations with discharge and with antecedent soil moisture conditions (Figure 8). With frequent rainfall and generally saturated soil conditions, Al concentrations varied little and tended to decrease with increasing discharge (October, 1984). During long dry periods (October 1985) Al concentrations were low, but storm events following this (November, 1985) gave large increases in Al concentrations although the highest discharge gave lower Al than the previous one.

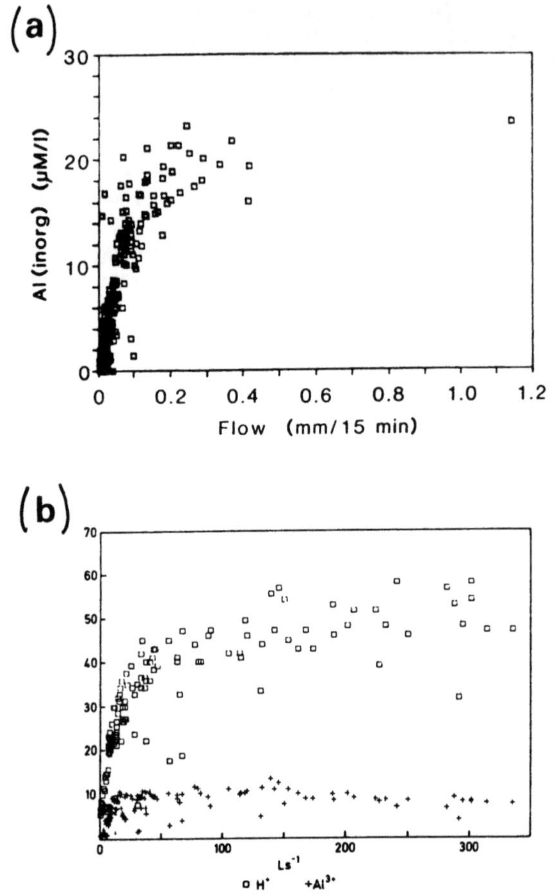

Figure 7. Aluminium concentrations in relation to the flow rate of streams (a) at Plynlimon, from Neal *et al*, 1989 and (b) at Birkenes, from Seip *et al*, 1989.

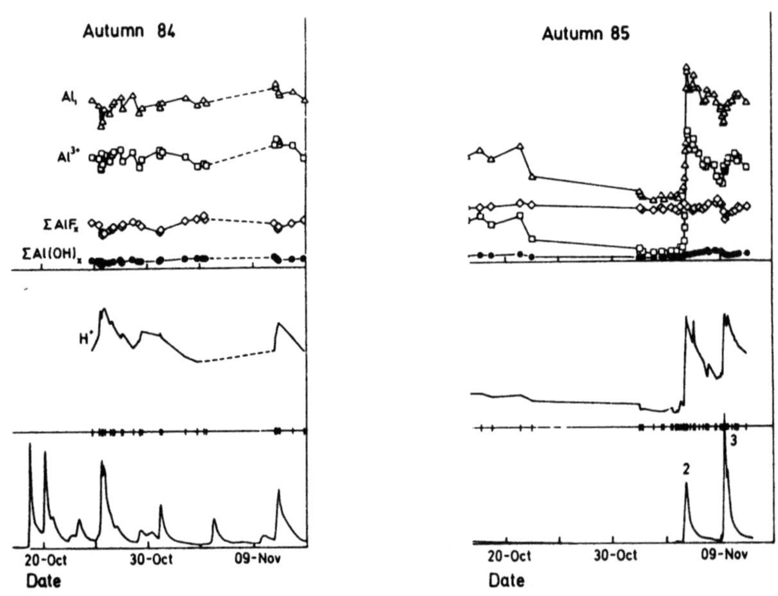

Figure 8. Stream discharge and Al concentrations in relation to time and antecedent moisture conditions for the stream draining the Birkenes catchment (from Seip *et al*, 1989).

The hydrogeochemistry of a 50ha catchment at the Svartberget Forest Research Station, Sweden, provides further detail to a complicated scenario (Townsend *et al*, 1990). During storm events, Al_i concentrations at the exit stream (weir A) peaked just before the hydrograph peak, but much of the Al in this stream was organically complexed (Figure 9). The chemistry along the stream from the headwaters to the bottom weir show some interesting relationships: during base flow conditions, the pH of the stream from a mire at the top of the catchment (site F) to weir A, 920m away, dropped from about 4.5 to about 6, with a decrease in both total and labile Al concentrations, as shown in Figure 10. (Some of this

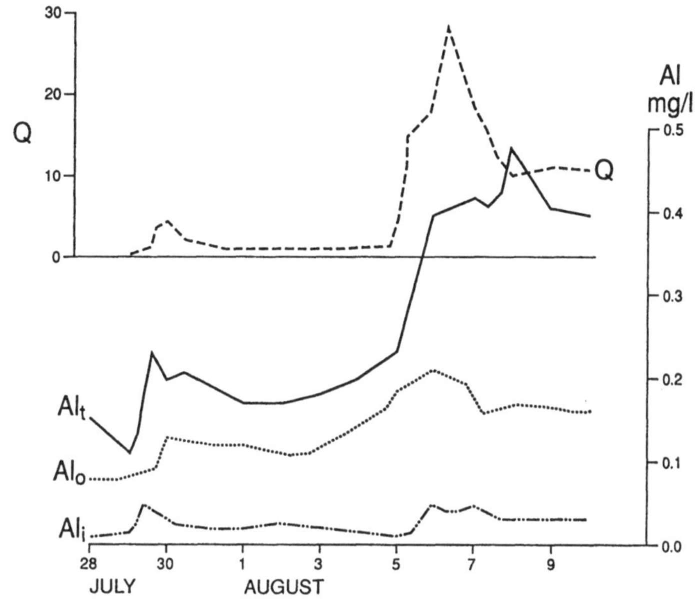

Figure 9. Aluminum fractions in relation to stream flow at the outlet weir of the Svartberget catchment, Sweden (from Townsend *et al*, 1990).

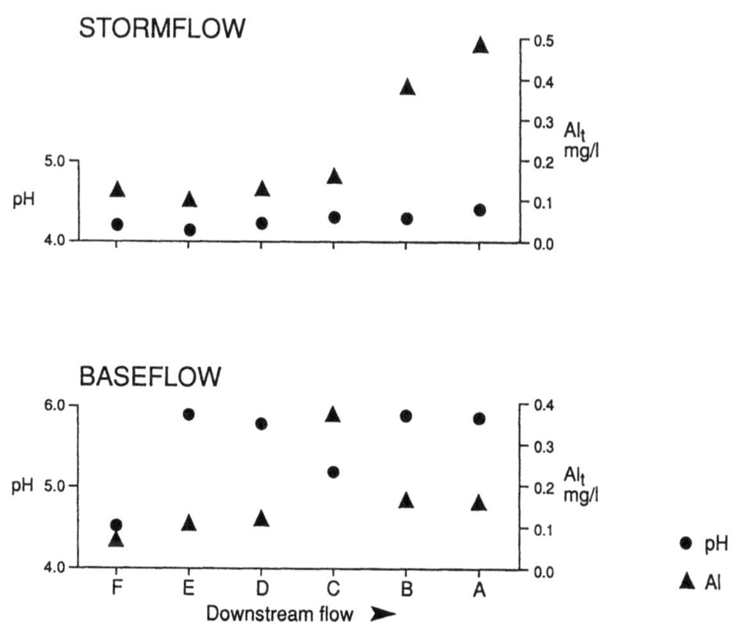

Figure 10. Aluminium and pH along the Svartberget stream during base flow and storm flow (from Townsend *et al* , 1990).

pH drop was attributed to degassing of CO_2 and H_2S from the stream water.) It seemed likely that Al had been adsorbed or precipitated onto the streambed materials between weirs C and A, and that the increase in Al concentrations in the lower part of the stream during the following storm event (Figure 10) was caused by its dissolution by lower pH water. This hypothesis was tested by leaching samples of the streambed gravel and sphagnum taken during the low flow conditions. The results of this experiment are shown in Figure 11. The sphagnum is a much better buffer than the streambed gravel, but both clearly contain considerable amounts of Al, which they release over time to sulphuric acid of pH4, and presumably would also release to low pH natural waters.

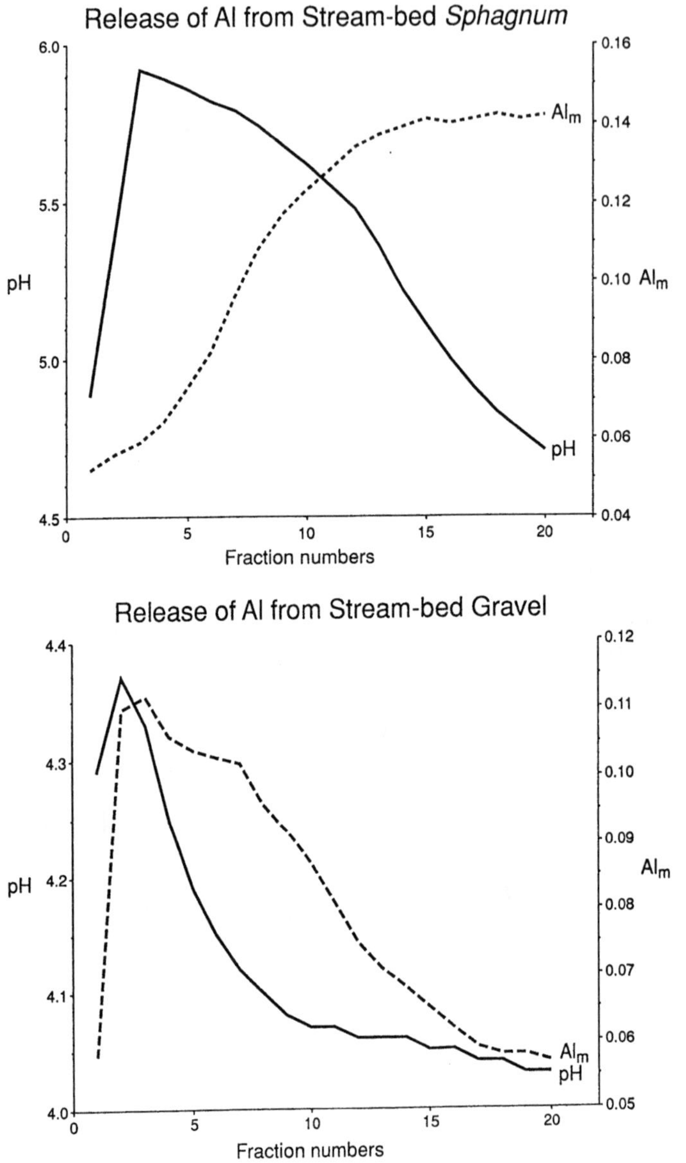

Figure 11. Release of entrained aluminium from (a) streambed *Sphagnum* and (b) streambed gravel from the Svartberget stream, by dilute sulphuric acid leaching (from Townsend, 1990).

4. THE INTERACTION OF CHEMISTRY AND HYDROLOGY

Studies of Al solubility in natural systems have been obsessed with comparisons with the solubility products of minerals, particularly gibbsite, that might be expected to control its solubility. At one time the proper data to do this were not available, because the speciation of Al in solution was unknown. With the fractionation schemes and speciation models now available, Al^{3+} concentrations can be determined and their activities calculated with reasonable confidence, but the data still do not fit simple solubility models (Neal, 1988). This is hardly surprising in view of the complexity of soil composition, and the slow rate of reactions of insoluble minerals; what is more surprising is that scientists seem unable to accept this. Rather, they should be looking for a more pragmatic approach based on rate processes and mixing models, and it is gratifying that this is at last being recognised (e.g. Neal *et al*, 1990).

A sequence of questions should be addressed to the system being considered, in order to discover the most appropriate chemical models that will apply.

(i) Is it a system at equilibrium? If so, what chemical equilibria are likely to be operating? One of the frustrations of the "gibbsite solubility" school is that solution activities for gibbsite, or amorphous Al hydroxide, are considered when there is no likelihood of these solid phases ever being present, let alone controlling Al concentrations in solution. Examples are the organic horizons of soils, systems dominated by Al sulphate salts derived from sulphate impaction onto a dry soil surface, systems dominated by soluble silica and silicate minerals, and systems dominated by metal-cation exchange. What one *can* say, is that at a given pH the gibbsite, or amorphous hydroxide, solubility relationship provides an *upper limit* to Al solubility, in that if Al concentrations exceeded this, then the hydroxide would precipitate. In most situations the Al concentration is, however, considerably below this, except in some cases at higher pH (>5.5) which can be explained by the formation of soluble Al complex ions formed with other ligands.

The only natural systems likely to be at equilibrium are ground waters that have been in contact with weathering rock for some time, so that base flow streams issuing from them may be saturated with the mineral phases of the regolith. It is, however, relatively easy to establish equilibrium in laboratory experiments with soils and finely divided mineral matter, and it would be necessary to do this to reveal many of the equilibrium systems mentioned above.

(ii) Most natural systems are unlikely to be at chemical equilibrium, and one then has to question the *type* of disequilibrium that is operating. First, there are simple chemical kinetic processes, such as the rate of dissolution of a mineral phase. Then, perhaps more importantly, there are dispersion phenomena where mass flow and diffusion processes operate together when solutions move through porous structured material such as soil, as discussed in a recent review (Bache, 1990). In these situations the Al concentrations found in solution could be anything from a close proximity to the equilibrium value, to the near zero value of the percolating rain. There is no simple way of predicting this, but both process and stochastic models are currently being developed to give some indication of expected concentrations in relation to pore size, solution flow velocity and the equilibrium value for the system (Townsend, 1990).

(iii) Finally, one has to question whether the observed results can arise from the mixing of waters of different chemical origins. This has been the approach of a number of recent modelling procedures. It can be illustrated by the Birkenes model, a convenient summary of which is given by Stone and Seip (1989). This has the virtue of being simple, in that apart from snowmelt, it has only two compartments, one for reactions with bare rock or shallow soil, and the other with deeper soil layers, and in each of these hydrology and chemistry is considered. Streamwater composition is then determined by the mixing of water from these two sources. For many components, the chemistry seems reasonable, but it is defective for Al, probably because the model assumes a gibbsite-type of relationship between H^+ and Al^{3+}, which these authors admit is wrong. The hydrology also is too simplistic, for although it is based on $H_2{}^{18}O$ isotopic tracing and the use of Cl as a conservative ion, it is limited to the sizes of the two reservoirs, and the rates of flow from them, when modelling the contrasts between base flow and storm flow.

A more detailed and realistic hydrology was presented by Bishop and Grip

(1990), who used spatially and temporally intensive measurements of hydraulic potential and soil water chemistry to define flow pathways during acid episodes at the Svartberget catchment. The majority of storm runoff moved through the upper soil horizons after they had become saturated, but prior to this these horizons contained large amounts of water which were virtually stationary because the ground was unsaturated. Thus the stormflow runoff consisted of much "old" water that had come to equilibrium with the acid organic soil layers.

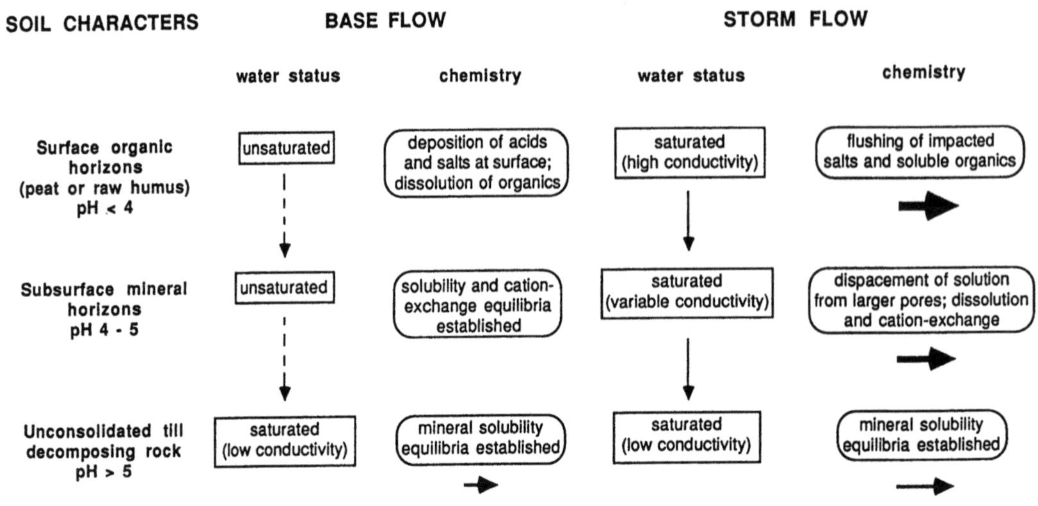

Figure 12. A diagrammatic scheme relating water chemistry to water status and composition of soil horizons. Arrows indicate flow, relative amounts roughly proportional to the size of the arrow.

A representation of the nature of the soil material, the water status of the soil for the contrasting conditions of base flow and storm flow, and the type of chemical reactions that may be occurring, is given in Figure 12. While being much simplified, this attempts to incorporate some of the important processes described above, which together determine the resulting water chemistry and in particular aluminium forms and concentrations that may be found in solution. This incorporates mixing of water from at least three sources, but in reality there may be many more depending on the detailed structure and chemistry of the soil. For reasons stated above, the highest Al_i concentrations can be expected to come from the mineral soil horizons, when the Al is mobilized by salts from the cation exchange system. The highest Al_o concentrations will come from the acid organic surface layers, particularly after periods of dryness, when the normally insoluble Al-humus complexes appear to dissolve to some extent in the acid waters.

5. CONCLUSIONS

1. One of the main effects of the impact of acid deposition onto land surfaces has been to increase the solubility of Al. This is because Al solubility is closely related to pH, and the acid depositions have decreased the pH of soils in all areas where there is no solid phase calcium or magnesium carbonate in the soils or in the underlying rocks.

2. Soils are made up of a wide variety of mineral and organic materials, and Al may exist in solution in a number of different ionic or molecular species. It is thus difficult to predict accurately solution Al concentrations at equilibrium for a specific material, except for fairly simple systems. A large number of chemical equilibria usually have to be taken into account, for some of which the appropriate thermodynamic data do not exist, so that in general reliable equilibrium calculations cannot be made.

3. Chemical equilibrium rarely exists in soil-water systems. Non-equilibrium systems result in very different Al concentrations in soil solutions, base flow stream waters, and storm flow stream waters. However, a consideration of the rates of flow, flow pathways, and the chemistry of the materials through which water is flowing, leads to both a qualitative *understanding* of observed Al concentrations and speciation, and a qualitative *predictive ability*, largely based on experience. Rigorous quantitative models for Al concentration and speciation in non-equilibrium systems are still a long way from being achieved.

REFERENCES

Andersen, S., Christophersen, N., Mulder, J., Seip, H.M. and R.D.Vogt (1990). Aluminium solubility in the various soil horizons in an acidified catchment. *Surface Water Acidification Programme, Final Report*, (J.Mason, ed.) The Royal Society, London, (in press).

Bache, B.W. (1974). Soluble aluminium and calcium-aluminium exchange in relation to the pH of dilute calcium chloride suspensions of acid soils. *Journal of Soil Science*,25, 320-332.

Bache, B.W. (1984). Soil-water interactions. *Philosophical Transactions of the Royal Society, London*, B 305, 393-407.

Bache, B.W. (1986). Aluminium mobilization in soils and waters. *Journal of the Geological Society, London*, 143, 699-706.

Bache, B.W. (1990). Solute transport in soils. In *Process Studies in Hillslope Hydrology* (M Anderson and T Burt, eds.) Wiley, Chichester (in press).

Bache, B.W. and G.S.Sharp (1976). Soluble polymeric hydroxy-aluminium ions in acid soils. *Journal of Soil Science*, 27, 167-174.

Backes, C.A. and E.Tipping (1987). An evaluation of the use of cation-exchange resin for the determination of organically-complexed Al in natural acid waters.*International Journal of Environmental Analytical Chemistry*, 30, 135-143.

Billet, M.F., Parker-Jarvis, F., Fitzpatrick, E.A. and M.S.Cresser (1990). Forest soil chemical changes between 1949/50 and 1987. *Journal of Soil Science*, 41, 133-145.

Birchall, J.D. and J.S.Chappell (1988). The solution chemistry of aluminium and silicon and its biological significance. 231-242 in *Geochemistry and Health* (I.Thornton ed.) Science Reviews Ltd., Northwood.

Bishop, K.H. and H.Grip (1990). The significance of flow pathways in an episodically acid stream. *Surface Water Acidification Programme, Final Report*, (J.Mason, ed,) The Royal Society, London, (in press).

Bull, K.R. and J.R.Hall (1986). Aluminium in the rivers Esk and Duddon, Cumbria and their tributaries. *Environmental Pollution (Series B)*, 12, 165-193.

Conyers, M. (1990). The control of aluminium solubility in some acidic Australian soils. *Journal of Soil Science*, 41, 147-156.

Driscoll, C.T. (1984). A procedure for the fractionation of aqueous aluminium in dilute acidic waters. *International Journal of Environmental Analytical Chemistry*, 16, 267-283.

Driscoll, C.T. and W.D. Schecher (1988). Aluminium in the environment. pp 59-120 in H.Sigel (ed.) *Aluminium and its role in biology*, Dekker, New York.

Driscoll C.T. and W.D Schecher (1990). The chemistry of aluminium in the environment. *Environmental Geochemistry and Health*, 12 (in press).

Driscoll, C.T., van Breemen, N. and J. Mulder (1985). Aluminium chemistry in a forested spodosol. *Soil Science Society of America, Journal*, 49, 437-444.

Edwards, A.C., Creasey, J. and M.S.Cresser (1984). The conditions and frequency of sampling for elucidation of transport mechanisms and element budgets in upland drainage basins. In Eriksson, E. (ed.) *Hydrochemical Balances of Freshwater Systems*, 187-202. International Association of Hydrological Sciences Publ. 150, Oxford.

Farmer, V.C., Adams, M.J., Fraser, A.R. and F.Palmieri (1983). Synthetic imogolite: properties, synthesis and possible applications. *Clay Minerals*, 18,459-472.

Ferrier, R.C., Anderson, J.S., Miller, J.D. and N. Christophersen (1989). Changes in soil and stream hydrochemistry during periods of snowmelt at a pristine site in mid-Norway. *Water, Air, and Soil Pollution* 44, 321-337.

Foy, C.D. (1984). Physiological effects of hydrogen, aluminium and manganese toxicities in acid soil. pp 57-97 in F. Adams (ed.) *Soil Acidity and Liming*, 2nd edn. American Society of Agronomy, Madison.

Hargrove, W.L. and G.W.Thomas (1981). Effect of organic matter on exchangeable aluminum and plant growth on acid soils. pp 151-166 in *Chemistry in the Soil Environment*, American Society of Agronomy, Madison.

Howells, G., Dalziel, T.R.K., Reader, J.P. and J.F. Solbe (1990). EIFAC water quality criteria for European freshwater fish: report on aluminium. *Chemistry in Ecology*, 4 (in press).

Johnston, A.E., Goulding, K.W.T. and P.R.Poulton (1986). Soil acidification during more than 100 years under permanent grassland and woodland at Rothamsted. *Soil Use and Management*, 2, 3-10.

Lee, Y.H. (1985). Aluminium speciation in different water types. pp109-119 in *Lake Gardsjon: An Acid Forest Lake and its Catchment* (F.Andersson and B.Olsson, eds.) Ecological Bullutins No. 37.

Lee, R., Bache, B.W.,Wilson, M.J. and G.S.Sharp (1985). The determination of cation-exchange capacity of some New Zealand podzolised soils. *Journal of Soil Science*, 36, 239-253.

Lydersen, E., Salbu, B., Poleo, A.B.S. and I.P.Muniz (1990). The influence of temperature on aqueous aluminium chemistry. *Water, Air, and Soil Pollution* (in press).

Mattson, S. and Y Gustafsson (1937). The laws of soil colloid behaviour XIX. The gel and sol complex in soil formation. *Soil Science* 43, 453-473.

Neal, C. (1988). Aluminium solubility relationships in acid waters - a practical example of the need for a radical reappraisal. *Journal of Hydrology*, 104, 141-159.

Neal, C., Reynolds, B., Stevens, P and M. Hornung (1989). Hydrogeochemical controls for inorganic aluminium in acidic stream and soil waters at two upland catchments in Wales. *Journal of Hydrology*, 106, 156-175.

Neal, C., Harriman, R., Christophersen, N., Ferrier, R.C. and R. McMahon (1990). Ion-exchange and solubility controls in acidified systems. *Surface Water Acidification Programme, Final Report*, (J.Mason, ed.) The Royal Society, London, (in press).

Paces, T. (1978). Reversible control of aqueous aluminium and silica during the irreversible evolution of natural waters. *Geochimica et Cosmochimica Acta*, 42, 1487-1491.

Reuss, J.O. (1983). Implications of calcium-aluminium exchange for the effect of acid precipitaion on soils. *Journal of Environmental Quality*, 12, 591-595.

Salbu, B., Bjornstad, H.E., Lindstrom, N.S., Lydersen, E. Breivik, E.M., Rambaek, J.P. and P.E.Paus (1985). Size fractionation techniques for the determination of elements associated with particulate or colloidal material in natural fresh waters. *Talanta*, 32, 907-913.

Schecher, W. and C.T.Driscoll (1987). An evaluation of uncertainty associated with aluminium equilibrium calculations. *Water Resources Research*, 23,525-535.

Seip, H.M., Andersen, D.O., Christophersen, N. Sullivan, T.J. and R.D. Vogt (1989). Variations in concentrations of aqueous aluminium and other chemical species during hydrochemical episodes at Birkenes, southernmost Norway. *Journal of Hydrology*, 108, 387-405.

Stone, A. and H.M.Seip (1989). Mathematical models and their role in understanding surface water acidification: An evaluation using Birkenes model as an example. *Ambio, 18*, 192-199.

Tamm, C.O. and L.Hallbacken (1988). Changes in soil acidity in two forest areas with different acid deposition: 1920s to 1980s. *Ambio*, 17, 56-61.

Tipping, E. and M.A.Hurley (1988). A model of solid-solution interactions in acid organic soils, based on the complexation properties of humic substances. *Journal of Soil Science*, 39, 505-519.

Townsend, G.S. (1990). Rate processes in aluminium mobilization from soils to surface waters. *Ph.D. thesis,* University of Cambridge.

Townsend, G.S., Bishop, K.H. and B.W.Bache (1990). Aluminium speciation during episodes. *Surface Water Acidification Programme, Final Report*, (J.Masom, ed.) The Royal Society, London, (in press).

Ulrich, B., Meyer, R. and P.K.Khanna (1980). Chemical changes due to acid precipitation in a loess-derived soil in Central Europe. *Soil Science* 41, 133-145.

Van Praag, H.J. and F. Weissen (1985). Aluminium effects on spruce and beech seedlings I. Preliminary observations on plant and soil. *Plant and Soil* 83, 331-338.

Young, S.D. and B.W.Bache (1985). Aluminium-organic complexation: formationm constants and an equilibrium model for the soil solution. *Journal of Soil Science,* 36, 261-269.

RESULTS OF RESEARCH INTO DECAY OF THE FIR (*ABIES ALBA* MILL.) IN THE PYRENEES. NEW DATA ABOUT NUTRITIONAL AND PHYSIOLOGICAL DISTURBANCES.

F. Fromard*, J. Dagnac **, T. Gauquelin** and V Cheret**
Institut de la Carte Internationale de la Vegétation(*)
and Laboratoire de Botanique et Biogéographie (**)
Université Paul Sabatier
39, allées Jules Guesde, 31062 Toulouse Cedex, France

ABSTRACT

In the Pyrenees, the decay of the fir (*Abies alba* Mill.) is spreading and consequently research has been undertaken to identify the main causes of this loss. The research is included in the French DEFORPA programme (DEpérissement FORestier attribué à la Pollution Atmosphérique).The aim of the present paper is to establish some relationships between obvious symptoms (such as needles yellowing and falling) and nutritional or physiological disturbances. The results obtained from mineral leaf analysis led to the following conclusions: there is a potassium deficiency in decaying fir stands growing on a schistous substratum whereas decaying firs on limestone soil suffer from manganese and iron insufficiency. Additionaly biochemical analysis indicates reduced CO_2 uptake by decaying trees and a conspicuous modification of the seasonal fluctuations of starch and soluble sugars.

1. INTRODUCTION

In the Pyrenees, firs have been showing signs of decay for several years, with the same symptoms as in other European countries but not as pronounced (Cheret, 1987; Cheret, Dagnac, Fromard, 1987). However, analysis of the data obtained from the forest observation network (Landmann, 1989) shows a real increase of fir decay: the percentage of trees having lost more than 25% of their needles increased from 2 to 10% between 1985 and 1987. The percentage of trees with yellowing needles was near 50% in 1986 but then decreased from 1987.For example, our investigations show that, in the Luchon region, 10% of the area covered by fir forests (350 ha/3500 ha) are affected.

The main aim of the research into fir decay was:

- to specify the Pyrenean decay symptoms and to estimate the extent of damage.
- to estimate the impact of environmental conditions including pollution, climate or nutritional deficits.

In this paper, we concentrate more specifically on the causes of the damage. Chemical analysis of needles and soil were perfomed in order to reveal possible nutritional imbalance able to explain the forest dieback.Physiological disturbances in yellowing trees were also analysed. Investigations were carried out in three areas of the Pyrenean massif: the Luchonnais, vallée d'Aure and Pays de Sault (figure 1) who show various climatic and edaphic features and different intensities of decay.

2. DECAY SYMPTOMS

The affected trees have yellowing needles, at first in the upper third, but progressively over the whole tree. As yet the current year's needles have always kept usual coloration.On a same shoot aged at least two years, healthy and yellowing needles may be found side by side.Defoliation may or may not follow yellowing. For example, trees with discolouration over 25% in 1984 showed, at the end of 1985, a characteristic defoliation, sometimes after a turning brown of the needles.But defiolation without preliminary discolouration has been so observed, the trees keeping needles for only four or five years. Some firs carry a flat top, expressing apex growth checking and a disorganized development of the secondary ramification.The foliage is then, in the upper third, quite transparent, revealing needle loss.

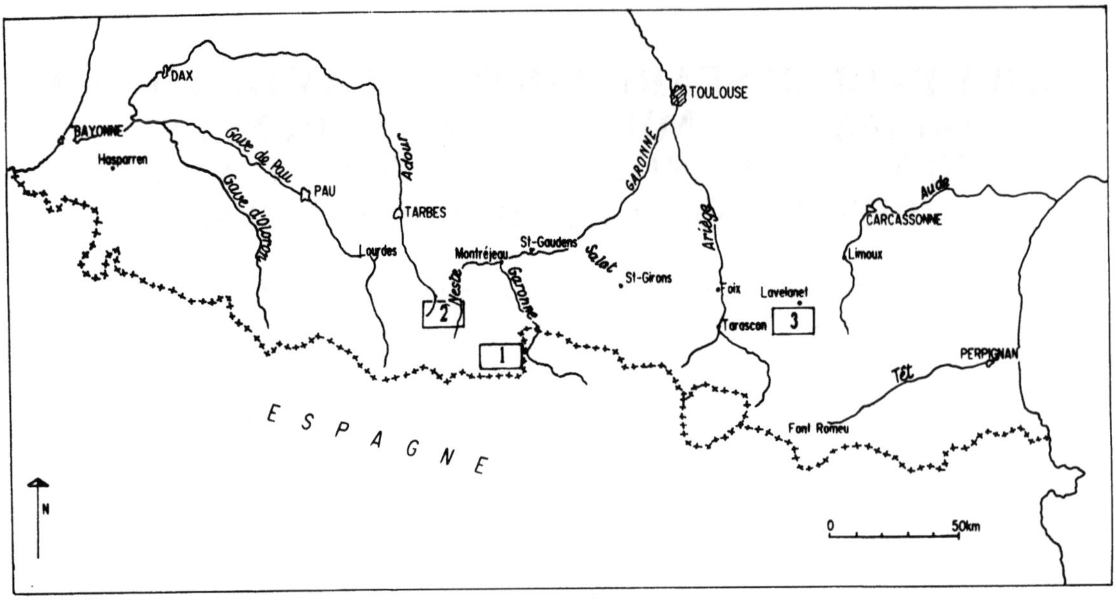

Figure 1. Location of study sites. [1] Luchonnais [2] Vallée d'Aure [3] Pays de Sault

Stress shoots are also frequenly noticed on damaged trees. It is then interesting to observe that the needles of these shoots are normal, even if the whole tree is strongly yellowing.

The most affected firs are most often mature dominant individuals but symptoms also exist among young trees.

They could be trees:

-with Mistletoe (*Viscum album* subsp. *abietii*) or without
-with abnormal branches produced by Rust (*Melampsorella caryophyllacearum*;) or without
-with injuries at the trunk base or without.

They form clumps of trees or are isolated in the midst of an apparently healthy population. they are situed as easily at the edges as in the centre of the forest.

As well as typical decay symptoms, browning needles appeared at the end of 1985, most often on the whole tree.This phenomen, not taken into account in our study is quite obviously linked to the exceptional climatic features of this period with a harsh winter and extreme drought between August and October.Among the affected trees, the proliferation of bark beetle (*Ips curvidens*) has increased decay. Large areas were affected, especially in the Pays de Sault, and many clear-cuts were then carried out.

3. NUTRITIONAL AND PHYSIOLOGICAL DISTURBANCES

"Foliar diagnosis" is based on a precise relation between nutrient concentration in the needles and tree growth.(Bonneau, 1988; Evers, 1972; Nihlgard, 1986).There is then, for a given element and a given species, an optimal level (maximal growth), a critical level (lower growth) and a deficiency level (strong growth reduction and some decay symptoms).The nutritional status dealing only with major nutrient concentrations was complemented by analysis of organic components . Starch, total soluble sugars and different metabolites (amino-acids, organic acids) were assayed in fir needles at different phenological dates.Comparison between healthy and yellowing trees shows possible metabolic disturbances .Variations in the chemical composition of the needles were also studied as a function of their age

3.1. Materials and methods

Foliar diagnosis requires accurate methodology concerning sample collection and a reliable analysis protocol which can be reproduced.

The factors affecting chemical composition are numerous:

- phenological state
- needle age classes
- social status of the tree: dominated/dominant

- location of the individual (edge or stand centre)
- needle position in the crown (from base to top)
- climatic conditions, important for some elements (e. g. calcium).

The analysis described below concerns needles sampled on healthy firs (blank) and yellowing firs in 84, 86 et 1987 in the Luchonnais and 1988 for the two other areas.

3.1.1. Sample collection

Trees were chosen in stands where healthy and yellowing specimens were mixed together.In the Luchonnais, 13 decaying trees and 7 healthy were chosen; in both Vallée d'Aure and Pays de Sault: 13 decaying and 13 healthy. Shoots were collected in the upper third, at the four exposures, by climbing trees.They were then grouped together in a single batch for each tree.For the chemical composition,the shoots were collected when the growth had stopped (October to December) when concentrations are the most stable.For the biochemical assays, four branches were collected between March and June.

In all cases, separate batches were made up from one-, two-, four- and six-year-old needles which were dried and ground.

3.1.2. Mineral and biochemical analyses

Nutrient levels were measured by absorption spectrophotometery after acid digestion of samples. Nitrogen was determined according to Kjeldahl's technique.Total soluble carbohydrates and starch were determined by colourimetry and chromatography . To measure CO_2 uptake and to follow metabolism products, healthy and yellowing fresh shoots were placed in a closed chamber with radio-labelled CO_2.

3.2.Nutritional Status

Three parameters must be considered :

- mineral content variations with age.
- comparison with the deficiency levels defined by Bonneau (1988) in a recent review.
- Ca/Mg and K/Mg ratios, regarded to give a better indication of the nutritional balance than just the concentrations of the elements (Nihlgard, 1989).

3.2.1. Nutritional status in the Luchonnais. (Table 1 and figure 2)

a. Mineral content variations with age.

- The magnesium, nitrogen and phosphorus levels were relatively stable in the different year classes with the highest figures in two-year-old needles (N, Mg) or in one-year-old and two-year-old needles.
- The potassium level was stable between 2 and 6 years, but highest in one-year-old needles.
- Calcium and manganese concentrations showed large fluctuations and a steady accumulation with age.

Among decaying firs, the same nutritional behaviour was observed but the nutrient levels were sometimes different. In 1984, the same variations were found in one-year-old needles of decaying and healthy trees.This shows that decay does not disturb the accumulation process (Ca, Mn) or distribution of mineral elements between needle year classes.

b-Nutrition and deficiency level.

The levels of the various elements assayed were as follows:

Nitrogen: the values were near the deficiency level (1 to 1.3) with some samples below this level. Nutrition is always mediocre and there was no significant difference between healthy and decaying firs.The figures decreased between 84 and 86 (from 1.35 to 1.18 for one-year-old needles).

Phosphorus: the values were always above the deficiency level (0.10%) whatever the age of the needles or their state of health.

Potassium: yellowing firs showed lower values than healthy trees (difference significant to 0.5 % and 0.1%). In decaying needles, the values were always below the deficiency level (0.35%). In healthy

Table 1
Nutrient concentrations in one year needles from the three sites
[healthy and yellowing fir trees]

		HEALTHY		YELLOWING		deficiency level	optimum content
Nutr[t]	Site	X	min max	X	min max		
N%	LU 84	1.29	1.09-1.49	1.29	0.96 -1.65		
	LU 86	1.12	0.82-1.50	1.15	0.90 -1.35		
	AU 88	1.27	1.11-1.52	1.23	1.12- 1.55	1 -1.3	1.5-1.9
	SA 88	1.13	0.93-1.28	1.12	1.04-1.20		
P%	LU 84	0.21	0.17-0.24	0.21	0.19-0.25		
	LU 86	0.24	0.15-0.45	0.24	0.13-0.48		
	AU 88	0.48	0.31-0.62	0.40	0.33-0.51	0.1-0.15	0.19-0.25
	SA 88	0.20	0.15-0.28	0.16	0.16-0.17		
K%	LU 84	0.53	0.35-0.81	**0.30**	0.20-0.40		
	LU 86	0.41	0.37-0.48	**0.27**	0.18-0.37		
	AU 88	**0.32**	0.20-0.48	**0.24**	0.09-0.50	0.4-0.5	0.6-0.8
	SA 88	**0.39**	0.20-0.60	**0.35**	0.28-0.46		
Ca%	LU 84	0.61	0.41-0.78	0.41	0.21-0.66		
	LU 86	0.86	0.48-1.30	0.72	0.55-0.95		
	AU 88	0.42	0.34-0.48	0.38	0.30-0.48	0.05-0.1	0.3-0.5
	SA 88	0.73	0.50-0.90	0.87	0.85-0.90		
Mg%	LU 84	0.13	0.08-0.20	0.15	0.10-0.28		
	LU 86	0.15	0.11-0.20	0.17	0.11-0.21		
	AU 88	0.13	0.08-0.18	0.12	0.10-0.15	0.06	0.1-0.14
	SA 88	0.11	0.08-0.15	0.11	0.10-0.12		
Mn%	LU 85	0.083	0.031-0.120	0.073	0.033-0.110		
	LU 86	0.041	0.020-0.060	0.041	0.020-0.077		
	AU 88	0.120	0.100-0.140	0.110	0.040-0.140	0.002	>0.005
	SA 88	0.038	0.010-0.072	**0.002**	0.001-0.003		
Fe%	LU 84	0.013	0.031-0.022	0.014	0.006-0.025		
	LU 86	-	- -	-	- -		
	AU 88	0.012	0.010-0.013	0.011	0.007-0.015	0.004	-
	SA 88	0.011	0.007-0.013	**0.005**	0.005-0.005		

LU 84 et LU 86: Luchonnais, 1984 and 1986; schist ; n=7 (healthy), n' (yellowing)=13.
AU 88: Vallée d'Aure 1988, schist; n= 5, n'=8.
Sa 88: Pays de Sault; limestone; n=10, n'=5.

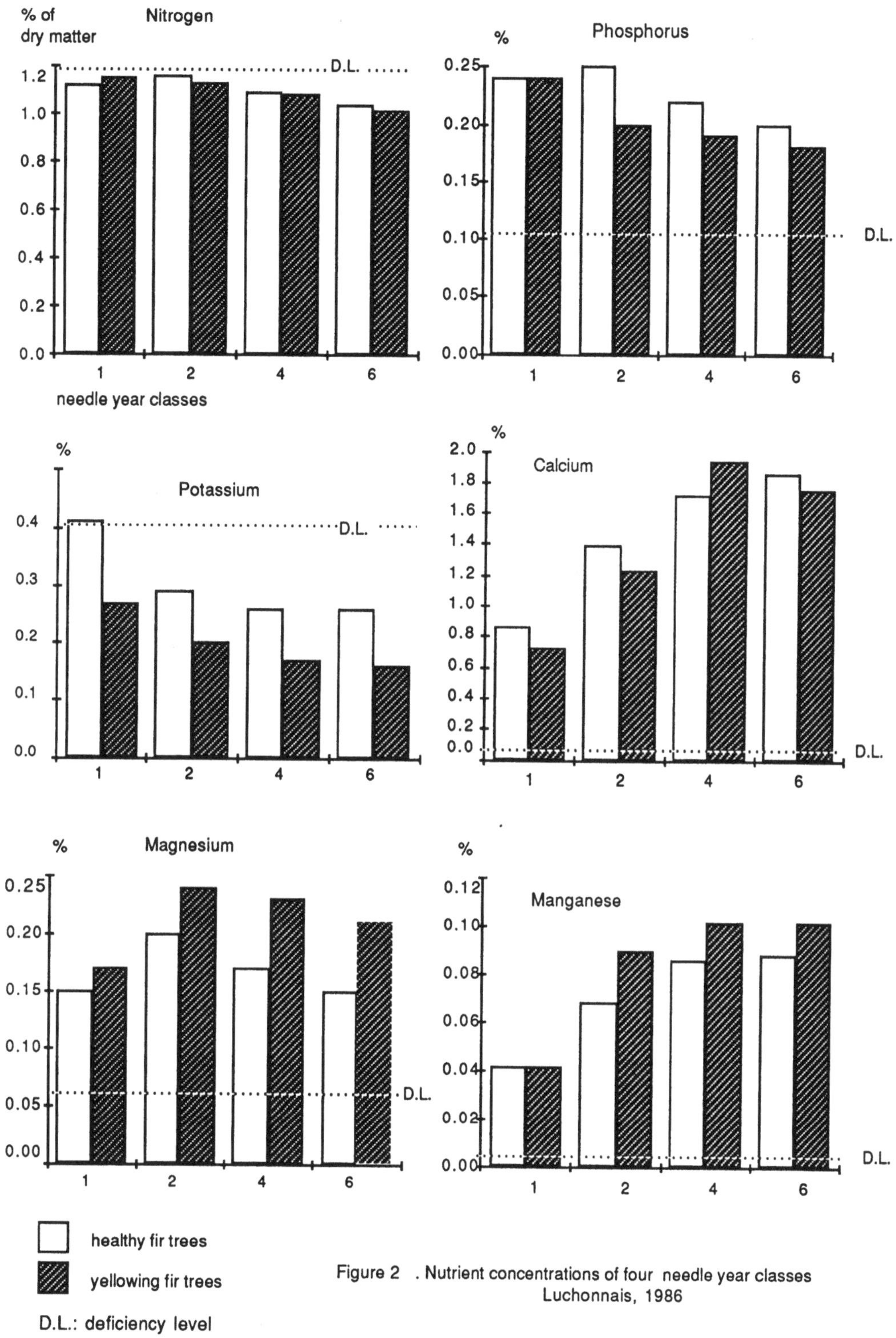

Figure 2 . Nutrient concentrations of four needle year classes
Luchonnais, 1986

needles the values were higher but lower than the optimal level (0.7).Between two collection years, the deficiency has not increased in yellowing needles, but values have decreased strongly for healthy trees.

Calcium: these was no significant difference between healthy and yellowing trees: the levels for one-year-old needles were near the optimal value and rose steadily with age. The deficiency noted in 1984 for one-year-old needles from yellowing trees was not confirmed in 1986.

Magnesium: the average level was far above the critical threshold defined at 0.06%. The levels were higher in yellowing specimens and those of '86 exceded those of '84.

Manganese: the levels were high and close to the optimum set at 0.005% for one-year-old needles.

The potassium deficiency therefore remains the main trait of yellowing firs in the Luchonnais: the levels of potassium being even lower in healthy specimens in 1986 than in the same trees two years before.

The study of the changes occurring in the K/Mg ratio strengthens these conclusions (figure 3): the ratios were lower than 2, considered to be a critical threshold, for yellowing trees at both sampling

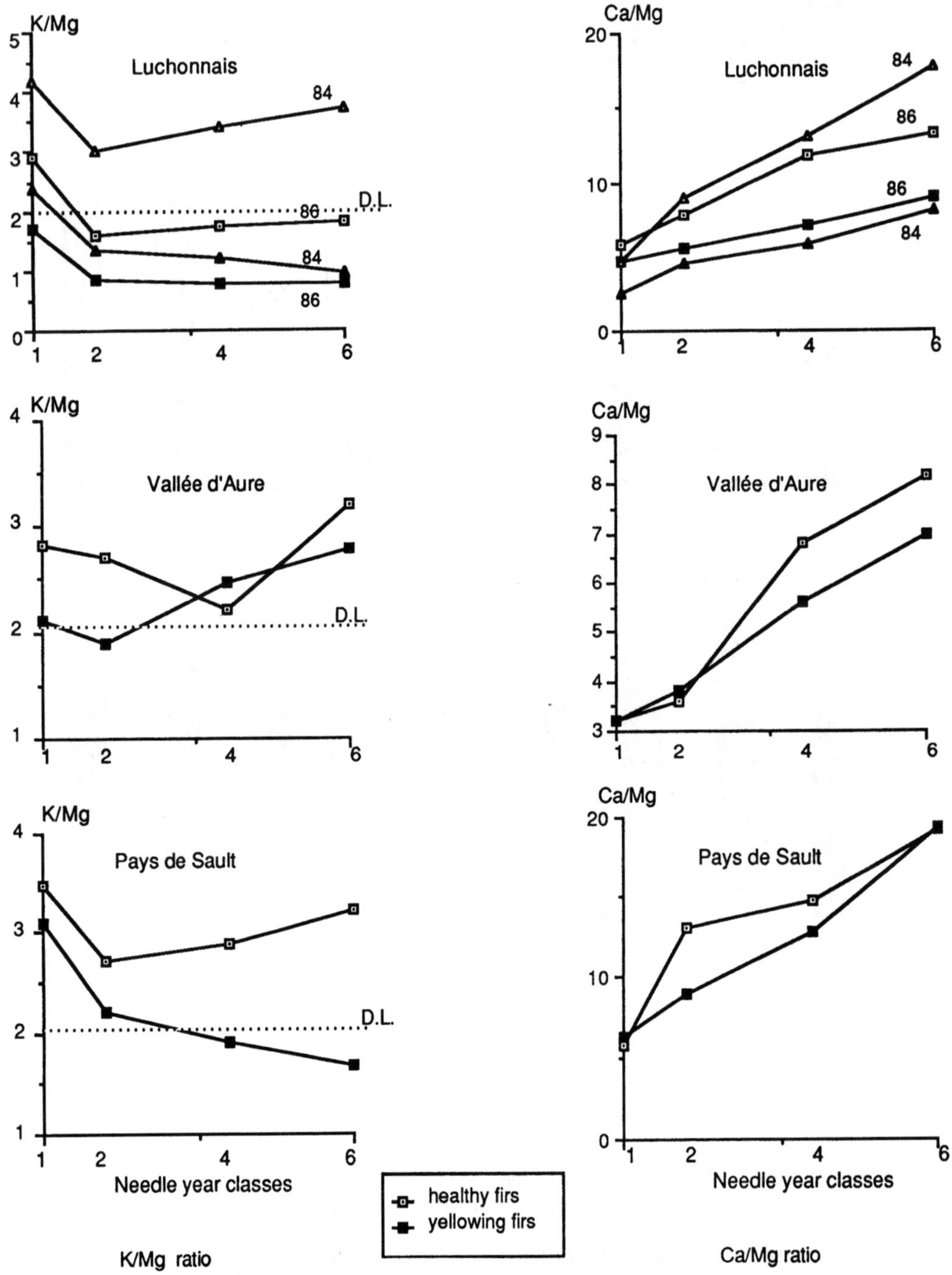

Figure 3: K/Mg and Ca/Mg ratios in healthy and yellowing fir tree needles.

periods as well as for heathly needles aged two years or more in 1986.The Ca/Mg ratios obtained in 1984 and '86 were comparable and the gap between healthy and yellowing tissues was maintained in spite of the rise in the level of calcium which could therefore be in relative deficit.

3.2.2 Foliar diagnosis in the Vallée d'Aure and Pays de Sault (table 1 , figures 4, 5)

The changes in concentration with the age of the needles now present a classic pattern: for

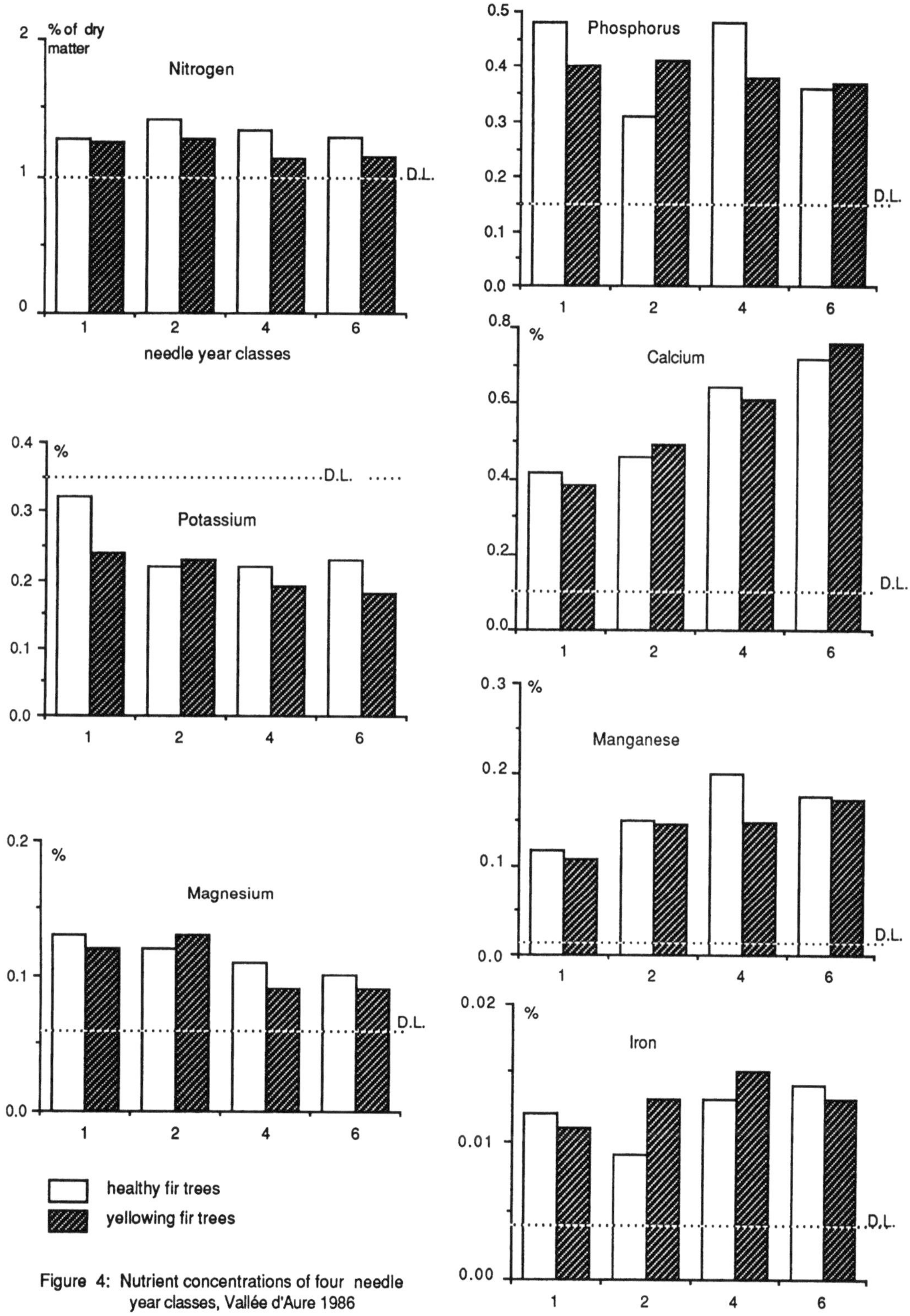

Figure 4: Nutrient concentrations of four needle year classes, Vallée d'Aure 1986

instance, a regular increase of the calcium content and a decrease of the potassium and magnesium contents, a slight variation of the nitrogen content and more irregular phosphorus and iron variations.

According to the deficiency thresholds and to the health of the trees, we can note:

-correct nutrition as regards calcium, magnesium and phosphorus for both tree classes grown on schistous and calcareous substrata,

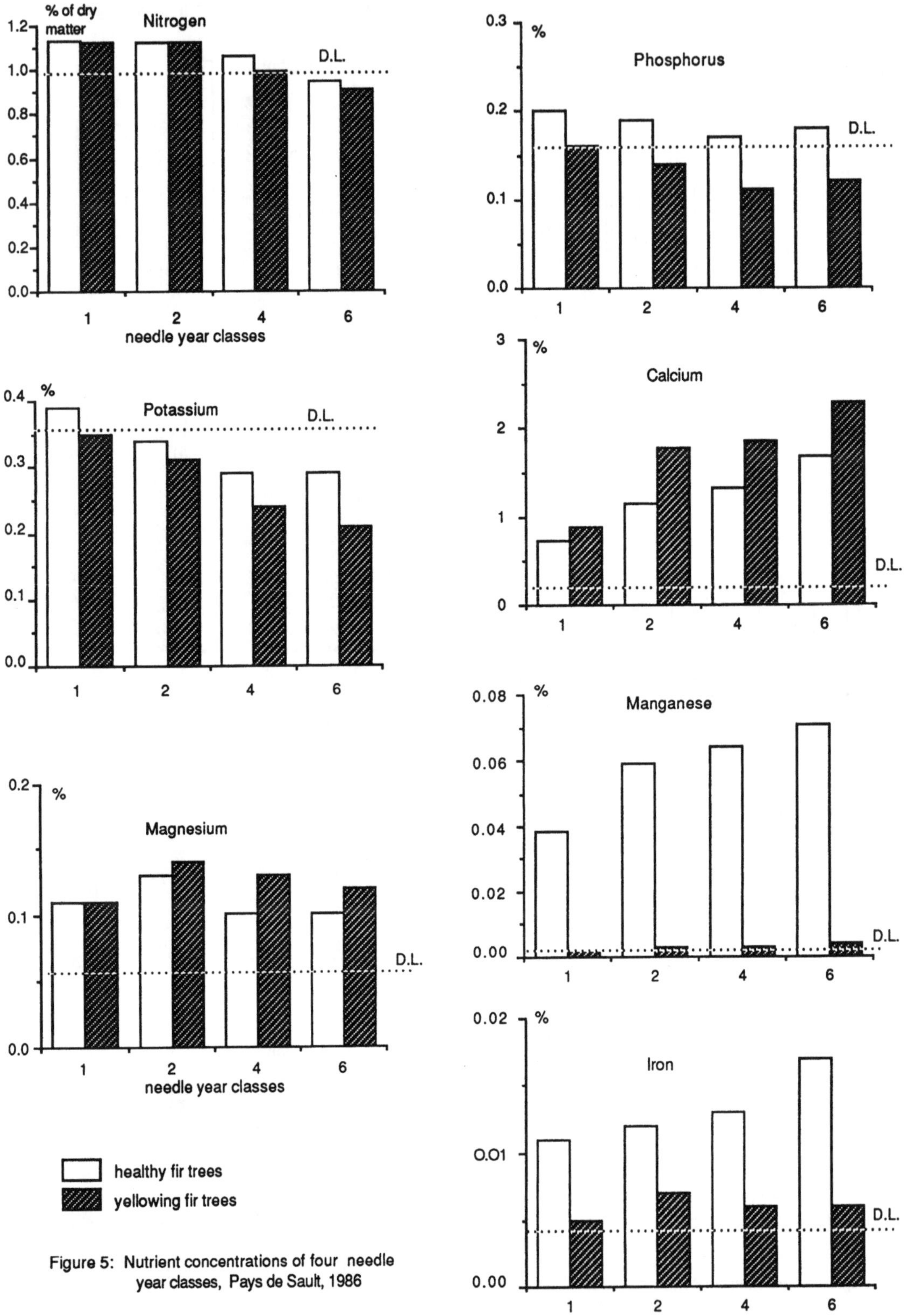

Figure 5: Nutrient concentrations of four needle year classes, Pays de Sault, 1986

-an average nutrition as regards nitrogen, especially on calcareous rocks, in which case the values range close to the deficiency level (1.10%)

-various disturbances in potassium, manganese and iron levels.

The optimum level for potassium is around 0.6%. All the values found here were below this point. On schistous substratum (Vallée d'Aure) all the K values were under the deficiency level (0.35) whether they concerned healthy (0.32% for one-year-old needles) or decaying subjects (O.24%) for which the deficiency was even more patent. Nutrition was correct for all other elements.

On calcareous substratum (Pays de Sault) the values found for one-year-old yellowing needles were close to the deficiency level (0.35), and slightly above for healthy samples (0.39).

The manganese deficiency level was estimated at 0.002%.The present results differ greatly according to the nature of the substratum.

-on schists, even though the concentrations were slightly weaker for decaying trees, they always remained below the deficiency level (D.L.) for both tree classes (>0.1%).

-on limestone, the manganese level in healthy needles was correct, although lower than that found on schsits (0.038% for one-year-old needles).It was much lower for yellowing trees where it was close to the D.L (0.002%).

As for iron,the D.L. was estimated at 0.004%; the values were good on schistous soils (0.012%) and for healthy subjects grown on calcareous rocks (0.011%). But for yellowing firs in calcareous areas, they come close to the D.L (0.005%).

Table 2

Nutrient composition of the soils of Pyrenean fir stand (horizons A1) (meq/100g)

	pH (water)	C/N	K	Ca	Mg	Mg/K	K/Ca
Luchonnais schist, n=27	4.6	15.5	0.28	8.8	0.06	3.8	0.03
Vallée d'Aure schist, n=2	4.4	22.8	0.5	1.07	1.01	2.02	0.46
Pays de Sault limestone, n=5	5.9	12.4	0.17	23.4	0.5	2.9	0.007

3.2.3 Discussion

The main deficiency brought to light by the study of the decay of firs in the Pyrenees is that of potassium. Yet, the response of the tree varies and does not seem strictly correlated to the type of substratum.

On the schists of the Luchonnais, the yellowing - deficiency link has been demonstrated (healthy firs: no deficiency; yellowing trees: potassium deficiency). There, the nutrition level is correct for all other elements, except for a noticeable calcium deficiency among decaying trees in 1984 which was not confirmed in 86. The study of the link between the main mineral nutrients also clearly underlines the existence of two tree classes (figure 3) - see in particular the K/Mg ratio under 2, a critical value for decaying subjects. The corresponding soils (tab. 2) are characterized by high acidity, insaturation of the exchange complex and a poor stock of exchangeable cations (Ca, Mg, K), yet without any actual deficiency of these elements.The nutritive possibilities of these soils are then weak. Correlation studies have shown, though, that the distribution of decaying areas seem to be independent of the soil's chemical characteristics (Cheret, 1987), but its potassium content is always below that found for instance in Vosges forest mountain soils.

In the Vallée d'Aure fir forests, the potassium deficiency is present in all the analysed subjects, but all of them do not present signs of decay. The nutrition is correct for all other elements.The K/Mg ratio does not show a clear difference between the classes; yet, its values are always close to the threshold from which an actual potassium deficiency is diagnosed (2.46 and 2 for young needles). It never reaches extreme values as in the Luchonnais (1 for two- to six-year-old needles).The corresponding soils, which are acid brown soils are also poor in potassium (0.1 to 0.9 meq/100 g for A_1 horizon).The K/Ca and Mg/K ratios (respectively >1 and close to 2) are however more favourable than those found earlier in the Luchonnais.

On calcareous substratum (Pays de Sault) the K/Mg ratios for healthy subjects are good for needles of all generations, as well as for the young (one and two-year-old) needles of decaying trees. The deficiency in potassium is less patent.On the other hand, a strong manganese deficiency can be noticed among yellowing subjects. The values encountered (0.002% and 0.004%) correspond to the D.L. defined by Bonneau (1988) for the one-year-old needles of coniferous trees. For healthy subjects, the levels found are high (0.038%) but remain below those found for all-class firs grown on schist. This manganese deficiency goes together - among the same subjects - with iron deficiency (threshold: 0.004%; values: 0.005%) as well as with a particulary high calcium level (0.87%). This phenomenon is a case of calcareous chlorosis already pointed out in Northern France beech and on Alps firs. The corresponding soils have a varying stock of exchangeable cations , but high levels of exchangeable calcium with Mg/K ratios under 2 for A_1 horizons, much higher for B horizons (2.8 to 6.8).The characteristic "soil chemistry / health state of the populations" correlation is here again difficult to establish, except for manganese: its values are much weaker in litter and humus that have developed on this calcareous substratum than those developed for instance on schists.

The results obtained in these types of fir forests, developed on various kinds of substratum do not permit a strict correlation to be established between the yellowing symptoms and a particular mineral deficiency of the soils.

This conclusion is not of course unexpected, as perturbations of different natures (diverse mineral nutrients, water deficiency) can induce identical symptoms and as other factors (structure and age of the populations, weather conditions, parasites) obviously act upon the general state of health of the forests and the trees.

Mere foliar diagnosis then, even if completed by the corresponding soil data, are generally insufficient to explain this kind of phenomenon. Our first study in the Luchonnais took into account all the ecological factors likely to act upon the correct or wrong functioning of the fir ecosystem; thus, it allowed us to establish better correlations between decay and the environnemental factors. It is also certain that the small number of samples analysed here forbids us to be too affirmative in our conclusions.

Yet, it is confirmed that:

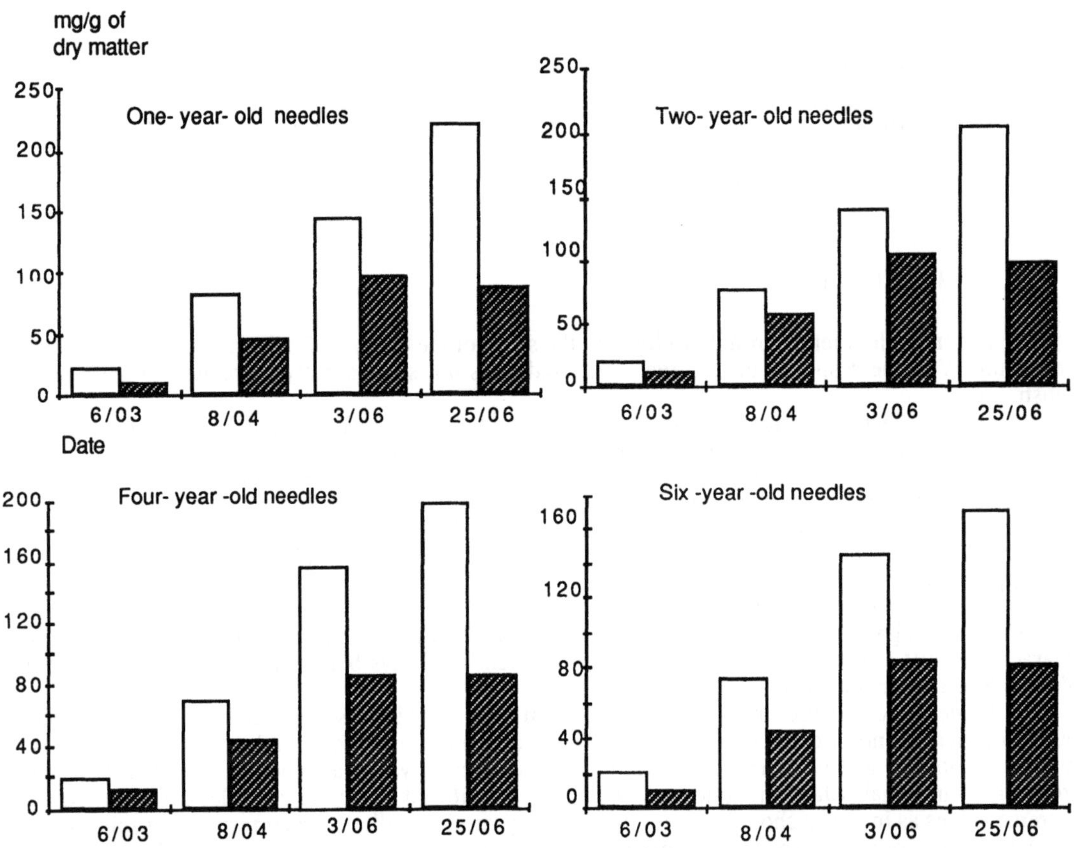

Figure 6 :Variation in the starch concentrations with time for needles of 4 different ages from Luchonnais site.

-potassium remains the main element to consider in the study of the decay of fir in the Pyrenees, and it can be deficient on schist as well as on limestone.

-on the contrary, magnesium is not a limiting factor

-lack of oligoelements like manganese and iron can induce or participate in chlorosis processes on limestone and can explain some cases of yellowing firs.

3.3 Biochemical analysis

3.3.1 Seasonal fluctuations in starch content (picture 5)

For all age classes (1- to 6-year-old needles) and for the two groups of firs studied, the starch content, direct photosynthesis product, increases from March to June. Storage is slower in yellowing needles than in healthy ones.This can be accounted for as the consequence of the CO_2 uptake decrease in diseased needles, as is shown in the experiment below.

The real reduced size of the branches of decaying subjects is also in close correlation with these observations,various authors having proved that the quantity of stored starch is positively correlated with growth activity.One week delay in the time when buds break has also been observed between healthy and diseased subjects.

3.3.2 Seasonal fluctuations of soluble sugar concentrations (figure 6)

From March to April a decrease occurs in total sugars, this is a normal phenomenon and corresponds to the beginning of the phase of starch accumulation from carbohydrates stocked by the

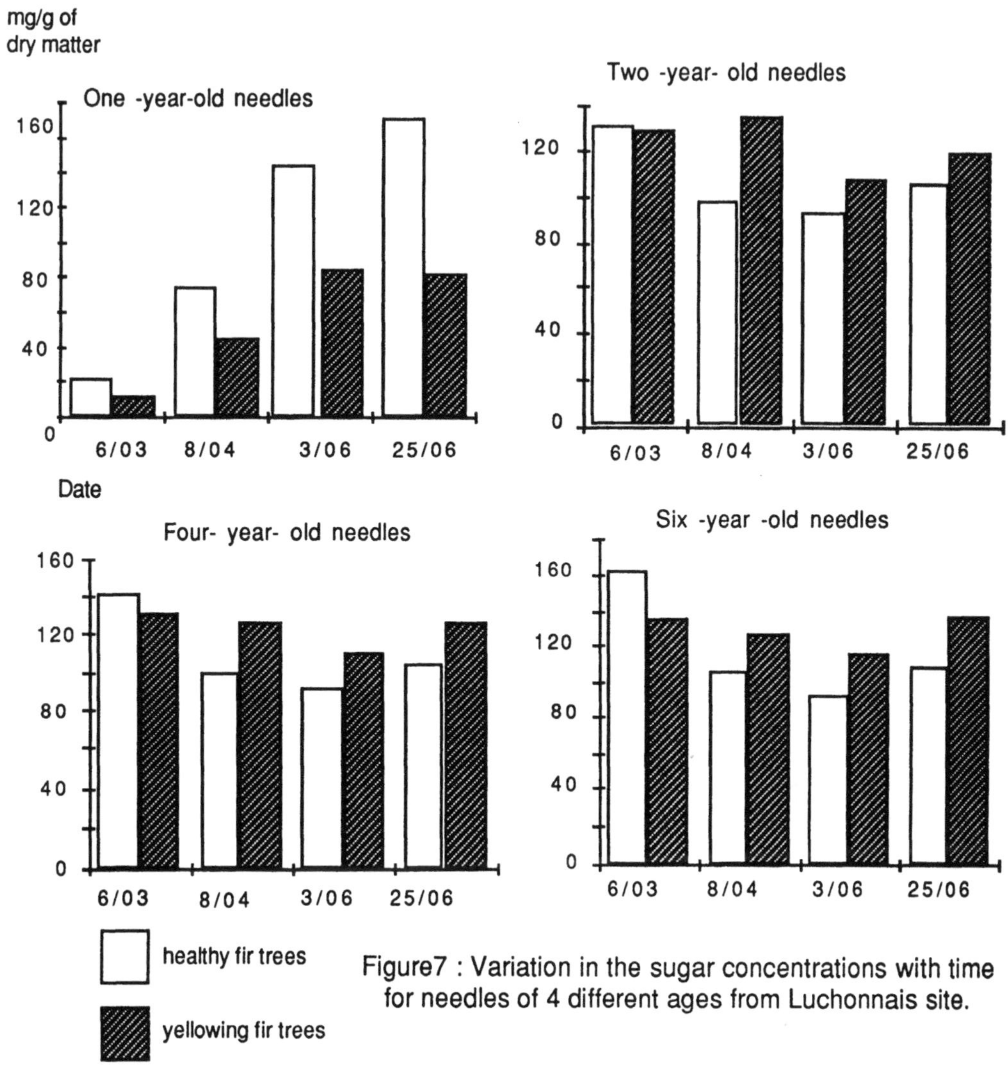

Figure7 : Variation in the sugar concentrations with time for needles of 4 different ages from Luchonnais site.

plant during the winter.In the diseased subjects, on the contratry, their concentration appreciably increases, and they present total soluble sugar contents higher than healthy trees.

In view of this, two hypotheses can be made:

-the sugars cannot be directly used for starch synthesis.

-the starch regeneration process occurs later in diseased subjects.

In both cases, this would arise from a disrupted or reduced activity in abnormal trees presenting reduced starch stockage.

For the same dates, the study of the variations in sucrose, glucose and fructose contents, shows different changes between healthy and diseased trees. For diseased subjects, we can note very irregular changes. The sucrose content, particulary, is always high in yellowing needles.

3.3.3 CO_2 uptake and variations of new metabolites

The experimental study of CO_2 uptake shows low assimilation of two- and four-year-old diseased needles. The study of new photosynthesis product distribution shows, in diseased subjects:

-amino-acid accumulation in needles of all ages

-total absence of organic acids also for needles of all ages

-decrease of soluble sugar content of young needles.

We can also note that the insoluble part is as plentiful in healthy as in declining trees, except in one-year-old needles.

3.3.4. Discussion

This physiological approach to forest decline in the Luchonnais shows:

- a clear diminution of CO_2 uptake in needles of 2 and 4 years, inducing poor starch accumulation.This also means that old needles will not fully participate in new needle growth. Branch

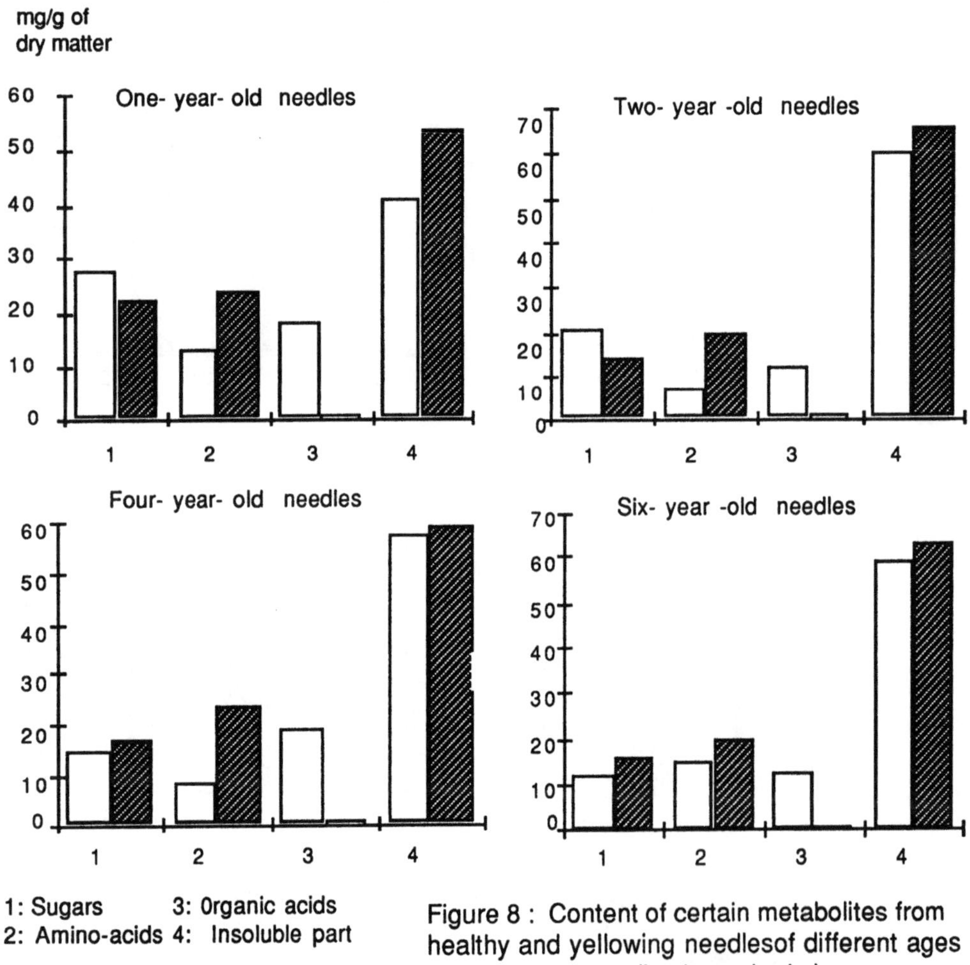

Figure 8 : Content of certain metabolites from healthy and yellowing needlesof different ages (Luchonnais site)

defoliation is an other factor which contributes to the reduction and the storage capacity of diseased subjects, also contributing to the poor growth of young needles.

- a global accumulation of amino-acids and very slow organic acid biosynthesis, caused especially by an excess proportion of NH_4^+ ions.

4. CONCLUSION

The decay of Pyrenean fir stands, which is characterized by yellowing of needles aged two years or more is linked to nutritional and physiological disturbances.

The K deficiency which appears in all decaying fir needles - and sometimes in healthy ones- is the main result of this phenomenon.

The low fertility of the soils of the fir stands is not the only factor able to explain forest decay. Other parameters must also be taken into account.

So, an experiment to define pollution conditions was run along side the study. The results are as follows:

-rainfall, collected in automatic raingauges is clearly acid (average pH: 4.0; limit values: 3.5 and 4.7) with high concentrations for sulfate (7.34 ppm) and nitrate (3.09 ppm) strongly correlated with H^+ concentration. The ammonium content of the rainfall is also high.

-throughfall sampled under fir stands contained higher concentrations of potassium and magnesium, and lower concentrations of ammonium, compared to incident water.

It is clear that there is a leaching phenomen which concerns the nutrient content of the needles (Horntvedt, 1979; Dambrine et Presvoto, 1989). Leached substances can be replaced by increasing the root uptake. In the Luchonnais site, where K contents of soils are very low, K deficiency in needles cannot be compensated.For Mg however the soil levels are sufficient to make up for the loss.

Likewise, most physiological disturbances observed could be related to an excessive ammonium supply from rainfall and subsequent foliar absorption :

-excessive amino-acid levels in yellowing trees can be explained as an "ammonium trap"

-yellowing, disturbances in photosynthesis and growth, are most likely symptoms of an excessive ammonium uptake for nutrient-deficient plants.

Atmospheric pollution can be considered as the main predisposing factor in forest decay phenomena, which can increase cation depletion and worsen nutritional imbalance. Climatic characteristics (rainfall deficits) can also be considered as a"weakening factor" in forest decay.

We are very much indebted to Dr Winterton for his contribution to the translation of this paper.

This research into the decay of firs in the Pyrenees was undertaken as part of a research contract with INRA Nancy who supervise the French DEFORPA programme.

REFERENCES

BONNEAU M. (1988). Le diagnostic foliaire .*Revue forestière française*, Special issue: *Diagnostics en forêt*, 19-28.

CHERET V. (1987). La sapinière du Luchonnais (Pyrénées Hautes-Garonnaises): Etude phytoécologique, recherche sur le phénomène de dépérissement forestier. *Thèse de Doct. Univ.*, Toulouse: 287 p.

CHERET V., DAGNAC J., FROMARD F. (1987) . Le dépérissement du sapin dans les Pyrénées luchonnaises. *Revue forestière française* , 34 (1):12-24.

DANBRINE E., PREVOSTO B. (1989) . Flux des éléments minéraux dans un écosystème forestier d'altitude soumis à la pollution atmosphérique.Relation avec le dépérissement*Journées de travail DEFORPA 23 Février, 2 et 3 Mars 1989, Nancy: Dépérissement forestier et pollution atmosphérique*, 4(8-2):1-34. INRA, Direction Générale, Paris.

EVERS F. H. (von) (1972) Die Jahrweisen Fluktuationen der Nährelementkonzentrationene in Fichtebnadeln und ihre Bedeutung für die Interpretation nadelanalytischer Befunde, *Allg. Forst. und Jagdzeitung,* 143 (3-4): 68-74.

HORNTVEDT R. (1979) . Leaching of chemical substances from tree crowns by artificial rain.*Mitt. I.U.F.R.O. Tagung 2.09 " Luftverunreinigung"*,Ljubljana, 1978 : 115-123.

LANDMANN G. (1989). Observation au sol de l'état sanitaire des forêts:évaluation critique des données 1983-1988. *Journées de Travail DEFORPA 23 Février,2 et 3 Mars 1989, Nancy: Dépérissement des forêts et pollution atmosphérique*, 1:1-16. INRA, Direction Générale, Paris.

NIHLGARD B. (1986). The seasonal variation of nutrient in needles of three differently exposed spruce stands.Proceed. from Workshop CEC "Air pollution research", report 4, 19-23 October 1986, Lokeberg:

NIHLGARD B. , (1989) . Nutrients and structural dynamics of conifer needles in South Sweden 1985-1987. *Medd.Nor. inst. skogforsk.* 42(1): 157-165.

INDIRECT EFFECTS OF ACID RAIN MEDIATED BY MINERAL LEACHING: AN EVALUATION OF POTENTIAL ROLES OF LEACHING FROM THE CANOPY

S. Leonardi
GSF-München, Institut für Biochemische Pflanzenpatholgie
D-8042 Neuherberg. Present address: Institut für Angewandte Pflanzenbiologie CH-4124 Schönenbuch

Abstract

The leaching of cations, in particular of K, Ca and Mg, plays a crucial role in leaf ionic relations. As a regular part of the biogeochemical nutrient cycling, the flux of leached cations needs to be balanced by an equivalent uptake and translocation. The plant's ability to adjust an increased flux due to acidic deposition determines the extent and duration of leaching-induced, altered ionic relations, first of all within the apoplastic space. - Basic principles and implications of cation leaching will be presented in order to elucidate this kind of an indirect effect of acid rain.

INTRODUCTION

"The removal of substances from plants by the action of aqueous solutions, such as rain, dew, mist and fog" is termed leaching (Tukey, 1970) and represents, on a whole plant level, a component of throughfall (Parker, 1983) and, on a leaf level, a process contributing to the overall balance of ions in the leaf (Pitman, 1988). Besides these ubiquitous processes, leaching may be accelerated by acidic precipitation (Tukey, 1980) and is then classified as an indirect effect of the potential impacts of "acid rain" (Tamm and Cowling, 1977). - On the analogy of the concept of the soil-plant-atmosphere continuum of plant water relations, the phenomenon of leaching may be thought as a flux of nutrients within a similar soil -plant-atmosphere continuum (i.e. availability - uptake/translocation/storage - leaching) , whose dynamics has been attributed the widely used term biogeochemical cycling. Thus, leaching is not a separate physiological process but rather a part of the overall nutrient balance of plants.

Regarding the accelerated leaching potential of acidic precipitation, several aspects have to be considered, if the effective leaching flux is to be assessed; they may be categorized as follows: i) physico-chemical properties of the precipitation, ii) physico-chemical conditions of the interface precipitation-leaf surface, iii) biological characteristics of the leaves and in particular of their surfaces. Some of these aspects will be addressed up on later to point out their crucial role in determining the quantity and quality of the leachates.

This paper is dedicated to Walter Flückiger who introduced to me the phenomenon of leaching. Part of the experimental work was done at his laboratory.

One major point of interest regarding the quantity of leached minerals emerged in the last decade, when several investigators hypothized that the leaching of cations could induce mineral deficiencies and that it thereby would contribute to the novel forest decline (Bosch et al., 1983; Prinz et al., 1982; Zech and Popp, 1983; Zöttl and Mies, 1983), especially to type 1 spruce decline, i.e. needle-yellowing at higher elevations of the German "Mittelgebirge" (FBW, 1986). Both ozone and frost have been proposed as additive or synergistic factors accelerating the leaching, especially of Mg. - Altered mineral relations of forest foliage including the ratios of mineral cations to nitrogen have been found in both broad-leaved (e.g. Flückiger et al., 1986) and conifer (e.g. Schulze, 1989) forests. Complex interactions between deposition inputs and the canopy (absorption/leaching) have been assumed to be involved in the promotion of observed nutrient imbalances.

In a recent paper, Roberts et al. (1989) have summarized some major results of the above hypothesis with special regard to the nutritional consequences in *Picea abies*. They concluded that "limited Mg availability uptake was more important than foliar leaching" (cf. Tomlinson, 1987, for a similar conclusion). However, if the nutrient availability in the rhizosphere and the uptake and translocation mechanisms are not sufficient to replenish leached pools of above-ground plant parts, leaching may well accelerate mineral deficiency (Kaupenjohann et al., 1988). Interpretation of average stand nutritional data and of calculated leaching fluxes may be misleading with respect to the affected trees, since the sources of leached material are single leaves and individual trees. Moreover, due to the great variability of atmospheric deposition - as the driving force of leaching - in time (e.g. Hendersen et al., 1977; Richter and Lindberg, 1988) and space (e.g. Beier and Gundersen, 1989; Draaijers et al., 1988), more attention should be drawn to temporal and spatial events. Of course, regional differences of site conditions including precipitation characteristics and impaction potential of the canopy are important determinating factors of throughfall fluxes (Olson et al., 1981).

It was shown (Haines et al., 1985a), that mineral leaching rates may vary considerably among forest species. An index of relative leachability was developed by Eaton et al. (1973) and used by Olson et al. (1981) for the comparison of the leaching of nutrients between hardwood and fir forests: regarding the mineral cations K, Ca and Mg, the leachability indices of the hardwood canopy were found to be two to seven times greater than those of the fir canopy. Similar findings were reported by Cronan and Reiners (1983) who compared a Balsam fir and beech-maple-birch vegetation. Most probably, the greater leachability of cations from broad-leaved species is due to the greater specific leaf area (area/weight). Accurate leachabiltiy indices cannot be calculated for European forest sites at the present time, since there are great uncertainties in determining effective leaching fluxes where dry and occult deposition are important inputs into forests and are very difficult to measure (Fowler et al., 1989a; Unsworth and Crossley, 1987) [1]. Nevertheless, the element specific leachability may be derived with the help of the 16-

1) This is particularly true for forest sites at higher elevations where capture of cloud droplets account for higher depositon rates (Unsworth and Crossley, 1987). Due to evaporation, the chemical concentrations of intercepted cloud water may further increase to values that are larger than those measured in cloud droplets themselves (Unsworth, 1984). It was shown by Joslin et al. (1988) and Waldman and Hoffmann (1988) that acidic cloud water may increase nutrient leaching substantially; the former

year-average data for two forest sites at Solling, FRG (Matzner, 1988): for both beech and spruce stands the leaching fluxes decreased in the order K > Ca > Mg. This is well in accordance with the data presented by Eaton et al. (1973) and Olson et al. (1981) who both worked in regions where the wash-off of dry deposition was estimated to be of minor importance in the throughfall flux.

It has been shown (Flückiger et al., 1988; Mitterhuber et al., 1989) that cations may also be leached easily from the bark tissue. Regarding the overall balance of mineral plant nutrition, the leached pools of bark tissue represent a sink/lacking source of the mineral translocation, especially in the next flush (cf. Ashmore et al., 1989, who found decreased Ca contents of leaves that were produced in the season subsequent to acid mist exposure).

In order to discuss further on the effects of mineral leaching upon whole plant physiology, an attempt is made to distinguish between direct and indirect effects of leaching. Depending on the replenishment of the leached pools, in particular of the apoplastic free space including the leaf surface, effects may last for only a short time (hours - days), or may persist.

1) *Direct effects:* associated with the removal of mineral elements from leaves.
- Changes in ionic relations of the leaves, particularly of the extracellular space.
- Altered compartmentation of leached minerals since the free space of the leaves is leached preferentially.
- Decreased buffer capacity, in particular of the extracellular compartment.

2) *Indirect effects:* consequences of the removal of mineral elements.
- Accelerated mineral cycling. In order to replenish the leached pools more minerals have to be taken up by the roots.
- Acidification of the rhizosphere due to increased cation uptake and associated proton exchange by the roots.
- Altered stomatal control since stomatal movements depend on ionic compartmentation and on extracellular ionic relations.
- Changed membrane permeabilities and cell wall elasticities since the structure and function of membranes and cell walls depend on their ionic micro-environment.
- Unspecific metabolic changes due to 1.2. and 2.4.
- Additional stressor in interactions with other environmental impacts.

Some of the listed indirect effects (2.3., 2.5., 2.6.) were classified by Tamm and Cowling (1977) as direct effects of acidic precipitation. Obviously, it is very difficult to attribute a certain effect to either "acid rain" or leaching. This particular point of interest will be discussed later.

---

authors estimated annual losses of K, Ca, and Mg of up to 5 %, 36 %, and 16 % of the standing foliar nurient pool, respectively. Forest sites at higher elevations with frequent occurence of clouds, mist and fogs are therefore particularly subjected to effects of acidic deposition. Interactions with gaseous air pollutants have been discussed to lead to the prevailing occurence of novel forest decline symptoms at high altitudes (Krause et al., 1986).

While the majority of reports on mineral leaching deals with the direct effects of element removal and subsequent changes of leaf contents, there are only few investigations concerning the indirect effects. Most attention have gained: i) accelerated mineral cycling (Johnson et al., 1985; Leonardi and Flückiger, 1988; Mecklenburg and Tukey, 1964), ii) acidification of the rhizosphere (Kaupenjohann et al., 1988; Leonardi and Flückiger, 1988, 1989a) and iii) altered stomatal control (Barnes et al., 1990; Evans et al., 1981; Leonardi and Flückiger, 1989a; Mengel et al., 1989). Only few data are available dealing with other indirect effects.

This paper will stress some basic principles of mineral leaching from leaves. Experimental results are presented to elucidate implications of acid rain effects mediated by mineral leaching.

MECHANISMS OF LEACHING

The sources of leachable minerals may be divided into:

- depositions on the leaf surface, derived either from plant internal sources (guttation, secretion, diffusion) or from atmospheric depostions
- ions of the water free space
- bound ions on exchange sites on the leaf surface and in the Donnan free space of the apoplast
- symplastic contents.

The first source is readily washed-off by natural precipitation (Leonardi and Flückiger, 1988), while ions from the other sources have to diffuse out of the leaf into the wetting solution. Therefore, the wettability of the leaf surface and the wetting potential of the precipitation are both important determinants for effective leaching fluxes (Haines et al., 1985b; Keever and Jacobson, 1983; Paoletti et al., 1989; Tukey, 1970) [2].

In Figure 1, the time course of the electrical conductivity (EC) of leaf diffusates (intact leaves of *Fagus sylvatica* immersed in H_2O) is depicted. The curve is clearly bi-phasic with a steep increase of EC within the first 30 minutes, reflecting the dissolution of surface deposited materials. Afterwards, a steady-state increase in EC results from a continuous efflux of electrolytes - most probably by diffusion - into the medium. With the help of an average specific conductance of 0.071 $\mu S\ cm^{-1}$ per $\mu eq\ L^{-1}$ the steady-state efflux into H_2O, derived from linear regression anlaysis of the dashed curve in Fig. 1, was calculated to approx. 0.9 $\mu eq\ g(f.w.)^{-1}\ h^{-1}$. As opposed to the diffusive pathway of efflux, 7.7 $\mu eq\ g(f.w.)^{-1}$ were calculated to be washed-off.

A similar experiment was done to illustrate exchange processes (i.e. third source of leachable minerals) between leaf cations and protons of the efflux medium: Figure 2. The time course of EC of leaf diffusates (intact leaves of *Fagus sylvatica* immersed in 1 mN H_2SO_4; same leaf sample set as used for the previous experiment) showed also an initial increase in EC. Thereafter, EC decreased steadily. This is due to the much higher specific conductance of the proton (0.35) as compared to those of mineral cations ($\pm$0.066); as a consequence, EC decreases for each proton of the efflux medium that exchanges with a cation from the leaf. - The steady-state exchange rate was calcu-

2) While a greater leaf wettability was generally found to increase leaching, Paoletti et al. reported a slight reduction of leaching upon addition of a surfactant to spray solutions.

Figure 1

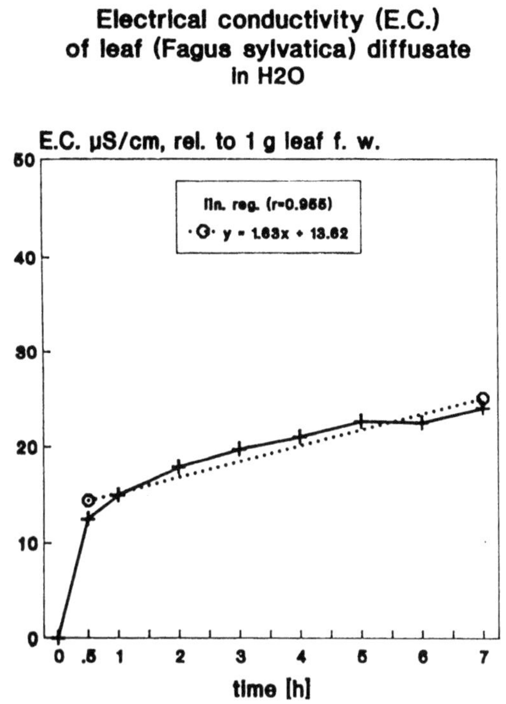

Figure 2

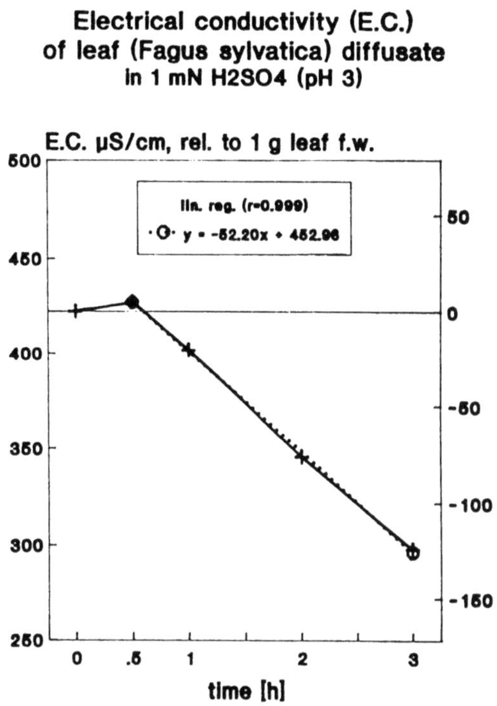

lated as follows: the slope of the dashed line in Fig. II was divided by the difference between the specific conductance of the proton and of cations (= 0.35 - 0.066) to yield µeq cations that were exchanged minus µeq ions that were diffused. Assuming that an equal amount of ions was diffused into both efflux media used, the 0.9/2 µeq cations $g(f.w.)^{-1}$ h^{-1} due to diffusion were added to come up to a net cation exchange rate of 7.8 µeq $g(f.w.)^{-1}$ h^{-1}. This figure is considered as a maximal exchange rate and it demonstrates that, under the chosen conditions, exchange is a far more effective process in leaching cations than that in the case of diffusion.

The calculated efflux of cations from leaves is summarized in Table 1. Additionally, measured values from the same assay of total efflux of K, Ca, and Mg are given for comparison. Splitting up the total efflux into three categories (dissolution of solid material, diffusion from the leaf, exchange against protons) demonstrates the relative contribution of each efflux mode: dissolution and exchange of cations in the acid solution are most significant in determining the leached quantity of cations.

The extent to which cations are leached in forests by exchange processes may vary greatly: reported percentages are 20 % (Cronan and Reiners, 1983), 27 % (Eaton et al., 1973), 28 % (Richter et al., 1983), 40-60 % (Lovett et al., 1985). In the experimental assay presented in Table 1, 25 % of the 40 µmol H^+ of the incubation medium were exchanged within 3 hours. But this value is thought to be a potential maximum of cation exchange since the assay was performed in a 1 mN H_2SO_4-solution of pH 3. Therefore, it is assumed that exchange rates in the field under ambient proton loads are lower. In a detailed study of canopy exchange processes in 26 forest stands of *Fagus sylvatica* in northwestern Switzerland (Leonardi & Flückiger, 1987), exchange of protons accounted for approx. 10 % of the leached

Table 1 Analysis of the efflux of K, Ca, and Mg from leaves of Fagus sylvatica

efflux from	into	by [a]	calculated tot. cations	measured K	Ca	Mg
			µeq g f.w.$^{-1}$			
intact leaf	- 1 mN H_2SO_4	1	8.53			
n = 44	(pH 3.0)	2	1.38			
3 hr assay		3	23.41			
		total	33.32	2.97	31.95	3.98
	- H_2O	1	3.84			
		2	1.38			
		total	5.22	1.81	5.95	0.76

[a] mode of efflux: 1 = dissolution of solid material
2 = diffusion from the leaf
3 = exchange against protons

cations. This study was done in late summer; so, the percentage is likely to increase as the season progresses.

However, the protons of the incident precipitation are not the only cations that may exchange with leaf cations. It was reported by Carlisle et al. (1966), Lovett et al. (1985) and Leonardi and Flückiger (1987) that N and in particular NH_4^+ was retained by the studied deciduous canopies and - calculated by Lovett et al. on an ion balance of retention and leaching - it was assumed that NH_4^+ may contribute to exchange processes with leaf cations. Indeed, in laboratory experiments with needles (Roelofs et al., 1985) and leaves (Leonardi and Flückiger, 1989), NH_4^+ was found to be very effective in leaching cations, in particular K.

Besides H^+ and NH_4^+ of the precipitation, the dissolution of the gaseous pollutants SO_2 and NO_2 in apoplastic water also yields protons (Pfanz and Heber, 1986) and may thereby increase leaching fluxes as has recently been discussed in experiments by Seufert and Evers (1990).

In any case, leached leaf cations have to pass the cuticular layer since a direct contact between the wetting precipitation and the apoplastic water through the stomata is not probable (Schönherr and Bukovac, 1972). Permeability coefficients of various cuticles for cations are greater for monovalent than for divalent cations and decrease in the order K > Mg > Ca (Garrec and Kerfourn, 1989; McFarlane and Berry, 1974; Riederer, 1989).

Leaching from the symplast (i.e. fourth source) is generally thought to be of minor importance (Tukey, 1970). However, leaching increases from injured and senescent tissues. A very illustrative example of increased leaching from injured leaves was reported by Skeffington and Roberts (1985): an aphid infection of Scots pine caused a large increase in leaching, particularly of K (cf. also Seufert (1988) for a similar observation; Table 2). Increasing values of leaching from the symplast were shown to correspond to increased permeabilities in senescing tissues (Eilam, 1965). Membrane permeabilities may also be increased by plant fumigation with ozone (Heath and Castillo, 1988) and by simulated acidic rain (Evans et al., 1981). Seufert (1988) concluded from his studies of throughfall analysis under *Picea abies* canopies exposed in open top-chambers that ambient air increased symplastic leaching. In own experiments with *Fagus sylvatica*, ozone-induced greater membrane permeabilities correlated with increased efflux rates, particularly of Mg (Leonardi and Langebartels, 1990). Since ozone is known to accelerate leaf senescence (Keller, 1988; Krause and Prinz, 1989), increased cation leachability is expected where high ambient levels of ozone promote premature senescence (cf. Brown and Roberts (1988) for an evaluation of confounding results regarding ozone-induced leaching due to NO_x-production during O_3 generation).

Table 2 Fluxes of leached cations as a percentage of the foliar nutrient pools

	K	Ca	Mg	H^+ [a]	H_2O [b]
	| - - - -	%	- - - - |	mmol m^{-2}	mm
a) field studies					
Solling [c]:	30	30	20	82	1032
Fichtel gebirge [d,e]:	8-20	2-7	1-7	46-57	998
b) semi OTC study [f]					
ambient	12.7	-0.1	0.1	45	584
filtered	(41) [g]	0.3	2.8	40	591
c) climatic chamber study [h]					
filtered	3-6	2-3	3-6	400	400

[a] H^+ deposited by precipitation or by spraying
[b] total amount of precipitation or spraying solution
[c] Matzner (1988); data: 1965-1985 average
[d] Horn et al. (1989); data: 1985
[e] Schulze et al. (1989); data: 1985
[f] Seufert (1988); data: 4/1986-4/1987; top irrigation pH 4
[g] abnormal leaching flux due to aphid attack
[h] Pfirrmann et al. (1989); data: 4/1988-9/1988; misting pH 3

QUANTITATIVE ASPECTS OF CATION LEACHING

It was recognized by Wood and Boramnn (1975) that leaching fluxes of cations estimated from throughfall analysis were greater than those measured in a laboratory assay. Moreover, the same assay used on two sample sets of greenhouse and field grown plants yielded a greater leachability of cations from leaves of the field grown plants (Leonardi and Flückiger, 1989b): the leachability was increased 2.3 x, 2.7 x, and 2.8 x for K, Ca, and Mg, respectively, from leaves of the field grown beech saplings.

The amounts of nutrient losses by foliar leaching are of crucial interest with regard to i) induction of or contribution to mineral imbalances/deficiencies of the foliage and ii) scaling-up from laboratory experiments to relevant field situations. Direct measurements of leaching fluxes under forest canopies are not possible and the interpretation of throughfall analysis needs several assumptions. On the other hand, adequate links between laboratory and field are still missing: concerning the leaching of cations, the discrepancy between laboratory and field data may rise up to one order of magnitude. This may be derived from the data shown in table 2: results from the most studied forest species in Central Europe, ***Picea abies***, are listed to focus the problem of scaling-up and -down (only longer-lasting (min. one season) experiments were taken for comparison).

Two main circumstances are thought to be responsible for the obvious discrepancies between field and laboratory [3] data of leaching fluxes:

1.) Since the cuticle (Berg, 1987) and in particular its polymeric cutin (Garrec and Kerfourn, 1989) is the main barrier of ion movements from the leaf interior to the outside, any abrasion and erosion of the cuticle would increase leaching (Berg, 1987; Eaton et al., 1973; Tukey and Tukey, 1963; Turunen and Huttunen, 1990). Damage to the cuticle is thought to be ubiquitous in nature since there are many physical, chemical and biological attacks affecting the integrity of the cuticle.

2.) Analysis of throughfall in forest is mostly done on a larger time-scale (months - years) than are leaching experiments performed in the laboratory (hours - weeks -months). Furthermore, the latter assays are done with young or mature leaves and - in the case of conifers - with a high percentage of current year foliage. This is in contrast to the field studies where throughfall analysis usually covers several seasons, thus including the stage of leaf senescence. The susceptibility to the loss of nutrients reach a peak at senescence (Cronan and Reiners, 1983; Tukey, 1970).

Additional sources of variability may be the managed nutrient status of the experimental plants (cf. Bosch et al., 1986; Mengel et al., 1987) as well as the chemical composition of the simulated acidic rain, in particular NH_4^+ concentration (cf. above), and the kind of application of spraying solutions (Laitat et al., 1989; Leonardi and Flückiger, 1989a).

QUALITATIVE ASPECTS OF CATION LEACHING

As already mentioned in the introductory chapter, the leachability of cations (on an equivalent basis) from forest canopies decreases in the order K > Ca > Mg. Additional evidence in support of this order comes from investigations of Baker and Attiwill (1987), Johnson et al. (1985), Klemm et al. (1989), Lindberg et al. (1986), Lovett et al. (1985), Pape et al. (1989). However, this rule is not applicable for every forest; Ca may well be leached more than K (e.g. Scherbatskoy and Klein, 1983; Schier, 1987).

On the contrary, most of the reviewed results of laboratory and greenhouse experiments indicate a decrease of the leachability of cations in the order Ca > K > Mg for all acidities of spraying solutions tested [4] (Glavac, 1987; Haines et al., 1985a; Klumpp and

3) cf. Jacobson (1984) for a discussion of contradictory experimental results

4) In general, there is an increased cation leaching with decreased pH of the treatment solution. But K doesn't seem to follow this rule strictly: Evans et al. (1981), Fairfax and Lepp (1975), Leonardi and Flückiger (1988, 1989a) and Paoletti et al. (1989) all reported decreased K leaching at lower pH.

Guderian, 1990; Mengel et al., 1987; Mitterhuber et al., 1989; Paoletti et al., 1989; Turner et al., 1989; Wood and Bormann, 1975). The same order of leachability was found in own experiments with leaves of *Fagus sylvatica*. However, this order is true for the amount of cations leached but not for the ion specific leachability.

If the latter is calculated with regard to the extracellular contents of the cations, the situation changes. This is shown in Table 3: 40 % of K of the Donnan free space could be leached in an efflux experiment, while only 4 % Ca and 5 % Mg were leached. This reflects a high leachability of K and as a consequence of K removal, the ratio of $[K^+]/([Ca^{++}+Mg^{++}])$ in the free space will greatly decrease. The leaching of K is further increased by its high mobility through the plant cuticle (cf. above). - In conclusion, greatest relative changes of leaf cation contents are expected for extracellular K. However, even minor intracellular changes might affect the cell's metabolism, in particular in the case of cytosolic free Ca whose concentration is strictly regulated and dependent on compartmentation.

Table 3 Leachability of K, Ca, and Mg from leaf discs of Fagus sylvatica as a fraction of cell wall contents [a]

	K	Ca	Mg
		µeq g d.w. $^{-1}$	
Cell wall contents	21.0±2.5	623.5±19.8	107.5±4.5
Leached	8.5±0.5	26.7± 3.4	5.6±0.5
(Leachability [b]	(40.3)	(4.3)	(5.2)

[a] original data from Leonardi and Langebartels, 1990
[b] percentage of element leached from cell wall content

ACID RAIN EFFECTS ON PLANT PHYSIOLOGY: DIRECT ACID EFFECT OR MEDIATED BY ION DISPLACEMENT BY LEACHING?

This chapter will not deal with morphological changes, particularly of leaf surfaces, due to acid rain, even if it is recognized that a morphological alteration (e.g. structure of cuticle) may affect physiological processes (e.g. water relations). Here, an attempt is made to differentiate acid rain effects on plant physiology between a real acid effect (i.e. change in pH, cf. Evans, 1984, for possible implications) and effects mediated by ion displacement from the water and Donnan free space. - To my knowledge, there is no specific literature on this topic. Neverheless, clarity on this question is needed in order to understand differential impacts of acid rain.

The most obvious changes between incident precipitation and net throughfall are decreases of proton [5)] and Nitrogen and increases of cation fluxes. Since these changes are balanced by corresponding, opposite changes within above-ground plant parts, ionic relations of affected tissues must change to a certain extent. Not only the

5) However, Hoffman et al. (1980) have shown that total acidity was conserved as rain passed through a forest canopy (*Quercus prinus*), although the strong:weak acid ratio declined.

amount of ions absorbed or leached determine the net changes, but also the extent to which changes may be buffered.

With regard to H^+, leaves are able to buffer incoming H^+ in order to keep the pH within a certain range. Penetrating H^+ are first buffered in the extracellular space by ion exchange with weak Brönsted bases (Cronan and Reiners, 1983) and within the Donnan free space with exchangeable cations. Indeed, the buffer capacity of isolated cell walls may account for more than half of the total leaf buffer capacity (Leonardi and Flückiger, 1988; Leonardi and Langebartels, 1990). From experiments with leaves of *Fagus sylvatica* it was derived that the amount of H^+ necessary to decrease the extracellular pH by one unit would be approx. 100 µmol per g leaf dry weight. Assuming no neutralization of H^+ by Brönsted bases and that 10 % of the incident H^+ would infiltrate into the apoplastic space, one would need 70 mm of precipitation with a pH of 3.0 to decrease the extracellular pH by one unit. But such conditions are quite unlikely; hence, significant pH changes in leaves are not expected to occur in the field. However, the situation may change if there are high levels of acidic gases such as SO_2 that contribute to the leaf's proton load (Pfanz and Heber, 1986).

On the other hand, leaching of cations lead to ion displacements first of all in the apoplastic space that change at least transiently the extracellular ionic relations. Impacts on those physiological processes that depend on extracellular ionic relations (e.g. membrane permeability and stomatal movements) are possible consequences. It was shown by Van Steveninck (1965) that apoplastic Ca displacement - with the help of EDTA as chelator - had a significant and reversible effect on increased membrane permeability. Schwarz (1985) reported impaired stomatal movements if epidermal strips were floated on a solution containing EGTA as Ca-chelator. Recently, it was demonstrated by Fink (1990), that acid rain-induced, apoplastic Ca displacement in needles of *Picea abies* is first observed in cell walls of the epidermis. Thus, even minor leaching fluxes might affect extracellular ionic relations without influencing the bulk nutrient status of the whole leaf.

Due to the lack of relevant data, one may only tentatively conclude that acid rain effects on physiological processes are mediated most probably by apoplastic ion displacement.

DIRECT EFFECTS OF CATION LEACHING

Once cations are leached from leaves, the leached pools have to be replenished. During this transient state of homeostasis, there are altered mineral relations in the affected tissues (Leonardi and Flückiger, 1988). Dependent on nutrient availability in the soil and on the functioning of uptake and translocation, accelerated uptake will replace the lossed nutrients (Mecklenburg and Tukey, 1964). Therefore, lasting deficiencies are not to be expected. On the contrary, homeostasis may even overshoot leading to increased nutrient contents in leaves of acid rain treated plants (e.g. Leonardi and Flückiger, 1989; Mengel et al., 1988; Skeffington and Roberts, 1985: all reported slightly higher nutrient contents in leaves of acid mist treated plants). However, if the requirements of an accelerated cycling are not fulfilled, developing nutrient deficiencies may well be strengthened by leaching processes.

Only little is known about affected compartmentation and local displacement of leached cations. But theoretical considerations and scarce experimental evidence (Fink, 1990; Leonardi and Langebartels, 1990) force the conclusion that cation compartmentation does alter due to leaching from selective compartments, i.e. apoplastic space, in particular leaf surface and epidermal cell walls.

It was shown that the apoplastic buffer capacity may be reduced by an acidic mist treatment of *Fagus sylvatica* (Leonardi and Flückiger, 1988). The extracellular buffer capacity largely depends on exchangeable cations on cell walls, particularly of Ca (Leonardi and Langebartels, 1990). In Figure 3, data pairs of exchangeable cell wall Ca and cell wall buffer capacity (µmol H^+ needed to reduce the initial pH to pH 3.0, per g leaf dry weight) are plotted: linear regression analysis revealed a significant correlation ($P<0.00001$) with a slope of 0.81, illustrating the dependence of extracellular buffer capacity on exchangeable cell wall Ca.

Figure 3

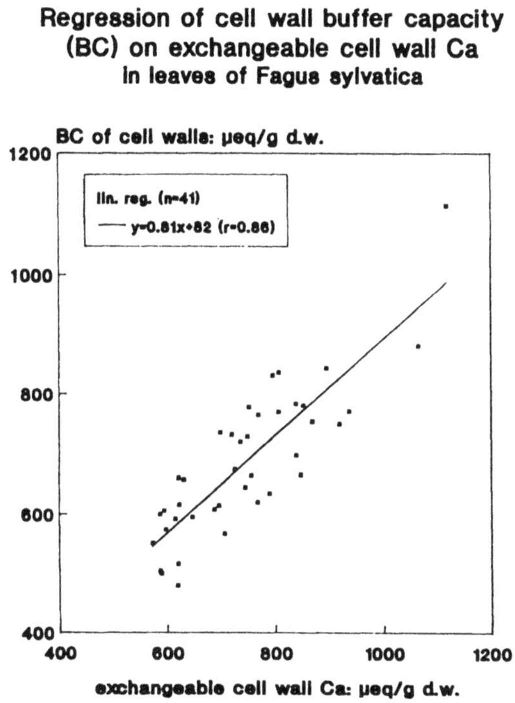

INDIRECT EFFECTS OF CATION LEACHING

As outlined above, leached nutrient pools have to be replenished by reabsorption from the soil. This was demonstrated by a radioactive tracer (Ca-45) experiment with *Phaseolus vulgaris* (Mecklenburg and Tukey, 1964) as well as in an experiment with seedlings of *Fagus sylvatica* (Leonardi and Flückiger, 1988). Due to the fact that far more cations than anions are leached (Tukey, 1970), the reabsorption of excess cations by the roots is balanced by a proton efflux (Marschner, 1983), thereby acidifying the rhizosphere. This indirect effect of cation leaching from the canopy was proposed by Miller (1984) and Ulrich (1983) and was confirmed experimentally by Kaupenjohann et al. (1988) using spruce saplings in nutrient solution and by Leonardi and Flückiger (1988, 1989a) using beech seedlings in nutrient solution as well as potted in soil.

Acidification of the rhizosphere may have significant impacts on the availability of Al, Mn, Fe, Pb, and Zn. Sarkar and Wyn Jones (1981) demonstrated that shoot Fe, Zn and Mn contents of *Phaseolus vulgaris* significantly correlated with the extractable levels determined in the rhizosphere. Laboratory (Rost-Siebert, 1985) and field (Meyer et al., 1988) investigations revealed evidence of decreased fine root growth with increasing Al:Mg ratios. Shortle and Smith (1988) have proposed aluminum-induced Ca deficiency involved in the decline of *Picea rubens* in northeastern United States. Regarding mycorrhizal interactions in the rhizosphere, conclusive data of the impact of acid rain (probably mediated by acidification of the rhizosphere) are not yet available: e.g., experiments with *Pinus taeda* seedlings exposed to rains of intermediate acidities revealed inhibited mycorrhiza formation (Shafer et al., 1985, at pH 3.2 and 4.8) or a greater number of mycorrhizal roots (Simmons and Kelly, 1989, at pH 3.8).

On an opposite plant-environment interface, i.e. stomata, there are acid rain effects, too. Evans et al. (1981), Leonardi and Flückiger (1988, 1989a), Leonardi and Langebartels (1990), Mengel et al. (1989), Neufeld et al. (1985) and Valentini et al. (1989) all reported increased transpiration rates of plants treated with simulated acidic mists. It was also shown that both stomatal and cuticular conductance may be increased. However, there are reports as well demonstrating decreased stomatal conductances due to acidic mist treatments (Barnes et al., 1990; Neufeld et al., 1985). At present, it is not known how stomatal control is affected. Several possibilities might be discussed: change in guard cell wall elasticity (Barnes et al., 1990; Bittisnich et al., 1987), proton-stimulated opening of stomata (Dittrich er al., 1979), altered ion fluxes in guard and epidermal cells (MacRobbie, 1988); all of these mechanisms may be affected by leaching processes due to acid rain.

To illustrate acid rain effects on stomatal conductance (g_s), the response of leaves of *Fagus sylvatica* to simulated acidic mist is shown in Figure 4: i) there is a marked temporal variation in g_s, starting with highly increased values followed by a steady decrease leading to lower g_s after 5 weeks of acidic mist treatment; ii) in a second experiment, there were no differences of g_s during day-time but significantly increased g_s at night (cf. Mansfield and Freer-Smith (1984), for similar findings with *Betula pendula* after fumigation with the acid gases SO_2 and NO_2). - It is concluded that acid rain affects the normal functioning of guard cells as well as the leaf's ability to withhold water (Barnes et al., 1990; Leonardi and Flückiger, 1989a; Mengel et al., 1989).

The permeability of cell membranes plays a crucial role in determining potential leaching fluxes from symplastic compartments. Since membrane integrity and function depend on both pH and ionic microenvironment, leaching induced changes of these parameters could affect the membrane permeability. There are only few reports dealing with this specific topic: Evans et al. (1981) showed with the help of isotope tests that cells within leaves of *Phaseolus vulgaris* were more permeable to SO_4^{2-} and H_2O and less permeable to Rb^+ after an acidic rain treatment of pH 2.7. In own experiments with leaf discs of *Fagus sylvatica* (Leonardi and Langebartels, 1990), no increased efflux rates of Dichlorofluorescein (DCF, a fluorescent tracer of membrane permeability), K^+, Ca^{2+} and Mg^{2+} were detected due to an acidic mist treatment of pH 3.5: Table 4. However, there were synergistic interactions between the acidic mist treatment and plant exposure to ozone with regard to the efflux rates of DCF, K^+ and Mg^{2+}.

Figure 4

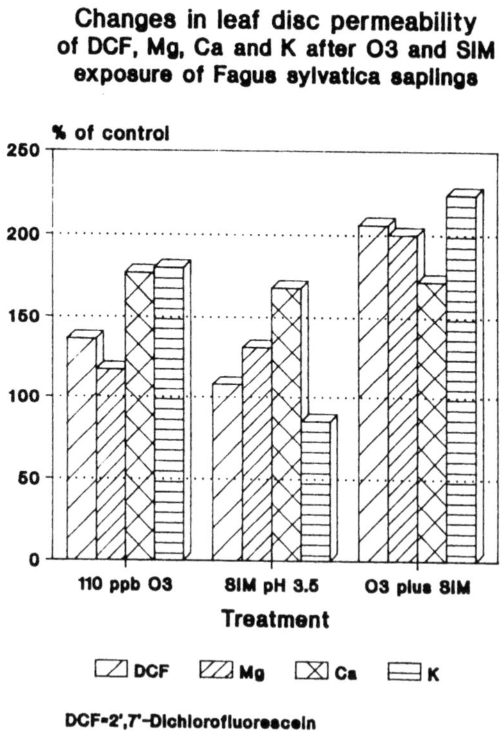

Table 4 Responses of stomatal conductance, g_s, of Fagus sylvatica exposed to simulated acidic mist.
all mean values ± s.e., g_s in mmol m^{-2} s^{-1}

i) [a] temporal variation of g_s

	1. week	2. week	3. week	4. week	5. week
Control [b] (n=12)	106.7 ± 7.3	116.5 ± 9.2	81.5 ±10.6	166.9 ± 8.5	237.0 ±11.2
pH 3.5 (n=12)	298.1 ±25.0	186.2 ±17.0	99.7 ± 8.8	80.7 ± 8.0	189.4 ±11.9
P (t-test)	<0.0001	0.01	-	0.0001	0.01

ii) [c] diurnal variation of g_s

	day-time	night-time
pH 5.0 (n=48)	100.3±11.0	21.3±1.1
pH 3.0 (n=40)	96.2± 8.2	31.7±1.2
P (t-test)	-	<0.0001

[a] Original data from Leonardi and Langebartels, 1990
[b] Control plants were not misted
[c] Original data from Leonardi and Flückiger, 1989a

The last example indicates the potential role of acid rain or induced leaching in interactions with other environmental stresses. But this vast field of interest shall not be dealt with here.

It is thought that altered ionic relations and affected membrane permeabilites may lead to several unspecific responses. As mentioned above, it remains unclear how biological effects of acidic precipitation (cf. reviews of Evans, 1984, and Tukey, 1980) are caused. For example, the reported decrease of frost hardiness of seedlings of *Picea rubens* due to acid rain treatments (Fowler et al., 1989b) was discussed to be caused by the interception of applied SO_4^{2-}, NO_3^-, H^+ and NH_4^+; the individual ion(s) responsible for the effect could not be identified. - The same conclusion is assumed to be true for most unspecific plant responses due to acid rain, including affected host plant-parasite interactions (Flückiger, 1987).

The lack of experimental evidence of effects due to cation leaching calls for more detailed studies in order to elucidate physiological and biochemical implications of the outlined indirect effects of acid rain.

CONCLUSION

Depending on the scale of interest (cell-leaf-plant-ecosystem), there are several ways to draw conclusions from the adduced evidence. The following is taken here: Leaching of cations as a regular, physiological process may develop to a plant-internal stress with multiple implications if the plant is no longer able to cope with it due to adverse environmental conditions.

Acknowledgments

I would like to thank J.D. Barnes, M. Bredemeier, L.S. Evans, J. Godt, P.M. Irving, G.H.M. Krause, E. Laitat, E. Matzner, A.R. McLeod, T. Pfirrmann, M. Riederer, G. Seufert, D.T. Tingey and D.P. Turner very much for their helpful advice and provision of literature. Also, many thanks to W. Heller and C. Langebartels for critical comments on the manuscript and to T. Kydd for editing my English.

References

ASHMORE, M.R., F.M. MCHUGH, R. MEPSTED, AND C. GARRETY. 1989. Effects of ozone and acid mist on saplings of Fagus syvlatica. In: *Air pollution and forest decline* (J.B. Bucher, and I. Bucher-Wallin, eds.). EAFV, Birmensdorf. pp. 384-386.

BARNES, J.D., D. EAMUS, AND K.A. BROWN. 1990. The infuence of ozone, acid mist and soil nutrient status on Norway spruce (Picea abies (L) Karst). I. Plant water relations. *New Phytol.*, in press.

BAKER, T.G., AND P.M. ATTIWILL. 1987. Fluxes of elements in rain passing through forest canopies in south-eastern Australia. *Biogeochemistry* 4, 27-39.

BEIER, C., AND P. GUNDERSEN. 1989. Atmospheric deposition to the edge of a spruce forest in Denmark. *Environ. Pollut.* 60, 257-271.

BERG, V.S. 1987. Plant cuticle as a barrier to acid rain penetration. In: *Effects of atmospheric pollutants on forests, wetlands and agricultural ecosystems* (T.C. Hutchinson, and K.M. Meema, eds.). Springer, Berlin. pp. 145-153.

BITTISNICH, D.J., L.O. ENTWISLE, AND T.F. NEALES. 1987. Acid-induced stomatal opening in Vicia faba L. and the role of guard cell wall elasticity. *Plant Physiol.* 85, 554-557.

BOSCH, C., E. PFANNKUCH, U. BAUM, AND K.E. REHFUESS. 1983. Über die Erkrankung der Fichte (Picea abies Karst.) in den Hochlagen des Bayerischen Waldes. *Forstw. Cbl.* 102, 167-181.

BROWN, K.A., AND T.M. ROBERTS. 1988. Effects of ozone on foliar leaching in Norway spruce (Picea abies (L.) Karst.): confounding factors dur to NO_x production during ozone generation. *Environ. Pollut.* 55, 55-73.
CARLISLE, A., A.H.F. BROWN, AND E.J. WHITE. 1966. The organic matter and nutrient elements in precipitation beneath a sessile oak (Quecus petraea) canopy. *J. Ecol.* 54, 87-98.
CRONAN, C.S., AND W.A. REINERS. 1983. Canopy processing of acidic precipitation by coniferous and hardwood forests in New England. *Oecologia* 59, 216-223.
DITTRICH, P., M. MAYER, AND M. MEUSEL. 1979. Proton-stimulated opening of stomata in relation to chloride uptake by guard cells. *Planta* 144, 305-309.
DRAAIJERS, G.P.J., W.P.M.F. IVENS, AND W. BLEUTEN. 1988. Atmospheric deposition in forest edges measured by monitoring canopy throughfall. *Water Air Soil Pollut.* 42, 129-136.
EATON, J.S., G.E. LIKENS, AND F.H. BORMANN. 1973. Throughfall and stemflow chemistry in a northern hardwood forest. *J. Ecol.* 61, 495-508.
EILAM, Y. 1965. Permeability changes in senescing tissue. *J. Exp. Bot.* 16, 614-627.
EVANS, L.S. 1984. Acidic precipitation effects on terrestrial vegetation. *Ann Rev. Phytopathol.* 22, 397-420.
EVANS, L.S., T.M. CURRY, AND K.F. LEWIN. 1981. Response of leaves of Phaseolus vulgaris L. to simulated acid rain. *New Phytol.* 88, 403-420.
FAIRFAX, J.A.W., AND N.W. LEPP. 1975. Effect of simulated 'acid rain' on cation loss from leaves. *Nature* 255, 324-325.
FBW. 1986. Forschungsbeirat Waldschäden/Luftverunreinigungen. 2. Bericht.
FINK, S. 1990. Histologische und histochemische Untersuchungen zur Nährstoffdynamik in Waldbäumen im Hinblick auf die "Neuartigen Waldschäden". III. Lokalisierungsversuche. 6. Statuskolloquium des PEF, March 6.-8.
FLÜCKIGER, W. 1987. Effects of pollution on natural communities. In: *Protection intégrée: quo vadis? - "Parasitis 86"* (V. Delucchi, ed.). Parasitis, Genf. pp. 331-349.
FLÜCKIGER, W., S. BRAUN, H. FLÜCKIGER-KELLER, S. LEONARDI, N. ASCHE, U. BÜHLER, AND M. LIER. 1986. Untersuchungen über Waldschäden in festen Buchenbeobachtungsflächen der Kantone Basel-Landschaft, Basel-Stadt, Aargau, Solothurn, Bern, Zürich und Zug. *Schweiz. Z. Forstwes.* 137, 917-1010.
FLÜCKIGER, W., S. LEONARDI, AND S. BRAUN. 1988. Air pollution effects on foliar leaching. In: *Scientific basis of forest decline symptomatology* (Cape, J.N., and P. Mathy, eds.). Air Poll. Rep. Ser., No. 15. CEC, Brussels. pp. 160-169.
FOWLER, D., J.N. CAPE, AND M.H. UNSWORTH. 1989a. Deposition of atmospheric pollutants on forests. *Phil. Trans. R. Soc. Lond. B* 324, 247-265.
FOWLER, D., J.N. CAPE, J.D. DEANS, I.D. LEITH, M.B. MURRAY, R.I. SMITH, L.J. SHEPPARD, AND M.H. UNSWORTH. 1989b. Effects of acid mist on the frost hardiness of red spruce seedlings. *New Phytol.* 113, 321-335.
GARREC, J.-P., AND C. KERFOURN. 1989. Effets des pluies acides et de l'ozone sur la pérméabilité à l'eau et aux ions de cuticules isolées. Implications dans le phénomène de dépérissement des forèts. *Environ. Exp. Bot.* 29, 215-228.
GLAVAC, V. 1987. Calcium-, Magnesium-, Kalium- und Zink-Gehalte in Blättern eines immissionsgeschädigten Rendzina-Buchenwaldes. *Allg. Forstz.* 42, 303-305.
HAINES, B., J. CHAPMAN, AND C.D. MONK. 1985a. Rates of mineral element leaching from leaves of nine plant species from a southern Appalachian forest succession subjected to simulated acid rain. *Torrey Botanical Club* 112, 258-264.
HAINES, B.L., J.A. JERNSTEDT, AND H.S. NEUFELD. 1985b. Direct foliar effects of simulated acid rain. II. Leaf surface characteristics. *New Phytol.* 99, 407-416.
HEATH, R.L., AND F.J. CASTILLO. 1988. Membrane disturbances in response to air pollutants. In: *Air pollution and plant metabolism* (S. Schulte-Hostede, N.M. Darrall, L.W. Blank, and A.R. Wellburn, eds.). Elsevier, London. pp.55-75.
HENDERSON, G.S., W.F. HARRIS, D.E. TODD, JR, AND T. GRIZZARD. 1977. Quantity and chemistry of throughfall as influenced by forest-type and season. *J. Ecol.* 65, 365-374.
HOFFMAN, W.A., S.E. LINDBERG, AND R.R. TURNER. 1980. Precipitation acidity: the role of the forest canopy in acid exchange. *J. Environ. Qual.* 9, 95-100.
HORN, R., E.-D. SCHULZE, AND R. HANTSCHEL. 1989. Nutrient balance and element cycling in a healthy and declining Norway spruce stand. In: *Air pollution and forest decline* (E.-D. Schulze, O.L. Lange, and R. Oren (eds.). Springer, Berlin. pp. 444-455.
JACOBSON, J.S. 1984. Effects of acidic aerosol, fog, mist and rain on crops and trees. *Phil. Trans. R. Soc. Lond. B* 305, 327-338.
JOHNSON, D.W., D.D. RICHTER, G.M. LOVETT, AND S.E. LINDBERG. 1985. The effect of atmospheric deposition on potassium, calcium, and magnesium cycling in two deciduous forests. *Can. J. For. Res.* 15,773-882.
JOSLIN, J.D. C. MCDUFFIE, AND P.F. BREWER. 1988. Acidic cloud water and cation loss from red spruce foliage. *Water Air Soil Pollut.* 39, 355-363.

KAUPENJOHANN, M, B.U. SCHNEIDER, R. HANTSCHEL, W. ZECH, AND R. HORN. 1988. Sulfuric acid rain treatment of Picea abies (Karst. L.): Effects on nutrient solution, throughfall chemistry, and tree nutrition. *Z. Pflanzenernähr. Bodenk.* 151, 123-126.

KEEVER, G.J., AND J.S. JACOBSON. 1983. Response of Glycine max (L.) Mevill to simulated acid rain. I. Environmental and morpholoical influences on the foliar leaching of ^{86}Rb. *Field Crop Res.* 6, 241-250.

KELLER, TH. 1988. Growth and premature leaf fall in American aspen as bioindications for ozone. *Environ. Pollut.* 52, 183-192.

KLEMM, O., U. KUHN, E. BECK, C. KATZ, R. OREN, E.-D. SCHULZE, E. STEUDLE, E. MITTERHUBER, H. PFANZ, W.M. KAISER, M. KAUPENJOHANN, B.-U. SCHNEIDER, R. HANTSCHEL, R. HORN, AND W. ZECH. 1989. Leaching and uptake of ions through above-ground Norway spruce tree parts. In: *Forest decline and air pollution* (E.-D. Schulze, O.L. Lange, and R. Oren, eds.). Springer, Berlin. pp. 210-237.

KLUMPP, A., AND R. GUDERIAN. 1990. Leaching von Magnesium, Calcium und Kalium aus immissionsbelasteten Nadeln junger Fichten (Picea abies (L.) Karst.). *Forstw. Cbl.* 109, 13-39.

KRAUSE, G.H.M., U. ARNDT, C.J. BRANDT, J. BUCHER, G. KENK, AND E. MATZNER. 1986. Forest decline in Europe: development and possible causes. *Water Air Soil Pollut.* 31, 647-668.

KRAUSE, G.H.M., AND B. PRINZ. 1989. Experimentelle Untersuchungen der LIS zur Aufklärung möglicher Ursachen der neuartigen Waldschäden. *LIS-Berichte*, Nr. 80. LIS, Essen.

LAITAT, E., R. IMPENS, AND I. RICHARDIN. 1989. Bases pour un diagnostic précoce de l'altération de l'état sanitaire d'épicéas dépérissants. In: *Man and his ecosystem* (L.J. Brasser, and W.C. Mulder, eds.). Elsevier Science Publisher, Amsterdam. pp. 219-224.

LEONARDI, S., AND C. LANGEBARTELS. 1990. Fall exposure of beech saplings (Fagus sylvatica L.) to ozone and acidic mist: effects on gas exchange and leachability. *Water Air Soil Pollut.*, in press.

LEONARDI, S., AND W. FLÜCKIGER. 1987. Short-term canopy interactions of beech trees: mineral ion leaching and absorption during rainfall. *Tree Physiol.* 3, 137-145.

LEONARDI, S., AND W. FLÜCKIGER. 1988. Der Einfluss einer durch saure Benebelung induzierten Kationenauswaschung auf die Rhizosphäre und die Pufferkapazität von Buchenkeimlingen in Nährlösungskultur. *Forstw. Cbl.* 107, 160-172.

LEONARDI, S., AND W. FLÜCKIGER. 1989a. Effects of cation leaching on mineral cycling and transpiration: investigations with beech saplings, Fagus sylvatica L. *New Phytol.* 111, 173-179.

LEONARDI, S., AND W. FLÜCKIGER. 1989b. Physiologische Auswirkungen der durch sauren Nebel induzierten Kationenauswaschung aus Buchenblättern. In: *Air pollution and forest decline* (J.B. Bucher, and I. Bucher-Wallin, eds.). EAFV, Birmensdorf. pp. 470-473.

LINDBERG, S.E., G.M. LOVETT, D.D. RICHTER, AND D.W. JOHNSON. 1986. Atmospheric deposition and canopy interactions of major ions in a forest. *Science* 231, 141-145.

LOVETT, G.M., S.E. LINDBERG, D.D. RICHTER, AND D.W. JOHNSON. 1985. The effects of acidic deposition on cation leaching from three deciduous forest canopies. *Can. J. For. Res.* 5, 1055-1060.

MACROBBIE, E.A.C. 1988. Control of ion fluxes in stomatal guard cells. *Botanica acta* 101, 140-148.

MANSFIELD, T.A., AND P.H. FREER-SMITH. 1984. The role of stomata in resistance mechanisms. In: *Gaseous pollutants and plant metabolism* (M.J. Koziol, and F.R. WHATLEY, eds.). Butterworth, London. pp. 131-146.

MATZNER, E. 1988. Der Stoffumsatz zweier Waldökosysteme im Solling. *Ber. d. Forschungszentrums Waldökosysteme/Waldsterben* 40.

MARSCHNER, H. 1983. General introduction to the mineral nutrition of plants. In: *Ency. Plant Physiol.* (new series) 15A (A. Läuchli and R.L. Bieleski, eds.). Springer, Berlin. pp. 5-60.

MCFARLANE, J.C., AND W.D. BERRY. 1974. Cation penetration through isolated leaf cuticles. *Plant Physiol.* 53, 723-727.

MECKLENBURG, R.A. AND H.B. TUKEY, JR. 1964. Influence of folial leaching on root uptake and translocation of calcium-45 to the stem and foliage of Phaseolus vulgaris. *Plant Physiol.* 39, 533-536.

MENGEL, K., H.-J. LUTZ, AND M.TH. BREININGER. 1987. Auswaschung von Nährstoffen durch sauren Nebel aus jungen intakten Fichten (Picea abies). *Z. Pflanzenernähr. Bodenk.* 150, 61-68.

MENGEL, K., A.M.R. HOGREBE, AND A. ESCH. 1989. Effect of acidig fog on needle surface and water relations of Picea abies. *Physiol. Plant.* 75, 201-207.

MEYER, J., B.U. SCHNEIDER, K.S. WERK, R. OREN, AND E.D. SCHULZE. 1988. Performance of two Picea abies (L.) Karst. stands of different stages of decline. V. Root tip and ectomycorrhiza development and their relation to aboveground and soil nutrients. *Oecologia* 77, 151-162.

MILLER, H.G. 1984. Deposition-plant-soil interactions. *Phil. Trans. R. Soc. Lond. B* 305, 339-352.

MITTERHUBER, E., H. PFANZ, AND W.M. KAISER. 1989. Leaching of solutes by the action of acidic rain: a comparison of efflux from twigs and single needles of Picea abies (L. Karst.). *Plant Cell Environ.* 12, 93-100.

NEUFELD, H.S., J.A. JERNSTEDT, AND B.L. HAINES. 1985. Direct foliar effects of simulated acid rain. I. Damage, growth and gas exchange. *New Phytol.* 99, 389-405.

OLSON, R.K., W.A. REINERS, C.S. CRONAN, AND G.E. LANG. 1981. The chemistry and flux of throughfall and stemflow in subalpine balsam fir forests. *Holarct. Ecol.* 4, 291-300.

PAOLETTI, E., R. GELLINI, AND E. BARBOLANI. 1989. Effects of acid fog and detergents on foliar leaching of cations. *Water Air Soil Pollut.* 45, 49-61.
PAPE, TH., N. VANBREEMEN, AND H. VANOEVEREN. 1989. Calcium cycling in an oak-birch woodland on soils of varying $CaCO_3$ content. *Pant Soil* 120, 253-261.
PARKER. G.G. 1983. Throughfall and stemflow in the forest nutrient cycle. *Adv. Ecol. Res.* 13, 57-133.
PFANZ, H., AND U. HEBER. 1986. Buffer capacities of leaves, leaf cells, and leaf cell organelles in relation to fluxes of potentially acidic gases. *Plant Physiol.* 81, 597-602.
PFIRRMANN, T., K.H. RUNKEL, AND H.-D. PAYER. 1990. Leaching losses from Norway spruce exposed to acidic mist and different levels of ozone during an environmental chamber experiment. Proc., Int. Congr. on forest decline research: state of knowledge and perspectives. Friedrichshafen, Oct. 2-6 1989. in press.
PITMAN, M.G. 1988. Whole plants. In: *Solute transport in plant cells and tissues* (Baker, D.A., and J.K. Hall, eds.), Longman Scientific & Technical, Harlow. pp. 346-391.
PRINZ, B., G.H.M. KRAUSE, AND H. STRATMAN. 1982. Waldschäden in der Bundesrepublik Deutschland. *LIS-Berichte*, Nr. 28. LIS, Essen.
RICHTER, D.D., D.W. JOHNSON, AND D.E. TODD. 1983. Atmospheric sulfur deposition, neutralization, and ion leaching in two deciduous forest ecosystems. *J. Environ. Qual.* 12, 263-270.
RICHTER, D.D., AND S.E. LINDBERG. 1988. Wet deposition estimates from long-term bulk and event wet-only samples of incident precipitation and throughfall. *J. Environ. Qual.* 17, 619-622.
RIEDERER, M. 1989. Simulation und Analyse des Verlustes anorganischer Ionen aus Blättern unter dem Einfluss von sauren Niederschlägen und Luftschadstoffen. Proc., 1. Statusseminar der PBWU zum Forschungsschwerpunkt Waldschäden, Feb. 27 - March 3 1989 (GSF-Bericht 6/89), 261-270.
ROBERTS, T.M., R.A. SKEFFINGTON, AND L.W. BLANK. 1989. Causes of type 1 spruce decline in Europe. *Forestry* 62, 179-222.
ROELOFS, J.G.M., A.J. KEMPERS, A.L.F.M. HOUDIJK, AND J. JANSEN. 1985. The effect of air-borne ammonium sulphate on Pinus nigra var. maritima in the Netherlands. *Plant Soil* 84, 45-56.
ROST-SIEBERT, K. 1985. Untersuchungen zur H- und Al-Ionentoxizität an Keimpflanzen von Fichte (Picea abies, Karst.) und Buche (Fagus sylvatica, L.) in Lösungskultur. *Ber. d. Forschungszentrums Waldökosysteme/Waldsterben* 12, pp. 219.
SARKAR, A.N., AND R.G. WYN JONES. 1988. Effect of rhizosphere pH on the availability and uptake of Fe, Mn and Zn. *Plant Soil* 66, 361-372.
SCHERBATSKOY, T., AND R.M. KLEIN. 1983. Response of spruce and birch foliage to leaching by acidic mists. *J. Environ. Qual.* 12, 189-195.
SCHIER, G.A. 1987. Throughfall chemistry in a red maple provenance plantation sprayed with "acid rain". *Can. J. For. Res.* 17, 660-665.
SCHÖNHERR, J., AND M.J. BUKOVAC. 1972. Penetration of stomata by liquids. *Plant Physiol.* 49, 813-819.
SCHULZE, E.-D. 1989. Air pollution and forest decline in a spruce (Picea abies) forest. *Science* 244, 776-783.
SCHULZE, E.-D., R. HANTSCHEL, K.S. WERK, AND R. HORN. 1989. Water relations of two Norway spruce stands at different stages of decline. In: *Forest decline and air pollution* (E.-D. Schulze, O.L. Lange, and R. Oren, eds.). Springer, Berlin. pp. 341-351.
SCHWARZ, A. 1985. Role of Ca^{2+} and EGTA on stomatal movements in Commelina communis L. *Plant Physiol.* 79, 1003-1005.
SEUFERT, G. 1988. Untersuchungen zum Einfluss von Luftverunreinigungen auf den wassergebundenen Stofftransport in Modellökosystemen mit jungen Waldbäumen. *Ber. d. Forschungszentrums Waldökosysteme* 44.
SEUFERT, G., AND F.H. EVERS. 1990. Schadgas-Ausschlussexperiment bei Fichte am Edelmannshof: Konzeption und erste Ergebnisse zum Stoffhaushalt. Proc., Int. congress on forest decline research: state of knowledge and perspectives. Friedrichshafen, Oct. 2 - 6 1989. in press.
SHAFER, S.R., L.F. GRAND, R.I. BRUCK, AND A.S. HEAGLE. 1985. Formation of ectomycorrhizae on Pinus taeda seedlings exposed to simulated acidic rain. *Can. J. For. Path.* 15, 66-71.
SHORTLE, W.C., AND K.T. SMITH. 1988. Aluminum-induced calcium deficiency syndrome in declining red spruce. *Science* 240, 1017-1018.
SIMMONS, G.L., AND J.M. KELLY. 1989. Influence of O_3, rainfall chemistry, and soil Mg status on growth and ectomycorrhizal colonization of Loblolly pine roots. *Water Air Soil Pollut.* 44, 159-171.
SKEFFINGTON, R.A., AND T.M. ROBERTS. 1985. The effects of ozone and acid mist on Scots pine saplings. *Oecologia* 65, 201-206.
TAMM, C.O., AND E.B. COWLING. 1977. Acidic precipitation and forest vegetation. *Water Air Soil Pollut.* 7, 503-511.
TOMLINSON, G.H. 1987. Acidic deposition, nutrient imbalance and tree decline: a commentary. In: *Effects of atmospheric pollutants on forests, wetlands and agricultural ecosystems* (Hutchinson, T.C., and K.M. Meema, eds.). Springer, Berlin. pp. 189-199.
TUKEY, H.B.,JR. 1970. The leaching of substances from plants. *Ann. Rev. Plant. Physiol.* 21, 305-321.

TUKEY, H.B.,JR. 1980. Some effects of rain and mist on plants, with implications for acid precipitation. In: *Effects of acid precipitation on terrestrial ecosystems* (Hutchinson, T.C., and M. Havas, eds.). Plenum Press, New York. pp. 141-150.

TUKEY, H.B.,JR, AND H.B. TUKEY, SR. 1963. The loss of organic and inorganic materials by leaching from leaves and other above-ground plant parts. In: *Radioisotopes in soil-plant nutrition studies.* International atomic energy agency, Vienna. pp. 289-302.

TURNER, D.P., D.T. TINGEY, AND W.E. HOGSETT. 1989. Acid fog effects on conifer seelings. In: *Air pollution and forest decline* (J.B. Bucher, and I. Bucher-Wallin, eds.). EAFV, Birmensdorf. pp. 125-129.

TURUNEN, M., AND S. HUTTUNEN. 1990. A review of the response of epicuticular wax of conifer needles to air pollution. *J. Environ. Qual.* 19, 35-45.

ULRICH, B. 1983. A concept in forest ecosystem stability and of acid deposition as driving force for destabilization. In: *Effects of accumulation of air pollutants in forest ecosystems* (B. Ulrich, and J. Pankrath, eds.). Reidel, Dordrecht. pp. 1-29.

UNSWORTH, M.H. 1984. Evaporation from forests in cloud enhances the effects of acid deposition. *Nature* 312, 262-264.

UNSWORTH, M.H., AND A. CROSSLEY. 1987. Consequences of cloud water deposition on vegetation at high elevation. In: Effects of atmospheric pollutants on forests, wetlands and agricultural ecosystems (Hutchinson, T.C., and K.M. Meema, eds.). Springer, Berlin. pp. 171-188.

VALENTINI, R., G. SCARASCIA-MUGNOZZA, P. DEANGELIS, AND R. MONACO. 1989. Short term effscts of simulated acid mist on gas exchange of Eucalyptos globulus. *Eur. J. For. Path.* 19, 200-205.

VAN STEVENINCK, R.F.M. 1965. The significance of calcium on the apparent permeability of cell membranes and the effects of substitution with other divalent ions. *Physiol. Plant.* 18, 54-69.

WALDMAN, J.D., AND M.R. HOFFMANN. 1988. Nutrient leaching from Pine needles impacted by acidic cloud water. *Water Air Soil Pollut.* 37, 193-201.

WOOD, T., AND F.H.BORMANN. 1975. Increase in foliar leaching caused by acidification of an artificial mist. *Ambio* 4, 169-171.

ZECH, H.W., AND E. POPP. 1983. Magnesiummangel, einer der Gründe für das Fichten- und Tannensterben in NO-Bayern. *Forstw. Cbl.* 102, 50-55.

ZÖTTL, H.W., AND E. MIES. 1983. Nährelementversorgung und Schadstoffbelastung von Fichtenökosystemen im Südschwarzwald unter Immissionseinfluss. *Mitt. Dtsch. Bodenkdl. Gesellsch.* 38, 429-434.

CONTRIBUTION TO FOREST DECLINE STUDIES; BIOCHEMICAL EFFECTS OF ATMOSPHERIC POLLUTION (ACID RAIN) IN *PICEA* TREES

V. R. Villanueva and Anne Santerre
Institut de Chimie des Substances Naturelles,
C.N.R.S. 91198 Gif-sur Yvette, France

ABSTRACT. We present and discuss results from a time-course cell metabolic comparative study conducted on 30 selected (healthy and diseased) forest trees (Picea abies) growing in the polluted area of the Donon Range (Vosges, France) and on trees growing: A) under controlled conditions in phytotronic chambers with and without subnecrotic levels of SO_2 (208 $\mu g.m^{-3}$) and B) in open-top chambers simulating the pollution conditions of the Donon Range. Mineral content (Al, Ba, Ca, Cu, Fe, Mg, Mn, K, Sr and Zn) in the same samples from forest trees was also determined in order to check possible correlations between deficiencies or excess of nutrients and differences in cell metabolite levels, particularly those of putrescine and polyamines which have been reported to accumulate in response to cell mineral deficiencies.
On the basis of the results obtained, a scheme is presented which summarizes a proposed sequence of metabolic events leading to the development of an adaptive metabolism which can allow trees to become healthy-resistant after passage through the diseased state. Putrescine, polyamines, tryptophan and sugars are proposed to be early markers of the biochemical and physiological effects of forest pollution.

INTRODUCTION

During the last 15 years, the loss of vitality of European forests surprised for its very rapid spread to new areas far from the industrialised zones. This phenomenom was called "forest decline", "Waldsterben" or "dépérissement forestier".
The main symptoms of the disease are necroses and discoloration, mainly chloroses and browning, together with early needle and leaf loss. Growth reductions in height and diameter are also noticeable, as well as an abnormal growth of buds. All the symptoms observed on the major forestry species, both on needle and leaf trees, are not specific of any known disease (Bonneau and Landmann, 1988). Moreover, some symptoms of injury seem to be responses to multiple stress conditions. Discussion has become much confused by the fact that over the past 15 years there have been several seasons of severe drought and extreme winters. Thus, widespread forest decline in Europe appears to be provoked by a combination of airborne pollutants and natural stresses, including climatic factors and pathogens (Hinrichsen, 1986).

In the last decades, an almost general consensus among scientists has emerged that forest dieback would not have occurred without the influence of air pollutants. The chronic impact of gaseous and dissolved pollutants, enhanced by the filtering action of tree canopies, has lowered the vitality of the forest ecosystems probably by their action on photosynthetic reactions, respiratory processes, biosynthesis and growth. Consideration of stand age and vigor is also necessary, as is an understanding of the dynamics of tree defenses and the differences between species and ages in defensive capacity (Loehle, 1988).

It is clear that an increased acidity of rain water can lead to accelerated loss of certain base cations, but there is insufficient reasons to believe that this alone could lead to declining health, for cations should still be retained within the nutrient cycles. There is nevertheless the possibility that the extreme acidity that can be occasionally encountered in cloud water may be such that permanent leaf damage, and consequent nutrient loss may result. Also an associated reduction in the ability of roots to recapture nutrients can be another possibility (Miller, 1986).

The present state of knowledge and theories on forest damage and eventual death of the trees observed in air polluted areas are at the basis of the French DEFORPA Program (Dépérissement des Forêts Attribué à la Pollution Atmosphérique). This program was established in 1984, following the observation of significant needle loss in conifer stands in the Vosges mountains.

Current research in this program includes analytical work aimed at detecting metabolic modifications ascribable to the mechanism(s) of action of pollutants on leaf metabolism. We are involved in this program, with a project dealing with an analytical biochemical study of cellular metabolism comparing needles from healthy and diseased *Picea* trees from the polluted area of the Donon Range (Vosges, France).

The primary aim of this work was to detect both qualitative and quantitative differences in the contents of cell metabolites that could then be used as biochemical markers of the physiological state for trees living in polluted areas, prior to any visible damage. Secondly, on the basis of the observed differences, we wanted to identify metabolic alterations caused by the effects of pollution in order to study, in a deeper and more detailed manner, the pathways affected or modified.

This paper presents a time-course biochemical comparative study (cell metabolites and mineral nutrients) of almost a year's metabolism cycle on 30 selected forest trees (*Picea*) as well as experimental studies under controlled conditions (Phytotron, Gif-sur-Yvette and open-top chambers, Montardon) using young *Picea* trees.

The results are presented and discussed in the light of the biochemical knowledge of the biological role of the different cell metabolites and mineral nutrients, their possible interrelationships, and the possible adaptive mechanism(s) following cell response to stress attack.

SAMPLING AND EXPERIMENTAL CONDITIONS

SAMPLING TECHNIQUES

Forest trees: With the help of botanists and foresters, 30 *Picea* trees (approximately 20 yr old) from the Donon Range were carefully selected and labelled for this study : 10 apparently healthy, 10 diseased and 10 very diseased. 100 needles from the third whorl of these trees were collected randomly and immediately frozen in dry ice for transport; seven samplings were carried out from March to August 1986. Needles of three different ages were collected : two years old and one year old needles were collected seven times, and needles of the year were only available for the four last samplings.

Phytotronic conditions: Young *Picea* trees (5 yr old) were grown in experimental chambers as described previously (Pavlides, 1983) in the presence or in the absence of subnecrotic levels of SO_2 (208 $\mu g.m^{-3}$). Once a month, 10 needles from each young tree were collected.

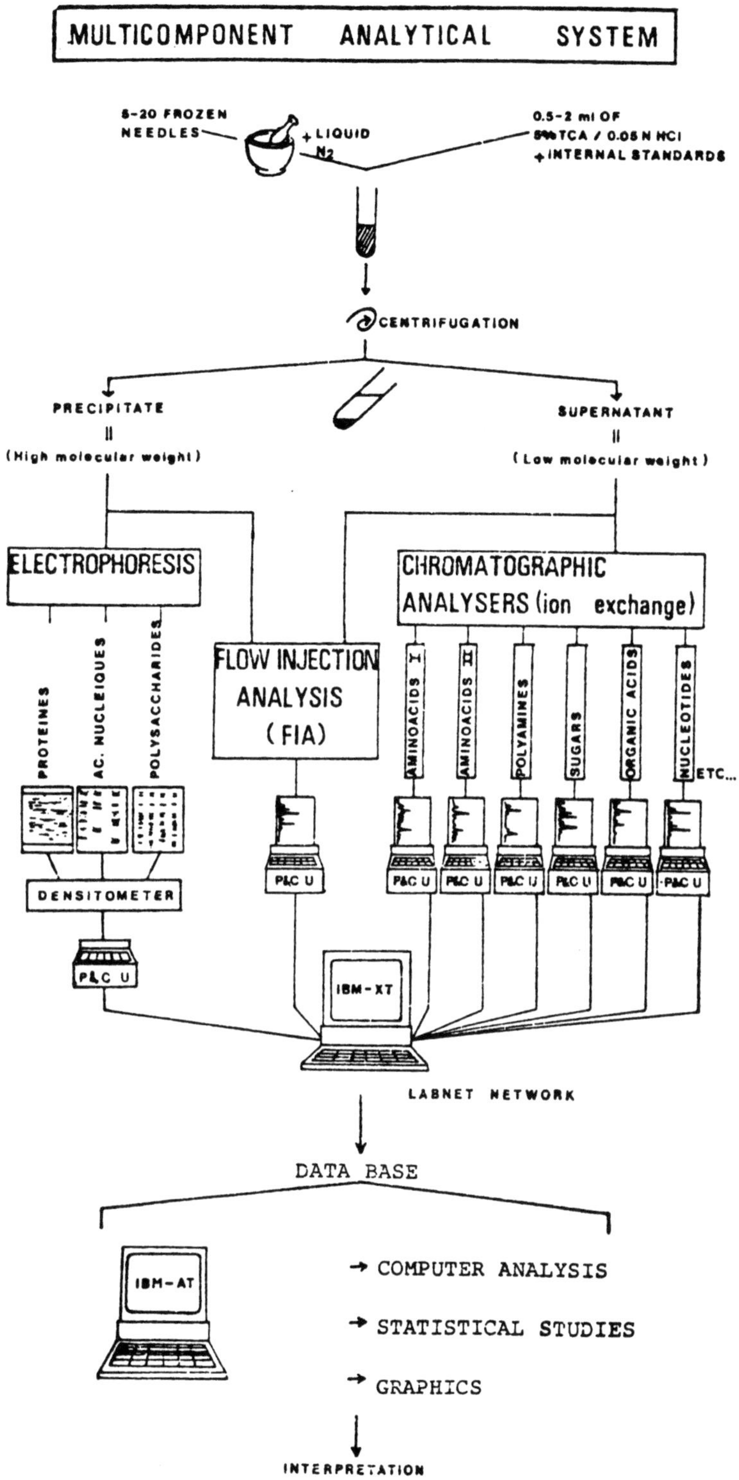

Fig. 1. Schematic representation of the computer-assisted multianalytical system. The event external control cable of a 4200 Integrator (Spectra-Physics) is used for the automation of chromatographs. The same Integrator serves for the quantitation of the chromatographic analysis. The whole system is interconnected with the Labnet network from Spectra-Physics. P&C U: Programmer and calculator unit.

Open top chambers: In addition, needle samples from three different Picea clones ("Istebna", "Gérardmer" and "Lac de Constance") growing in open-top chambers were collected. Each of these open-top chambers, situated at Montardon (South of France) contained 15 young Picea trees (5 yr old). These chambers, as already described (Bonte et al., 1986) define five different conditions : a) air; b) filtered air; c) filtered air + SO_2; d) filtered air + O_3; e) filtered air + (SO_2 + O_3); trees grown out of the chambers were also entered in this study. The levels of SO_2 and O_3 injected into the chambers were those mesured in the polluted area of the Donon Range. Samples were taken once a month from April 1987 to March 1988.

SAMPLE PREPARATION AND ANALYSIS

a) Cell Metabolites

After being manually powdered in a mortar with liquid nitrogen, needles were extracted with a solution of 5% trichloroacetic acid (TCA) in 0.05M HCl, containing appropriate internal quantitation standards. After centrifugation, aliquots of the crude supernatant were used directly for the analysis of low molecular weight components, and the precipitate, after washing and solubilization, was used for the analysis of high molecular weight components.

Likewise, several analyses were carried out on needles from diseased trees. The needles were separated only by their color: green, yellow or brown.

The computer-assisted multicomponent analytical system previously described (Villanueva et al., 1987) was employed. Aliquots of the supernatant, obtained as described above, were injected directly into ion exchange chromatographic analyzers for the determination of polyamines, aminoacids, sugars ... etc. The precipitated fraction was analyzed by electrophoresis (Figure 1).

Protein concentration was determined using the method of Bradford (1976) with bovine serum albumin as a standard and total sugar by the orcinol method (Montreuil and Spik, 1963).

b) Mineral Nutrients

Before mineralisation, needles were powdered and dried under vacuum in an exsiccator (on KOH) to constant weight.

Considering the large number of samples to prepare, and the small quantities of material for each sample, mineralisation was performed by concentrated nitric acid digestion (Santerre et al., 1990) in a microwave apparatus (Microdigest 300, Prolabo, France): 50 mg of dry needle powder was digested in 4 ml of nitric acid (15 min at 20% of maximum Microdigest 300 power), and the liquid residue was transfered and ajusted to 10 ml with permuted water. ICP-AES analysis (Mermet et al., 1988) was then performed directly, without dilution, on a Philips PV8060 system equiped with a polychromator and associated computer modules (Philips, Bobigny, France), permitting rapid and simultaneous measurements of up to 32 elements in solution. Ten elements were finally accurately measured in our samples of Picea needles, namely Al, Ba, Ca, Cu, Fe, Mg, Mn, K, Sr, Zn. The other twenty-two elements were present at non-reliable levels (near detection limits) or could not be detected.

Calibration was achieved with an ICP multi-element standard solution of 19 elements (Merck), to which known amounts of five other elements (K, Na, Ca, Mg and Hg) were added to complete it with all the desired standard elements. Mo could not be added because it provoked precipitation of other elements. Analytical data result from the average of 3 integrations.

COMPARATIVE STUDIES BETWEEN HEALTHY AND DISEASED PICEA TREES FROM THE POLLUTED AREA OF THE DONON RANGE.

A- CELL METABOLITES

Each of the 540 samples entering this study was processed according to our methodology (see Figure 1) in order to fractionate it into high and low molecular weight material. The use of our computer-assisted analytical system, based on electrophoresis and chromatography, has allowed us to analyze a large number of cell components: proteins, nucleic acids, polysaccharides, aminoacids, sugars, carboxylic acids, polyamines and nucleotides.

Analysis by Electrophoresis:

Macromolecular material (proteins and nucleic acids) was analysed by this technique. Results clearly showed marked qualitative and quantitative differences between samples from healthy and diseased trees. This preliminary analysis indicates that a more complete and detailed study by electrophoresis on healthy and diseased tree samples may well provide considerably useful information which could correlate with the physiological state of the trees. However we have not yet attempted this as electrophoresis, being a very elaborated technique, would have consumed most of our efforts and time in application to the study mentioned. The initial aim of our work was rather to analyse the largest possible number of metabolites to determine which could give indications so as to be used to detect metabolic modifications ascribable to the biochemical and physiological effects of pollution.

Chromatographic analysis: Analysis by "family" of metabolites.

Analysis of sugars:

Figure 2 shows the mean content and evolution profiles of total acid-soluble sugars for the healthy and diseased physiological groups

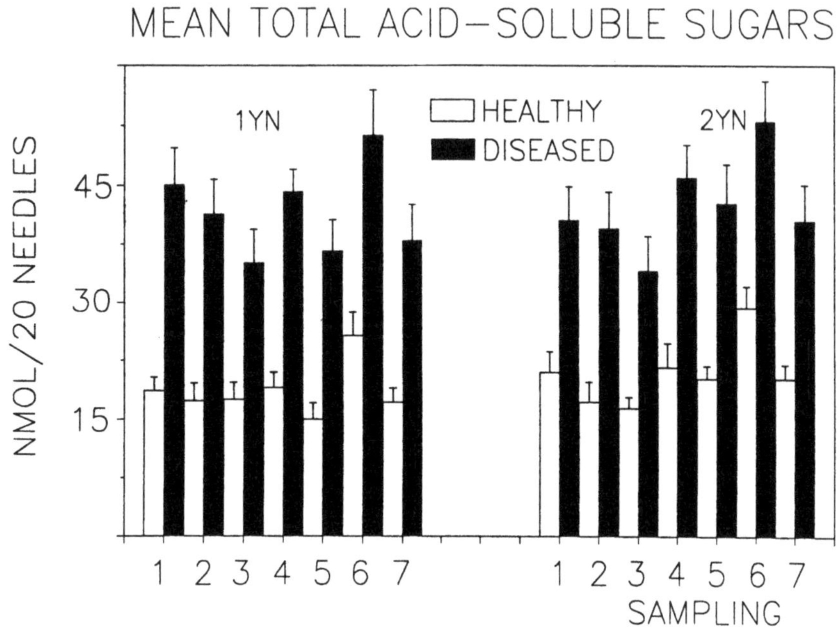

Fig. 2. Mean (10 Picea trees /classs, + - SEM) content evolution of total acid-soluble sugars in healthy and diseased trees. Needles aged 1 (1YN) and 2 years old (2YN) were collected seven times (March to August). Note that diseased trees contain higher levels of sugars than healthy ones.

of trees, while Figure 3 shows that of the total acid-insoluble sugars. These differences in sugar metabolism between healthy and diseased trees have also been observed in the case of Scots pine exposed to fluorine and SO_2 pollution (Mejnartowicz and Lukasiak, 1985). These authors suggested that the increased amount of monosaccharides in the needles of diseased trees may contribute to their greater injury by gaseous emissions, as an increase in the level of fructose and glucose increases the turgor of cells, and in turn contributes to the opening of the stomata through which pollutants can enter and disturb the carbohydrate metabolism. On the contrary, the smaller amounts of glucose and fructose in the seemingly healthy trees may lower the osmotic potential of cells in the needles, which could contribute to the closure of the stomata and a reduction in the amount of pollutants entering the leaves.

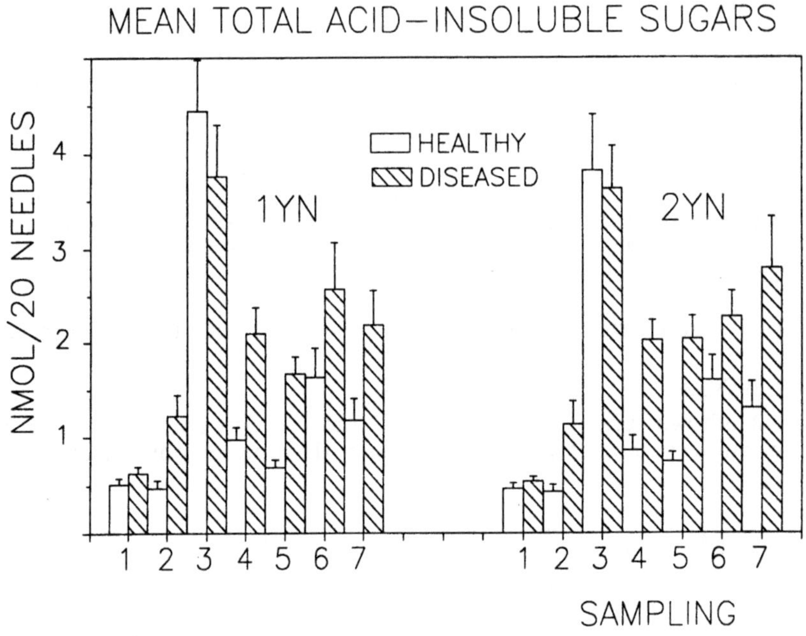

Fig. 3. Mean (10 Picea trees/class, + - SEM) content evolution profile of acid-insoluble sugars in healthy and diseased trees. See fig. 2 for symbols and commentaries.

Analysis of aminoacids:

Large differences in the content of three aminoacids : histidine, tryptophan and arginine, were also observed in our study. Figure 4 illustrates the case of tryptophan. These kinetic profiles suggest an abnormal utilization by diseased trees of this compound which is known to be the precursor of indole-acetic acid, a plant growth substance (Luckner, 1984).

There is also an important difference between the kinetic profiles of arginine in healthy and diseased Picea trees. For instance, as can be seen in Figure 5, profiles in one year old needles, of healthy trees (10 individuals) are rather compact and homogeneous, while in the case of diseased trees (10 individuals) a large dispersion of the individual profiles is observed, probably due to differences in the nature and degree of their diseased state.

Analysis of putrescine and polyamines:

Probably the most spectacular results from this work concern the content and evolution of putrescine (Figure 6) and polyamines

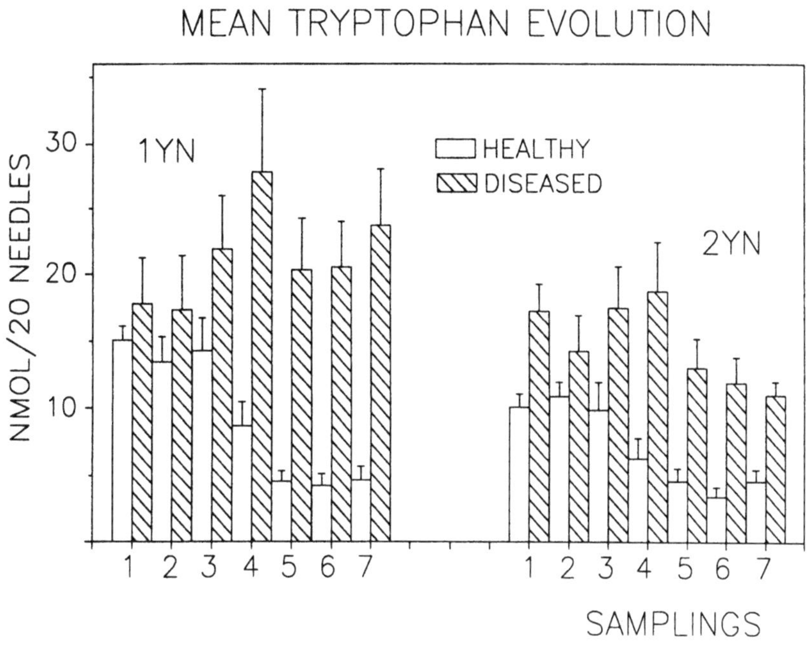

Fig. 4. Mean tryptophan values (+-SEM) for healthy and diseased trees in 1 (1YN) and 2 (2YN) years old needles. Note that while healthy trees seem to use their tryptophan content (particularly during june to august), at the contrary, diseased trees seem to accumulate it.

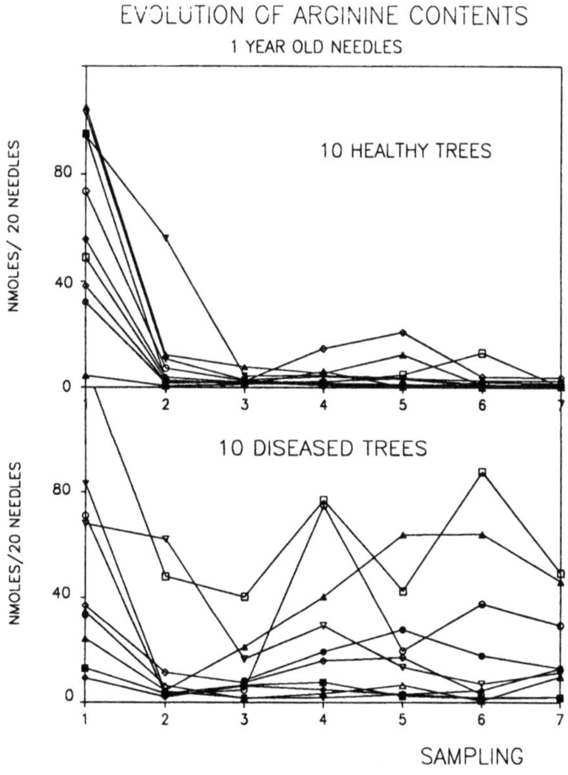

Fig. 5. Individual arginine content profiles of one year old needles from healthy and diseased trees. While profiles of healthy trees are rather homogeneous, in the case of diseased trees a large dispersion is observed, probably due to differences in the nature and degree of their diseased state.

(Figure 7), biogenetically related compounds derived from arginine. The results clearly show marked differences in the levels and in the evolution of the content of these compounds according to the physiological state of the trees.

Other cell metabolites also showed differences in their level and evolution according to the physiological state of the group of trees but these differences were less marked.

Analysis of needles according to their color:

We have checked the influence of the color of the needles (yellow, green or brown) during the samplings of the diseased trees,

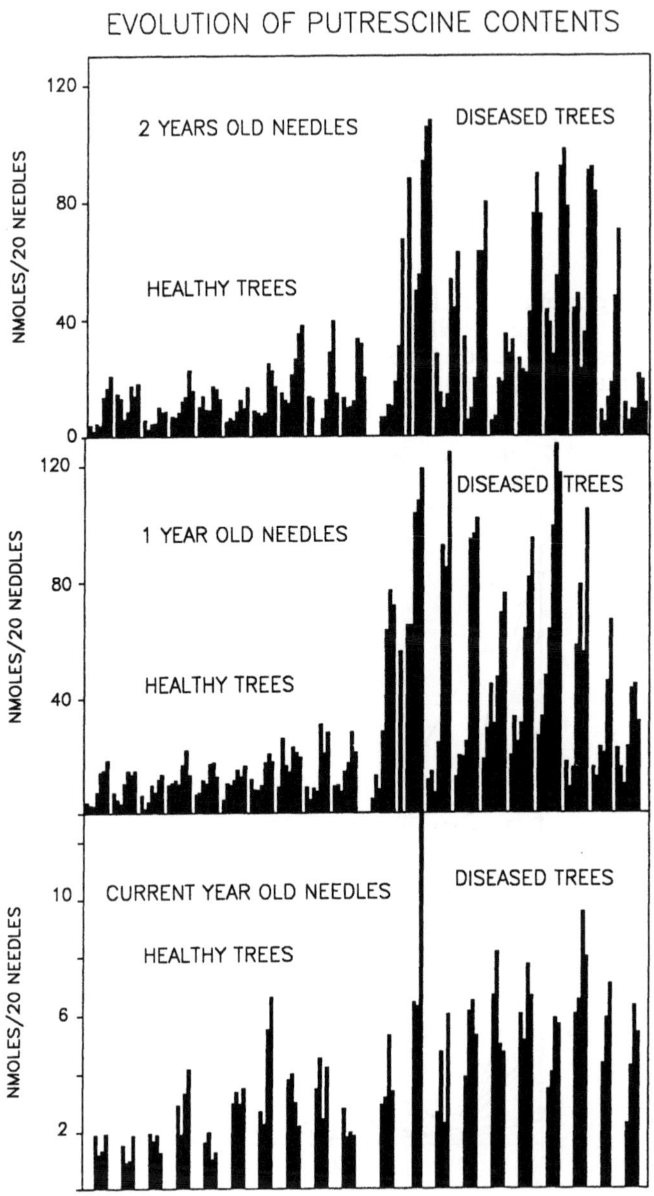

Fig. 6. Comparative histograms showing the difference of putrescine levels for the ten healthy and the ten diseased Picea trees studied. Seven samplings (March to August) were collected for 1 and 2 years old needles and four samplings (June to August) for current year old needles.

for its contribution to the final content and evolution of the different metabolites considered. During the fifth and the sixth samplings on the selected Picea trees, needles from the diseased trees group were analysed separately according to their colour. Tryptophan, histidine, sugars, putrescine, polyamines and protein contents were determined in green, yellow and brown neddles. Yellow needles, in the cases of putrescine, total polyamines, histidine and tryptophane, contain slighly higher levels of these metabolites as compared to the green needles from the same sample. For sugars, contents of yellow and green needles are similar. Brown needles contain variable amounts of the metabolites considered . In any case, independently of the colour of the needles, the physiological state of the tree appears to be important and responsible for the differing contents of the cell

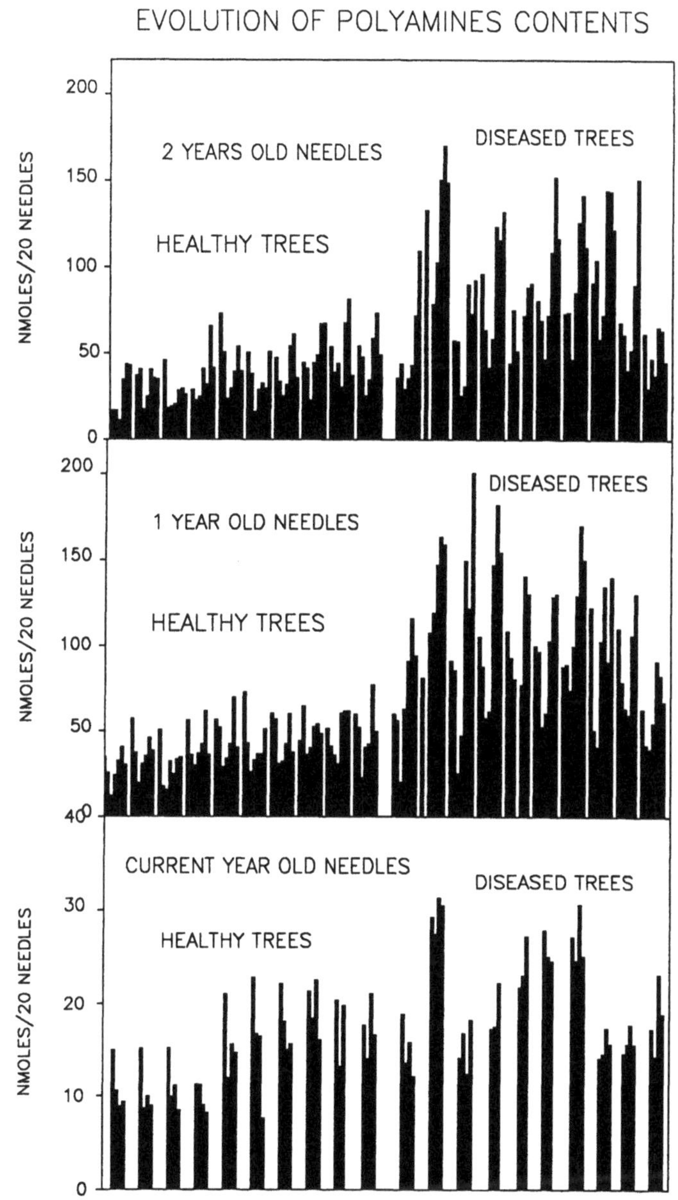

Fig. 7. Comparative histograms showing the difference of polyamine levels for the ten healthy and the ten diseased Picea trees studied. Seven samplings (March to August) were collected for 1 and 2 years old needles and four samplings (June to August) for current year old needles.

metabolites. Moreover, as can be seen in Figure 8, there are large differences in the metabolite contents of green needles whether they come from a diseased tree or from a healthy one.

B- MINERAL NUTRIENTS

In connection with the above results concerning cell metabolites, particularly putrescine and polyamines, we wanted to measure if deficiencies or excesses of some nutrients occurred in needles from diseasing trees.

Figures 9 and 10 show the mean (ten trees/class) kinetic (7 samplings from March to August for one and two years old neeldes, 4 for current year needles) nutrient profiles of the *Picea* trees studied. Levels of ten mineral constituents (Al, Ba, Ca, Cu, Fe, K, Mg, Mn, Sr, Zn) are presented as a function of the physiological group (healthy, diseased) or needle age (current year needles, one year old and two years old). Although the differences are not always clear-cut the following general tendencies can be observed:

1 - Nutrients whose content evolution profile changes according to needle age:

a) Al, Ba, Ca, Sr, Mn and Fe levels increase with needle age.
b) Mg, K and Cu levels decrease.
c) Zn levels increase in healthy trees and decrease in diseased trees.

2 - Nutrients whose content evolution profile changes according to the physiological state of the trees:

a) Ba and Sr are higher in diseased trees.
b) Mg is higher in healthy trees.

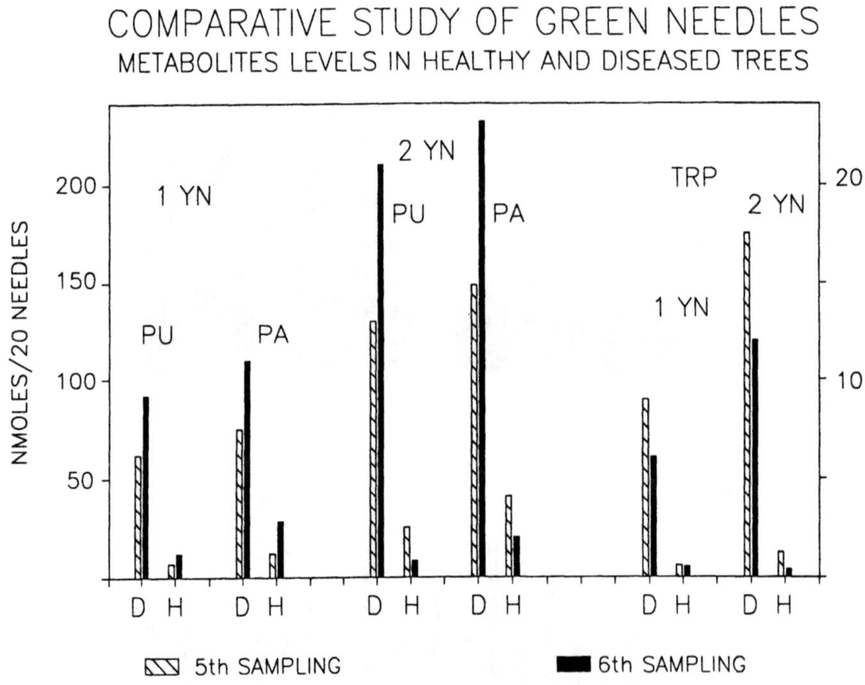

Fig. 8. Comparative histograms of putrescine (PU), Polyamines (PA), and tryptophan (TRP) content in 1 (1 YN) and 2 (2 YN) years old GREEN needles from healthy and diseased trees. This study was done during the 5th and 6th samplings. Note the large difference of metabolites contents between green needles whether they come from a diseased or a healthy trees. Thus, even if a diseased tree is green like a healthy one, its needle metabolites content should allow to diagnostic its physiological state.

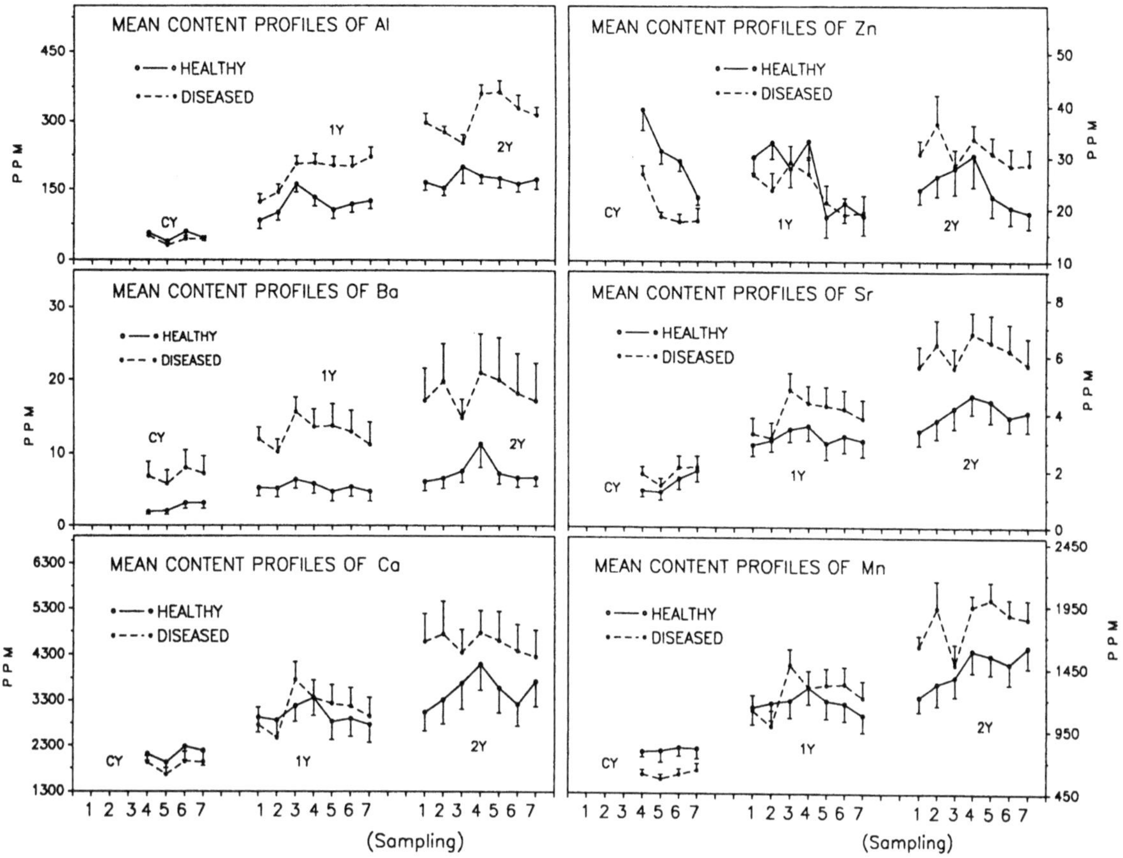

Fig. 9 and 10. Comparative time-course mineral content of current year (CY), 1 year (1Y) and 2 years (2Y) old needles from healthy and diseased polluted forest trees. A gradient, in the levels of the nutrients, can be observed from current year to 2 years old needles according to the physiological sate of the tree.

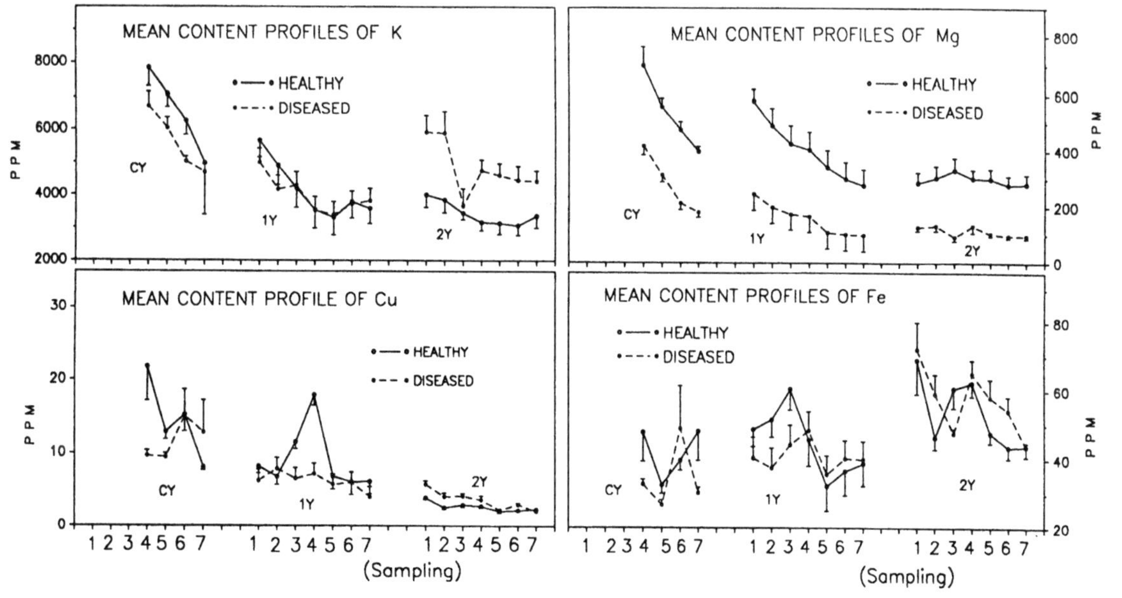

3 - Nutrients whose content evolution profile is reversed according both needle age and physiological state of the trees:

a) Al, Ca, and Mn : Diseased trees contain lower levels of these elements in the youngest needles. With the age, these minerals accumulate more rapidly in the diseased trees than in the healthy ones to finally clearly show higher levels in diseased trees.

Although less clear-cut, Fe contents seem to follow the same tendency.

b) K and Zn show a particular behavior. In the youngest needles, healthy trees contain higher levels than diseased ones.

In older needles the contrary is observed but not for the same reason. For K, in both healthy and diseased trees, levels decrease but with different rates, so that finally diseased trees show higher contents than healthy ones. In the case of Zn, while diseased trees increase their levels with needle age, healthy trees decrease them. The Cu pattern profile, although less clear, resembles that of K.

Third or fourth samplings (in June) were key points for most of the constituents studied (maxima or minima). This corresponds to a period of very active and fast growth of shoots and needles.

Al, Ba, Ca, Mn and Sr were present in needles from diseased trees in higher levels than in seemingly healthy trees. On the contrary, Mg levels were quite high in healthy trees in comparison with the levels in diseased ones. An obvious reason is that green healthy needles contain higher levels of chlorophyll, a magnesium complex. This could partially explain the differences observed. Profiles of the contents of aluminium are higher in needles from diseased trees than in healty ones. Al accumulation can also be detected in current year needles and increases regularly. It is known that high levels of Al are toxic for the cells, for example by interaction with calmodulin-dependent enzymes (Williamson and Ashley, 1982). Thus, high levels of Al in diseased trees could be associated with disease symptoms. Ulrich (1986), proposed that acid rains may acidify the soil and develop a toxicity related to aluminium solubilisation in soil solution. Al can be absorbed by plants to the leaves before roots are destroyed.

We observe that the evolution of contents of Sr and Ba is parallel. Previous results, on the behaviour of these divalent cations together with Ca, have indicated that they are competively absorbed by plants (Hutchin and Vaughanb,1968; Miller et al.,1970)

Measurements of K, Mg, Ca, Zn, Ba and Mn in needles of different branch whorls for a given needle year and of different needle years for a given branch whorl, of healthy and diseased spruce trees from Southwest Germany have been reported. Results are similar to ours and showed that spruce diseases are not necessarily associated with a malnutrition with this elements. Uptake and translocation of elements by the roots or phloemmobility also do not appear to be causally related to the expression of disease symptoms (Krivan and Schaldach, 1985). Mengel et al. (1986) showed that under experimental conditions of seven weeks, the amount of cations (Mg^{2+}, Mn^{2+}, Zn^{2+}, Ca^{2+}, K^+ were studied) leached by acid fog (pH=2.75) had effectively higher decrease rates than with normal fog (pH=5), but the loss of nutrients was low in comparison with the normal content of these cations in spruce needles. In fact cation concentrations found in needles after this seven week experiment rather increased than decreased.

POLYAMINE LEVELS AND MINERAL DEFICIENCIES OR EXCESS

It has been reported (Richard and Coleman, 1952; Sinclair, 1967; Priebe et al., 1978; Smith, 1983 and 1985) that stress, caused especially by K or Mg deficiencies or excess of several other inorganic elements, can induce in higher plants a universal response of increase of putrescine concentration. Putrescine is synthetised in K-deficient plant leaves to take the place of this cation which is the one required in largest amount in cells for normal growth. In the present study, measurements of levels did not lead to the expected results as diseased trees showed lower levels of K, as compared to healthy ones only in current year needles but no longer in one and two years old needles in which diseased trees showed higher levels of K than healthy trees. Then, the putrescine accumulation observed in diseased trees (Villanueva and Santerre, 1989; Santerre et al., 1990) does not appear to be linked with a K-deficiency.

We have already said that Mg-deficiencies and needles yellowing where probably linked through chlorophyll destruction. Mg normal concentration in cells is important but not for the same reason as K, whose deficiency can influence changes in cellular pH and ionic equilibrium. Smith (1983) has suggested that Mg is probably involved in nucleic acid synthesis and that polyamines may substitute for Mg^{2+} in this synthesis when it is deficient.

It is also likely that putrescine may be formed in higher plants in response to reduced pH of the medium to help cells to maintain their ionic equilibrium and cell homeostasis (Morel et al., 1980). Increase of arginine decarboxylase activity in acid feeding experiments suggested that putrescine may be formed in response to a reduction in pH (Young and Galston, 1983). In this hypothesis putrescine formation was therefore, rather a system for maintenance of a constant cytoplasmic pH (Villanueva et al., 1987; Villanueva and Santerre, 1989).

At this stage, two aminoacids, arginine and methionine, appeared to play a very important role as interrelated precursors of metabolic compounds whose biochemical and physiological roles are well recognized (see Figure 11). For instance, in plants, arginine is the main storage form of soluble N (Durzan, 1968) ; it is also, together with methionine, the biogenetic precursor of polyamines, which play an essential role in the macromolecular metabolism and in the different processes leading to cell division, cell growth and cell differentiation ; methionine, after activation by ATP, gives S-adenosyl methionine (SAM) which is the universal methyl donor for transmethylation reactions, the aminopropyl donor for polyamines biosynthesis and the precursor of the plant hormone : ethylene. Since ethylene tends to be a senescence inducer (Adams and Yang, 1979) and polyamines to be senescence inhibitors (Kaur-Sawhney and Galston, 1979), the fate of SAM would seem to be crucial. Once decarboxylated, SAM is committed to the polyamine pathway. In this connection, it is of importance that in etiolated pea epicotyls, the conversion of Pr-Phytochrome to Pfr immediately inhibits the formation of ethylene (Goeschl et al., 1967) and promotes the formation of polyamines through elevation of the activity of arginine decarboxylase (Dai and Galston, 1981). This interesting system abounds with feedback controls, since polyamines inhibit ethylene formation (Apelbaum et al., 1981), and ethylene inhibits polyamine formation. Thus each pathway, once initiated, tends to shut off the other (Galston, 1983).

Of relevant importance are the reports that stress ethylene formation determines plant sensitivity to ozone (Mehlhorn and Wellburn, 1987) and that plant defense genes are regulated by ethylene (Ecker and Davis, 1987).

A detailed differential study of the metabolic pathways of these two aminoacids, between healthy and diseased trees, would certainly help in understanding the multiple steps followed by a cell to adapt its metabolism for survival in stress conditions.

COMPARATIVE METABOLIC STUDIES UNDER CONTROLLED CONDITIONS

A- PHYTOTRON, Gif-sur-Yvette

Preliminary work, using two groups of *Picea* trees (5 yr old) was carried out under controlled conditions. One group was in a chamber with filtered air and the other in a chamber with filtered air supplemented with SO_2 in subnecrotic concentrations (208 $\mu g.m^{-3}$). Once a month, for a period of 4 months, similar analyses as those made in

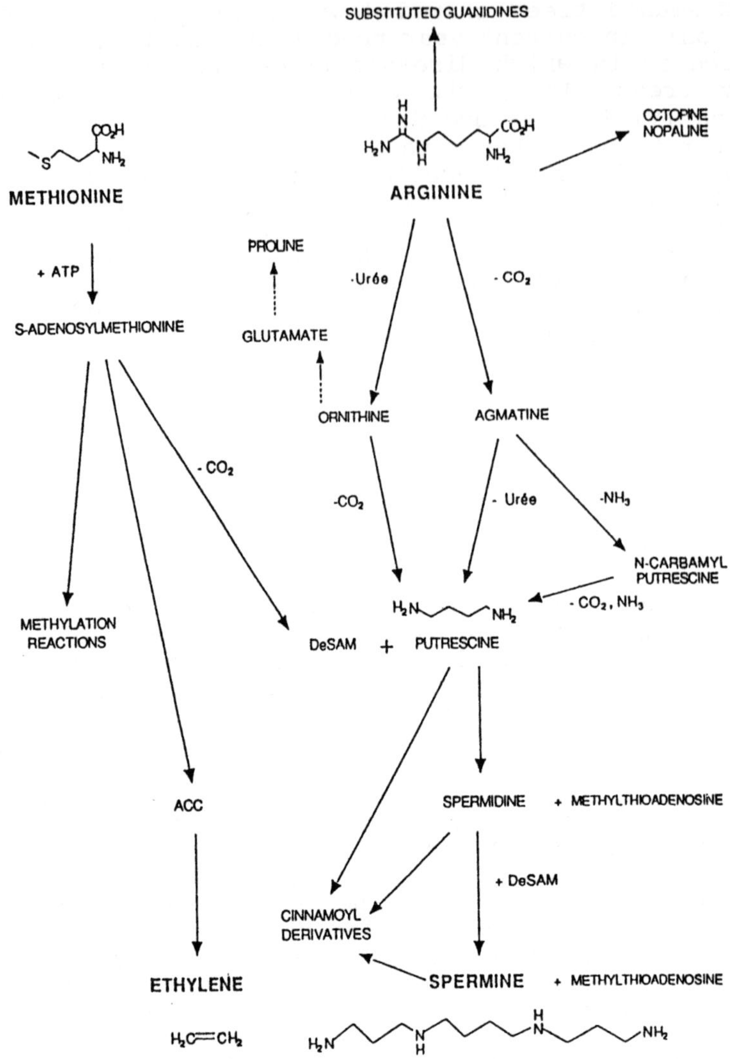

Fig. 11. The interrelated metabolic steps of Methionine and Arginine lead, through important intermediates, to two kind of very different compounds: polyamines and ethylene. Polyamines intervene in processes engaged in cell division, cell grow and cell differentiation and ethylene in maturation and senescence processes.

the case of the forest trees were carried out. As exemplified by the kinetic evolution of free sugars (Figure 12, two sets of experiments), the trees exposed to SO_2 present higher metabolic levels at the beginning (as the diseased forest trees) which then start to decrease and finally (four months later) they appear to contain lower metabolic levels as compared to the tree growing in absence of SO_2 (as healthy-resistant forest trees).

B- OPEN TOP CHAMBERS: Montardon

A similar study as that described above was carried out using open top chambers situated at Montardon (South of France). In these chambers, the levels of SO_2 and O_3 introduced were the same as those daily measured in the Donon Range (Vosges, France). Young *Picea* trees (5 yr old) were also used here. Twelve samples from April 1987 to March 1988 were collected and studied as in the case of forest trees.

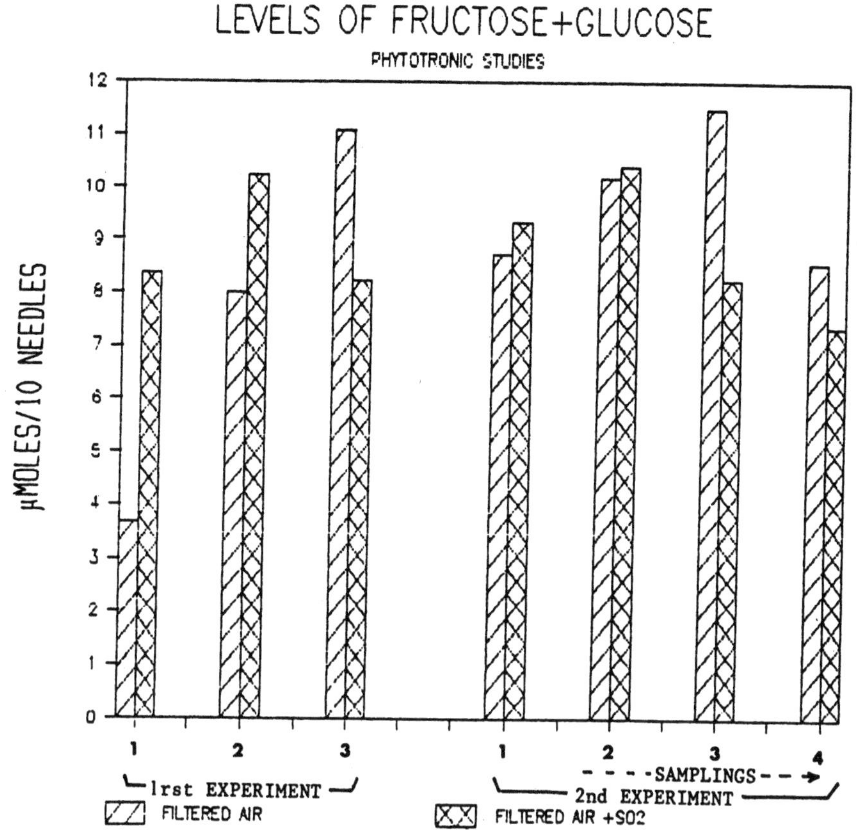

Fig. 12. Evolution of glucose and fructose content in 1 year old needles from young *Picea* trees (five years old) growing in controlled conditions (Phytotron, Gif-sur-Yvette) in the presence or absence of subnecrotic levels of SO_2 (208 $\mu g.m^{-3}$). Note that trees in presence of SO_2 raise, at the begining, their sugar content and levels are higher as compared to those growing in filtered air. Thereafter sugar content seems to start to decrease to finally show lower levels. A similar process could take place in forest trees submited to chronic pollution (see also figs. 13 and 14).

Unexpectedly, the low levels of SO_2 and O_3 in the Donon Range during the period of the study and thus used in the experimental work of open-top chambers at Montardon, was insufficient to induce marked metabolic differences among the different conditions. Only a first point, in April, consecutively to a relatively high SO_2 atmospheric pollution in March showed metabolic differences (see Figure 13). We also clearly observed (and this for the three clones "Lac de Constance" ,"Istebna" and "Gérardmer") a difference in the rate of the kinetic cellular metabolism according to the development of the trees among them.

ADAPTIVE METABOLISM AND HEALTHY-RESISTANT TREES

One of the general findings of this study is that diseased trees contain higher levels of all the metabolites studied than healthy trees. This could be the result of a metabolic response of the cell to counteract the biochemical and physiological effects of the stress(es).

On the basis of our experimental work and the different biochemical observations, we have elaborated the scheme shown in

Figure 14 (Villanueva , 1988), which summarizes a proposed sequence of events in the cell exposed to pollution. The scheme takes into account : a) the heterogeneity of the degree of sensitivity of the tree population to the stress(es) and b) the passage of the healthy trees through the "diseased" state (primary response) before becoming healthy-resistant. This means that, under stress, the cell reacts

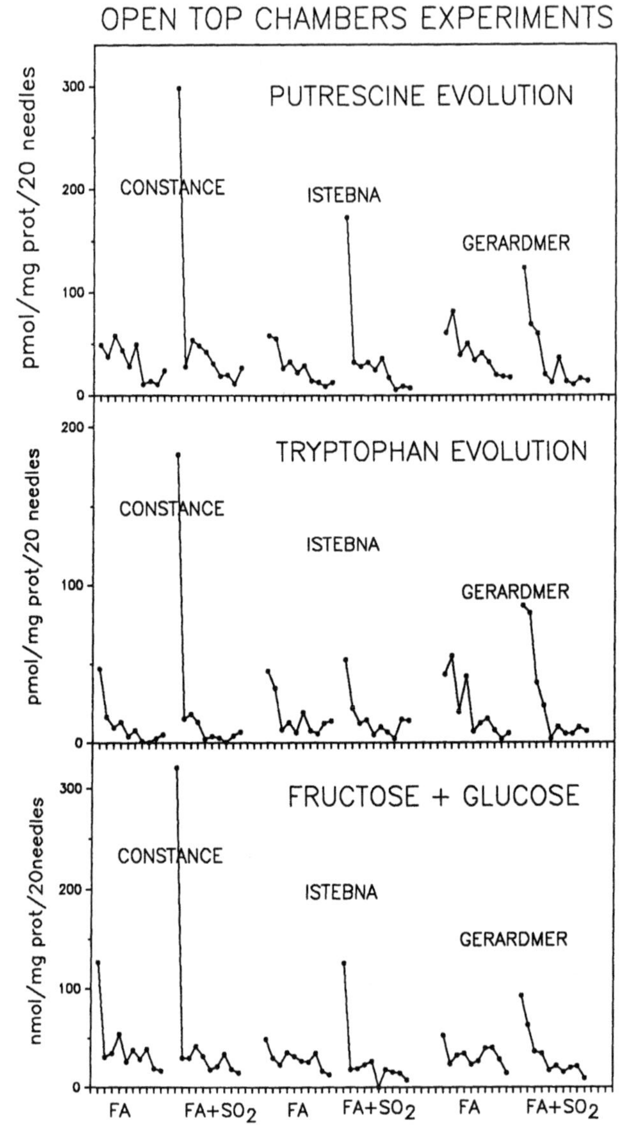

Fig. 13. Compared profile evolutions of putrescine, tryptophan and sugars in young _Picea_ trees growing in filtered air (FA) and in filtered air added with SO_2 ($FA+SO_2$) in open-top chambers (OTC). The OTC were situated at Montardon (South of France) and reproduced the real atmospheric polluted conditions mesured at the Donon range (near Strasbourg). Unexpectedly, during the period of this experiment there was only, at the beginning (Fev/March), a relatively high SO_2 level of pollution at the Donon range and also in the OTC. This could provoque the response observed for the first point in the profile level of putrescine, tryptophan and sugars. The rest of the year the levels of pollutants were to low to induce marked metabolic changes. Noteworthy is also the difference in the kinetic response according to the clone tree (Constance, Istebna and Gerardmer).

immediately (short-term kinetic response) to defend itself by trying to counteract the initial harmful effects. The intracellular level of the different metabolites will rise (as observed in diseased trees and also in the experiments under controlled conditions at the Phytotron and in open top chambers, Figures 12 and 13). At this stage the effects are not always necessarily visible on plants, but longer

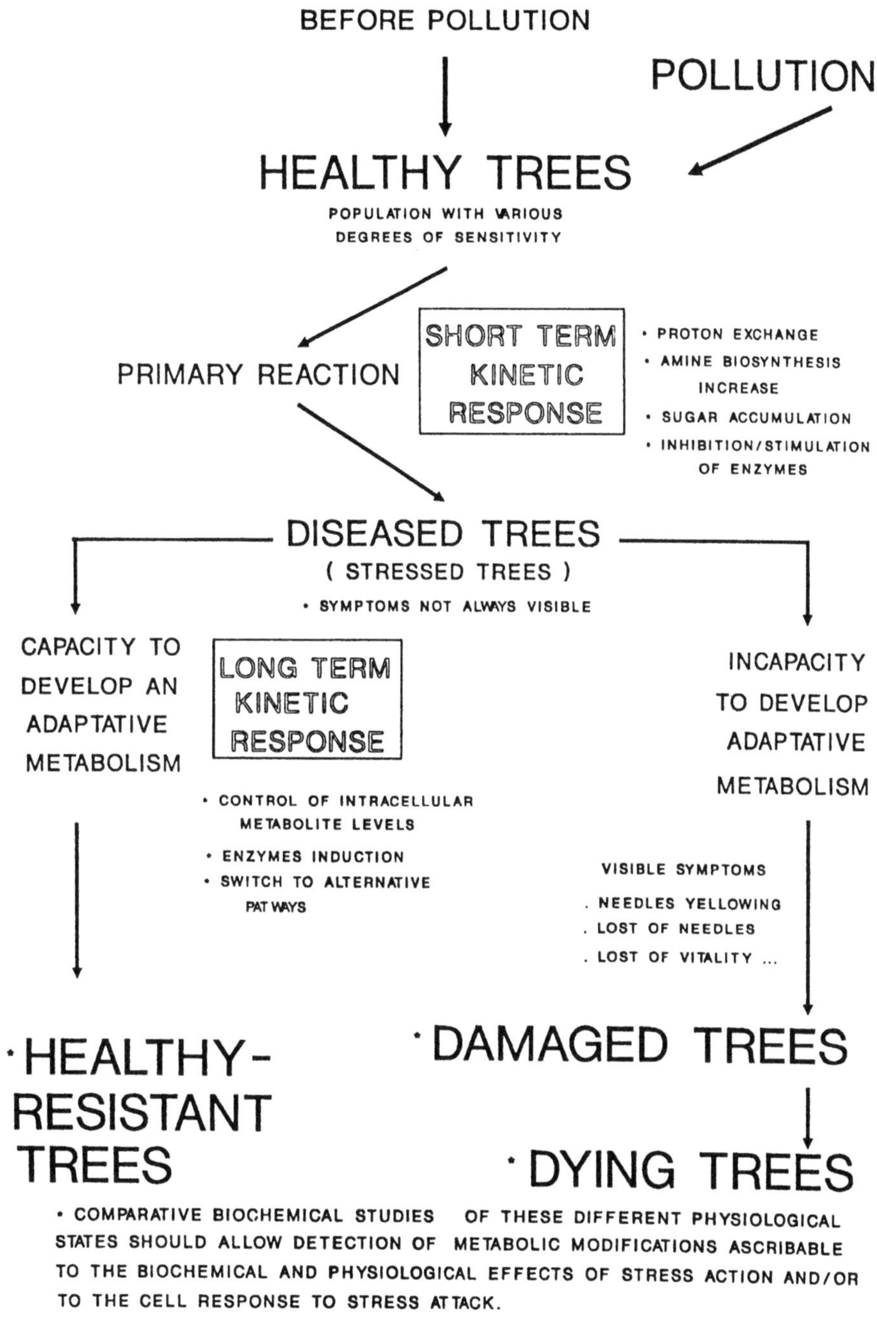

Fig. 14. Proposed sequence of events leading to healthy-resistant trees in a polluted forest submitted to chronic pollution (see also figs. 8, 12 and 13). Two processes must be distinguished: a short term kinetic response (an immediated response of cell defense mecanism) and a long term kinetic response (a possibility to develop an adaptive metabolism to continue living in stress conditions) .

exposure to the stress could finally, by the same mechanism by which the cell responds, cause damaging effects to the plant. Trees must in this case, as an adaptive mechanism, modify their usual metabolic pathways (long-term kinetic response), to become healthy-resistant to be able to continue living in polluted areas. Those trees which can not adapt their metabolism will remain in what we call short-term kinetic response and, as mentioned above, the metabolic disturbances, after leading to visible injuries, will finally kill the plants.
In this context it is also of interest to point out that the clear difference between healthy and diseased trees, in their content of amines, is of special significance in the case of acid stress. An increase of polyamine biosynthesis is the result of the stimulation by the acidic environment of arginine decarboxylase (Young and Galston, 1983), the key enzyme which produces putrescine, the polyamine precursor, from arginine via agmatine. Amine biosynthesis increase is then one of the primary metabolic responses of plants to acid stress.

Also, more experimental work concerning the long-term kinetic response, as part of adaptive metabolism, could help us to understand how the cells manage to find their "own solution" to survive in polluted areas.

CONCLUSIONS

The present biochemical work is one of the most extensive ever reported in this field, considering the number of individual trees and samples used and the duration of the time-course kinetic analysis. It was an almost non-destructive, as only few needles (a hundred) had to be taken from each tree. This was made possible by 1) the use of the analytical system developped in our laboratory, which allows analysis of a large number of metabolites from a single crude sample and 2) the considerable progress riched in the instrumentation for mineral analysis (the Philips apparatus we used is able to simultaneously measure 52 elements in a single sample) and sample preparation (the microwave apparatus from Prolabo allowed us to process the 540 samples at a rate of 20 min each).

The results presented here have clearly shown a marked difference in the levels and in the evolution of several metabolites, and these findings can be well correlated with the physiological state of the trees. The most noteworthy results are those found for putrescine, tryptophan, free sugars, bound sugars, arginine and polyamines. Such differences are observable when comparing green needles from healthy trees with either yellow or green needles from diseased trees. Hence, even if a diseased tree is green, its needle contents will allow determination of its physiological state (Figure 8). These metabolites, in particular putrescine (Figure 6), tryptophan (Figure 4), sugars (Figures 2 and 3) and polyamines (Figure 7) can thus serve as **early biochemical markers** of the nocive effects of cell pollution. They could also help to differentiate apparently healthy trees which have started to suffer damages due to pollution and will not be able to live in polluted areas, from those healthy-resistant trees, which are able to live in polluted areas.

Results obtained suggest that, for biochemical studies, experiments in controlled conditions at the Phytotron can allow clearer interpretations than those from open top chambers. This could be due to the fact that under the phytotronic conditions the different parameters can be standardized better and maintained constant, while in open top chambers too many factors, often episodically, can intervene.

At this point, it appears that although acid rain could accelerate loss of some mineral nutrients, this fact alone can not cause forest decline as cations should be retained within the nutrient cycles. As shown here and by others (Krivan and Schaldach, 1985; Mengel et al., 1986) diseased trees even contained higher levels, of most of the minerals studied, when compared to healthy ones.

There is no doubt that forest decline is a very complex biochemical process, whose visually observed symptoms are not only due to the effects of the atmospheric pollution (Schütt and Cowling, 1985) but also to cell response and cell adaptation (Villanueva and Santerre, 1989). Moreover, we must keep in mind that pollutants can also act as a drastic agent of selection, causing degradation of the existing plant communities, reducing their specific composition. As a result, only that part of the original population which was able to adapt to the polluted environment will remain.

ACKNOWLEDGMENTS

This work was supported in part by the French Ministries of Agriculture (Direction des Forêts), of Environment (SRETIE), and of Research and by the Commission of European Communities, DG XII of CEC (DEFORPA Program).

REFERENCES
Adams, D.O. and Yang S.F.: 1979, Proc. Natl. Acad. Sci. USA **76**, 170.
Apelbaum, A., Burgoon, A.C., Anderson, J.D., Lieberman, M., Ben-Arie, R. and Matoo, A.K.: 1981, Plant Physiol. **68**, 453.
Bonneau, M. and Landmann, G. : 1988, La Recherche **19**, N°205, 1542.
Bonte, J., Cantuel, J., and Malka, P. : 1986, Programme DEFORPA **3**, 13.
Bradford, M.M. : 1976, Anal. Biochem. **72**, 248.
Dai, Y.R., and Galston, A.W. : 1981, Plant Physiol. **67**, 266.
Durzan D.J. : 1968, Can. J. Bot. **46**, 909.
Ecker, J.R., and Davis, R.W. : 1987, Proc. Natl. Acad. Sci. USA 84, 5702
Galston, W. : 1983, BioScience **33**, N°6, 382.
Goeschl, J.D., Pratt, H.K., and Bonner, B.A. : 1967, Plant Physiol. **42**, 1077.
Hinrichsen, D. : 1986, Ambio **15**, 258
Hutchin, M.E. and Vaughanb, E.: 1968, Plant Physiol., **43**, 1913.
Kaur-Sawhney, R., and Galston, A.W.: 1979, Plant Cell Environm. **2**, 189.
Krivan V. and Schaldach,G.: 1985, Projekt Europäisches Forschungszentrum für Massnahmen zur Luftreinhaltung (PEF), 5, 163.
Loehle, C. : 1988, Can. J. For. Res. **18**, 208.
Luckner, M. : 1984, Secondary Metabolism in Microorganisms, Plants and Animals, Springer, Berlin.
Mehlhorn, H., and Wellburn, A.R. : 1987, Nature **327**, 417.
Mejnartowicz, L.E. and Lukasiak, H. : 1985, Eur. J. For. Path. 15, 193.
Mengel, K., Breininger, M.Th. and Lutz, H.J.: 1986, CEE Air Pollution Research Report **4**, 25.
Mermet, J.M., Robin, J. and Trassy, C.: 1988, Techniques de l'Ingénieur , Fascicule P2719, 1.
Miller, H.G.: 1986, CEE Air Pollution Research Report **4**, 10.
Montreuil, J. and Spik, G.: 1963, Microdosage des glucides totaux, Fasc. 1, 21, Faculté de Sciences de Lille.
Morel, C., Villanueva V.R., and Queiroz, O.: 1980, Planta, **149**, 440.
Pavlides, D. : 1983, Physiol. Vég. **15**, 187.
Priebe, A., Klein, H. and Jäger, H.J.: 1978, J. of Exp. Bot., 29, 1045.
Richard, F.J. and Coleman, R.G.: 1952, Nature, **170**, 460.
Santerre, A., Villanueva, V.R. and Mermet, J.M.: 1990, ICP Inform. Newslet. **15**, 355.

Santerre, A. Markiewicz M. and Villanueva, V.R.: 1990, Phytochemistry (in press).
Schütt, P. and Cowling, E.B.: 1985, Plant Desease, **69**, 548.
Sinclair, C.: 1967, Nature, **213**, 214.
Smith, T.A.: 1983, Recent Adv.Phytochem., **18**, 7.
Smith, T.A.: 1985, Ann. Rev. Plant Physiol., **36**, 117.
Ulrich,B.: 1986, Z.Pflanzerernähr. Bodenkd., **149**, 702.
Villanueva, V.R. : 1987, Programme DEFORPA **3**, 141.
Villanueva, V.R. : 1988, Proc. of the 2nd Internationnal Symposium on Air Pollution and plant metabolism. GSF-Report **9/87**, 97.
Villanueva, V.R., Mardon, M., Le Goff, M.Th., and Moncelon, F.: 1987, J. Chromatog., **393**, 97.
Villanueva V.R. and Santerre A.: 1989, Water, Air and Soil Pollution, **48**, 59.
Young, N.D., and Galston, A.W. : 1983, Plant Physiol. **71**, 767.
Williamson, R.E. and Ashley, C.C.: 1982, Nature, **296**, 647.

ACID WATERS IN THE UNITED KINGDOM: EVIDENCE, EFFECTS AND TRENDS

Dr. A. Tickle,
The Centre for Environmental Technology, Imperial Collge, U.K.

ABSTRACT

Acidification of surface waters has now been demonstrated unequivocally in Britain and over a far wider area of the country than was hitherto thought. Incontrovertible palaeoecological proof from almost all areas of upland Britain has in particular underlined the extent, and more importantly, the recent (post-1850) timescale of acidification. It was this advance in understanding together with mounting evidence of loss from fish stocks that led to the decision by the electricity supply industry to invest in pollution control measures. However the extent to which these reductions will influence the pH of surface waters is a key question. Although encouraging signs have been observed that effects may be reversible in some areas, this will probably not be the case for all systems. Modelling studies in particular point to the fact that current emission reduction plans are insufficient to halt, let alone reverse, acidification in many areas.

1. INTRODUCTION

Large scale effects of acidic deposition were first recognised for surface waters in the late 1960's (Oden, 1968) where in southern Norway and Sweden the pH of lakes and rivers was falling, to the detriment of fish stocks. By the end of the next decade it became clear that a severe and often terminal decline had occurred in fish populations over widespread areas of Scandinavia (Drablos & Tollan, 1980).

In the UK, acid emissions (principally SO_2 and NO_x) from fossil fuel burning and other combustive processes have risen markedly throughout most of this century. Correspondingly, acidic and other potentially damaging components in rain, particularly hydrogen (H^+) and sulphate (SO_4^{2-}) ions, are also likely to have increased. But in spite of the warning signs from Scandinavia and the fact that the potential link between acid deposition and acid waters in the UK had been identified some twenty years previously (Gorham, 1958), recognition of acidification problems has been slow.

Deterioration in water quality was first identified in the late 1970's in Galloway and the Trossachs (DAFS, 1978). At this time these reports of acidic waters with sparse or absent fish populations were regarded as a local problem, possibly caused by land use changes - in particular the widespread planting of conifers in upland catchments. An obvious requirement, therefore, of research undertaken since that date has been to separate the relative contributions of natural acidity from forest soils and acidity scavenged by conifers from the atmosphere and to distinguish these from effects of acidic deposition in unafforested catchments. Equally important to our understanding is the need to identify the extent of acidification over the country, its effect on biological systems, and to gain some idea of the timescale over which it has occurred.

In response, research efforts have intensified over the last decade; in the last five years particularly, detailed and often long-term studies by co-operating research groups principally funded by the Department of the Environment (DoE), the Central Electricity Generating Board (CEGB) and British Coal (under the auspices of the Surface Waters Acidification Programme [SWAP] of the Royal Society of London and the corresponding societies of Norway and Sweden), the Natural Environment Research Council (NERC) and the Nature Conservancy Council (NCC) have accumulated a large body of information in answer to these questions. With this evidence at hand, it is now possible to give a realistic overall perspective of acid waters in the UK and also evaluate current and future trends.

2. EVIDENCE FROM SURFACE WATERS

Records of water quality held prior to the recognition of acidification problems have not, for the most part, helped to identify potentially significant changes in acidity or alkalinity. Most are in terms of acidity (pH) and, being composed of infrequent or irregular sampling regimes, are usually inadequate for the detection of long-term (ten year or so) trends. Despite this caveat, a working group reporting to the DoE (Acid Waters Review Group, 1986) identified roughly a third of the waters for which good records were available as being vulnerable to acidification (Table 1.). No evidence however was found of a uniform drop in surface pH over the country as a whole.

Table 1. **Surface waters identified as highly vulnerable to acidification by the Acid Waters Review Group (AWRG, 1986).**

Water	Area	Source of record
Beiste Brice Mhullaich Thull	Northwest Scotland	Freshwater Fish Laboratory, (DAFS).
River Annalong Silent Valley Reservoir Ben Crom Reservoir	Northern Ireland	Department of the Environment (NI)
River Duddon Ennerdale Water Thirlmere Crummock Water	Cumbria	North West Water
River Lymington River Beaulieu River Plym East & West Dart River Dart Burrator Reservoir Siblyback Reservoir	Southern England Southwest England	Southern Water South West Water
Alwen Conwy Seiont Tryweryn Mynach Teifi Ystwyth Irfon (Abergwesyn) Irfon (Dol-y-coed) Wye	North Wales Mid-Wales	Welsh Water

Further evidence for changes in water quality are detailed below and come from contemporary studies designed specifically to identify possible trends due to acidification. Almost without exception these studies are drawn from catchments in upland Britain underlain by hard, slow-weathering rocks with little buffering capacity, and as such have been identified as highly susceptible to acidification by Kinniburgh & Edmunds (1984). Figure 1. (redrawn from their study) broadly shows the extent of potential risk once the pattern of acid deposition is superimposed.

Figure 1. The susceptibility of UK surface and groundwaters to acid deposition.

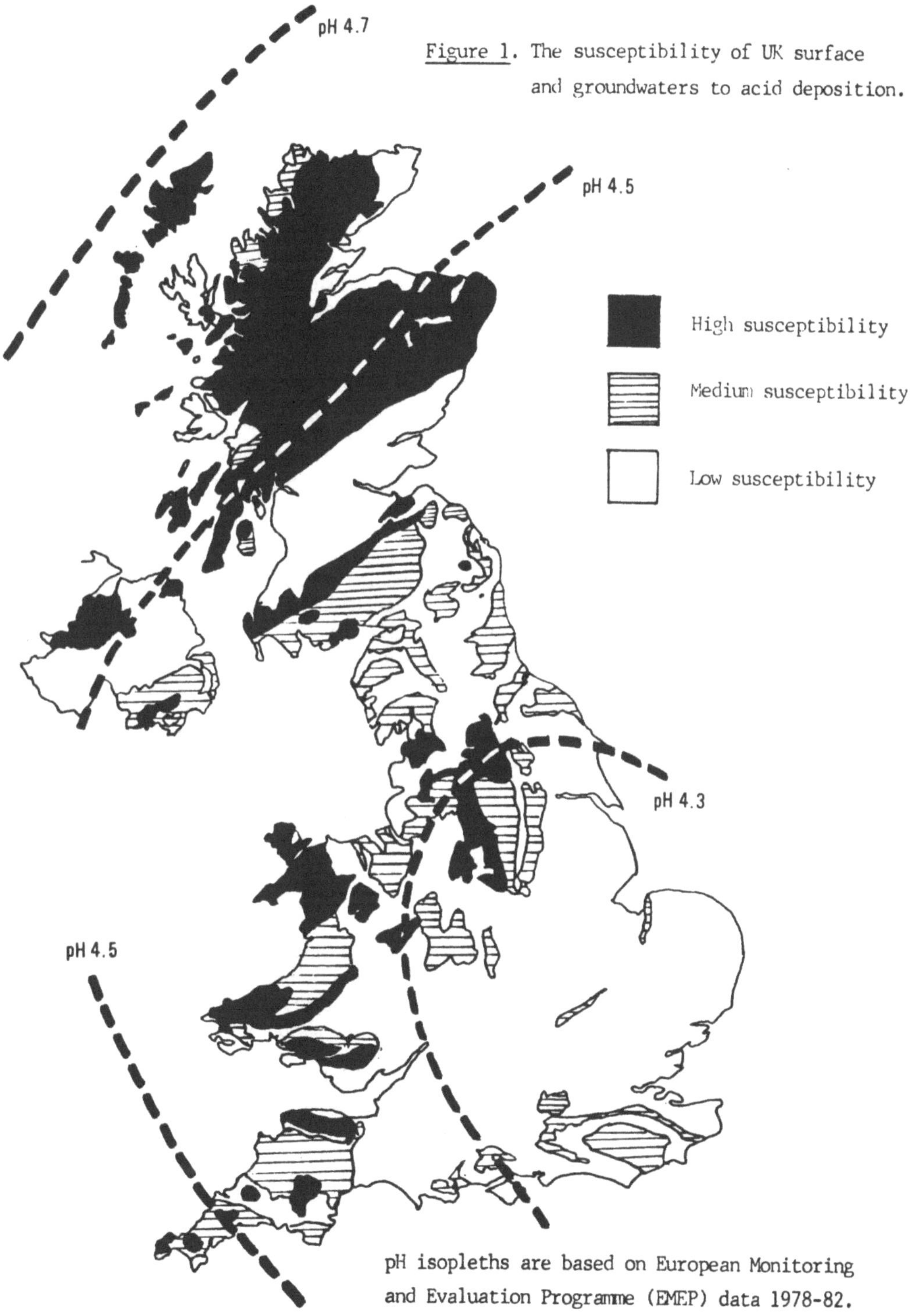

pH isopleths are based on European Monitoring and Evaluation Programme (EMEP) data 1978-82.

2.1 Scotland

Upland Scotland has the largest contiguous area susceptible to acidification in the UK and much of it receives highly acid deposition. The first studies of acidification were reported from the Loch Ard area in the Trossachs (Harriman & Morrison, 1982) where total levels of deposited sulphur (wet and dry) are high, exceeding 4.0 grammes of sulphur (S) per square metre per year (g.S $m^{-2}yr^{-1}$). In a comparison of forested and non-forested catchments it was shown that streams in the planted zone were always more acid and contained higher concentrations of aluminium and manganese. In addition both forested and non-forested catchments showed a highly significant correlation between SO_4 levels from deposition and their acidity. Further work on forested and non-forested catchments showed that trees collect air pollutants, giving rise to increased stream concentrations of acidity and SO_4 (Harriman & Wells, 1985). In the same paper the extent of acidification of some lochs was shown, calculated using an 'acidification' equation (Henriksen, 1982), giving indications of highly acidified waters in the Trossachs (Lochs Katrine & Chon) and Galloway (L. Grannoch and L. Fleet). Additionally, further surveys of 92 streams from all over Scotland (Wells *et al.*, 1986) showed a highly significant correlation between these 'acidification' values and excess (non-marine) sulphate in their waters.

In 1987 the results of long-term surveys of water quality (1978-79 and 1983-84) in upland Galloway were published by Harriman *et al.* (1987a). This is a sensitive granite area with high deposition levels, 0.08 g.$H^+m^{-2}yr^{-1}$ and 3.14 g.S $m^{-2}yr^{-1}$ as sulphate (in 1979), although part of this load comes from marine sources. Sixteen of the 22 lochs and 12 of the 21 streams had a mean pH less than 5.0. All the lochs and 20 of the streams had less than 100 microequivalents per litre ($\mu eq.l^{-1}$) of calcium, a buffering level that could be regarded as critically sensitive (Dickson, 1986) in view of the high sulphur deposition. They also presented data for pH change in seven lochs over the last twenty years or less. Although many of the changes could be associated with afforestation, a drop of 1.9 pH units (1958-1987) took place in Neldricken, a loch unaffected by forestry. Regression analysis of further data showed streams with similar buffering levels would be 0.7 pH units higher draining moorland compared to semi-mature forest catchments.

Work since then has extended beyond those areas regarded as highly sensitive (Harriman & Wells, 1985) to transitional regions in terms of their median pollutant loads (Rannoch Moor; the Cairngorms) or remote sites receiving little deposited acidity or SO_4 (northwest Scotland). Correlations between acidification indices and excess SO_4 have still been shown to exist at these more distant locations. There was also significantly greater SO_4 in streams draining mature conifer forests than moorland catchments - even in areas receiving little or no deposited SO_4 (Wells & Harriman, 1987). They concluded that acid deposition provides a major contribution to acidification in upland Scotland, with forests being a contributing factor in scavenging atmospheric acidity. However their work on pristine areas does suggest an influence of afforestation alone in increasing soil acidity.

2.2 Northwest England

2.2.1 Cumbria

The English Lake District contains extensive areas of acid sensitive lithologies - Eskdale Granites, Borrowdale Volcanics and Skiddaw Slates; in addition high levels of acidity (0.06 - 0.12 g.$H^+m^{-2}yr^{-1}$) and sulphur (2-4 g.S $m^{-2}yr^{-1}$) have been deposited over long periods of time (Sutcliffe, 1983). Mean records of rainfall acidity over a similar timescale (1954-1979) range between pH 4.1 - 4.4, dropping in individual samples as low as pH 3.6. Despite this and the seeming wealth of records held by the Freshwater Biological Association (Carrick & Sutcliffe, 1982, 1983), evidence for acidification trends in surface waters is patchy and has been difficult to substantiate.

Initial symptoms of problems were seen in the early 1980's with a decline in fishery status being reported for the River Esk followed by extensive fish mortalities on the Esk and Duddon in 1980 (Prigg, 1983). Patterns of water quality have been monitored in these and other sensitive rivers since that date by North West Water and others (Crawshaw, 1984; Bull & Hall, 1986). Prior to that programme of research, little data existed for Cumbria apart from that collected by the FBA. These data have now however been subjected to repeated examination for possible evidence of acidification (Sutcliffe and others, 1982-1988). These studies have shown that many extremely acid ('very soft water') upland tarns exist, often with nil or negative alkalinities and many more lose their alkalinity in winter. As late as 1986 Sutcliffe and Carrick stated that although the frequency of acid episodes may have increased in the 1970's, there has not been an increase in water acidity in the past 35-56 years. It was also suggested that the absence of coniferous forest might be a factor distinguishing Cumbria from other parts of the UK where acidification had been identified. More recently however it has been admitted that acidification may be occurring (albeit very slowly) and that if levels of acidic deposition rise beyond current levels, further acidification of vulnerable tarns might be expected (Sutcliffe & Carrick, 1988).

2.2.2 Pennines

Attention has only recently turned to the upland streams and waters of the southern Pennines. They are underlain by acid Millstone Grits and currently receive some of the highest levels of deposited acidity, sulphur and nitrogen in the UK (Review Group on Acid Rain, 1987). The region has also been strongly contaminated by general industrial pollution in the past (Press *et al.*, 1983). It is unsurprising therefore that in a comparative study with Lakeland streams (Diamond *et al.*, 1987) the Pennine waters were more acidic. They also tended to have higher concentrations of aluminium at any given pH. Calculations from the data of Turnpenny *et al.* (1987) also show mean concentrations of labile monomeric aluminium to be seven times higher in Pennines streams than in those Mid or North Wales. Information collected by Anderson *et al.* (1988) also shows that half the reservoirs in the region for which records are available, currently have a mean pH < 5.5. Unfortunately no further information on pH trends can be drawn from available records of surface water.

2.3 Wales

Acid susceptible areas occur principally in North and Mid-Wales over ancient sedimentary and metamorphic rocks. Although rainfall in general is less polluted over Wales than Scotland or northern England (Warren Spring Laboratory, 1987) it can give rise to large amounts of deposited acidity and sulphur. Examples include Plynlimon with 0.06 $g.H^{+}m^{-2}yr^{-1}$ (RGAR, 1987) and the Berwyns with 3.1 $g.S\ m^{-2}yr^{-1}$ (Press *et al.*, 1986). Mean acidity of rainfall is about pH 4.4 - 4.5 in sensitive areas, with bulk deposition (includes both wet and dry contributions) sometimes falling as low as pH 3.5 (Welsh Water, 1988).

For the most part reports and studies detailing surface water effects have paralleled those from Scotland. Large surveys of acidic streams were undertaken first in Mid-Wales between 1981-83 after reports of poor fish survival on the Rivers Twyi and Camddwr above Llyn Brianne (Stoner *et al.*, 1984). A major potential source of acidity in the streams was found to be atmospherically-derived anions and "excess" sulphate levels were also some 30% higher in afforested streams. Indeed, the increase in concentration of mobile acid anions from passage through vegetation canopies is thought to be more important than throughfall acidity in influencing reactions in the acid upland soils of Wales, and in acidification of surface waters (Hornung *et al.*, 1990). An earlier report (Stoner & Gee, 1985) also confirmed higher levels of acidity and aluminium for the afforested Twyi. The same study showed similarly depressed pH and elevated aluminium levels in several afforested upland lakes where fish catches had declined considerably since the 1960's.

Comparison of records in the Wye Catchment (Ormerod & Edwards, 1985) reported a fall in mean pH at both afforested and moorland sites between 1964-70 and 1983-85. The decline in pH for unafforested sites, although small, was suggestive of a potentially significant trend. Figures for agricultural liming in Wales were also presented showing that a decline in application by 55% by 1970 could possibly be contributing to the trend observed.

Thus although acidification in Welsh waters has been accelerated by conifer afforestation in some areas and possibly by a reduction in liming, it has principally been as a result of acidic deposition (Stoner & Harriman, 1988).

3. EVIDENCE FROM LAKE SEDIMENTS

The difficulty of establishing trends in acidity from existing data for rivers and streams has been demonstrated. However reconstruction of historic pH levels for standing waters can be achieved from records of fossil diatom assemblages usually present in lake sediments (Renberg & Hellberg, 1982; Battarbee, 1984). This elegant technique has been used by Battarbee and his colleagues since 1981 to establish systematically the evidence, or otherwise, of pH change in lochs, llyns and lakes throughout Scotland, Wales and England. The studies have been particularly instrumental in distinguishing acidification due to atmospheric contamination from other postulated effects.

The first sites studied were the low pH lakes of Galloway (Flower & Battarbee, 1983) where analyses of the diatom records demonstrated a marked acidification from about 1840, a fact that could not be satisfactorily explained by afforestation or other changes in land use. Acidic deposition - by default - therefore seemed responsible (Battarbee *et al.*, 1985). Over the rest of Scotland diatom records from Loch Laidon on Rannoch Moor (Flower et al, 1988), Loch Tinker in the Trossachs (Battarbee, 1987), Loch Tanna (Arran) and Lochnagar in the Cairngorms (Battarbee *et al.*, 1988b) have shown the same trends, despite in some cases receiving substantially lower levels of atmospheric deposition.

Studies on sites in Wales since 1984 (Fritz *et al.*, 1986 *et seq.*) have demonstrated the same decline in pH since industrial times, and as with the Scottish data are often accompanied by losses or declines in fish stocks. Land use changes were identified in Wales for some lakes; some may have accelerated acidification but in all cases the actual start in the pH decline had already occurred. In Cumbria, despite lack of early evidence (Haworth, 1985), clear pH declines due to atmospheric deposition have now been identified for Scoat, Greendale and Low Tarns (Haworth *et al.*, 1987). This study and others (Jones *et al.*, 1986) now convincingly refute theories of long-term natural soil acidification (Pennington, 1984).

Finding natural lakes in other parts of northern England has been difficult so attempts to reconstruct pH histories have been made using sediments from Watersheddles Reservoir in the South Pennines (Anderson *et al.*, 1988). Although a variety of problems in the sample core prevented pH reconstruction, the diatom flora was found to be consistently representative of acid waters throughout, leading to the conclusion that the water was acidified below pH 5.0 at the time of the reservoir's construction in 1877. This finding is in agreement with the pollution history of the area (Press *et al.*, 1983).

In almost all of the waters examined in Scotland, England and Wales the diatom floras are remarkably stable in their composition until 1850 or thereabouts. This points strongly to the fact that contamination from burgeoning industrial sources since that date has been primarily responsible for the subsequent acidification and diatom changes. However the link to fossil fuel combustion is indisputably made by the increases in trace metals (lead and zinc) and carbonaceous particles ('soot') seen in the sediment cores over the same post-1850 timescale. This trend has been shown at all but the most distant of sites (northwest Scotland) examined by Battarbee and his colleagues indicating the large geographical extent of industrial contamination.

Current interest now lies in areas distant from sources of pollution, with sites being examined in northwest Scotland, the south and southwest of England, Eire and Northern Ireland (Battarbee and others, ongoing). However their findings thus far indicate that few water bodies in acid susceptible areas south of Loch Ness will be immune from acidification (Battarbee *et al.*, 1988b).

4. BIOLOGICAL EFFECTS

4.1 Effects on fish

Mortalities of fish due to short term acid episodes ('acid flushes') associated with meltwater or spate conditions have probably been occurring naturally in upland streams prior to the advent of acidic deposition. However reports of fish kills (Prigg, 1983; Milner & Hemsworth, 1984) and other effects of acidity on fish are now more common (Harriman & Morrison, 1982; Stoner *et al.*, 1984). Laboratory tests quickly established that acid waters with low calcium and high levels of aluminium adversely affected fish survival, particularly at early stages in the life cycle. Salmonids are particularly affected with salmon being more sensitive than trout. Many acid waters where salmonid fish are now sparse or absent have met these toxicological criteria. Many more are thought to be susceptible when spate conditions markedly lower acidity and increase aluminium concentrations (Prigg, 1985). All these effects are difficult to monitor *in vivo* so many surveys have used experimental techniques, such as caging fish in the stream water under study, to gauge survival over a defined period of time. Widespread effects have now been demonstrated throughout sensitive areas of the UK.

In Scotland a general decline in salmon catch has been significantly correlated with upland afforestation (Egglishaw *et al.*, 1986) but other declines in populations are unlinked to forestry. This is the case for Galloway where acid deposition is reported to be the major cause of changes in fishery status (Harriman *et al.*, 1987a). Also in this area, eel densities in the Loch Fleet system have been reported to be low (Turnpenny *et al.*, 1988), despite the species' much higher tolerance to acidity when compared to salmonids. Mortalities in experimental field tests have also occurred in studies in the Cairngorms (Harriman *et al.*, 1987b), an area regarded as transitional in terms of its vulnerability to acidification.

Historical data from Wales show that catches of salmon (particularly by rod) have declined on twenty major rivers between 1966-75 and 1976-85 (Ormerod & Gee, 1990). The pattern of catch decline is consistent with that noted for Scotland by Egglishaw *et al.* (*ibid.*). Modelling hindcasts on a regional basis for Wales (Ormerod *et al.*, 1988b) also suggest a considerable reduction in salmon abundance and trout survival time over the period 1844-1984. Estimates of the potential loss to Welsh fisheries from acidification have been put between £5-25 million, for salmon and sea trout alone (Milner & Varallo, 1990).

Analyses of stream water quality and fishery status in Wales (Turnpenny, 1985) have shown salmonid density, biomass and condition to be adversely affected at higher levels of acidity. It was suggested these trends were associated with heavy metal toxicities, rather than pH *per se* or limited food availability. Other studies in Wales have implicated aluminium toxicity (Stoner & Gee, 1985) and in particular, an experimental addition of sulphuric acid and/or aluminium sulphate to a stream in the Llyn Brianne catchment (Ormerod *et al.*, 1987b) where mortality of salmonids due to aluminium and low pH were 50-87% compared with 7-10% for low pH alone.

However studies do agree on the importance of aluminium in Northwest England, being implicated both in fishkills and decline in Cumbria (Diamond *et al.*, 1987) and low densities of fish in Pennine streams (Diamond *et al.*, *ibid.*; Turnpenny, 1985). A later report (Turnpenny *et al.*, 1987) does however point to the overall restricted distribution of migrant species (including eels) in the industrialised English rivers where pollution and impoundments create barriers. Additionally for Cumbria, Diamond *et al.* suggest from initial analysis of data from the Duddon and Esk that upstream migration of salmonids is inhibited when pH falls below pH 5.5. Finally, effects of forestry on fish have not been identified from Pennine or Cumbrian data sets.

4.2 Effects on invertebrates, birds and other species

Although restricted distributions of stream-dwelling invertebrates due to low pH and other related factors are well known (Sutcliffe & Carrick, 1973), the further link between diversity reduction and acidification has been difficult to prove. Recent studies have, however, been able to reveal strong and consistent relationships between acid-base status and the faunal composition of upland streams (Ormerod & Wade, 1990). Impoverished distributions of invertebrates (particularly mayflies and stoneflies) have been noted in Scotland and Wales (Harriman & Morrison, 1982; Stoner *et al.*, 1984) and northwest England, particularly in the south Pennines (Diamond *et al.*, 1987). Mechanisms are far from understood, but in a realistic field experiment (Ormerod *et al.*, 1987b) aluminium caused a decline in the mayfly, *Baetis rhodani*, with drift densities of other invertebrates being enhanced by both aluminium and low pH. Despite this demonstration of acute effects, long-term effects of acidification on invertebrate communities may be in terms of change rather than loss, with more acid-tolerant species replacing sensitive ones; this has been demonstrated in models of faunal change through time for a number of streams at Llyn Brianne (Ormerod *et al.*, 1988b). The lack of correlation found between pH and invertebrate biomass (Morrison & Battarbee, 1988) lends support to the notion of species change rather than loss.
Invertebrates are an important food source for species at higher trophic levels such as fish and birds, although such secondary effects on fish are disputed by the data of Turnpenny *et al.* (1987). However lower abundance of invertebrates in acidic streams have been shown to have

serious effects on bird populations,particularly dippers in Wales (Ormerod et al, 1986). In low pH water lack of prey rich in calcium may also be contributing to dipper decline through effects on eggshell-thickness and weight (Ormerod *et al.*, 1988a). Recent studies have also shown effects on otter populations in Wales (Mason & McDonald, 1987) possibly acting through decline in fish populations. Indirect effects are also implicated in the decline of natterjack toads in southern England, whose breeding pools have suffered marked drops in pH in the last fifty years (Flower & Beebee, 1987). In addition acidification has been experimentally shown to affect frogs (Cummins, 1987).

Details of effects on plants, however, are scarce for the UK as a whole, although a survey of macrophytes in fifty Welsh lakes (Lowther, 1988) showed reduced taxon richness at the most acidic sites. However, limited historical data available for half the sites surveyed revealed no gross species changes over the past two decades. More oblique evidence (from experimental additions of lime at Llyn Berwyn and Llyn Hir) has shown altered distribution and abundance of genus such as *Sphagnum*, *Nardia*, *Fontinalis* and *Lobelia* in response to changing water chemistry (Ormerod & Wade, 1990). It has also been suggested that loss of aquatic macrophytes could lead to further effects on invertebrates (Ormerod *et al.*, 1987a).

5. CURRENT AND FUTURE TRENDS

Although levels of acidic deposition are still high over many parts of the UK, there has been a significant countrywide reduction in wet deposited acidity and a small decrease in wet deposited sulphate in the last ten years (RGAR, 1987). This drop, which is principally due to the sharp decline in SO_2 emissions since 1970, has been particularly noticeable in Scotland (Harriman & Wells, 1985). Correspondingly the impetus of some studies has changed direction and signs of reversibility in systems are now being sought.

The only evidence to date has come from Scotland. Deposited acidity and sulphate in the Loch Ard area have dropped by some 50% or over which has caused a small but significant increase in the acid neutralising capacity of one monitored stream, equivalent to an increase of 10-15 $\mu eq.l^{-1}$ over ten years (Wells & Harriman, 1987). However no detectable fall in stream sulphate can be found, suggesting that a substantial reservoir of sulphate has built up in the catchment soils. The time taken for this to flush out may cause a considerable lag in improvements. A second study (Battarbee *et al.*, 1988a) seemingly points to the potential for a more rapid recovery. Re-analysis of diatom cores from two lochs in Galloway show a trend of decreasing acidity from 1981-86 and surface water pH in one loch, Loch Enoch, has increased slightly.

The improvement detected by Battarbee *et al.* had in fact already been predicted from a modelling study in a nearby catchment above Loch Dee (Cosby *et al.*, 1986). But this amelioration, principally caused by the sharp post-1970 fall in SO_2 emissions, will only be a temporary respite according to the model; current (1984) sulphate deposition levels are still sufficient to continue depleting the buffering capacity of the catchment, leading to a further long fall in stream pH. The same model, MAGIC (Model of Acidification of Groundwater In Catchments), has been applied to part of the Llyn Brianne catchment in Wales (Whitehead *et al.*, 1987) and shows a similar long-term decline in acidity, if the same sulphate deposition trends are applied. Both studies show that acidification can only be halted by a 50% reduction from 1984 levels of deposition. MAGIC has also been extended to model biological responses to acidification (Ormerod *et al.*, 1988b); when this is applied on a regional basis to Wales, the same 50% reduction in deposition gave no marked improvement in trout density or survival times. Indeed, the regional model for Wales predicts (Whitehead *et al.*, 1990) that only a 90% reduction in deposition of sulphate would return stream conditions to pre-acidification levels.

Under the terms of the European Community (EC) Large Combustion Plants Directive (LCPD 88/609/EEC) the UK is required to reduce large plant (>50 MW) emissions of sulphur dioxide (SO_2) 60% by 2003 and oxides of nitrogen (NOx) 30% by 1998, both from 1980 levels. The effects of these future reductions on deposition have been modelled for the Llyn Brianne catchment (Metcalfe & Derwent, 1990): for both the pollutants the reduction in deposition was roughly half that of the percentage reduction in emissions. Although the control of nitrogen oxides will be strengthened after 1992 by a further EC Directive (89/458/EEC) limiting vehicle emissions, it is still clear from these studies that unless deposition levels are cut back more sharply, the buffering capacity of many catchments is unlikely to recover, thereby preventing a large amelioration in water quality.

In view of the fact that such sharp declines are not envisaged at present and soil sulphate pools may thus still be increasing, several Water Boards and other groups have resorted to ameliorative methods to combat acidification, although usually only on a small and restricted experimental basis. The most widespread of these techniques has been the direct liming of lakes and waters which has been carried out in Scotland (CEGB, 1988), Cumbria (Diamond *et al.*, 1987) and Wales (Underwood *et al.*, 1987; Welsh Water, 1988). All have been ameliorated sufficiently for successful restocking of fish, although no increases in invertebrate diversity seem apparent as yet (Prigg, 1987). Obviously direct liming must be repeated at regular intervals (proportional to the water residence time), which can be expensive for remote locations. Some liming of catchments has been carried out (Loch Fleet; Llyn Brianne), which will increase the period for which treatment is effective although source-area liming may be ineffective at some flows. But more importantly, the effects of adding large quantities of lime to the ombrogenous mires and blanket bogs - typical in naturally acid upland ecosystems - has already been shown to be detrimental to *Sphagnum* (Mackenzie, 1989) with the possibility of further consequences for invertebrates, amphibians, small mammals and birds (Ormerod, 1989).

Other work at Llyn Brianne has investigated differing management practices to overcome acidification, particularly in afforested catchments (Welsh Water, 1988). Attempts so far involving creation of stream-side buffering zones have been ineffective (Stoner & Harriman, 1988). New forestry practices which increase buffering base flows are also being investigated which theoretically could improve stream acidity problems (Whitehead *et al.*, 1986).

Notably, Welsh Water were the first to publish guidelines for acceptable conifer afforestation to protect streams of low hardness. Similar codes of practice have now been introduced for several other water areas in the UK but despite being given limited ratification by the Forestry Commission (Forestry Commission, 1988), these guidelines as yet have no legal status. Confusingly however, the Commission still states that firm planting guidelines cannot be given on the acidification issue, as it still disputes the proof of a significant forest effect *per se* (Nisbet, 1990).

6. CONCLUSION

Acidified waters are now a widespread phenomenon throughout areas of hard, slow-weathering rocks in the UK (Table 2.). Trends in acidity have been difficult to identify from past records of surface waters, but studies conducted over the last decade in Scotland, Wales and to a lesser extent England have now shown acid deposition to be the major cause of freshwater acidification and declines in fish populations. Additional scavenging of acidity from the atmosphere by conifers has also exacerbated the problem in many upland areas.

Evidence from diatom analysis of lake sediments has been instrumental in clarifying the timescale, pattern and extent of acidification. These studies have demonstrated that lakes in sensitive areas of Scotland, Northern England and Wales have been acidified to a similar extent and over the same (post-1850) timescale. These drops in lake pH have often been accompanied by declines in fishery status. Palaeoecological studies have also underlined the fact that acidification processes and their associated effects on biological systems occur in the absence of afforestation or any other land-use changes. It was this research in particular, together with mounting evidence of loss of fish stocks, that led to the decision in 1986 by the electricity supply industry to retrofit three large UK power stations with flue gas desulphurisation technology in an attempt to control acid emissions.

In 1988 the UK further agreed to control emissions of acidic gases by signing the EEC Large Combustion Plants Directive, which binds the UK to reducing high-level emissions of sulphur dioxide (SO_2) 60% by 2003 and nitrogen oxides (NO_x) 30% by 1998, both from a base year of 1980. However current plans by the electricity industry to limit SO_2 emissions, which include a range of measures from retrofitting power stations with flue gas desulphurisation, the construction of new gas-powered capacity (which does not emit SO_2), to the use of low-sulphur fuels, look unlikely to meet the stipulated targets of the Directive as long as demand for electricity continues to rise (Sweet & Tickle, 1990). However a further directive placing

Table 2. Affected waters throughout the UK.

Water	Status	Source of Record
Scotland		
NorthWest		
Lochan Dubh	acidified	Battarbee *et al.*, 1988b
Cairngorms		
Lochnagar	acidified	Battarbee *et al*, 1988b
Dubh Loch	"	"
Loch nan Eun	"	"
Allt-a-Mharcaidh	vulnerable	Harriman *et al*, 1987b
Rannoch Moor		
Loch Laidon	acidified	Flower *et al*, 1988
Trossachs		
Loch Tinker	acidified	Battarbee, 1987
Loch Chon	"	"
Keltie Water	fishless	Mason, 1987
Loch Katrine	vulnerable	Harriman & Wells, 1985
River Forth (upper)	fish decline	"
Tanna		
Loch Tanna	acidified	Battarbee *et al*, 1988b
Galloway		
Loch Riecawr	fish decline	Harriman & Wells, 1985
Loch Fleet	fishless	Turnpenny *et al*, 1988
Loch Enoch	acidified*	Battarbee *et al*, 1985
Loch Grannoch	"	Flower & Battarbee, 1983
Round Loch of Glenhead	"	"
Loch Arron	fishless	Harriman *et al*, 1987a
Loch Narroch	"	"
Loch Neldricken	"	"
Loch Valley	"	"
Loch Dee	acidified*	Flower *et al*, 1987
Loch Skirrow	"	"
Northwest England		
Cumbria		
Greendale Tarn	acidified	Haworth *et al*, 1987
Scoat Tarn	"	"
Low Tarn	"	"
Devoke Water	"	Battarbee *et al*, 1988b
River Duddon	fish loss	Prigg, 1985
River Esk	"	"
River Liza	vulnerable	"
Great Langdale Beck	"	"
Seathwaite Beck	"	"
River Glenderamackin	vulnerable	Prigg, 1985
Broadcrag Tarn	"	Sutcliffe & Carrick, 1988
Brownrigg Moss Tarn	"	"
Chapel Hill Tarn	"	"
Gillercomb Tarn	"	"
Greenup Edge Tarn	"	"
Grey Knotts Tarn	"	"
Haystacks Tarn	"	"
High Scawdel Tarn	"	"
Kirkfell Tarn	"	"
Launchy Tarn	vulnerable	Sutcliffe & Carrick, 1988
Low Birker Tarn	"	"
Parkgate Tarn	"	"
Rough Crag Tarn	"	"
Standing Crag Tarn	"	"
Angle Tarn (Langdale)	"	"
Beckhead Tarn	"	"
Eel Tarn	"	"

Table 2. (contd.)

Water	Status	Source of Record
Northwest England		
Cumbria		
Floutern Tarn	"	"
Foxes Tarn	"	"
High House Tarn	"	"
Innominate Tarn	"	"
Leverswater	vulnerable	Sutcliffe & Carrick, 1988
Red Tarn (Langdale)	"	"
Scales Tarn	"	"
Seathwaite Tarn	"	"
Siney Tarn	"	"
Stony Tarn	"	"
Tarn-at-Leaves	"	"
Three Tarns (Bowfell)	"	"
Pennines		
Watersheddles Reservoir	acidified	Anderson *et al*, 1988
Wales		
Mid-Wales		
River Twyi	fish decline	Stoner & Gee, 1985
River Camddwr	"	"
Syfydrin	"	"
Pendam	"	"
Blaemelindwr.	"	"
Llyn Hir	acidified*	Fritz *et al*, 1986
Llyn Bewyn	fish loss	Underwood *et al*, 1987
Llyn Gynon	acidified	Stevenson *et al*, 1987a
Gwynedd		
Llyn y Bi	acidified*	Fritz *et al*, 1987
Llyn Dulyn	"	Stevenson *et al*, 1987b
Llyn Eiddew Bach	"	Patrick *et al*, 1987a
Llyn Llagi	"	Patrick *et al*, 1987b
Llyn Cwm Mynack	"	Kreiser *et al*, 1987
Llyn y Adar	vulnerable	Battarbee *et al*, 1988b
Llyn Edno	"	"
Southern England		
Hampshire		
Woolmer Pond	acidified	Flower & Beebee, 1987

NB. The following waters have now been restored by liming and restocked with fish: Loch Dee, Loch Fleet, parts of the Esk catchment, Llyn Hir, Llyn Berwyn.

* signifies effects on fish additionally.

stringent limits on vehicle emissions from 1992 should help in reducing UK NO_x emissions, although once again, upward trends in vehicle numbers are likely to offset these reductions partially.

A key question at present therefore, is the extent to which envisaged downward trends in emissions will influence the pH of surface waters. Although few studies have addressed this crucial area, there are signs from Sweden, and now more recently in the Galloway region of Scotland, that the capacity for a rapid reversibility of lake acidification does exist (Forsberg *et al.*, 1985; Battarbee *et al.*, 1988a).

However modelling studies applied on a regional basis to Wales suggest that a 50% reduction from current day sulphate levels will only halt alkalinity falling further in a large number of sensitive systems (Whitehead *et al.*, 1990). It seems probable that differences in the size of the soil sulphate pool held in these systems (which range from blanket peats to mineralized forest soils) could account for large differences in the timescale and potential extent of their recovery.

But, in general terms for the UK as a whole, it has been noted that only a 90% reduction in acid deposition would restore something approaching pristine conditions in the majority of affected waters (AWRG, 1989). Given the non-linear relationship between modelled UK emission reductions and resultant deposition, such a recovery will not be achieved without concerted and drastic reduction measures implemented throughout northwest and central Europe. Renegotiation of current international conventions on long-range transport of air pollutants, based on defined critical loads (Nilsson & Grennfelt, 1988), is likely to be the way forward in reaching such reductions.

ACKNOWLEDGEMENTS

This paper is based on a report originally published in 1988 by Greenpeace UK. Besides those acknowledged in that report, I would also like to thank Rick Battarbee, Keith Bull, Mike Hornung, John Innes, Steve Ormerod, Paul Stevens, Andy Turnpenny and Bob Wilson for helping to make available more recently published material on the subject.

REFERENCES

Anderson, N.J., Patrick, S.T., Appleby, P.G., Oldfield, F., Rippey, B., Richardson, N., Darley, J. & Battarbee, R.W. (1988) An assessment of the use of reservoir sediments in the southern Pennines for reconstructing the history and effects of atmospheric pollution. Research Paper No. 30, Palaeoecology Research Unit, University College London.

AWRG (1986). *UK Acid Waters Review Group: Acidity in the United Kingdom Fresh Waters.* Department of the Environment/ Department of Transport, South Ruislip.

AWRG (1989) *Acidity in United Kingdom Freshwaters: United Kingdom Acid Waters Review Group Second Report.* HMSO, London.

Battarbee, R.W. (1984) Diatom analysis and the acidification of lakes. *Phil. Trans. R. Soc.* B 305, 451-477.

Battarbee, R.W. (1987) Palaeolimnology programme - UK. In: *Surface Water Acidification Programme mid-term review conference, Bergen, Norway.* Royal Society, London.

Battarbee, R.W., Flower, R.J., Stevenson, A.C. & Rippery, B. (1985) Lake acidification in Galloway: a palaeoecological test of competing hypotheses. *Nature* 314, 350-352.

Battarbee, R.W., Flower, R.J., Stevenson, A.C., Rippey, B., Harriman, R. & Appleby, P.G. (1988a) Diatom and chemical evidence for reversibility of Scottish lochs. *Nature* 332, 530-532.

Battarbee, R.W., Anderson, N.J., Appleby, P.G., Flower, R.J., Fritz, S.C., Haworth, E.Y., Higgitt, S., Jones, V.J., Kreiser, A., Munro, M.A.R., Natkanski, J., Oldfield, F., Patrick, S.T., Richardson, N.G., Rippey, B. & Stevenson, A.C. (1988b). *Lake Acidification in the United Kingdom 1800-1986. Evidence from Analysis of Lake Sediments.* Ensis, London.

Bull, K.R. & Hall, J.R. (1986) Aluminium in the Rivers Esk and Duddon, Cumbria, and their tributaries. *Environ. Pollut. Ser. B* 12, 165-193.

Carrick, T.R. & Sutcliffe, D.W. (1982) *Concentrations of major ions in lakes and tarns of the English Lake District.* (Occasional publication no. 16.) Freshwater Biological Association, Ambleside.

Carrick, T.R. & Sutcliffe, D.W. (1983) *Concentrations of major ions in streams and catchments of the River Duddon (1971-1974) and Windermere (1975-1978), English lake District.* (Occasional publication no. 21.) Freshwater Biological Association, Ambleside.

CEGB (1988). *Loch Fleet Project. A report of the intervention phase (2) 1986-87.* (eds. Howells, G. & Dalziel, T.R.K.). Central Electricity Research Laboratory, Leatherhead

Cosby, B.J., Whitehead, P.G. & Neale, R. (1986) A preliminary model of long-term changes in stream acidity in south western Scotland. *J. Hydrol.* 84, 381-401.

Crawshaw, D.H. (1984) The effect of acid run-off on the chemistry of streams in Cumbria. (TSN/84/3.) Rivers Division, North West Water Authority, Warrington.

Cummins, C.P. (1987) Factors influencing the occurrence of limb deformities in common frog tadpoles raised at low pH. *Annales de la Societe Royale Zoologique de Belgique* 117, (Suppl. 1), 353-364.

DAFS (1978). *Freshwater Fisheries Laboratory, Triennial Review of Research 1976-78.* Department of Agriculture and Fisheries for Scotland, Edinburgh.

Diamond, M., Crawshaw, D.H., Prigg, R.F. & Cragg-Hine, D. (1987) Stream water chemistry and its influence on the distribution and abundance of aquatic invertebrates and fish in upland streams in Northwest England. In: *Acid Rain: Scientific and Technical Advances* (eds. Perry, R.M., Bell, J.N.B. & Lester, J.N.) 481-488. Selper, London.

Dickson, W. (1986) Some data on critical loads for sulphur on surface waters. In: *Critical Loads for Nitrogen and Sulphur* (ed. Nilsson, J.) SEPB, Solna.

Drablos, D. & Tollan, A. (eds.) (1980) *Ecological Impact of Acid Precipitation - Proceedings of an International Conference, Sandefjord, Norway.* SNSF Project, Oslo.

Egglishaw, H., Gardiner, R. & Foster, J. (1986). Salmon catch decline and forestry in Scotland. *Scott. geogr. Mag.* 102, 57-61.

Flower, R.J. & Battarbee, R.W. (1983) Diatom evidence for recent acidification of two Scottish lochs. *Nature* 305, 130-133.

Flower, R.J. & Beebee, T.J.C. (1987). *The recent palaeolimnology of Woolmer Pond, Hampshire, with special reference to the documentary history and distribution of the natterjack toad,* Bufo Calamita L. Ensis, London.

Flower, R.J., Battarbee, R.W. & Appleby, P.G. (1987). The recent palaeolimnology of acid lakes in Galloway, south west Scotland: diatom analysis, pH trends and the role of afforestation. *J. Ecol.* 75, 797-824.

Flower, R.J., Battarbee, R.W., Natkanski, J., Rippey, B. & Appleby, P.G. (1988) The recent acidification of a large Scottish loch located partly within a National Nature Reserve and a Site of Special Scientific Interest. *J. appl. Ecol.* 25, 715-724.

Forestry Commission (1988) *Forest and water guidelines.* Forestry Commission, Edinburgh.

Forsberg, C., Morling, C., & Wetzel, R.G. (1985) Indications of the capacity for rapid reversibility of lake acidification. *Ambio* 14, 164-166.

Fritz, S.C., Stevenson, A.C., Patrick, S.T., Appleby, P.G., Oldfield, F., Rippey, B., Darley, J. & Battarbee, R.W. (1986) Palaeoecological evaluation of the recent acidification of Welsh lakes. I, Llyn Hir, Dyfed. Research Paper No. 19, Palaeoecological Research Unit, University College London.

Fritz, S.C., Stevenson, A.C., Patrick, S.T., Appleby, P.G., Oldfield, F., Rippey, B., Darley, J., Battarbee, R.W., Higgitt, S.R. & Raven, P.J. (1987) Palaeoecological evaluation of Welsh lakes. VII, Llyn y Bi, Gwynedd. Research Paper No. 23, Palaeoecological Research Unit, University College London.

Gorham, E. (1958) The influence and importance of daily weather conditions in the supply of chloride, sulphate and other ions to fresh waters from atmospheric precipitation. *Phil. Trans. R. Soc.*, B 241, 147-178.

Harriman, R. & Morrison, B.R.S. (1982) The ecology of streams draining forested and non-forested catchments in an area of central Scotland subject to acid precipitation. *Hydrobiologia* 88, 251-263.

Harriman, R. & Wells, D.E. (1985) Causes and effects of surface water acidification in Scotland. *Journal of Water Pollution Control* 84, 215-222.

Harriman, R., Morrison, B.R.S., Caines, L.A., Collen, P. & Watt, A.W. (1987a) Long-term changes in fish populations of acid streams and lochs in Galloway, south west Scotland. *Water, Air & Soil Pollution* 32, 89-112.

Harriman, R., Wells, D.E., Gillespie, E., King, D., Taylor, E. & Watt, A.W. (1987b) Stream chemistry and salmon survival in the Allt a'Mharcaidh, Cairngorm. In: *Surface Water Acidification Programme mid-term review conference,Bergen, Norway*. Royal Society, London.

Haworth, E.Y. (1985) 'The highly nervous system of the English Lakes': aquatic ecosystem sensitivity to external changes, as demonstrated by diatoms. *Rep. Freshwater biol. Ass.* 53, 60-79.

Haworth, E.Y., Atkinson, K.M. & Riley, E. (1987) Acidification in Cumbrian waters: past and present distribution of diatoms in local lakes and tarns. Report to the DoE, Freshwater Biological Association, Ambleside.

Henriksen, A. (1982). Alkalinity and acid precipitation research. *Vatten* 38, 83-85.

Hornung, M., Reynolds, B., Stevens, P.A. & Hughes, S. (1990) Water quality changes from input to stream. In: *Acid Waters in Wales* (eds. Edwards, R.W., Gee, A.S. & Stoner, J.H.). Kluwer, Dordrecht.

Jones, V.J., Stevenson, A.C. & Battarbee, R.W. (1986) Lake acidification and the land-use hypothesis: a mid-post-glacial analogue. *Nature* 322, 157-158.

Kinniburgh, D.G. & Edmunds, W.M. (1984) *The susceptibility of UK ground waters to acid deposition*. Report to the Department of the Environment. British Geological Survey, Wallingford.

Kreiser, A., Patrick, S.T., Stevenson, A.C., Appleby, P.G., Rippey, B., Oldfield, F., Darley, J., Battarbee, R.W., Higgitt, S.R. & Raven, P.J. (1987). Palaeoecological evaluation of the recent acidification of Welsh lakes. X, Llyn Cwm Mynach, Gwynedd. Research Paper No. 27, Palaeoecological Research Unit, University College London.

Lowther, R. (1988) Macrophyte assemblages in Welsh Lakes. Unpublished report, University of Wales Institute of Science and Technology.

Mackenzie, S. (1989) The ecological impact of catchment liming on ombrotrophic bog communities. Unpublished report to the Nature Conservancy Council.

Mason, B.J.(1987) Description and instrumentation of SWAP field sites in Britain. In: *Surface Water Acidification Programme mid-term review conference, Bergen, Norway*. Royal Society, London.

Mason, C.F. & McDonald, S.M. (1987) Acidification and otter *Lutra lutra* distribution on a British river. *Mammalia* 51, 81-87.

Metcalfe, S.E. & Derwent, R.G. (1990) Llyn Brianne - Acid deposition modelling. In: *Acid Water in Wales* (eds. Edwards, R.W., Gee, A.S. & Stoner, J.H.). Kluwer, Dordrecht.

Milner, N.J. & Hemsworth, R.J. (1984) Fisheries investigation on Afon Glaslyn fish kill, September 1984. Welsh Water internal report 84/2.

Milner, N.J. & Varallo, P.V. (1990) Effects of acidification on fish and fisheries in Wales. In: *Acid Water in Wales* (eds. Edwards, R.W., Gee, A.S. & Stoner, J.H.). Kluwer, Dordrecht.

Morrison, B.R.S. & Battarbee, R.W. (1988) Effects on Freshwater Flora and Fauna (Excluding Fish). In: *Acid Rain and Britain's Natural Ecosystems* (eds. Ashmore, M., Bell, N. & Garretty, C.). Imperial College Centre for Environmental Technology, London.

Nilsson, J. & Grennfelt, P. (eds.) (1988) *Critical Loads for Sulphur and Nitrogen*. Miljorapport 1988:15. Nordic Council of Ministers, Copenhagen.

Nisbet, T.R. (1990) *Forests and surface water acidification.* Forestry Commission Bulletin 86. HMSO, London.

Oden, S. (1968) *The acidification of air precipitation and its consequences in the natural environment.* Energy Committee Bulletin 1, Swedish Natural Sciences Research Council, Stockholm.

Ormerod, S.J.(ed.) (1989) Some ecological implications of liming to counteract surface-water acidity. Ecology sub-group report, Llyn Brianne Acid Waters Project, Llanelli.

Ormerod, S.J. & Edwards, R.W. (1985) Stream acidity in some areas of Wales in relation to historical trends in afforestation and the usage of agricultural limestone. *J. Environ. Management* 20, 189-197.

Ormerod, S.J. & Gee, A.S. (1990) Chemical and ecological evidence on the acidification of Welsh lakes and rivers. In: *Acid Waters in Wales* (eds. Edwards, R.W, Gee, A.S. & Stoner, J.H.). Kluwer, Dordrecht.

Ormerod, S.J. & Wade, K.R. (1990) The role of acidity in the ecology of Welsh lakes and streams. In: *Acid Waters in Wales* (eds. Edwards, R.W, Gee, A.S. & Stoner, J.H.). Kluwer, Dordrecht.

Ormerod, S.J., Allinson, N., Hudson, D. & Tyler, S.J. (1986) The distribution of breeding Dippers (*Cinclus cinclus* Aves) in relation to stream acidity in upland Wales. *Freshwater Biology* 16, 501-507.

Ormerod, S.J., Wade, K.R. & Gee, S.A. (1987a) Macro-floral assemblages in upland Welsh streams in relation to acidity and their importance to invertebrates. *Freshwater Biology* 18, 545-557.

Ormerod, S.J., Boole, P., McMahon, C.P., Weatherley, N.S., Pascoe, D. & Edwards, R.W. (1987b) Short-term experimental acidification of a Welsh stream: comparing the biological effects of hydrogen ions and aluminium. *Freshwater Biology* 17, 341-356.

Ormerod, S.J., Bull, K., Cummins, C., Tyler, S.J. & Vickery, J.A. (1988b) Egg mass and shell thickness in dippers *Cinclus cinclus* in relation to stream acidity in Wales and Scotland. *Environ. Pollut.* 55, 107-121.

Ormerod, S.J., Weatherley, N.S., Varallo, P.V. & Whitehead, P.G. (1988a) Preliminary empirical models of the historical and future impact of acidification on the ecology of Welsh streams. *Freshwater Biology* 20, 127-140.

Patrick, S.T., Fritz, S.C., Stevenson, A.C., Appleby, P.G., Rippey, B., Oldfield, F, Darley, J., Battarbee, R.W., Higgitt, S.R. & Raven, P.J. (1987a) Palaeoecological evaluation of the recent acidification of Welsh lakes. VIII, Llyn Eiddew Bach, Gwynedd. Research Paper No. 24, Palaeoecological Research Unit, University College London.

Patrick, S.T., Stevenson, A.C., Fritz, S.C., Appleby, P.G., Rippey, B., Oldfield, F, Darley, J., Battarbee, R.W., Higgitt, S.R. & Raven, P.J. (1987b) Palaeoecological evaluation of the recent acidification of Welsh lakes. IX, Llyn Llagi, Gwynedd. Research Paper No. 27, Palaeoecological Research Unit, University College London.

Pennington, W. (1984) Long-term natural acidification of upland sites in Cumbria: evidence from post-glacial lake sediments. *Rep. Freshwat. biol. Ass.* 52, 28-46.

Press, M.C., Ferguson, P. & Lee, J.A. (1983) Two hundred years of acid rain. *Naturalist* 108, 125-129.

Press, M.C., Woodin, S.J. & Lee, J.A. (1986) The potential importance of an increased nitrogen supply to the growth of ombrotrophic *Sphagnum* species. *New Phytol.* 103, 45-55.

Prigg, R.F.(1983) Juvenile salmonid populations and biological quality of upland streams in Cumbria with particular refernce to low pH effects. (BN 77-2-83.) Rivers Division, North West Water Authority, Warrington.

Prigg, R.F. (1985) Faunal and chemical observations on the River Glenderamackin above Mungrisdale following a probable acid event. (Technical Note NC 299) Rivers Division, North West Water Authority, Carlisle

Prigg, R.F. (1987) Trout and invertebrates in Spothow Gill, Eskdale before and after limimg. Tech. Note NC353, Rivers Division, North West Water Authority, Carlisle.

Renberg, I. & Hellberg, I. (1982) The pH history of lakes in southwestern Sweden as calculated from the subfossil diatom flora of the sediments. *Ambio* 11, 30-33.

RGAR (1987). *Acid deposition in the UK 1981-1985: A second report of the United Kingdom Review Group on Acid Rain.* Warren Spring Laboratory, Stevenage.

Stevenson, A.C., Patrick, S.T., Fritz, S.C., Rippey, B., Oldfield, F., Darley, J., Higgitt, S.R. & Battarbee, R.W. (1987a). Palaeoecological evaluation of the recent acidification of Welsh lakes. IV, Llyn Gynon, Dyfed. Research Paper No. 20, Palaeoecological Research Unit, University College London.

Stevenson, A.C., Patrick, S.T., Fritz, S.C., Rippey, B., Appleby, P.G., Oldfield, F., Darley, J., Higgitt, S.R., Battarbee, R.W. & Raven, P.J. (1987b) Palaeoecological evaluation of the recent acidification of Welsh lakes. VI, Llyn Dulyn, Gwynedd. Research Paper No. 22, Palaeoecological Research Unit, University College London.

Stoner, J.H., Gee, A.S. & Wade, K.R. (1984) The effects of acidification on the ecology of streams in the upper Tywi catchment in west Wales. *Environ. Pollut. Ser. A.* 35, 125-157.

Stoner, J.H. & Gee, A.S. (1985) Effects of forestry on water quality and fish in Welsh streams and lakes. *J. Wat. Engnrs. Sci.* 39, 27-46.

Stoner, J.H. & Harriman, R. (1988) Forests and freshwater acidification. In: *Acid Rain and British Forests.* Royal Society, London.

Sutcliffe, D.W. (1983) Acid precipitation and its effects on aquatic systems in the English Lake District (Cumbria). *Rep. Freshwater biol. Ass.* 51, 30-62.

Sutcliffe, D.W. & Carrick, T.R. (1973) Studies on mountain streams in the English Lake District. I: pH, calcium and the distribution of invertebrates in the River Duddon. *Freshwater Biology* 3, 437-462.

Sutcliffe, D.W. & Carrick, T.R. (1986) Effects of acid rain on waterbodies in Cumbria. In: *Pollution in Cumbria* (ed. Ineson, P.) ITE symp. no. 16. Institute of Terrestrial Ecology, Grange-over-Sands.

Sutcliffe, D.W. & Carrick, T.R. (1988) Alkalinity and pH of tarns and streams in the English Lake District (Cumbria). *Freshwater Biology* 19, 179-189.

Sweet, J. & Tickle, A. (1990) Memorandum submitted by Greenpeace. In: *The Flue Gas Desulphurisation Programme.* Fourth report from the Energy Committee, House of Commons 371. HMSO, London.

Turnpenny, A.W.H. (1985). The fish populations and water quality of some Welsh and Pennine streams. Central Electricity Research Laboratory Report TPRD/L/2859/N85. CERL, Leatherhead.

Turnpenny, A.W.H., Sadler, K., Ashton, R.J., Milner, A.G.P. & Lynam, S. (1987) The fish populations of some streams in Wales and Northern England in relation to acidity and associated factors. *J. Fish Biol.* 31, 415-434.

Turnpenny, A.W.H., Dempsey, C.H., Davis, M.H. & Fleming, J.M. (1988) Factors limiting fish populations in the Loch Fleet system, an acidic drainage system in south-west Scotland. *J. Fish Biol.* 32, 101-118.

Underwood, J., Donald, A.P. & Stoner, J.H. (1987) Investigations into the use of limestone to combat acidification in two lakes in west Wales. *J. Environ. Management* 24, 29-40.

Wells, D.E., Gee, A.S.& Battarbee, R.W. (1986) Sensitive surface waters - a UK perspective. *Water, Soil & Air Pollution* 31, 631-668.

Wells, D.E. & Harriman, R. (1987) Acidification studies in Scottish catchments: the effects of deposition, catchment type and runoff. In *Acid Rain: Scientific and Technical Advances* (eds. Perry R., Harrison, R.M., Bell, J.N.B. & Lester, J.N.). Selper, London.

Welsh Water (1988). *Acid Waters Project. An Investigation into the Effects of Afforestation and Land Management on Stream Acidity.* Welsh Water, Llanelli.

Whitehead, P.G., Neal, C. & Neale, R. (1986) Modelling the effects of hydrological changes on stream water acidity. *J. Hydrol.* 84, 353-364.

Whitehead, P.G., Neale, R. & Paricos, P. (1987) Modelling stream acidification in upland Wales. In: *Systems Analysis in Water Quality Management* (ed. Beck, M.B.). Pergamon, Oxford.

Whitehead, P.G., Musgrove, T.J. & Cosby, B.J. (1990) Hydrochemical modelling of acidification in Wales. In: *Acid Waters in Wales* (eds. Edwards, R.W, Gee, A.S. & Stoner, J.H.). Kluwer, Dordrecht.

WSL (1987) *United Kingdom Acid Rain Monitoring Results for 1986.* Warren Spring Laboratory, Stevenage.

THE INFLUENCE OF REDUCTIONS IN ATMOSPHERIC SULPHATE DEPOSITION ON ION LEACHING FROM PODZOLIC SOILS UNDER HARDWOOD FOREST

N.W. Foster and P.W. Hazlett
Forestry Canada, Ontario Region, Sault Ste, Marie, Ontario, Canada

ABSTRACT

Annual fluxes of SO_4^{2-} through the forest canopy and podzolic soil at the Turkey Lakes Watershed (TLW) in central Canada decreased in response to reductions in atmospheric SO_4^{2-} deposition. Despite these reductions, there was no additional release of previously adsorbed SO_4^{2-} from the soil. In general, Ca^{2+} leaching from soil at the two sites examined increased, despite the lower SO_4^{2-} fluxes, because of increased leaching of NO_3^- from the soil. In the absence of changes in NO_3^- cycling, changes in SO_4^{2-} alone can reduce base leaching from the TLW soil. Controlling emissions to reduce SO_4^{2-} inputs to forested watersheds will reduce SO_4^{2-} leaching from podzolic soils similar to those examined.

INTRODUCTION

Average sulphate (SO_4^{2-}) deposition in eastern Canada between 1982 and 1986 ranged from 200 to 700 mol_c ha^{-1} yr^{-1} (Pearson and Percy 1990). The Turkey Lakes Watershed (TLW) in central Ontario received an estimated 800 mol_c ha^{-1} yr^{-1} of SO_4^{2-} in the late 1970s (Kelso and Jeffries 1988) and a measured 4-yr mean of 620 mol_c ha^{-1} yr^{-1} of wet SO_4^{2-} in the early 1980s (Sirois and Vet 1988). Calculations of the annual ion budget and laboratory examinations of SO_4^{2-} contents and retention in soil all suggest that the TLW soils retained none of this atmospherically deposited SO_4^{2-} (Foster *et al.* 1986). Atmospheric SO_4^{2-} contributed significantly to leaching of base cations from the podzolic soils of TLW (Foster 1985).

Sulphate deposition in eastern North America has been declining in the past decade (Dillon *et al.* 1988) and is likely to decline further as emission controls are established. A logical consequence of reduced deposition might be lower SO_4^{2-} concentrations in soil solution. Alternatively, SO_4^{2-} adsorbed by soil iron and hydrous aluminum oxides under formerly higher rates of SO_4^{2-} deposition might be released into the more-dilute current soil solutions, thereby acting to maintain SO_4^{2-} levels closer to those observed under the higher deposition regimes. The former scenario would have a greater impact on moderating leaching of base cations from podzolic soils than would the latter one.

In 1986 and 1987, TLW received significantly less SO_4^{2-} from the atmosphere than in previous years. The objective of this report was to demonstrate how leaching of strong-acid anions and base cations from two TLW podzolic soils responded to the reductions in SO_4^{2-} inputs to the forest.

MATERIALS AND METHODS

Study Area

Soil solutions were collected from two study sites at the TLW (lat. 47°03'N, long. 84°24'W) in central Ontario within uneven-aged 'old-growth' stands dominated by sugar maple (*Acer saccharum* Marsh.), with lesser components of yellow birch (*Betula alleghaniensis* Britton). Upland soils consist of a stony, silty-loam ablation till over a compacted sandy basal till at a depth of 0.5 m. Soils at both sites are Orthic Humo-Ferric and Ferro-Humic Podzols (Canada Soil Survey Committee 1978) derived from greenstone and granite. The average January temperature is -10°C, the average July temperature is 18°C and annual precipitation averages 1200 mm, with 400 mm falling as snow between December and April. Detailed information on watershed characteristics and forest structure are contained in Jeffries *et al.* (1988).

Solution Collection

Soil solutions were obtained from 1985 to 1988 in a 0.02-ha plot at 400 m elevation (southeastern aspect) within the Little Turkey Lake snowmelt-runoff study site (basin 30). During the snow-free period, piping water and percolating water (saturated flow) were collected, but no more frequently than every second week, from below the forest floor (F horizon) and within the effective rooting zone (30 cm depth) with four to six (15 by 30 cm) plastic tension-free lysimeters (Jordan 1968).

Solutions collected weekly from 1982 to 1988 on a 1-ha study site at 390 m elevation (southwestern aspect) within basin 31 were also used to assess solution chemistry in greater detail. Basin 31 contained up to four bulk-precipitation, 36 throughfall, and 24 forest-percolate tension-free lysimeters; six mineral-soil solution samplers were also present. Precipitation and throughfall were collected in 20-cm-diameter, continuously open collectors, either plastic funnel-bottle sets or aluminum rain gauges. Ceramic cups (4.8 cm outer diameter) (Soilmoisture Equip. Corp.) collected soil solution by extraction at -20 kPa tension 60 cm below the mineral-soil surface. Before use, the ceramic cups were washed with H_2SO_4 and rinsed with distilled water until the chemical composition of the water passing through the cup equaled that of a distilled-water blank. More frequent solution collections were made at both sites during the snowmelt period in the spring.

Sulphate, chloride (Cl^-), nitrate (NO_3^-), and ammonium (NH_4^+) in water were measured with a Technicon Autoanalyzer II system by the methylthymol blue, mercuric thiocyanate, cadmium reduction and sodium nitroprusside methods, respectively. Sulphate and Cl^- in highly colored samples were measured by means of ion chromatography. Cations were analyzed with an atomic-absorption spectrophotometer by means of flame-emission spectrophotometry for potassium (K^+) and sodium (Na^+) and atomic-absorption spectrophotometry for calcium (Ca^{2+}) and magnesium (Mg^{2+}). Hydrogen (H^+) and bicarbonate (HCO_3^-) were calculated from glass-electrode pH measurements and the results of total-alkalinity electrometric titration, respectively.

Ion fluxes were calculated as the product of concentration and measured solution volumes (or estimated volume for mineral-soil solutions). Estimated solution volumes at basin 31 were calculated according to the water-balance equation (Thornthwaite and Mather 1957). Annual fluxes were not estimated at basin 30 because of the infrequency of collection by tension-free lysimeters during the growing season. The data was summarized on the basis of a water year from 1 November to 31 October to accomodate snowpack accumulation and snowmelt in the same period.

RESULTS

In the 1986 and 1987 water years, SO_4^{2-} deposition in bulk precipitation at TLW averaged 413 mol_c ha^{-1} yr^{-1} (Fig. 1), or 69% of the average levels observed from 1982 to 1985 and in 1988. In the 1986 water year, lower SO_4^{2-} concentrations in bulk precipitation (35 vs 48 μmol_c L^{-1} from 1982 to 1985) were coincident with lower SO_4^{2-} concentrations in soil solution from basin 31 and percolating water from basin 30 (Table 1). Fluxes of SO_4^{2-} through the canopy and soil of basin 31 in 1986 decreased in response to reductions in SO_4^{2-} deposition (Fig. 2).

In the 1986 water year, thawing of the snowpack resulted in less water, H^+, and SO_4^{2-} movement and more NO_3^- and Ca^{2+} leaching through the soil than was observed for the 1983 to 1985 period (Fig. 3, 4, 5). During the 1986 growing season (mid-May to mid-October), water, SO_4^{2-}, NO_3^-, H^+, and Ca^{2+} transport through the soil increased (Fig. 3,4, 5). In contrast, the inputs to and outputs from the soil of Cl^- and base cations other than Ca^{2+} did not deviate greatly, in any season, from those during the 1983 to 1985 period (Fig. 4,5).

Sulphate and NO_3^- represented ~35% and 50%, respectively, of the total anion charge in soil water from both watersheds in the water years before 1986 (Fig. 6). The proportional (Fig. 6) and absolute (Table 1) decreases in mean annual SO_4^{2-} concentrations in soil water in the 1986 water year from the levels in other years, were greater in watershed 31. Likewise, on a yearly basis, less NO_3^-, Ca^{2+}, total bases and SO_4^{2-} were leached from the forest floor in the 1986 water year; however, more NO_3^- and Ca^{2+} were leached from the effective rooting zone (60-cm depth) of basin 31 (Table 2). In basin 30, the mean annual base concentrations in soil water from 1986 to 1988 were only 80% of those observed in 1985 (Table 1). It is likely, therefore, that fewer bases would be leached from the soil in the 1986 through 1988 water years in response to lower NO_3^- and SO_4^{2-} concentrations in 1986, lower SO_4^{2-} concentrations in 1987, and lower NO_3^- concentrations in 1988, respectively (Table 1).

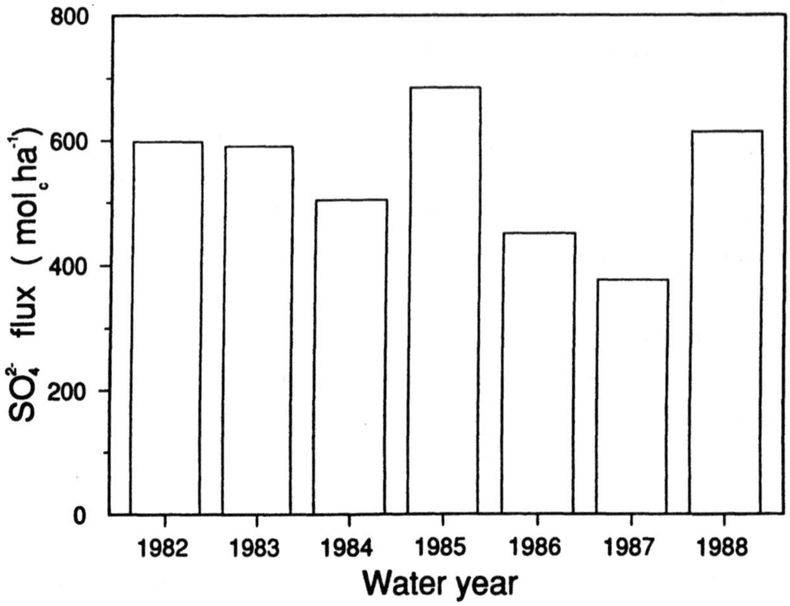

Figure 1. Annual (1 Nov. to 31 Oct.) fluxes of sulphate in bulk precipitation at the Turkey Lakes Watershed.

Table 1. Mean annual[1] ion concentrations (μmol_c L^{-1}) in soil percolate in two tolerant-hardwood basins of the Turkey Lakes Watershed, central Ontario.

	Mean annual ion concentration in soil percolate (μmol_c L^{-1})							
	Basin 31				Basin 30			
Ion	1983	1984	1985	1986	1985	1986	1987	1988
SO_4^{2-}	113	112	95	61	131	114	114	142
NO_3^-	117	132	169	324	151	123	171	74
HCO_3^-	2	5	14	2	40	34	21	29
Cl^-	14	16	14	16	11	11	9	9
Ca^{2+}	152	178	194	270	255	225	264	207
Bases[2]	243	267	279	378	366	298	338	271
Anions	246	265	292	403	333	282	315	254

[1]based on a "water year" from 1 Nov. to 31 Oct.

[2]Σ Ca^{2+}, Mg^{2+}, K^+, Na^+

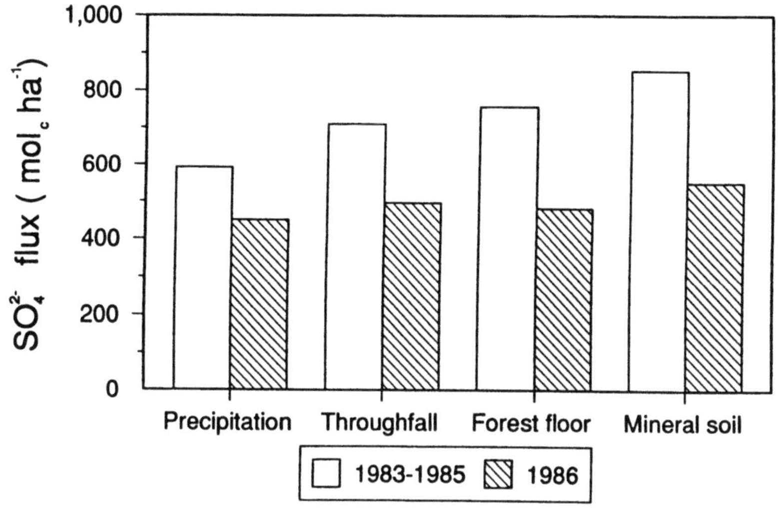

Figure 2. Annual (1 Nov. to 31 Oct.) fluxes of sulphate in solutions from a mature tolerant-hardwood ecosystem in watershed 31.

DISCUSSION

Input/Output Budgets For Sulphate

There is considerable uncertainty associated with the estimates of ion leaching from the mineral soil. Chloride ions are considered to be conservative because they do not react with the soil. For Cl^-, the output from the mineral soil should therefore be balanced with inputs to the soil and the ratio of outputs to inputs should be close to unity. In fact, the calculated annual Cl^- ratios for basin 31 were between 0.91 and 0.94 (1984 excepted). Assuming that the measured inputs of Cl^- to the mineral soil are accurate, the water-balance approach provides a reasonable, although slightly low, estimate of annual Cl^- outputs from the soil. The annual output/input ratios for SO_4^{2-} in the soil of basin 31 varied from 1.12 to 1.14 between the 1983 and 1986 water years. A consistent net loss of SO_4^{2-} from the soil each year, therefore, was determined.

The TLW soil in basin 31, therefore, may be currently losing a small proportion of previously adsorbed SO_4^{2-} or SO_4^{2-} from some of the much larger reserves of organic S (see Foster *et al.* 1986, Foster and Nicolson 1988). Similarly, SO_4^{2-} budgets for a 10-year period in small undisturbed watersheds with podzolic soils and tolerant hardwood forests at the Hubbard Brook Watershed in New Hampshire indicated a net annual loss of SO_4^{2-} (Hornbeck *et al.* 1987). However, Fuller *et al.* (1985) concluded that the podzols they examined under a tolerant hardwood forest in upper New York state, were neither losing nor gaining SO_4^{2-}, since soil-solution SO_4^{2-} was in equilibrium with adsorbed SO_4^{2-}. David and Mitchell (1987) have documented the rapid immobilization of added SO_4^{2-} in organic layers and adsorption in mineral horizons of Adirondack podzolic soils in field experiments with ^{35}S in the form of SO_4^{2-}. They concluded that native soil S was being released in quantities equal to the retention of added S by soil, since the soil was not accumulating S. Other podzolized forest soils have been shown to adsorb SO_4^{2-} (e.g., Neary *et al.* 1987) and retain added SO_4^{2-} (Singh *et al.* 1980, Wright *et al.* 1988).

The lower atmospheric deposition of SO_4^{2-} in the 1986 and 1987 water years resulted in reduced leaching of SO_4^{2-} from the soil of basin 31 and lower SO_4^{2-} concentrations in percolating waters from basin 30. Reduced SO_4^{2-} deposition in 1986 and 1987 may, in part, be associated with an overall reduction in SO_4^{2-} emissions in eastern North America (Dillon *et al.* 1988). However, in the 1988 and 1989 (not shown) water years, SO_4^{2-} deposition was similar in magnitude to pre-1986 levels. It is more likely, therefore, that the TLW experienced, in 1986 and 1987, a shift in dominant air-mass movement from a southerly to a westerly direction, thereby temporarily replacing pollution-laden air with cleaner air.

If lower atmospheric deposition of SO_4^{2-} had an influence on the release or retention of SO_4^{2-} by reserves of organic S and adsorbed SO_4^{2-} in the soil, the reduction in SO_4^{2-} levels in soil solution in 1986 would not be proportional to the reductions in deposition. Similar proportional decreases (i.e., current level divided by the average level from 1983 to 1985) from the mean fluxes recorded in water years 1983 to 1985 in basin 31 were observed in SO_4^{2-} leaching from organic

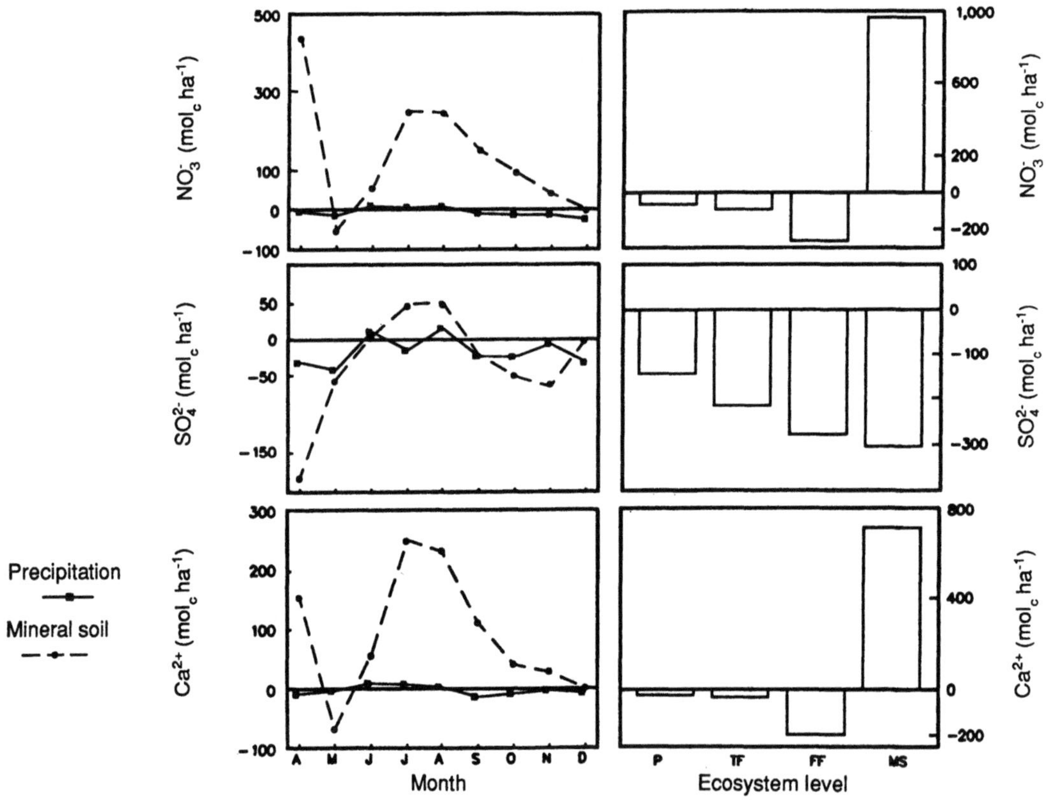

Figure 3. Monthly (lines) and annual (bars) deviation of 1986 fluxes of NO_3^-, SO_4^{2-} and Ca^{2+} from mean 1983 to 1985 values for watershed 31. (P = precipitation, TF = throughfall, FF = forest floor, MS = mineral soil)

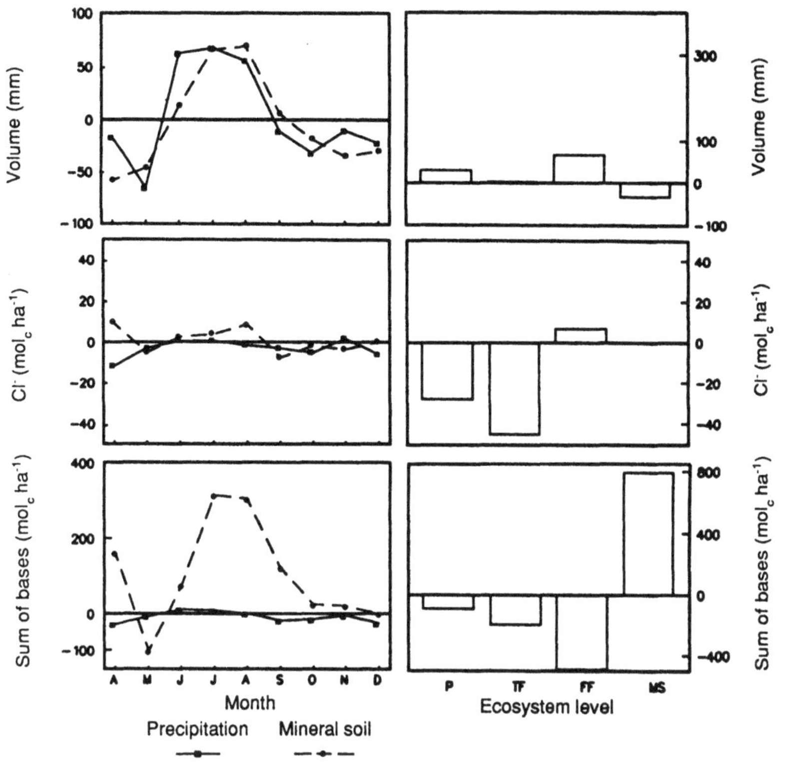

Figure 4. Monthly (lines) and annual (bars) deviation of 1986 fluxes for water volume, Cl^- and sum of bases (Ca^{2+}, Mg^{2+}, K^+ Na^+) from mean 1983 to 1985 values for watershed 31. (P = precipitation, TF = throughfall, FF = forest floor, MS = mineral soil)

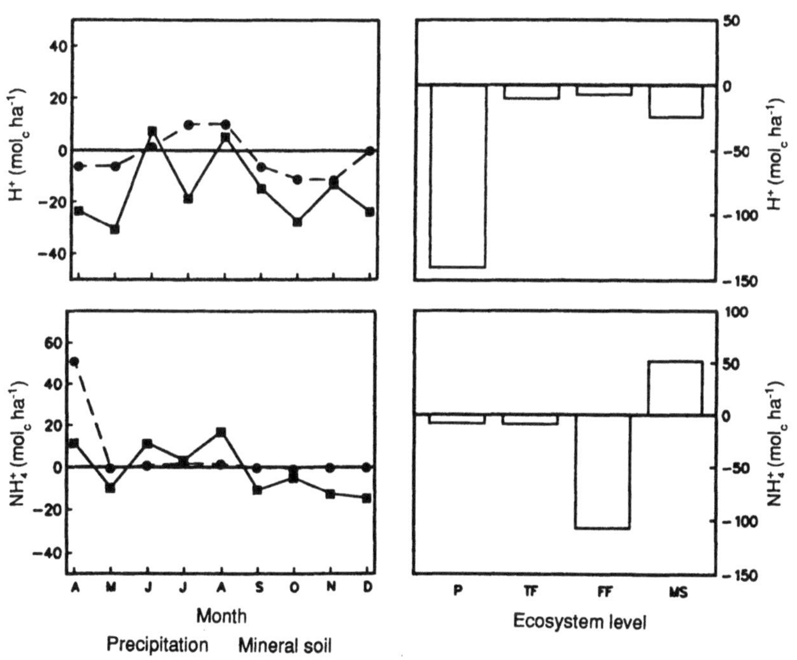

Figure 5. Monthly (lines) and annual (bars) deviation of 1986 fluxes of H^+ and NH_4^+ from mean 1983 to 1985 values for watershed 31. (P = precipitation, TF = throughfall, FF = forest floor, MS = mineral soil)

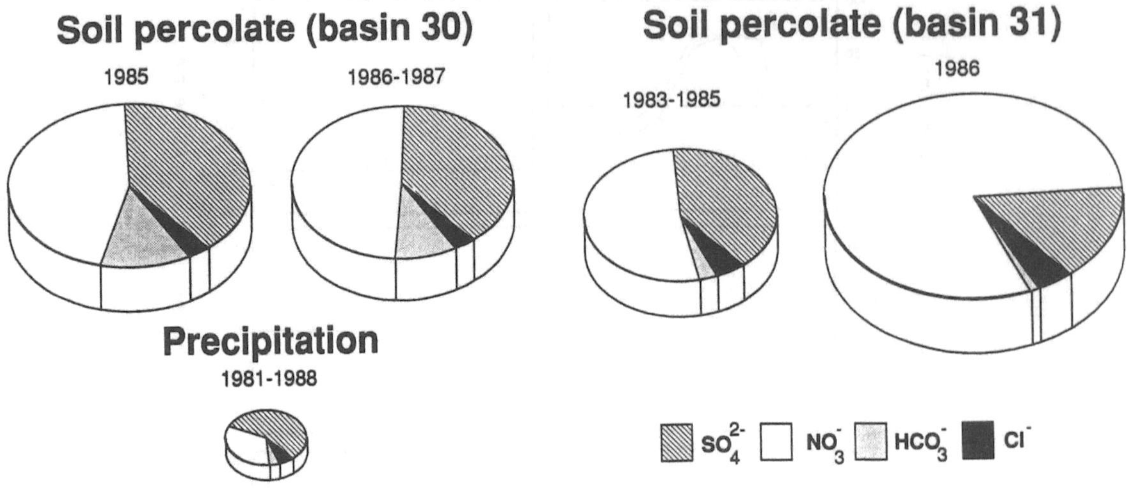

Figure 6. Relative composition of anions in solutions from watersheds 30 and 31, Turkey Lakes Watershed, central Ontario.

Table 2. Mean annual ion fluxes ($mol_c\ ha^{-1}$) in solutions from basin 31 (based on a "water year" from 1 Nov. to 31 Oct.).

	Avg. annual ion flux ($mol_c\ ha^{-1}$)					Avg. annual ion flux ($mol_c\ ha^{-1}$)				
Ion	1983	1984	1985	83-85	1986	1983	1984	1985	83-85	1986
	Precipitation					Forest floor				
SO_4^{2-}	591	504	685	593	450	819	732	723	711	482
NO_3^-	306	308	370	328	263	924	814	438	673	462
HCO_3^-	19	51	105	58	<1	-	-	-	-	19
Cl^-	75	82	77	78	50	112	111	142	114	129
Ca^{2+}	138	192	213	181	158	1584	1423	1349	1360	1255
Σ Bases[1]	286	360	374	340	251	2682	2598	2157	2311	1990
Σ Anions	991	945	1237	1058	763	1855	1657	1303	1498	1092
Volume (mm)	1261	1165	1295	1240	1270	974	935	1024	917	1044
	Throughfall					Mineral soil				
SO_4^{2-}	721	703	706	710	496	924	822	818	855	552
NO_3^-	360	313	320	331	237	1114	1046	1341	1167	2133
HCO_3^-	82	293	260	212	2	27	40	139	69	18
Cl^-	162	124	148	145	100	102	128	134	121	121
Ca^{2+}	303	349	408	353	320	1354	1348	1640	1447	2159
Σ Bases[1]	886	1080	1050	1005	813	2093	2033	2381	2169	2964
Σ Anions	1325	1433	1434	1397	835	2167	2036	2432	2212	2824
Volume (mm)	1146	1045	1120	1104	1107	840	749	842	810	776

[1]Σ Ca^{2+}, Mg^{2+}, K^+, Na^+

layers (0.68), from the effective rooting zone (0.65) and for below-canopy deposition as throughfall (0.70). On the basis of individual rain-storm events, close positive relationships between SO_4^{2-} fluxes in throughfall and forest-floor percolate ($r = 0.84$) and in throughfall and soil solution ($r = 0.60$) have been reported (Nicolson, unpublished data). Both seasonal and annual comparisons indicate that there was no net release of adsorbed SO_4^{2-} from the soil into solution in response to reduced concentrations and fluxes of SO_4^{2-} entering the soil. In other words, the decrease in SO_4^{2-} leaching from the soil was more or less in proportion to the decrease in SO_4^{2-} inputs.

Sulphate Adsorption/Desorption

Changes in SO_4^{2-} solution chemistry are consistent with the relatively low adsorbed SO_4^{2-} content and weak affinity for SO_4^{2-} exhibited by the mineral horizons of this soil, especially at the concentration levels observed in the soil water (Foster *et al.* 1986). The TLW soil contained and retained less SO_4^{2-} than many of the 20 soils (spodosols, inceptisols, ultisols) examined by Harrison *et al.* (1989) in the Integrated Forestry Study.

The net amount of SO_4^{2-} leaching from the rooting zone of the soil in basin 31 in the 1986 water year (70 mol_c ha^{-1}) was actually less than the annual net average from 1983 to 1985 (~140 mol_c ha^{-1}). The change in net SO_4^{2-} leaching could be interpreted as suggesting that either desorption of SO_4^{2-} slowed when SO_4^{2-} inputs lessened or that microbial immobilization/mineralization changes resulted in less net SO_4^{2-} release in 1986. On the other hand, the change in net SO_4^{2-} export was small and these interpretations are tentative, since SO_4^{2-} reactions in this particular soil are incompletely understood and uncertainties exist in estimating SO_4^{2-} output. It appears at least that dilution of forest-floor percolate concentrations by the reduction in SO_4^{2-} inputs was insufficient to increase SO_4^{2-} desorption from the mineral soil and thereby alter the output/input ratio for SO_4^{2-}. In absolute amounts, 230 mol_c ha^{-1} (>75% of the reduction in SO_4^{2-} output from the soil of basin 31 in the 1986 water year in relation to that from 1983 to 1985) could be explained by the decrease in SO_4^{2-} input as a result of atmospheric deposition.

Sulphate retention can also be increased by acidification of soil solutions (Nodvin *et al.* 1986). Nodvin *et al.* (1988) have hypothesized that SO_4^{2-} retention in a tolerant hardwood podzolic soil is enhanced when accelerated nitrification induces protonation of surface adsorption sites. Links between acidification and SO_4^{2-} adsorption are complex and can vary temporally. For example, the increased nitrification in the soil of basin 31 in 1986, by increasing acidification of the soil solution (see Fig. 5), should have favoured additional retention of SO_4^{2-} during the growing season, but this was not observed. In the dormant season, soil solutions lost less acidity in 1986 than expected (Fig. 5) because retention of acidity from nitrification counteracted reduced deposition of atmospheric acidity. Sulphate levels in solution during the dormant period of 1986 declined sharply from those in other years despite a small net decrease in soil-solution acidity. However, the sharp reduction in nitrification in the soil of basin 30 in 1988 was accompanied by an increase in SO_4^{2-} concentrations in soil solution that was greater than that expected as a result of changes in SO_4^{2-} deposition alone.

Theory and soil-equilibration experiments both suggest that SO_4^{2-} adsorption is reversible; a likely consequence of reducing atmospheric SO_4^{2-} deposition would be increased leaching of SO_4^{2-} during the desorption period (Johnson and Reuss 1984). For example, in a whole-catchment manipulation in Norway, exclusion of acidic precipitation decreased SO_4^{2-} concentrations in runoff but SO_4^{2-} output from the basin exceeded atmospheric inputs in both wet and dry deposition (Wright *et al.* 1988). Output/input differences might be explained by SO_4^{2-} desorbed from soil in response to the lower SO_4^{2-} concentrations in solution that resulted from exclusion of SO_4^{2-} in precipitation. Perhaps these soils contained greater quantities of recently adsorbed SO_4^{2-} than the TLW soils and were in equilibrium with current levels of SO_4^{2-} in acidic precipitation.

Nitrate

Precipitation in the 1986 and 1987 (not shown) water years also contained less NO_3^- than was measured from 1982 to 1985. Reduced NO_3^- supply to the soil should also favor reduced cation leaching. The rates of NO_3^- addition in precipitation at TLW in these years, however, were considerably less than rates of NO_3^- release from organic-N reserves in the soil (Table 2).

Lower precipitation during the dormant period of the 1986 water year was associated with reduced SO_4^{2-} levels in soil solution and increased NO_3^- leaching from the soils in basin 31. Lower SO_4^{2-} leaching was largely related to reduced inputs in precipitation, but excess NO_3^- was related to increased leaching from within the soil. Increased natural tree mortality on the study plot in basin 31 is likely to have reduced the forest's demand for N (Foster *et al.* 1989). Because of reduced N uptake by the vegetation, more NH_4^+ was available to be nitrified in the soil and NO_3^- leaching was subsequently increased. Greater-than-normal precipitation during the growing season of 1986 also contributed to NO_3^- leaching from the soil.

The mature hardwood forest at TLW appears to be in a dynamic state with respect to N cycling between vegetation and soil. In the 6 years examined in this report, mean annual concentrations of NO_3^- in soil solution varied by 200% in basin 30 and by almost 300% in basin 31 (Table 1). Nitrate concentrations in soil solution are largely controlled by the difference between mineralization and immobilization of inorganic N in soil by microbes and the demand for N by vegetation. Atmospheric inputs of inorganic N also contribute to variation in NO_3^- levels in soil solution.

Nitrate concentrations made a significant contribution to the overall total anion composition of soil solutions in both basins. In basin 30, for which tree mortality was not observed, year-to-year changes in mean NO_3^- concentrations were sufficient to alter the total anion composition of soil water significantly.

Base Leaching

Despite the dominance of acidic cations on the exchange complex of the TLW soils, soil base saturation, which averaged 25% in the surface 15 cm of mineral soil, exceeded suggested critical levels (10-15%; Reuss 1983) below which release of aluminum into soil solution is greatly enhanced. Acidic cations in throughfall, therefore, displaced base cations (chiefly Ca^{2+}, which was the dominant base on the exchange sites of the soil) into soil solution. Base cation leaching from the TLW soils was promoted by external inputs of SO_4^{2-} and NO_3^- from the atmosphere and by internal production of NO_3^- from organic N. The significant positive correlation between NO_3^- and Ca^{2+} concentrations in soil solution, especially during the dormant period, has been reported previously for TLW soils (Foster 1985). Calcium movement, therefore, seems particularly dependent on the presence and mobility of NO_3^- ions.

In precipitation and throughfall, the reduction in deposition of strong-acid anions was coincident with a reduction in H^+ deposition (Fig. 5). The impact of reduced SO_4^{2-} deposition was a decrease in leaching of Ca^{2+} and other bases from the forest floor. However, in the 1986 water year, increased leaching of NO_3^- from the mineral soil in basin 31 offset the effect of reductions in strong-acid anions, so that Ca^{2+} leaching was greater than that reported in other years. In basin 30, on the other hand, the lower concentrations of strong-acid anions in precipitation decreased their level in soil solution and thereby contributed to a reduction in the concentrations of Ca^{2+} and/or the sum of bases in percolating water. In the 1987 water year, SO_4^{2-} was the only anion in precipitation with lower concentrations and contents. Reductions in concentrations of SO_4^{2-} and of summed cations, but not of Ca^{2+} in comparison with 1985 levels, were observed in soil percolate in 1987. Decreases in SO_4^{2-} leaching, in the absence of changes in NO_3^- cycling, can reduce base leaching from the TLW soil. Although the decrease in atmospheric deposition of SO_4^{2-} to TLW in the 1986 and 1987 water years appears to have been temporary, these results imply that controlling emissions to reduce SO_4^{2-} inputs is likely to reduce SO_4^{2-} leaching from the soil and thereby slow the rate of soil acidification. Changes in the leaching of Ca^{2+} and other base cations may differ from those predicted by examining SO_4^{2-} alone because of strong year-to-year differences in NO_3^- leaching from the soil.

ACKNOWLEDGMENTS

The assistance of S. Curtis and W. Johns, with sample collection, and of L. Chartrand, S. Gibbs and D. Kurylo, with chemical analysis of samples, is gratefully acknowledged.

REFERENCES

Canada Soil Survey Committee. 1978. The Canadian System of Soil Classification. Dep. Agric., Ottawa, Ont. Publ. 1646. 164 pp.

David, M.B. and Mitchell, M.J. 1987. Transformations of organic and inorganic sulfur: importance to sulfate flux in an Adirondack forest soil. *J. Air Pollut. Contr. Assoc.* 37:39-44.

Dillon, P.J., Lusis, M., Reid, R. and Yap, D. 1988. Ten-year trends in sulphate, nitrate and hydrogen deposition in central Ontario. *Atmospher. Environ.* 22:901-905.

Foster, N.W. 1985. Acid precipitation and soil solution chemistry within a maple-birch forest in Canada. *For. Ecol. Manage.* 12:215-231.

Foster, N.W., Morrison, I.K. and Nicolson, J.A. 1986. Acid deposition and ion leaching from a podzolic soil under hardwood forest. *Water Air Soil Pollut.* 31:879-889.

Foster, N.W. and Nicolson, J.A. 1988. Acid deposition and nutrient leaching from deciduous vegetation and podzolic soils at the Turkey Lakes Watershed. *Can. J. Fish. Aquat. Sci.* 45 (Suppl. 1):96-100.

Foster, N.W., Nicolson, J.A. and Hazlett, P.W. 1989. Temporal variation in nitrate and nutrient cations in drainage waters from a deciduous forest. *J. Environ. Qual.* 18:238-244.

Fuller, R.D., David, M.B. and Driscoll, C.T. 1985. Sulfate adsorption relationships in forested spodosols of the northeastern U.S.A. *Soil Sci. Soc. Am. J.* 49:1034-1040.

Harrison, R.B., Johnson, D.W. and Todd, D.E. 1989. Sulfate adsorption and desorption reversibility in a variety of forest soils. *J. Environ. Qual.* 18:419-426.

Hornbeck, J.W., Martin, C.W., Pierce, R.S., Bormann, F.H., Likens, G.E. and Eaton, J.S. 1987. The northern hardwood forest ecosystem: ten years of recovery from clearcutting. USDA For. Serv., Northeastern Forest Experiment Station, Broomhall, PA. Rep. NE-RP-596. 30p.

Jeffries, D.S., Kelso, J.R.M. and Morrison, I.K. 1988. Physical, chemical and biological characteristics of the Turkey Lakes Watershed, central Ontario, Canada. *Can. J. Fish. Aquat. Sci.* 45(Suppl. 1):3-13.

Johnson, D.W. and Reuss, J.O. 1984. Soil-mediated effects of atmospherically deposited sulphur and nitrogen. *Phil. Trans. R. Soc. Lond.* B. 305:383-392.

Jordan, C.F. 1968. A simple tension-free lysimeter. *Soil Sci.* 105:81-86.

Kelso, J.R.M. and Jeffries, D.S. 1988. Response of headwater lakes to varying atmospheric deposition in north-central Ontario, 1979-85. *Can. J. Fish. Aquat. Sci.* 45:1905-1911.

Neary, A.J., Mistry, E. and Vanderstar, L. 1987. Sulphate relationships in some central Ontario forest soils. *Can. J. Soil Sci.* 67:341-352.

Nodvin, S.C., Driscoll, C.T. and Likens, G.E. 1986. The effect of pH on sulfate adsorption in a forest soil. *Soil Sci.* 142:69-75.

Nodvin, S.C., Driscoll, C.T. and Likens, G.E. 1988. Soil processes and sulfate loss at the Hubbard Brook Experimental Forest. *Biogeochemistry* 5:185-199.

Pearson, R.G., and Percy, K. (Eds.) 1990. The 1990 Assessment Document. Dep. Environ., Terrestrial Effects Subgroup. Res. Monitoring and Coordinating Comm., Downsview, Ont. 96 p.

Reuss, J.O. 1983. Implications of the calcium-aluminum exchange system for the effect of acid precipitation on soils. *J. Environ. Qual.* 12:591-595.

Singh, B.R., Abrahamsen, G. and Stuanes, A. 1980. Effect of simulated acid rain on sulfate movement in acid forest soils. *Soil Sci. Soc. Am. J.* 44:75-80.

Sirois, A. and Vet, R.J. 1988. Detailed analysis of sulfate and nitrate atmospheric deposition estimates at the Turkey Lakes Watershed. *Can. J. Fish Aquat. Sci.* 45(Suppl. 1):14-25.

Thornthwaite, C.W. and Mather, J.R. 1957. Instructions and tables for computing potential evapotranspiration and the water balance. p. 185-311 in: Climatology Vol. X, No. 3.

Wright, R.F., Lotse, E. and Semb, A. 1988. Reversibility of acidification shown by whole-catchment experiments. *Nature* 334:670-675.

CROWN STRUCTURE AND TREE VITALITY

A. Roloff
Institute of Silviculture II, Buesgenweg 1, Göttingen, F.R.G.

Abstract

In the present paper firstly general methodical problems of vitality assessments of deciduous trees are discussed and also existing disparities or contradictions are pointed out if assessments based on "leaf loss" and those based on crown structures are compared. The necessity of a consideration of branching is substantiated and methods which have been developed until now are presented.

By a new approach based upon branching structures, a long-term, chronic decrease of vitality can be recognized. Thereby, it is a practicable method to detect forest decline and to discriminate it from short-term influences, e.g. drought damages.

1. Introduction

It is still very difficult to determine vitality of deciduous and coniferous trees. This is due to that in most inventories only parameters such as "percentage leaf loss" and leaf or needle colouring are considered, either because of lack of further knowledge or in some cases, because there is a reluctance to change from commonly used methods.

Therefore, in the following paper the branching and the crown structure of trees in the assessment of tree vitality is considered. This is also important for the interpretation of aerial photographs.

Among deciduous trees European beech (Fagus sylvatica) is of particular interest, as it's distribution ranges throughout

Central Europe and in many countries it represents the most important broadleaved tree species, furthermore it grows on extremely different sites.

2. Stress symptoms of tree crowns: "leaf loss" versus crown structure

The consideration of crown structures in the assessments of tree vitality has become increasingly important as scientists are now aware of the problem of "leaf loss" has been shown in many studies (ROLOFF 1986, 1989a,b, FLÜCKIGER et al. 1986, 1989, PERPET 1988, GIES et al. 1989, MÖHRING 1989, RICHTER 1989, WESTMAN 1989): the number of leaves and above all the leaf size underlie considerable annual fluctuations, for example as a result of drought and insect damage or flowering, fructification, respectively, and thus is to some degree inappropriate for the vitality assessment of deciduous trees. On the other hand of course the foliage also has to be taken into account. The correlation between "leaf loss", that is crown transparency, and fructification has been well demonstrated (FLÜCKIGER et al. 1989). However, the leaf size even within the same crown of a beech tree underlies such a great variation, that it is difficult to show a statistical significance between different trees (ROLOFF 1986, GIES et al. 1989).

For this reason it is not surprising if considerable disagreements between so called "leaf loss" and the crown structure, which will be discussed later, do occur if vitality assessments of the same beech trees are compared (fig. 1). The assessments may differ by up to 2 damage or vitality classes and agreement is only achieved for about 50% of the assessed trees (HESS.FORSTL.VERS.ANST. 1988, ATHARI & KRAMER 1989a).

It would therefore be of great advantage if the term "leaf loss" could be replaced by a different term (for example crown transparency), which does not lead to the misconception that we are talking about shedded leaves. As it is commonly known for the inventory of forest decline: A deciduous tree showing a "leaf loss" of 30% does not mean that 30% of the leaves were shed, but that they simply had never existed at the beginning of the vegetation period (ROLOFF 1986).

In this context, there is one more aspect which should be mentioned, as it was conspicuous especially in the most recent

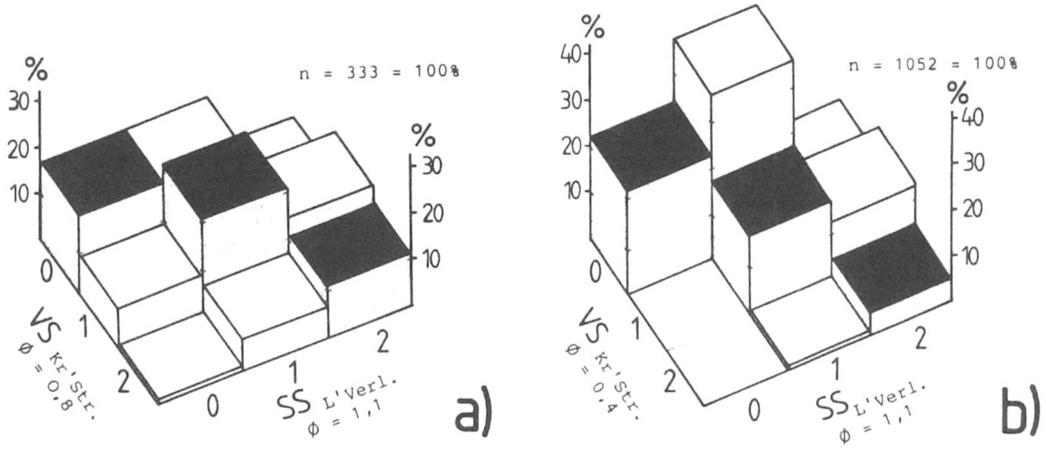

Fig. 1: Disagreements between a vitality assessment of beech trees based upon leaf loss damage classes (SS L'Verl.) and crown structure vitality classes (VS Kr'Str.) after investigations of a) ATHARI & KRAMER 1989a,b and b) HESS. FORSTL.VERS.ANST. 1988; the black columns mark the share of trees which are in the same vitality/damage class in accordance with both methods of assessment

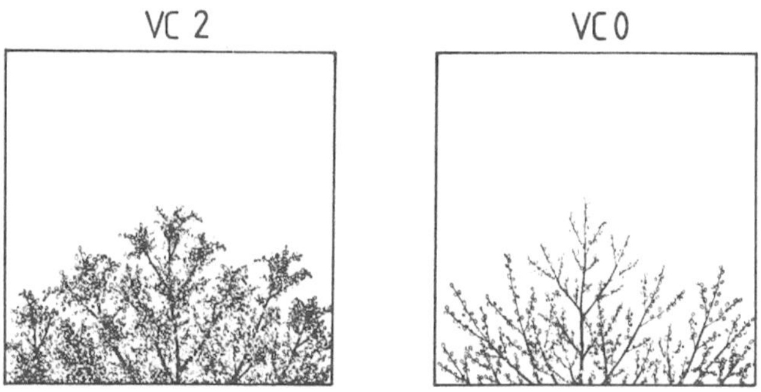

Fig. 2: Disagreement between a vitality assessment based upon crown transparency and crown structure (vc = vitality class according to ROLOFF 1989a): in many tree species (here: wild cherry) the crown becomes more transparent with better growth

investigations. There are a great variety of tree species, in which the crown with increasing shoot lengths, that is better growth, becomes more transparent (fig. 2). In this case a vitality assessment on the basis of crown transparency versus crown structure is bound to produce exactly opposite results (ROLOFF 1989a,b).

A problem inherent to present inventories for the determination of damage progression, is discussed by MÖHRING (1989). By means of a series of photographs taken over a period of 5 years of the same crown parts of beech trees, he could show that due to die-off and breakage of branches, damage trend is represented inadequately by "leaf loss". Thus, no change of damage class is noticed when the dying-off and shedding of branches happens simultaneously (fig. 3). If the number of branches which break off is greater than the number of branches which die-off, a damage decrease is determined. However, if the number of branches which die-off is

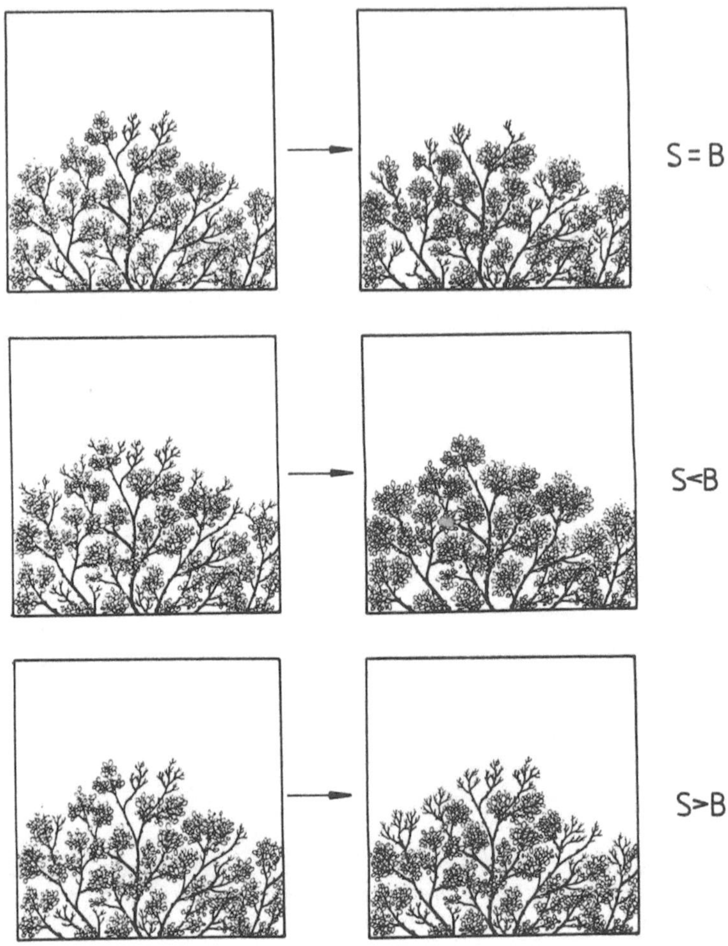

Fig. 3: Damage trend is often represented inadequately by "leaf loss" because of different time course of die-off (S) and breakage (B) of branches

greater than that of the branches shed, a "damage" increase is recorded. In any case, the true damage progression can not be assessed by "leaf loss".

3. Changes in the crown structure with decreasing vitality

In the following, tree vitality is discussed in terms of growth potential, which in trees is expressed in shoot growth. Although various branching structures within one tree crown have long been known (BÜSGEN 1927, THIEBAUT 1981, 1988), their significance as a vitality indicator has only recently been discovered (ROLOFF 1985).

3.1. Shoot Morphology

Carefully observing the branching pattern of a hardwood tree first of all closely packed grooves upon the shoot surface can be recognized conspicuously (fig. 4). These are the so-called 'shoot-base scars' which were unfortunately ignored for a long time. They are the scars of the bud scales, which closely packed formerly encased the young shoot primordia and hence they mark the boundary between two year's growth exactly to a millimetre. In beech there is the good luck that the rind does not turn to bark as for example in oak, but remains more or less smooth. For this reason it becomes possible to retrace the development of any hardwood branching pattern for many years, partly - for example in beech - over decades and in this way to reconstruct its growth.

Further investigating the branches in most tree species can be distinguished two kinds of shoots: short shoots and long shoots (fig. 5). Short shoots only are a few millimetres or centimetres long, only have 3 to 5 leaves and do not ramify in the following

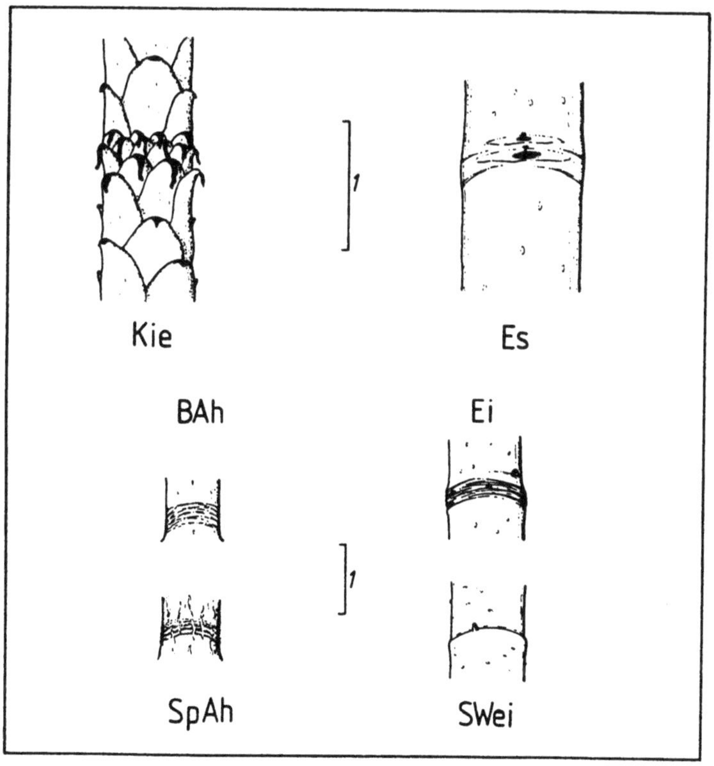

Fig. 4: Shoot base scars of Scotch pine (Kie), European ash (Es), Sycamore (BAh), Norway maple (SpAh), oak (Ei) and common willow (SWei) (scales in cm)

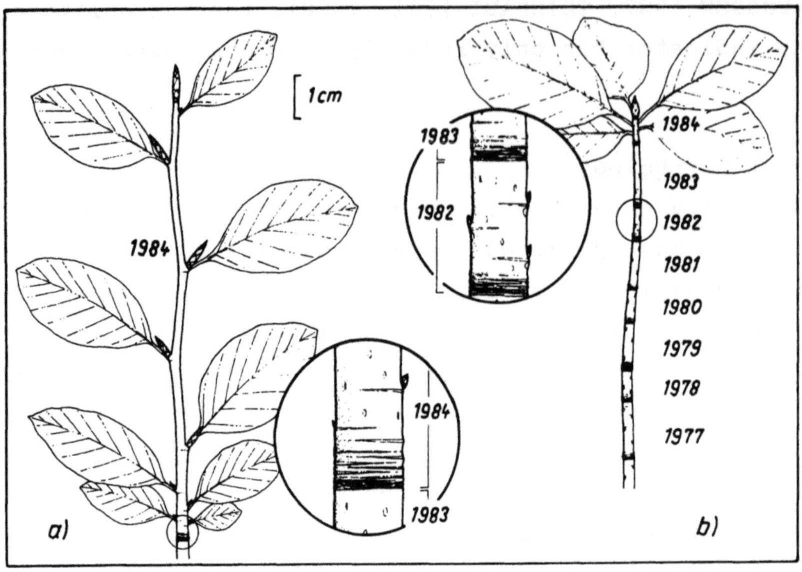

Fig. 5: Shoot morphology of beech. (a) long-shoot with shoot-base scar (circle), alternate distichous leaves and lateral buds, terminal bud; detail: shoot-base scar, the boundary between the annual shoots of 1983 and 1984 (b) 9-year-old short-shoot chain without ramification and with terminal cluster of leaves; detail: annual shoot boundaries, clearly marked by the shoot-base scars; dormant lateral buds ("reserve buds")

years, because they laterally produce only very little dormant buds. The terminal bud of short shoots, however, either produces a short shoot again the following season and there are formed successive short shoots ('short shoot chains'), or else it anew turns into forming a long shoot. Long shoots are clearly longer, show more leaves and ramify in the following years.

Finally, it is still important to know that in beech, oak and many other broadleaved tree species only those primordia are shooted in spring, which were developed in the bud during the previous year: this is called 'determinate growth'. The result of it is that a dry summer cannot affect the shoot lengths before the following year.

Owing to mature age, in any tree species the annual growth of the treetop shoots and therefore the height increment decreases after passing a culmination point. That reflects the decreasing vitality. The lengths of the treetop shoots are to interpret as a vitality sign, because the strategy of a forest forming tree species to conquer new airspace, has to occur especially in the very top of the tree. On the other hand, the shoot lenghts in the inner, lower and lateral crown areas mainly depend on competition and lighting conditions and therefore are unsuitable for the assessment of the vitality of a whole tree.

3.2. Model of growth phases

Fig. 6 shows how a typical ramification originates from a leader shoot of a vigorous beech. It is similar in most other hardwoods only with little modifications. This exploration-phase produces the branching structure which is best known and most frequently found: the terminal and the upper lateral buds yearly develop into long shoots, from the lower lateral buds arise short shoots and the lowermost at last, don't shoot at all but remain lasting for years as very little dormant buds preserved for unusual events.

In every annual leader shoot the lengths of the lateral younger shoots decrease from top to bottom, the developing ramification is turned upwards or forwards. In this way, there is developed an obviously storey-like branch system and we can distinguish the annual shoot boundaries (marked by the interruptions of the black lines) even from a distance (by the steps of the branching pattern, by the abrupt change from long lateral shoots to short shoots). This exploration-phase is the normal appearance of the leader shoots in healthy vigorous trees until an old age, because only in this way the treetop can fulfil its main purpose for the tree steadily to conquer new airspace, to

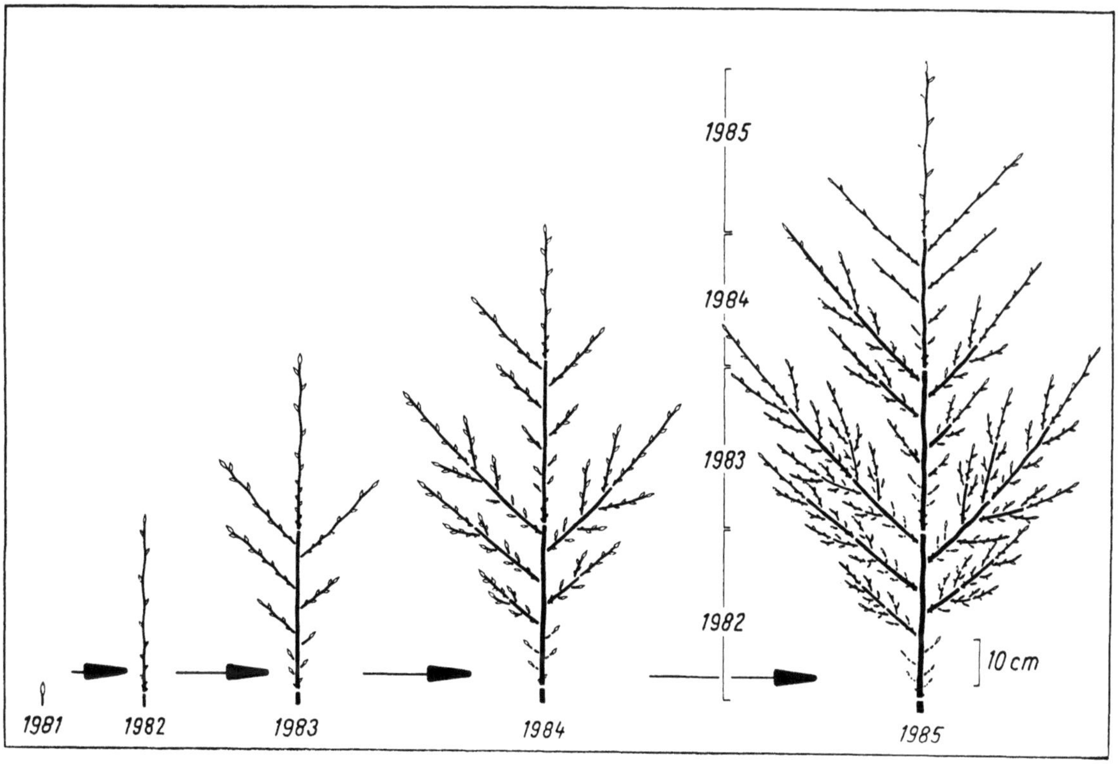

Fig. 6: Detailed illustration of the 4-year-old growth of a typically ramified beech (exploration phase)

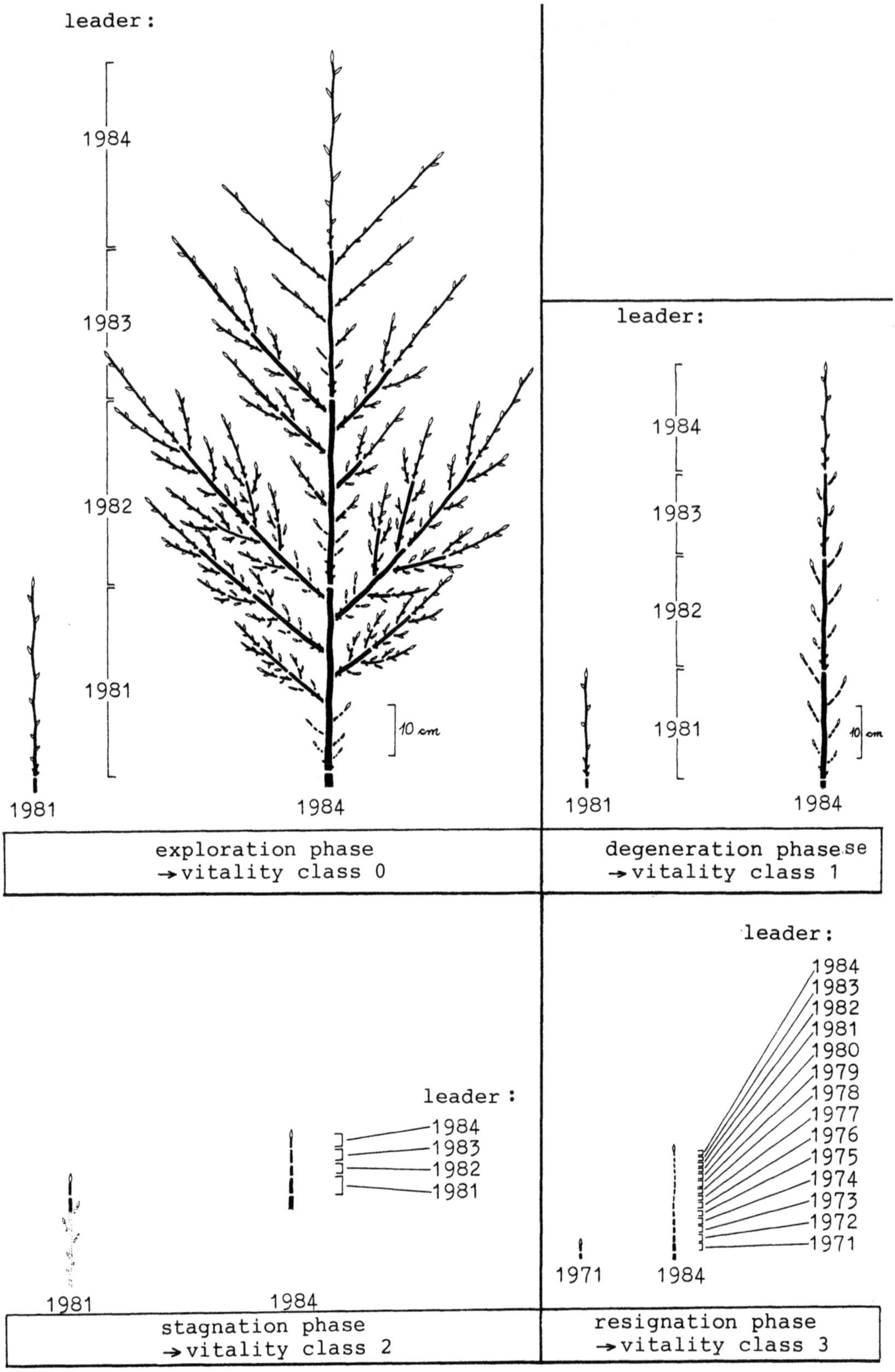

Fig. 7: Growth phases of beech (model): on the left the state of the leader in 1981, on the right its ramification after 3 years (1984) (annual boundaries marked by interrupted black lines)

fill it up with lateral shoots and to be successful against rival trees.

In the degeneration-phase (fig. 7), however, the terminal bud still may develop but into shorter long shoots, but from nearly any lateral bud (from the uppermost, too) arise short shoots almost without exception. Thereby an obvious impoverishment of the branching pattern takes place and there are formed spears in the periphery of the crowns, which are to perceive from a long distance, too.

In the course of a further decreasing vitality, even the terminal bud changes into developing short shoots. In this stagnation-phase no more ramification occurs because short shoots do not ramify. On account of the short annual lengths of these shoots, the length increment of the branch or the height increment of the tree stagnates.

If this stagnation-phase persists longer than a few years (if it does not have temporary feature), the branch or - if concerning the treetop shoots - the whole tree dies. This is called the resignation-phase.

Already as a result of their disadvantageous mechanical-static features (a dense cluster of leaves at the end of very delicate shoots), the short shoot chains cannot grow to any length or any age in the upper crown area exposed to the wind. Now secondary factors decide on the exact time of dieback. There are formed claws in the crown periphery typical for this stage, because the short shoot chains getting longer stretch to the light.

After these remarks, it should be understandable that the different phases of growing are due to decreasing shoot lengths and, therefore, reflect a decreasing vitality.

Similar growth-phases are to distinguish in every other investigated broadleaved tree species, only with little species-dependant modifications, e.g. in oak, maple, ash, birch, willow and others (ROLOFF 1989a).

Basing on these growth-phases the following vitality class system was developed.

3.3. Vitality classes

Healthy undamaged trees of the vitality class 0 (fig. 8) show treetop shoots in the exploration-phase: both the main axes and partly the lateral twigs consist of long shoots. For this reason a regular netlike branching pattern is developed, which reaches deep to the interior of the crown. The crowns are equally

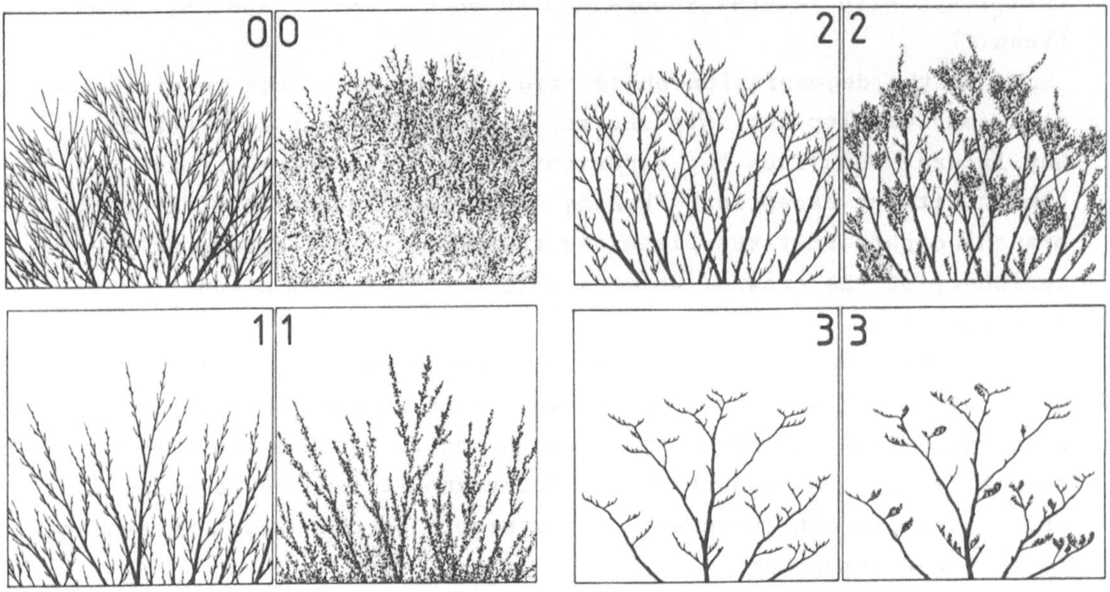

Fig. 8: Vitality classes in beech (view of the upper crown in winter and summer; basing on the growth phases of fig. 7)

closed and domed and do not show any greater gap unless just a stronger intervention has happened in the stand, because such a gap is closed quickly again by the intensive ramification. Just so the newly conquered airspace is quickly filled by the harmonical branching pattern. In the summer a dense foliage arises without any greater gap.

Weakened trees of the vitality class 1 show treetop shoots in the degeneration-phase. Thus, there are formed spears rising above the canopy. On these spears the leaves are arranged densely and all around (at the top of the lateral short shoots or short shoot chains). The crowns make a frazzled impression on the outside and show a fastigiate appearance, because the airspace between the spears is not completely filled by leaves and twigs, there arises a spikey outline of the crowns. Inside the crown the branching pattern and therewith the foliage, too, is quite dense, because it still dates from so to speak 'better times'. In this vitality class straight percurrent mean axes of the treetop branches are still dominant, but the crowns no more look as intact as in class 0 because of the spears shooting out of the canopy.

In obviously declining and devitalized trees of the vitality class 2 the treetop shoots begin to turn into building short shoots in the stagnation-phase. In the leafless state, it could be designated as the claw stage, because the short shoot chains in the outside of the crowns grow longer, are predominant and stretch claw-like to the light. These short shoot chains growing too long break off in the summer in thunderstorm and heavy rain and strew the forest ground in declining beech stands. Under normal circumstances beech and other tree species, too, get rid of a part of their unimportant twigs in the inner and lower crown parts in this way. But if the treetop shoots themselves are now devitalized, the self-pruning of twigs progresses into the outskirts of the crown, the crowns are getting thin from inside outwards. The cause for that is not premature leaf fall as often stated falsely, but broken short shoot chains, lack of ramification and dead buds and twigs. The still existing ramification is accumulated bushily and lumpily in the periphery of the crowns. This causes summer and winter brushy crown structures and greater gaps. There still are hardly straight percurrent branches in the crown periphery.

In considerably damaged or dying trees of the vitality class 3 the crowns finally fall apart by breaking off of larger branches and dieback of whole crown parts. The tree only seems to consist of more or less a lot of subcrowns dispersed rather accidentally in the airspace and forming whip-like structures. The treetop often is dying back or it already is dead, because the treetop shoots have grown in the resignation-phase.

It should be pointed out that an assessment of vitality by this way is possible in trees of any age. Then they should be called vitality classes. If someone wants to call them damage classes, he has to take into account the age of the assessed trees and, e.g. in beech, he should not assess the vitality of trees older than 150 years because of the natural aging trend.

3.4. Drought damages, relationship of the presented vitality criteria to environmental factors

After the previous explanations, it should be self-evident that such a vitality class system based on criteria of the branching structures especially goes back to long-term, chronic diminution of vitality.

E.g., drought damages can influence branching only temporarely and do not lead to a fundamental variation of the branching structure. Drought damages are to identify even after years by the help of the shoot base scars, because the treetop shoots are abruptly very short in the year after a dry summer (because of the determinate growth) and recover their original values quickly again (fig. 9).

On the other hand, drought damages can cause a fundamental passing thinness of the foliage because of the premature leaf-fall

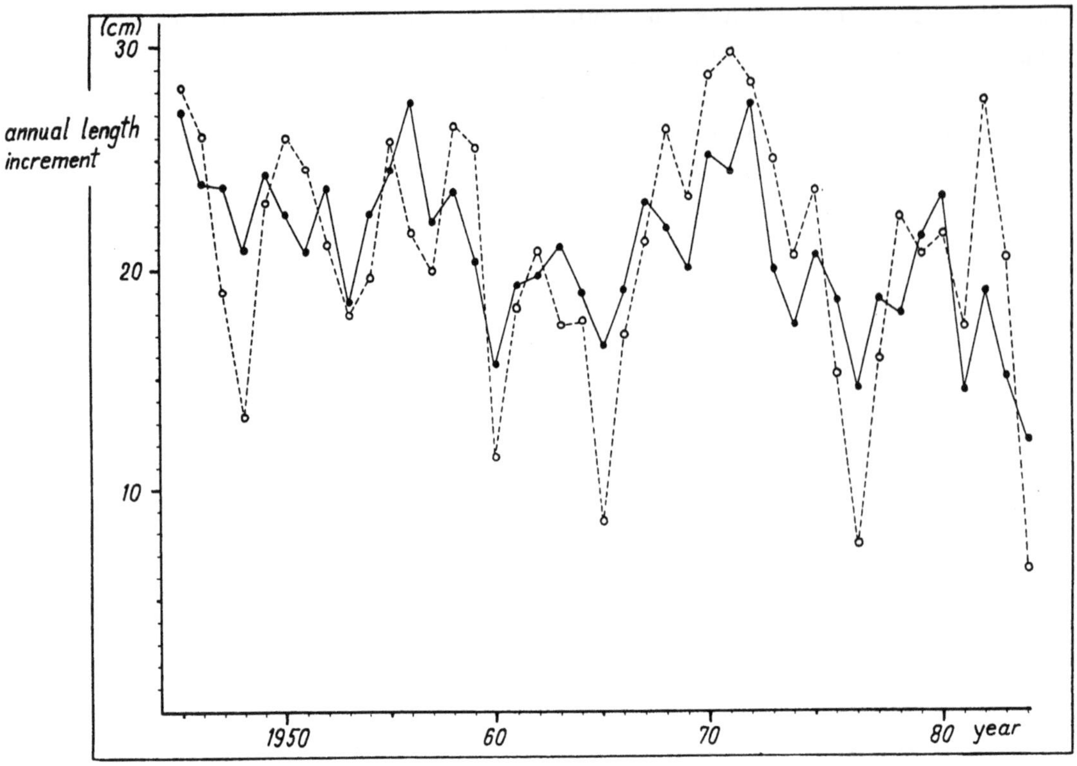

Fig. 9: Annual length increament of the leader shoots of two 120-year-old beech trees, one of which clearly shows drought damages (very short shoots formed abruptly in the year following an extremely dry summer and afterwards quick recovery), whereas the other remained nearly unaffected; ——●—— beech with only moderate reaction to the droughts, ----O---- beech damaged by the droughts of 1947, 1959, 1964, 1976 and 1983

during the current summer and smaller and less leaves in the following season . But it is impossible to recognize temporary short shoots of the past caused by drought from a distance, e.g., from the forest ground, and they cannot effect a fundamental modification of the branching structure (fig. 10), whereas a chronic decrease of the vitality (connected with a long-term decrease of the shoot lenghts in the treetop) cause a conspicuously different branching structure in hardwoods (fig. 11).

Of course, the best time for an exact assessment of vitality with the help of the presented method is after leaf-fall, in autumn and winter.

As only long-term influences can effect a fundamental modification of the branching pattern, the presented method above all is suitable to identify 'forest decline' in hardwoods, because the recent rise of decline in our forests is supposed to be caused by already long time lasting influences. Thereby, modifications of

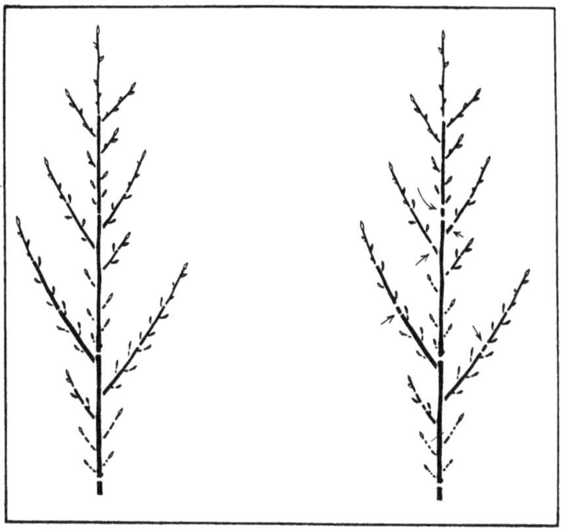

Fig. 10: A "hidden" drought damage in the ramification on the right (s. arrows) does not lead to a fundamental modification of the branching structure, but to a "delay" of one year compared with the unaffected one on the left

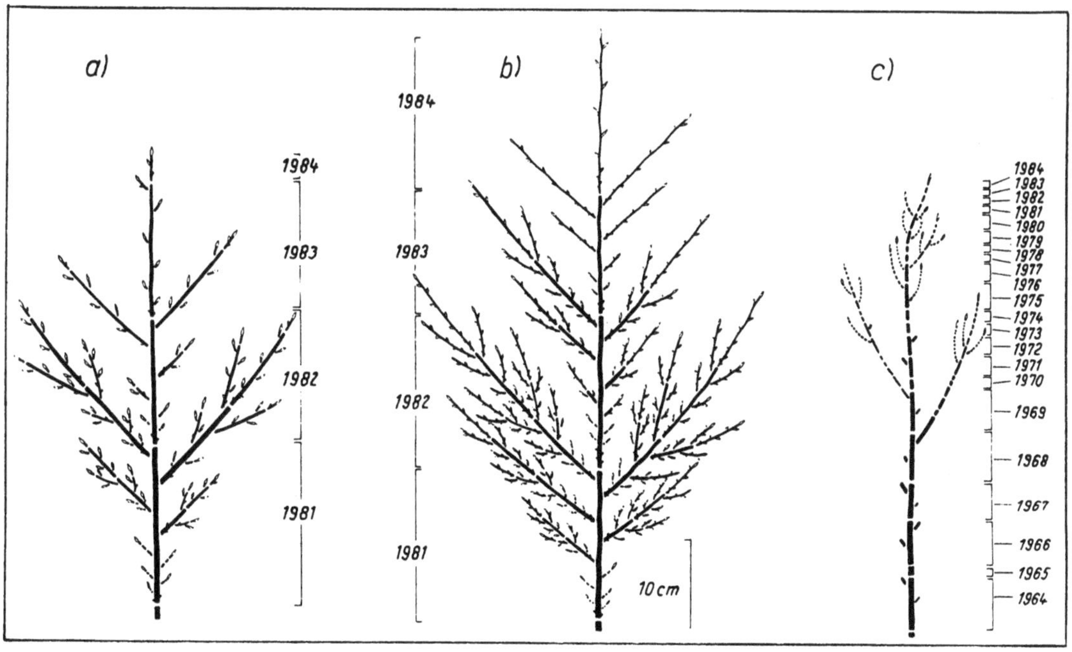

Fig. 11: Comparison of typical treetop branching patterns of a healthy beech (b), a drought damaged (a) and a chronically declining beech (c)

the branching structure are not exclusively specific to air pollution, but to long-term negative factors affecting tree growth as for example air pollution. A study trip to high polluted areas in Yugoslavia has shown that air pollution unambiguously causes the presented modifications of the branching structure, on the other hand, e.g., a long-term salt stress along the main roads can have the same result.

It is similar in almost all other symptoms, which are known up to the present day: they can be caused by air pollution, but may also be a result of other stress factors.

A legitimate assumption is that the presented modifications of the branching structure are related to similar modifications of the root system, because recent research has shown the high correlations between both these components of the system tree (ROLOFF & RÖMER 1989).

Further investigations in beech surprisingly have clarified, how long the decline in this important European hardwood species has already lasted (fig. 12). Only the vigorous, healthy trees of the vitality class (vc) 0 follow the age trend of the yield tables, whereas trees of the vc 1 to 3 decline since many years. Transforming the absolute values of the leader increment into values in per cent of vc 0 (fig. 13), emphasizes that trees of vc 1 + 2 are in most cases declinging since about 10 to 15 years,

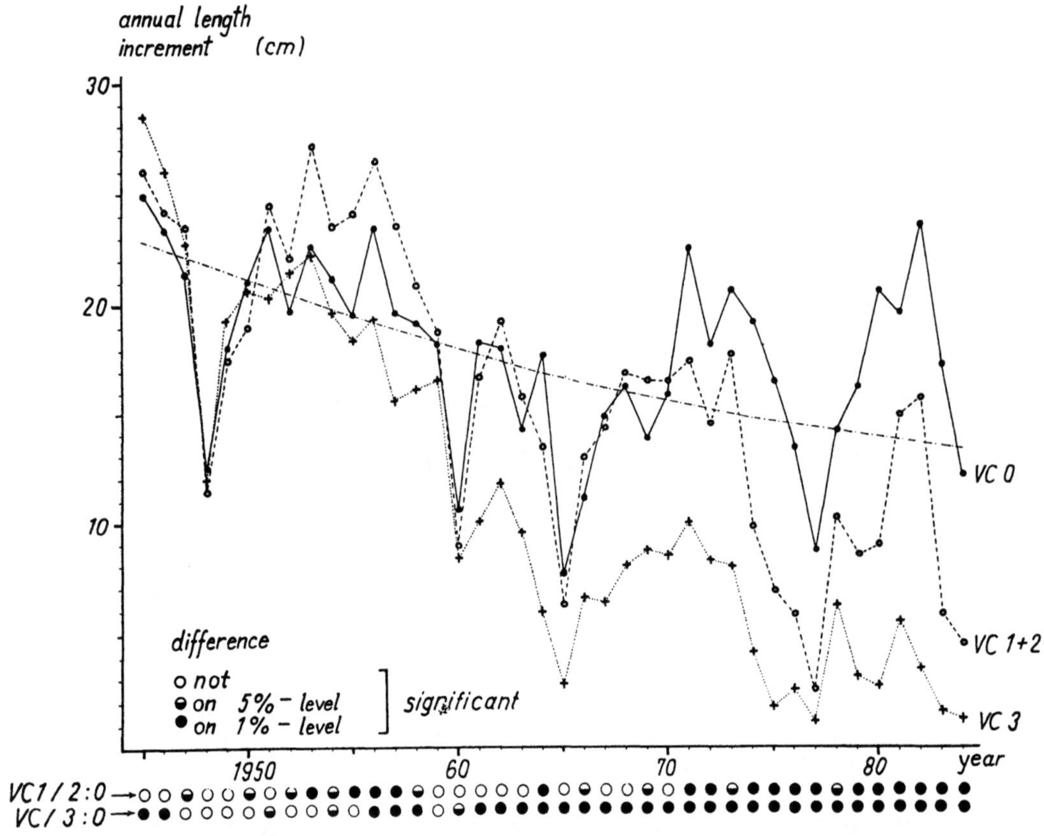

Fig. 12: Annual length increment of the leader shoots of 100- to 160-year-old beech trees during the last 40 years (classified by vitality-class, vc = vitality class) -·-· expected trend according to the yield tables (vc 0: n = 140, average age 131 y.; vc 1 + 2: n = 279, average age 132 y.; vc 3: n = 141, average age 134 y.)

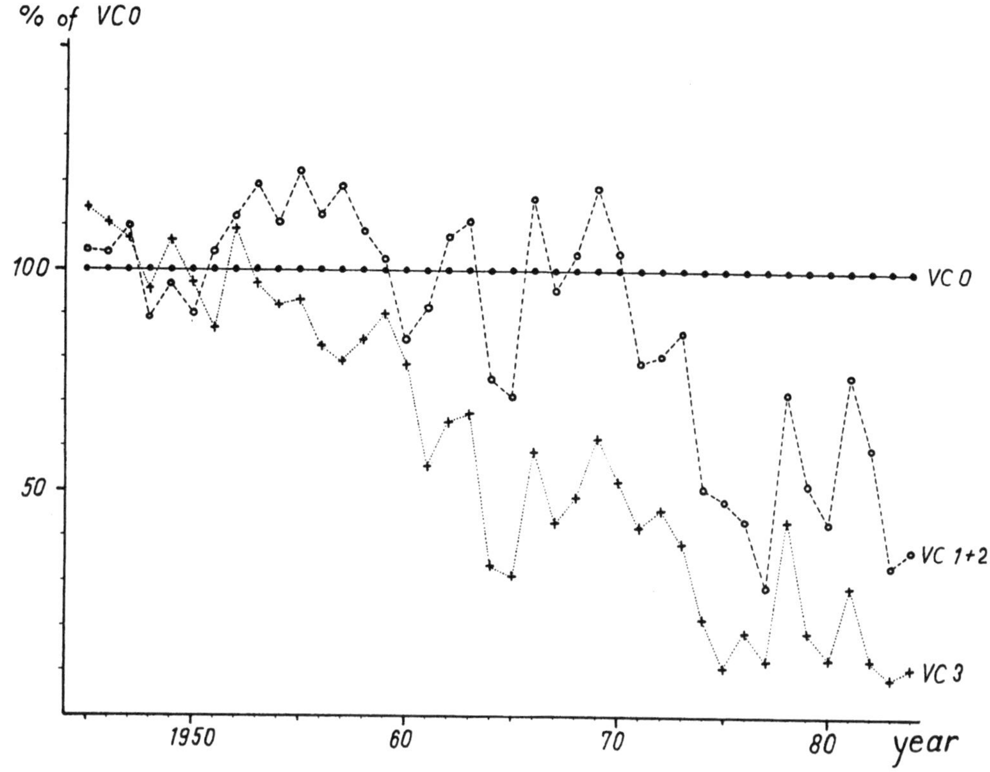

Fig. 13: Annual lenght increment of the leaders as in fig. 2, but the vitality classes (vc) 1 + 2 and 3 in % of the annual value of vc 0 (vc 0 in each year as 100%)

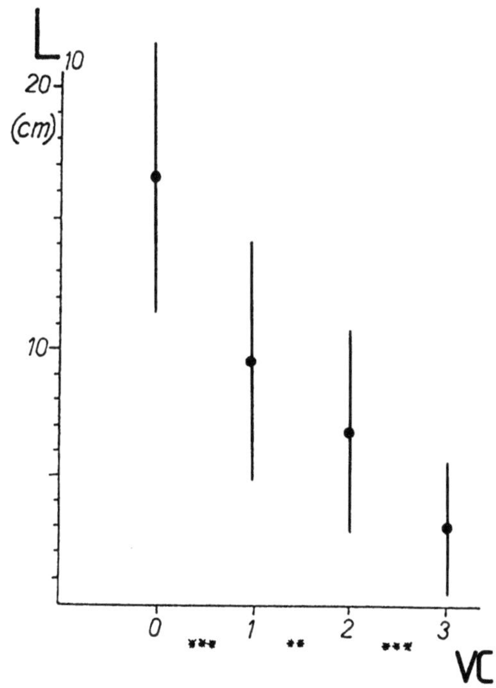

Fig. 14: Average annual lenght increment of the leaders of different vitality classes (vc) in the last 10 years (L_{10}) (mean value, standard deviation, statistical significance of the difference)

trees of vc 3 since about 20 to 25 years. Another possibility to show these long-term differences between the vigour of trees of different vc.es is the average annual lenght increment of the leader shoots during the last 10 years (fig. 14). This diagram, again, emphasizes the expressiveness of the presented vitality class system.

This system based on crown structures can be used successful in aerial photographs, too, and thereby it becomes possible to cover a larger area in a short time (fig. 15).

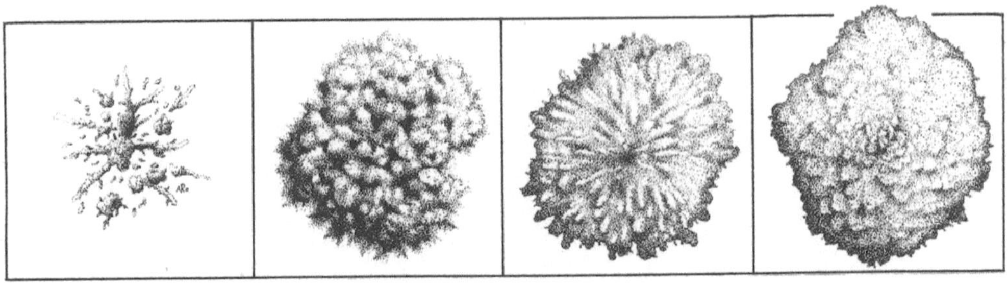

Fig. 15: Vitality classes (according to fig. 8) in aerial photographs (beech)

Recent research in 18 other hardwoods (and in pine) has shown the possibility to assess the vitality of other tree species of the Northern hemisphere in the same way as reported here for beech as an example. There are to take into account only little species-dependant modifications. This research has been finished now for

Silver Birch, Pubescent Birch (ROLOFF 1989a, WESTMAN 1989), Norway maple, Sycamore, Sugar maple, Horse-chestnut, Black alder, Hornbeam, European Ash, American Beech, Scotch pine, Wild cherry, Sessile oak, English oak, False acacia, Common willow, Small-leaved lime and Large-leaved lime (ROLOFF 1989a).

This vitality class system based upon branching structures should be used in future more frequently for vitality assessments of trees, because thereby it becomes possible to recognize a chronic decline and maybe better than by a 'leaf loss in per cent'.

The vitality class key for beech has been approved and applied in a variety of investigations (LONSDALE 1986, 1988, DOBLER et al. 1988, HESS.FORSTL.VERS.ANST. 1988, GIES et al. 1989). PERPET (1988) suggests for the vitality class 1, being the most important for an early diagnosis, an even more differentiated key, with low, medium and strong spear-like development of branches.

Interestingly enough, on fertilized sites (set up in the

early 80's) in the Solling mountains with plots which have either been acidified, limed or left untreated, a differentiation of the crown structures of beech trees is just beginning to show (fig. 16). This can be substantiated by the investigation of SCHENK et al. (1989). In the meantime vitality class keys based on the branching structure have been developed for other deciduous tree species (ROLOFF 1989a, WESTMAN 1989).

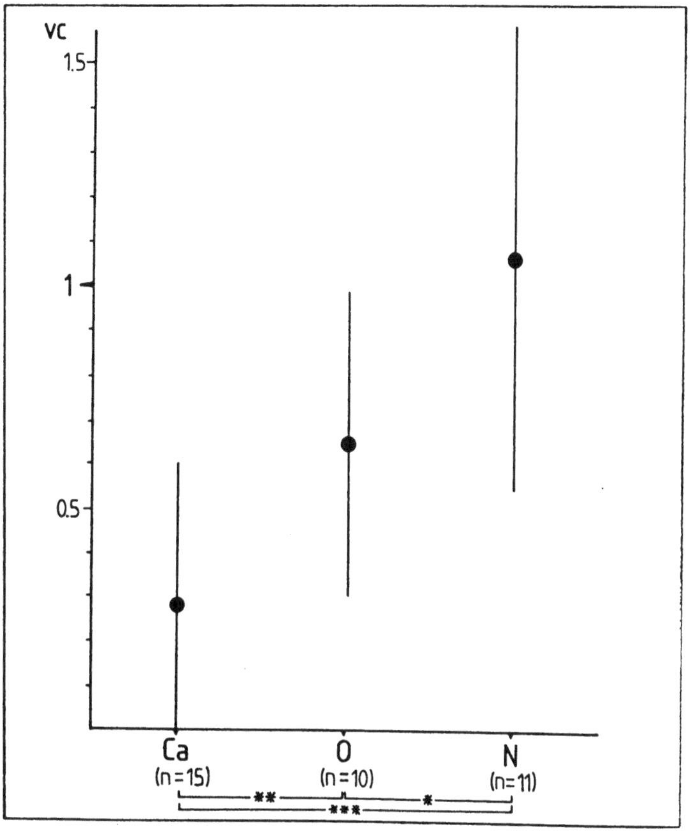

Fig. 16: Vitality classes (vc) after ROLOFF (1988) of beech trees on fertilized sites in the Solling mountains (set up in the early 80's) with acidified (N), limed (Ca) and untreated (0) plots

4. Effects on radial growth, relationships with root development and genetical as well as silvicultural consequences.

Although a close correlation between vitality class (on the basis of crown structure) and radial growth of branches of the crown does exist (fig.17, ROLOFF 1989a), it is difficult to correlate with radial growth at breast-height (GÄRNTNER & NASSAUER 1985, WAHLMANN et al. 1986, PERPET 1988, MAHLER 1988, ATHARI & KRAMER 1989a,b, FISCHER & ROMMEL 1989), in particular if the space of the trees investigated is not considered. Possible reasons for this generally unexpected discrepancy shall not be discussed here (refer to ROLOFF 1986, 1989a) However, it must be pointed out,

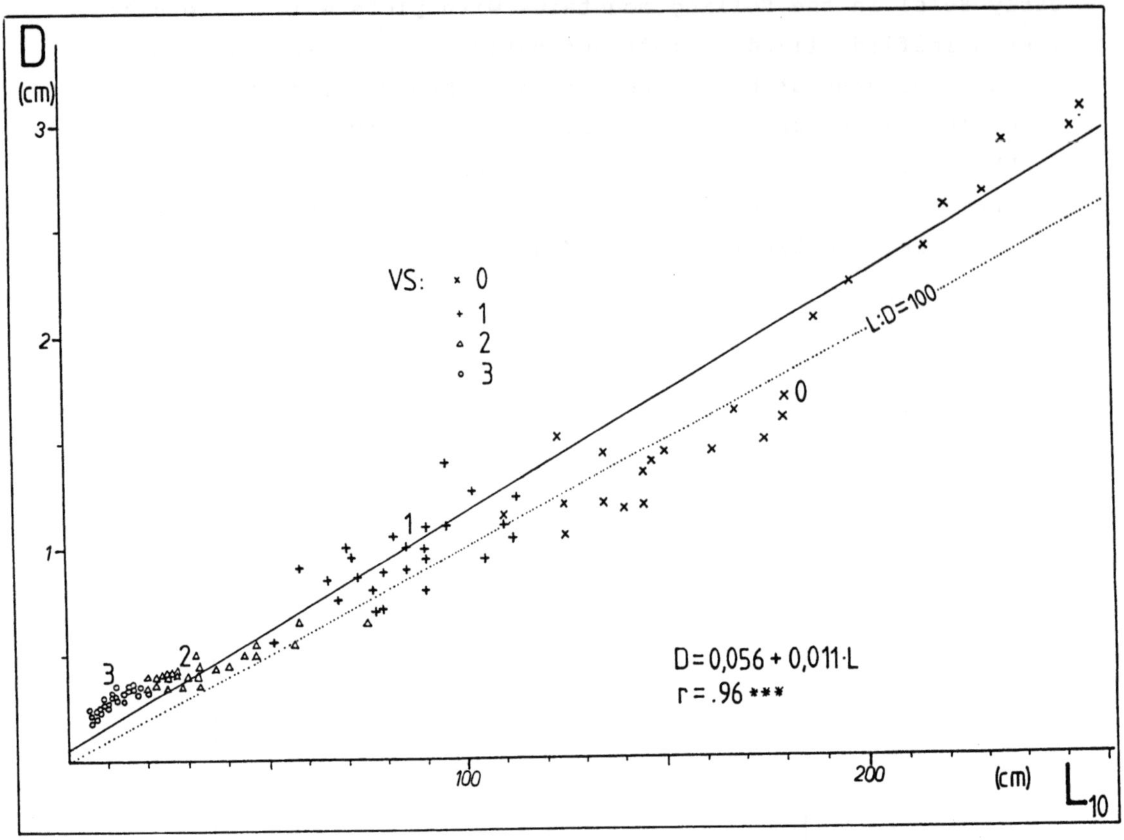

Fig. 17: Correlation between 10 year's growth of shoots (L 10) and their basal diameter (D) at the 10-year-old bud scale scar (vc = vitality class after ROLOFF 1985)

that the growth at breast-height as a vitality indicator is not only problematic in the case of beech. On the other hand, close relationships have been found if the space of the trees were taken into account and if investigated trees had the same vitality class based on both "leaf loss" and crown structure.

Of particular interest of course should be the correlation between changes of the crown and root growth, which unfortunately require considerable expenditure of research. For this reason results about close correlations between crown and root growth of beech trees are quite scarce (ROLOFF & RÖMER 1989, VINCENT 1989) and primarily reflect the vast need of research in this area.

Genetical consequences are presented by MÜLLER-STARCK & HATTEMER (1989). In beech trees he found (on the basis of crown structure) a decreasing genetic variety with decreasing vitality of the tree (heterozygotic grade, genetic diversity, respectively).

Accordingly, only beech trees of high genetic variety survive stress without suffering obvious damage.

Finally, changes of the crown structure of stressed beech trees may also have silvicultural consequences (DOBLER et al. 1988, HESS.FORSTL.VERS.ANST. 1988). As the canopy becomes more transparent after thinning, it is difficult to selectively regulate natural regeneration of beech stands. A modification of thinning approaches and methods have been suggested.

5. Final conclusions and further research requirements

A multitude of investigations carried out in recent years have confirmed the significance of changes of the crown structure of deciduous trees (primarily beech trees) with decreasing vitality. These changes should be considered to the same extent at least as "leaf loss" and can certainly improve the assessment of single standing trees or entire stands to a large degree. This becomes even more urgent since crown structures can be determined from aerial photographs (RUNKEL & ROLOFF 1985).

It is necessary to reconsider the inventory methods of forest decline. They only appear to be sensible if they are carried out in longer time intervals and also during the winter season.

Ecological morphology is a new, promising approach to explain modifications of tree growth and crown development and to detect its causes. E.g., by the help of the 'shoot base scars', it becomes possible to reconstruct the crown development over the last ten years, partly over decades. In every investigated (broad leaved) tree species (and in Scotch pine), there are to discriminate four growth phases: exploration, degeneration, stagnation and resignation phase, which are due to (statistically significant) decreasing shoot lenghts with the result of fundamental modifications of the branching structure. Especially in the leaf-less state, these different branching structures in the very treetop are to perceive from a distance (and in aerial photographs, too) and are the basis of a vitality assessment by four vitality classes, which were developed for 18 hardwoods and Scotch pine.

It is essential to carry out further research primarily in the following fields.

- Further elucidation of the discrepancies between radial growth at breast-height and radial growth of branches within the crown of declining trees.
- Investigations into correlations between crown - and root development as well as their morphology.

Because of the lack of an exact boundary between different stress symptoms, further investigations referring to forest decline can only be promising if considering simultaneously soil, root and crown parameters of the same trees.

7. References

ATHARI, S. & H. KRAMER (1989a) Problematik der Zuwachsuntersuchungen in Buchenbeständen mit neuartigen Schadsymptomen. *Allg. Forst- u. Jagdztg.* 160: 1-8

ATHARI, S. & H. KRAMER (1989b) Beziehungen zwischen Grundflächenzuwachs und verschiedenen Baumparametern in geschädigten Buchenbeständen. *Allg. Forst- u. Jagdztg.* 160: 77-83

BÜSGEN, M. (1927) Bau und Leben unserer Waldbäume. 3. Aufl.Jena

DOBLER, D., K. HOHLOCH, B. LISBACH & M. SALIARI (1988) Trieblängen-Messungen an Buchen. *Allg. Forstzschr.* 43: 811-812

FISCHER, H. & W.-D. ROMMEL (1988) Jahrringbreiten und Höhentrieblängen von Buchen mit unterschiedlicher Belaubungsdichte in Baden-Württemberg. *Allg. Forstzschr.* 44: 264-265

LÜCKIGER, W., S. BRAUN, H. FLÜCKIGER-KELLER, S. LEONARDI, N. ASCHE, U. BÜHLER & M. LIER (1986) Untersuchungen über Waldschäden in festen Buchenbeobachtungsflächen der Kantone Basel-Landschaft, Basel-Stadt, Aargau, Solothurn, Bern, Zürich und Zug. *Schweiz. Zschr. Forstwesen* 137: 917-1010

FLÜCKIGER, W., S. BRAUN, S. LEONARDI, L. FÖRDERER & U. BÜHLER (1989) Untersuchungen an Buchen in festen Waldbeobachtungsflächen des Kantons Zürich. *Schweiz. Zschr. Forstwesen* 140: 536-550

GÄRTNER, E. J. & K. G. NASSAUER (1985) Aktuelles zur Waldschadenssituation in Hessen. *Allg. Forstzschr.* 40: 1265-1271

HESSISCHE FORSTLICHE VERSUCHSANSTALT (1988) Waldschadenserhebung 1988. Selbstverlag Hann.Münden

LONSDALE, D. (1986) Beech health study. *For. Comm. Res. Devel. Paper* 149: 1-15

LONSDALE, D. & I. T. HICKMAN (1988) Beach health study. *Forest Res.* 23: 42-44

MAHLER, G., J. KLEBES & B. HÖWECKE (1988) Holzkundliche Untersuchungen an Buchen mit neuartigen Waldschäden. *Allg.Forst- u. Jagdztg.* 159: 121-125

MÖHRING, K. (1989) Wuchsstörungen und Absterben in den Kronen einiger Buchen im Solling. *Allg. Forstzschr.* 44: 113-116

MÜLLER-STARCK, G. & H. H. HATTEMER (1989) Genetische Auswirkungen von Umweltstreß auf Altbestände und Jungwuchs der Buche. *Forstarchiv* 60: 17-22

PERPET, M. (1988) Zur Differentialdiagnose bei der Waldschadenserhebung auf Buchenbeobachtungsflächen. *Allg. Forst- u. Jagdztg.* 159: 108-113

ROLOFF, A. (1985) Schadstufen bei der Buche. *Forst- u. Holzwirt* 40: 131-134

ROLOFF, A. (1986) Morphologie der Kronenentwicklung von Fagus sylvatica L. (Rotbuche) unter besonderer Berücksichtigung möglicherweise neuartiger Veränderungen. *Ber. Forschungszentr. Waldökosysteme* 18: 1-177

ROLOFF, A. (1989a) Kronenentwicklung und Vitalitätsbeurteilung ausgewählter Baumarten der gemäßigten Breiten. Frankfurt
ROLOFF, A. (1989b) Entwicklung und Flexibilität der Baumkrone und ihre Bedeutung als Vitalitätsweiser. *Schweiz. Zschr. Forstwesen* 140: 775-789, 943-963
ROLOFF, A. & H.-P. RÖMER, (1989) Beziehungen zwischen Krone und Wurzel bei der Buche (Fagus sylvatica L.). *Allg. Forst- u. Jagdztg.* 160: 200-205
RUNKEL, M. & A. ROLOFF (1985) Schadstufen bei der Buche im Infrarot-Farbluftbild. *Allg. Forstzschr.* 40: 789-792
SCHENK, J., W. STICKAN & M. RUNGE (1989) Phänologie und Blattmorphologie von Buchen (Fagus sylvatica L.) unter dem Einfluß von Kalkung oder N-Düngung. Poster Intern. Kongreß Waldschadensforschung Friedrichshafen/Bodensee 2.-6.10.89
THIEBAUT, B. (1981) Formation des rameaux. In: E. Tessier du Cros (ed.): Le Hêtre. Paris: 169-174
THIEBAUT, B. (1988) Tree growth, morphology and architecture, the case of beech: Fagus sylvatica L. In: Comm. Eur. Communities (ed.): Scientific basis of forest decline symptomatology. Brüssel: 49-72
VINCENT, J.-M. (1989) Feinwurzelmasse und Mykorrhiza von Altbuchen (Fagus sylvatica L.) in Waldschadensgebieten Bayerns. *Eur. J. For. Path.* 19: 167-177
WAHLMANN, B., E. BRAUN & S. LEWARK (1986) Radial increment in different tree heights in beech stands affected by air pollution. *IAWA Bull.* 7: 285-288
WESTMAN, L. (1989) A system for regional inventory of damage to birch. Poster Intern. Kongreß Waldschadensforschung Friedrichshafen/Bodensee 2.-6.10.89

NEW FOREST DAMAGES IN CENTRAL EUROPE: DEVELOPMENT, CAUSES AND POLICIES

Hans Essmann, Freiburg, F.R.G. and
Willi Zimmermann, Zurich, Switzerland

Abstract

Beginning of the 1980's the public in Central Europe first perceived that forests in wide areas had been damaged severely and seemed to die rapidly. The term of the *Waldsterben* (forest die-back) has been created.

In the following the development of new forest damages, present extent, the causes, the measurement methods of the damages and the policies defending the damages will be outlined and discussed. Peoples' special attitude to forests as well as a quite similar political treatment of the forest damage problem in Austria, the Federal Republic of Germany and Switzerland justifies comprehensive treatise under the term 'Central Europe'.

1. Introduction

1.1 Area of investigation

This investigation concentrates on three countries: Federal Republic of Germany (FRG), Austria (A), Switzerland (CH). For simplicity, they are together referred to as Central Europe. This geographical distinction of forest decline is not exclusive; it was chosen however, based upon:

1. comparable political approach to the problem and
2. author's familiarity with the damage situation

in the three listed countries.

Through a detailed analysis of the phenomenon of forest decline, it is evident that its spatial occurence and extent is quite variable. One must recognize that these differences are often reduced to the size of single stands.

The pattern of fully intact through heavily damaged forests and forest stands becomes more complex since the different tree species are differently affected. In addition, during the last decade deciduous trees, especially oak (*Quercus robur and Quercus petraea*) and European beech (*Fagus sylvatica*), have been increasingly influenced by the new damaging factors. Concurrently in some regions, conifers, i.e. Norway Spruce, (*Picea abies*) began a slight trend toward improvement. Two parameters can be defined which exhibit somewhat regularity in contrast to the varying spatiality and occurence. The forest stand altitude seems to have an effect since frequently the percent of damaged trees increases with increasing altitude. Generally, it is correct to state that montane forest stands and areas above 800 meters are strongly affected by new forest damages. Furthermore, the exposition to the main wind direction appears to be a factor. Both components, altitude and exposition, support and strengthen the early proposed hypothesis that the probable cause of new forest damages is primarily long distance, widely dispersed air pollutants (BMFT, 1985).

Nevertheless, it must be emphasized here: The additional parameters such as soil, climate, water availability etc. cannot fully clarify the complex mosaic of damage occurence in order to justify an accepted cause & effect theory of the new forest damages. These possible explainations don't even suffice to reduce the many damaged forest areas into a few large catagories. In summary: A discourse over the development and present situation of forest decline in Central Europe such as the one presented here, can only offer a coarse synopsis of the quantitative aspects of the phenomenon.

The situation changes, however, when the qualitative aspects of 'forest die-back', it's social and political dimensions, are examined and analysed. In this regard, an outlined account of the forest damage problematic in Germany, Austria, and Schwitzerland is useful. First, strongly influenced through the *Zeitgeist* of the Romantic period, the people of these three countries developed a deep inner attraction to the forest, stronger than any other material aspirations and which is without parallel in Europe (Rozsnyay, 1990). The character of the concept *positive Waldgesinnung,* for which no adequate English word exists, encompasses an emotional, almost mystical excessive attitude of many people toward the forest.

The close bond with the forest primarily explains why in Centeral Europe the first waves of broad public information about damaged or even dieing forests in the early 1980's created turmoil in the public opinion which was unmatched in other countries. Reports of equally serious environmental damage have neither before nor since then found a comparable echo.

Corresponding to the strong public interest in forest damages, the growing public concern, and political activism the pressure was great for the political decisonmakers to act quickly.

In addition to this emotional aspect, a summary of these three countries is meaningful since the political decision-making process to fight 'forest die-back' has proceeded similar. Based on the early and generally accepted hypothesis that air pollutants were the main cause for the new type of forest damages, the steps against *Waldsterben* were concentrated on air pollution policies. Both premises, *positive Waldgesinnung* and good comparability of the political efforts form a solid basis for this presented analysis. Relevant differences between the three countries, present and past, are noted accordingly.

1.2 'Forest die-back' or (only) new forest damages?

In the general public, the term *Waldsterben* originated through a series of articles in the news magazine *Der Spiegel* (1981 a, b) about the large-scale forest damage in Czechoslovakia and the Black Forest. The topic of 'forest die-back' quickly caught the attention of forest authorities, forest scientists and politicians. Until then, this type of public information was not common so that the forest damage was viewed as an 'abrupt event' (Niesslein, 1986). Even in the scientific community, specialists used this term and searched for possible explainations based on an 'abrupt occurence' (Poston/Stewart, 1978). Forest experts were already concerned for some years about a definite increase of damaged trees in their forests (Schärer/Zimmermann, 1984). However, since the 'new type' of damage symptoms was observed mainly on the silver fir (*Abies alba*), the phenomenon was referred to as *Tannensterben* (Roeder, 1979). The media soon full-filled the conclusions of many, namely within an few years, extensive areas of forest would die. Thus, in the early 1980's, the word *Waldsterben* depicted the real concerns and fears of most citizens.

At the present date, the prognosed, large-scale death of forested areas has not occured, as a result the term *Waldsterben* is seldom used in scientific journals, and then only in parenthesis. As an alternative, the term 'new forest damages' (*neuartige Waldschäden*) is used (Schlaepfer/Haemmerli, 1990). This term reflects the unchanged recognition that the forest is and continues to be heavily damaged through external influences. At the same time, it denotes the generally accepted causality theory, namely that of damaging influences or processes of this type and quantity until now unobserved. 'New' is the broad geographical distribution of the damage symptoms of individual tree species followed within a few years by an appearance on several other tree species and a prolonged damage (Niesslein/Voss, 1985). Finally, the word 'damages' clearly does not imply a sudden demise of the forests which was believed just six years ago.

2. Development and extent of forest damages

2.1 Measurement methods and their significance

For only six years, forest damage has been measured by a standardized technique. Previous methods were local and compiled only from subjective approximations of damage symptoms. A standardized inventory allows spacious as well as chronological comparisons. The inventory methods are based upon mathematical-statistical sampling. Due to financial and personal restraints, full inventories are not possible.

Since the initial forest damage inventories in 1983, using regionally different, later in 1984 standardized methods, a spatially balanced sampling grid was developed for the forests. In heavily damaged areas, a finer grid was used. At each coordinate cross-point, a given number of sample trees (max. 20) are systematically chosen and clearly marked. Using this procedure, the health of approximately 240,000 trees (A: 70,000; FRG: 160,000; CH: 8,000) is registered. The results are then projected for the entire forested area of each country.

During the annual forest damage inventory, between July and September, the condition of each sample tree is registered. Two main criteria are distinguished:

1. the relative needle or leaf loss divided in 5 % damage classes (FRG and CH) or in light penetration levels (A);
2. the degree of needle or leaf yellowing also registered in classes (BML, 1989; BUWAL, 1989; Neumann / Pollanschütz, 1988).

The divison of sample trees in the 5 damage classes based on these criteria is as follows:

Table 1: Classification of Forest Damage

Damage Classes		Needle/Leaf Loss Light Penetration in %	
FRG/CH	A	FRG/CH	A
0	1: without any signs of damage	<10	<20
1	2: slightly damaged (warning stage)	11-25	21-40
2	3: moderately damaged	26-60	41-60
3	4: heavily damaged	>60	>60
4	5: dead	--	--

If a moderate-heavy yellowing on a sample tree appears, it is placed in a higher damage class. For example, trees which due to needle or leaf loss are classifed in class 2, exhibit however, moderate yellowing, are regraded to class 3. In order to balance this picture, occular approximations of the insect and fungus infestation are recorded and added to the damage class information.

New findings have resulted in a different assessment of the various damage classes as in the first inventories. Particularily when using the needle/leaf loss criteria, the natural variability of this characteristic must be considered. These variatons can cause healthy trees to be placed in the first damage/light penetration class. This is especially important by the species silver fir (*Abies alba*) and Scotch pine (*Pinus sylvestris*). Therefore, the first class can only be interpreted as a warning or transition class, whereas definite tree damage is assumed by stage 2/3 that is moderately damaged (BML, 1989). The classes 3/4 (heavily damaged) and 4/5 (dead) as well as 2/3 - 4/5 are often represented together. The last number is currently the actual data with which the quantitative extent of forest damages described.

The validity and reliability of the forest damage inventories have been viewed very critically. In actuality, it has been shown that needle or leaf loss is to some extent naturally variable and does not solely indicate tree damage. The greatest objection, however, lies in the recording of symptoms which do not directly signify the cause (Essman et al., 1989); thus far, the essential data concerning the sample tree sites as well as the climate, soil and water availability have only been measured in a few cases. Such information allows a differential analysis to what extent tree lighting and yellowing is caused by site limiting factors or by non-site factors such as air pollutants (BMFT, 1985). These inadequacies are recognized and could be improved through extensive surveys of forest soil conditions which are now being planned.

A second, stochiastical problem arises in that all sample trees underlie normal forest management. Consequently, if a sample tree is cut, the object of observation, measurement and above all comparison disappears and must be replaced with a previously uncontrolled tree. As long as the number of these exchange remains small, the constant replacement of trees has no statistical significance. Critical, however, remain changes due to management or otherwise unforseeable interference (i.e. storm throw) resulting in the loss of a large contingency of sample trees on several sample plots. Consequently, due to these tree losses, the number of trees in the highest damage catagories (3/4 - 4/5), mostly over age 60, is smaller than if these dynamic processes did not occur.

2.2 Development and level of forest damages

The last forest damage inventories in Ceneral Europe showed moderate to heavy damage (classes 2/3 - 4/5) on an average of 3.6 % (A: 1988) to 15.9 % (FRG: 1989) of the forested areas (CH: 12 % in 1988). The low percentage in Austria compared to Germany and Switzerland is accredited to the high light penetration percent, 41 % for class 3, whereas the comparable class 2 in Germany and Switzerland starts with 26 % light penetration (see table 1). When Austrian values are conformed to the measurement system of German and Switzerland (also valid for ECE), they increase to approximately 15 %.

After the peak of the forest damages in 1985 (FRG: 19.2 %, after 9.7 % in 1983 for classes 2 - 4), 1986 (A: 5 % in classes 3 - 5) and 1987 (CH: 15 % after 8 % in 1985), the present data show a slight reduction or stagnation in the forest damage. An explanation for these reduction or stagnation trends appears that the damage factors still exist, however, other positive acting factors such as climate or the gradual reduction of SO_2 emissions could be counteracting the damaging influences.

Table 2: Forest Damage in Central Europe 1988

Country	Needle/Leaf Loss in % of Forested Area		
	no damage signs 0/1	lightly damaged 1/2	definite damage 2/3-4/5
West Germany (FRG)	47.6	37.5	**14.9**
Austria (A)	71.2	25.2	**3.6**[1]
Switzerland (CH)	57.0	31.0	**12.0**

[1] small since damage stage 2/3 first recorded at 40%

Currently, the damage class 1/2, as already stated is considered as a warning or transition stage and in 1988 accounted for approximately two-thirds of the total results. During the same time, forested area without any damage symptoms decreased. The regional development of the forest damages, based on the main tree species and on tree age was so variable that prognosis of future damage trends should only be made with precaution. Nevertheless, some careful generalizations can be made:

1. In stands over 60 years of age, the percent of moderately to heavily damaged classes (classes 2/3 - 4/5) continues to increase compared to the general trend (from 1988 - 89 in FRG: up 2.5 % to 28 %).

2. The condition of Norway Spruce improved slightly in the classes 2/3-4/5 (from 1988-89 in FRG: 15 % to 14 %; in CH: 15 % to 11 %) in contrast to the silver fir which remains strongly affected (1989 in FRG: 44 %; in CH: 23 %).

3. The damage on the most important broad-leaf species remains high or in some cases is increasing; the oak (*Quercus robur, Quercus petraea*) is the second most damaged species ('oak die-back' is now being termed) (A: 18 %; FRG: 32.4 %; CH: 11 %). Here also stands over age 60 are most affected.

4. Regionally, the occurence and extent of the new forest decline in Central Europe has become stronger differentiated; the situation has somewhat eased, at the same time, however, strongly hit areas can be clearly defined, for example, in the higher altitudes of the *Mittelgebirge* and the Alps.

2.3 About the causes

The forest damage inventories provide information about the state and development of the visible forest damage, but not about the causes (BUWAL, 1989). Although, since the beginning of the broad public information in Central Europe about *Waldsterben* in the early 1980's, the financial means for research into the causes of forest decline have increased (in FRG since 1982 approx. 600 projects have been funded with 277 million DM), no absolute certainty prevails about them. The only generally accepted assumption is that not a single cause, rather several diverse in nature exist (Schlaepfer, 1988).

Several biotical as well as abiotical man-made causes for the forest damage are plausible. A majority agreement is also reached about the assumption that 'forest die-back' would not be explainable without the influence of air pollutants (Schütt, 1988). A minority claims that nutrient deficiency through a permanent forest utilization without successive fertilization is responsible for the new forest damages. In the near future, scientists will not be able to provide proof for the accuracy of either hypothesis. In comparison to the initial phases of forest decline, scientists have clearly become more cautious through differentiating research results and interpretating them more stringently.

While the influences of biotic and abiotic factors on the health of forest trees and the direct effect of certain damaging gases (the 'classic smoke damage') on forests have been studied for years, the indirect influences from widely dispersed air pollutants and their transformations are new and little studied factors. It is assumed with little controversy that these air pollutants are damaging in two ways (BML, 1989): They act as acidic percipitation directly on the needles and leaves, thus impairing normal metabolic processes. In addition, through acidic and nitric depositions, the chemical and biological structure of the forest soils change. The nutrient supply via the roots is altered and consequently the health and growth of the trees. Measurements have shown that the forest ecosystems are more strongly affected by air pollutants than adjacent open fields. Through the larger crown surface area, the pollutants are filtered from the air and consequently deposited in higher concentrations on the forest floor.

The acidification is only one indicator for the grave changes in forest soils. Evidence lies for example in the change of concentration and supply of exchangable calcium and magnesium. Over large areas it appears that the supply of plant essential calcium and magnesium has been so strongly reduced that sufficient plant nutrition is not obtained. Especially in the higher altitudes of the *Mittelgebirge* and the Alps, the needle yellowing is accepted as a visible sign for indirect, unidentified changes in the forest soils.

Currently, the input of air pollutants seems to polarize forest growth: On the one side, they damage the already poorly nutriented soils, particularily in the higher altitudes and cause nutrient deficiencies which are recognized as new forest damages. On the other hand, NO_x depositions on once nitrogen poor soils are causing growth spurges. Both phenomenon are cited to support the hypothesis that the main cause of the deteriorating forest conditions are the human induced emissions.

3. Policies

3.1 Forest damage and public opinion

A broad public discussion about 'forest die-back' started approximately in the early 1980's. The increasing number of reports on the 'deadly ill' forests led the discussions to a peak in 1983 - 84. A disquieted and emotionally, deeply concerned public unprecidently pressured the political decision-makers to take measures immediately (see section 3.2). *Robin Wood*, one of the most famous groups which protested against the forest *Waldsterben* as well as other citizen initatives were founded at this time.

The well-developed consciousness of the majority of citizens in conjunction with a *positive Waldgesinnung* allowed a good platform for a public discussion of the forest decline problematic (Essmann/Niesslein, 1986). Polls reported over a long period of time, 'forest die-back' as the number one environmental problem. Although the public's sensitvity concerning environmental problems has remarkably increased since, other problems such as climate change, tropical deforestation, the pollution of rivers and seas, poisioned food and soil have diverted the public's attention. Reports of *Waldsterben* have nearly vanished from the newspapers front pages. The annually published forest damage inventories hardly grasp public attention anymore.

Among other reasons this resulted from the hasty prognosis of severe forest damage in which large areas of Central Europe would become entirely deforested within a few years, in reality, the damage of the forest was not so spectacular. The damaging processes still continue, however, with much less public attention. The ceasing attention for forest damage may be traced as well to the general phenomenon no event, however shocking or alarming it may be, is able to hold the public interest permanently (Luhmann, 1971). Irregardless if any measures have been taken or not to solve the problems connected to an event, after some time of intensive public discussion, their interest fades and turns to other more current events. However frightening a problem may be, in the public's perception it is no longer there (objectiviation, accustomization, supression, forgetfulness). Accordingly, the pressure on decision-makers to act will cease (ESPE, 1986).

The public debate about 'forest die-back' peaked at a time when information on the dimensions of the damage were hardly available, not to mention knowledge about the causes which led to which effects. Scientists and forest experts paved the way for the hypotheses and lately theories, (i.e. virus theory, fungus theory etc.) about possible causes of the wide-spread forest damages (Niesslein/Voss, 1985).

3.2 The political strategies

The first political appeals to tackle the forest damages were made in 1982 (Bericht Waldsterben, 1984; BMFT, 1985). However, they were vaguely formulated and concentrated on the problem of 'acid rain'. The problem of the 'forest die-back' was made a political issue only late in 1983 and early 1984. Within a short period, numerous appeals had been registered concerning the forest damages (BMFT, 1985). In addition to parliamentary sessions, other political institutions such as political parties, environmental organizations and concerned citizens appealed to politicians. These activities were supported by demonstrations with thousands of participants demanding that the health of the forests be maintained. Although science provided more assumptions than evidence about the causes of the 'forest die-back', political appeals, demands and manifests called mainly for a reduction of harmful emissions at the same time emphazising the need for research and financial support in the forest sector.

From the political viewpoint, the beginning of the 'forest die-back' is characterized by an unusually large and long-lasting interest of the mass media due to the integration of science in the discussion of its causes, the broadly-based political demands and the emphasis on the environmental impact of the new forest damages. Based on the statements of scientists, public officials and mass media presented a very drastic picture about the dimensions and the causes and effects of *Waldsterben*, making a discriminate assessment hardly possible for any of the concerned politicians. Furthermore, during this time national elections were held and thus, the problem of *Waldsterben* rapidly became a relevant political issue for all fractions.

3.3 The political decison-making process

The governments of the Central European countries reacted immediately to the manifold of political demands, supported by the strong public pressure: As a first step, the public was informed about the state of knowledge concerning forest damage and air pollution. In 1984, the Swiss Federal Department of Internal Affairs stated that "the appearance and spreading of the current forest damage gives evidence that causes other than natural are decisive for the forest damage. A primary cause must definitely be century long, steadily increasing air pollution" (BUWAL, 1984). According to another source, "in Central Europe new types of forest damage are occuring simultaneously and have continued to develop. Consequently, it seems that the same factor or group of factors are responsible for its complex of causes. With a high degree of probability they are air polluting chemicals and their metamorphoses"

(BMFT, 1985). From these statements evolved the general environmental objective to cut the present pollution levels to those in the 1950s or 1960s.

Starting in 1983, large initiatives for new laws and changes of existing laws were undertaken. Based on the hypothesis that stated air pollution to be the main causing factor of the forest damage, the new policies aimed at reducing pollutants. Accordingly, the major emitors were first approached. The most important legal measures were:
- decrees for huge heating units to reduce the emission of SO_2, NO_x and dust (in FRG investments of 28 billion DM)
- intensification of critical emission standards
- regulations for industrial and commercial plants to adapt to the latest technical standards of emission control
- intensification of the security standards of industrial plants with potentially harmful environmental impacts.

The second important target of the anti-air pollution policies was the automobile. The general goal was the introduction of a car with reduced exhaust emissions. This was aimed through an intricate system of direct, fast acting fiscal benefits and regulations. The most important measures were:

- reduction and finally ban on leaded fuel
- tax benefits and regulations (CH) for cars with a three-way catalytic converter
- reduction of sulfur in gasoline and diesel fuels
- intensification of emission standards
- speedlimits on highways (Austria and Switzerland only).

Further measures are being implemented or are still being discussed. They are indispensible since thus far, the aspired reduction of air pollution has not been achieved, although, a reduction in SO_2 has. Yet, recent research shows that many other air polluting chemicals, particularily nitrogen oxide, amonia, hydrocarbons and ozone are contributing to the new forest damages. These emissions have not decreased rather increased. The increase of nitrogen oxides is due on the one hand to the steadily increasing traffic on the other hand few cars are equipped with a three-way catalytic converter. Almost no technical measures have been taken to reduce air pollutants from heavy transport trucks. Future financial incentives and reduction on legally allowable standards for air pollutants will help to achieve this goal.

In the forest economy, measures have been taken which in addition to the anti-air pollution policies should improve the resistance of the forest ecosystems, maintain the stability of forest soils and consequently mitigate the damaging processes. Forests which are stressed by acute chronic pollution are also susceptible to naturally occuring stress factors. Air pollutants, therefore, intensify the common and well-known problems of forest protection such as excessive game density, insects, storms etc.

An improvement of the situation is being pursued by the following measures:

- maintance and regeneration of ecologically sustainable mixed stands, rich in species diversity
- avoidance of large clear cuts
- intensive tending of stands
- forest protection based on integrated plant protection, use of forest management techniques which protect soils and stocks
- reduction of game density appropriate to the local natural situations
- fertilization of extremely poor soils on severely damaged sites.

All these forestry measures are sponsored more or less generously with public funds. Direct compensation for the damage due to air pollution, has not been paid to the affected forest enterprises. The legal basis to apply the responsibility principle for pollutors is not sufficient. Moreover, it is not possible to claim the state vicariously for those who are responsible for the damage.

3.4 Present political discussion

Due to the decreasing public interest for the topics linked to 'forest die-back', and the rising criticism against one-sided theories, it became more difficult for politicians to enforce specific measures for an immediate air pollution reduction. This is the main reason why the annual compulsory exhaust gas controls or the introduction of regulated catalitic converters for cars were delayed. Further measures which had been planned in the anti-air pollution concepts were postponed and have not yet been implemented (i.e. heavy truck tolls).

A specification of all dispositions made would make a considerable list, yet the actual outcome of political decisions made in relation to the forest damage would only be partly covered. In Central Europe, parliaments and goverments have actually been adding measures for possible solutions into already existing programs rather than creating independent programs against "forest die-back". Due to the relationship between current air pollution and the new forest damages, the topic of *Waldsterben* has had a substantial impact on environmental policy. The framework for these new policies was easier to make since alarms about an obvious 'forest death' were parallel to discussions about better environmental protection (i.e. endorsement of the Swiss Environmental Protection Law). During this stage, discussions on forest damage accelerated the political process in the parliaments and led to much faster than usual agreements on important environmental questions avoiding further delays in passing other environmental legislation.

Political debates about forest damages have been positively altering the content and implementation of environmental decisions. Within only a few years, anti-air pollution and other environmental protection regulations have been passed. Due to the increased public awareness spurred by the 'forest die-back', it became possible for the public administration to integrate rather strict codes of conduct and procedures into these regulations. In all three countries the legal basis for anti-air pollution programs has been established which through new laws can meet the demands on environmental groups. The political importance of *Waldsterben* as stimulus for environmental issues must be ranked as high as all other protection measured passed to fight it.

The annual damage reports of the last two or three years indicate a slightly decreasing trend (damage classes 2/3 - 4/5), however, large areas of damaged forest remain. Nonetheless, this topoic has nearly disappeared from the agendas of intermediary organizations as well as those of the political organizations and the mass media. In Central Europe, the phenomenon of forest damage is no longer a relevant political topic. This is seen in the staggering or even questioned discussions on speed limits to reduce traffic (especially in Switzerland). Although forest damage was the primary reason for the introduction of speed limits, it has ceased to be an argument in current discussions. Currently, the main argument for speed reduction lies in safety and general environmental aspects, namely noise and air pollution.

One of the reasons for the lacking interest in topics of 'forest die-back' is also to be found with science. In the beginning of public discussions, science played a vital role, partly due to the (financial) research interests, partly due to the strong pressure of the mass media. During this emotional and strongly environmental protection oriented time, some scientists gave statements which were based on vague assumptions, rather than on scientific facts. These generalizations and sometimes 'exotic' theories about the causes of forest damage were much more eagerly grasped by the mass media than the differential scientific findings of recent research which do not provide direct answers for the over simplified question about the causes.

Regarding the political dimension of this problem, the situation is completely different now than 5 - 6 years ago. Political decison-makers agreed upon decisons and measures meeting all expectations of the major interest groups to a great extent or even entirely. These expectations have been treated formally, however not sufficiently, apparently the 'accustimation, suppression and forgetfulness-behavior' of the concerned groups has been stronger. Three main results of these 'satisfaction policies' are:

- the forest economy and forest research have recieved an unprecidented amount of federal funds (Zimmermann, 1989)
- environmental organizations and closely related political parties were temporarily satisfied with rapidly passed and comparatively rigorous and encompassing legal provisions on environmental protection
- the economic interest groups, especially the automobile industry, were spared from drastic measures, since the emphasis of the environmental policy was laid on technical and economically interesting long-term solutions rather than immediate regulations.

Most of the forest policy measures induced through *Waldsterben* are now in a phase of implementation. They are executed without much public attention, since only small sectors of the public are directly affected by these measures. Problems of implementation are dealt by those who are directly concerned, either personally or through their respective lobby or responsible administration. If no consenses can be achieved, the conflicts will be settled at the juridicial and not at the political level.

Literature

Bericht Luftreinhalte-Konzept (1986) Bundesblatt 1986 III 269 - 369

Bericht Waldsterben (1984) Parlamentarische Vorstöße und Maßnahmenkatalog, Bundesblatt 1984 III 1129 - 1421

BML/Bundesministerium für Ernährung, Landwirtschaft und Forsten, Hrsg. (1989) Waldzustandsbericht. Ergebnisse der Waldschadenserhebung 1989, Bonn.

BUWAL/Eidgenössische Forstdirektion und Eidgenössische Anstalt für Wald, Schnee und Landschaft WSL (1989) Sanasilva-Waldschadensbericht 1989, Bern und Birmensdorf.

BMFT/Der Bundesminister für Forschung und Technologie, Hrsg. (1985) Umweltforschung zu Waldschäden, 3. Bericht, Bonn.

Caprez/Fischer/Stadler/Weiersmüller (1987) Wald und Luft. Verlag Paul Haupt, Bern und Stuttgart.

Der Spiegel, (1981a) Etwas stirbt immer, Nr. 29.

Der Spiegel, (1981b) Da liegt was in der Luft (Serie), Nr. 47/48/49.

Eidgenössisches Departement des Innern (1984) Waldsterben und Luftverschmutzung, Bern.

Espe, H. (1987) Waldschadensbericht 1986. Eine Analyse der Pressereaktionen. *IIUG-report* 87-2, Wissenschaftszentrum Berlin, Berlin.

Essmann, H. et al. (1989) Policy. Aspects of Natural Resource Monitoring. Proceedings of IUFRO/FAO Congress 'Global Natural Resource Monitoring and Assessments', Venice.

Essmann, H./Niesslein, E. (1986) Einflußströme für politische Entscheidungen zum Waldsterben. *Allgemeine Forstzeitschrift* 12.

Luhmann, N. (1971) Öffentliche Meinung, in: N. Luhmann, Politische Planung, Opladen.

Mayer, H. (1990) Entwicklungsperspektiven der Wälder unter Immissionsbelastung, in: Rossmanith, H.P., Waldschäden - Holzwirtschaft, Wien.

Neumann, M./Pollanschütz, J. (1988) Waldzustandsinventur 1988 - Der Wald hat sich heuer erholt. *Österreichische Forstzeitung* 11.

Niesslein, E. (1986) The impact of atmosperic deposition on forests from policy aspects. Proceedings of the 18th IUFRO World Congress (Division 4), Ljubljana.

Niesslein, E./Voss, G. (1985) Was wir über das Waldsterben wissen, Köln.

Niesslein, E. et al. (1986) Ökonomische und politische Folgen des Waldsterbens, Forschungsbericht KfK-PEF 18, Karlsruhe.

Österreichische Akademie der Wissenschaften, Hrsg. (1989) Die Bedrohung der Wälder. Schäden, Folgeerscheinungen und Gegenmaßnahmen, Wien.

Poston, T. and Steward, J. (1978) Catastrophe Theory and its Applications, Piman Publishing Ltd. London.

Roeder, V. (1979) Immissionen - Hauptursache für das Tannensterben? *Holz - Zentralblatt* 105.

Rozsnyay, Z. (1990) Wandel der Beziehungen zwischen Wald, Mensch und Gesellschaft. *Forst und Holz* 2.

Schärer/Zimmermann (1984) Politische und rechtliche Betrachtungen zum Thema Waldsterben in der Schweiz: Eine Standortsbestimmung. Beiheft zu *Schweizerische Zeitschrift für Forstwesen* Nr. 73.

Schlaepfer, R. (1988) Waldsterben. Eine Analyse der Kenntnisse aus der Forschung. Berichte Nr. 306, Eidgenössische Anstalt für das forstliche Versuchswesen, Birmensdorf.

Schlaepfer, R., Haemmerli, J. (1990) Das 'Waldsterben' in der Schweiz aus heutiger Sicht. *Schweizerische Zeitschrift für Forstwesen* 4.

Schütt, P. (1988) Waldsterben - Wichtung der Ursachenhypothesen, *Forstarchiv* 59.

Zimmermann, W. (1985) Waldsterben in der Schweiz: Die Leistungen der staatlichen Institutionen. *Forstarchiv* 56.

THE EFFECT OF LIMING AGRICULTURAL LAND ON THE WATER QUALITY OF THE RIVER ESK, CUMBRIA

Mark Diamond, David Hirst, Linton Winder, David H. Crawshaw* and Raymond F. Prigg
National Rivers Authority, North West Region,
PO Box 12, Warrington, U.K.
*North West Water Ltd., Warrington, U.K.

Abstract

A land use study indicated that a reduction in agricultural liming may have been a major factor in the development of acid episodes and consequent fish kills in the River Esk and River Duddon.

A field study was carried out to determine whether the re-introduction of catchment liming on the River Esk could improve water quality sufficiently to prevent fish kills. Lime was applied to previously limed areas comprising 10 % of the catchment and a range of determinands including flow, calcium, aluminium, total humic substances, conductivity and invertebrate abundance were measured both before and after liming. The River Duddon was also monitored as a control.

There was a general improvement in water quality both at the River Esk and the River Duddon during the study period. In addition there were substantial and unexpected changes in the water quality of the River Esk associated with liming.

This study demonstrates the benefits of continuous monitors in the study of long term changes in water quality.

Introduction

Agricultural liming is the spreading of powdered limestone, largely calcium carbonate, on land to reduce soil acidity. It has been used in this way since Roman times, but its use to mitigate the acidification of surface waters is a recent development. In contrast to direct liming treatment (Fraser & Britt, 1982; Weatherley, 1988) there is little information available on the effects of agricultural or catchment liming on surface waters . This is probably because direct liming of surface waters has an immediate effect and is said to be more cost-effective (Warfvinge and Sverdrup, 1984). However, catchment liming may be an effective component of the treatment of acidified surface waters when the ground is particularly susceptible to the leaching of aluminium (Hultberg and Andersson, 1981), where spates are likely and where lakes have a high flushing rate (Swedish Ministry of Agriculture, 1982).

This paper describes the results and preliminary analysis of a catchment liming exercise carried out on the River Esk in the South-West of the English Lake District, Cumbria. The work was stimulated by the observation of acid related mortalities of salmonids, simultaneously in the River Esk and the adjacent River Duddon in June 1980 and again in September 1983 (Diamond et al., 1987). Such mortalities are thought to be caused by high concentrations of aluminium (> 0.25 mg/l) which are toxic at low pH (< 5.5) and low calcium concentration (< 1 - 2 mg/l) (Reviewed by Warren et al., 1986). Preliminary investigations demonstrated that many parts of the Esk and Duddon catchments had a water chemistry and biology indicative of acidification and there had been a decline in the numbers of returning migratory salmonids (Crawshaw et al., 1987). It was known that the application of lime to land in Great Britain had declined rapidly since the early 1960's (Ormerod and Edwards, 1985) and an investigation into the past application of lime to the Esk Valley indicated a five-fold reduction since 1976 when the liming subsidy was removed (Robinson, 1984). Therefore, it was decided that an attempt would be made to improve water quality in the River Esk by liming all of the agricultural land in the catchment which had been limed in the past (generally the flatter land immediately adjacent to the river) and monitor the effects on the water quality and biology. The River Duddon catchment, where liming for normal agricultural purposes continued at a reduced rate, was used as a control.

Description of the Sites

The River Esk and River Duddon are situated in South-West Cumbria in the English Lake District (Fig. 1). The Esk rises on the southern slope of Scafell Pike and is a relatively short river being some 23.6km to the tidal limit. At the gauging station at Cropple How (National Grid Reference SD 131 977), 3.4km above the tidal limit, its catchment area is 70.2 square km and its average daily flow is approximately 5.6 cubic m/s. The Duddon rises 6km to the south-east of the source of the Esk at the top of Wrynose Pass. The distance to the tidal limit is 22km and at the gauging station at Duddon Hall (National Grid Reference SD 196 897), 1.5km above the tidal limit, its catchment area is 78.2 square km and the average daily flow is approximately 6.2 cubic m/s.

The underlying rocks in both catchments are of low buffering capacity and are largely Borrowdale Volcanics in the Duddon catchment and Eskdale Granites in the Esk catchment. In both

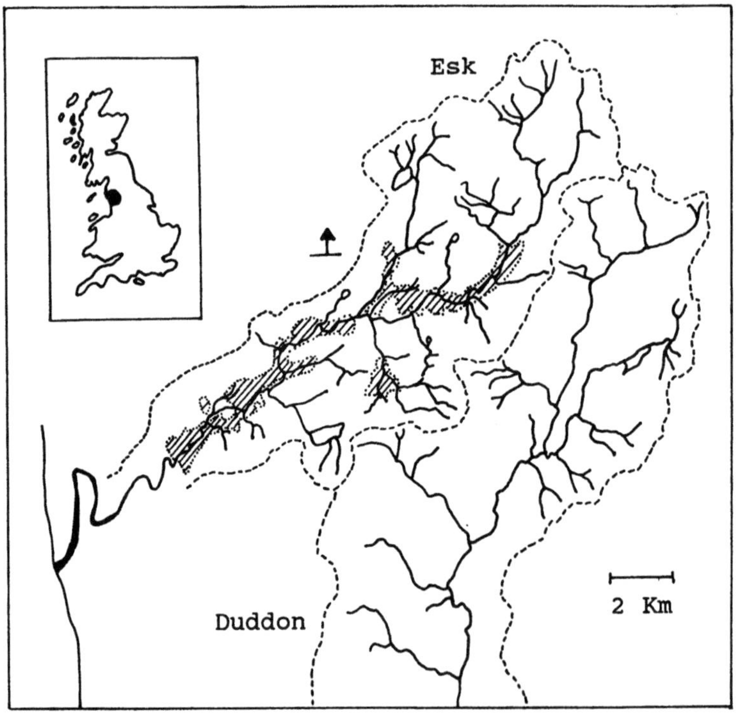

Figure 1. The Esk and Duddon catchments. Hatching within the Esk catchment indicates area within which lime was applied.

catchments the soils are described as humic rankers, podzols and brown earths (Millward et al., 1978). Most of the land immediately adjacent to the rivers is semi-improved agricultural grassland. On the valley slopes moorland predominates, and the vegetation consists largely of grasses, bracken and *Sphagnum* moss.

In both valleys the land is used mainly for the grazing of sheep and, to a much smaller extent, cattle. A major change in farming practice resulted from the withdrawal of the general liming subsidy in 1976. Between 1940 and 1976 annual application rates of 13 tonnes per hectare to intake land were common throughout the Esk and the Duddon Valleys. Between 1976 and 1983, 40% of this land received no lime and the remainder received just one application at the reduced rate of 5 tonnes per hectare (Robinson, 1984).

A desk-study of the calcium budget of the Esk catchment indicated that the restoration of the pre-1976 liming rates could halt the apparent deterioration in conditions which had led to the fish mortalities in 1980 and 1983 (Crawshaw, 1984).

Methods

Continuous Water Quality Monitors and Automatic Samplers

The continuous monitoring of water quality, water level and upstream fish movement commenced at Duddon Hall on the River Duddon in April 1983 and at Cropple How on the River Esk in May 1983. The monitors took measurements at quarter hour intervals of pH, river level, electrical conductivity, temperature and upstream fish movement. All readings were recorded on a solid state data logger. Sensors were calibrated at three week intervals.

Automatic water samplers at each of these sites were triggered when the water level reached a pre-set height. Composite samples, made up of four half-hourly sub-samples, were taken at two hour intervals for 48 hours. These samples, which were taken to monitor changes in water chemistry during episodes of high flow, were analysed for pH, alkalinity, calcium, magnesium, aluminium (soluble and total after acidification of the sample), conductivity, chloride and humic substances.

Catchment Liming

The powdered limestone was provided to the following specification: Sedimentary rock consisting largely of calcium carbonate and containing not more than 10% of magnesium (expressed as MgO) by weight, of which 100% will pass through a sieve of 5mm, not less than 95% will pass through a sieve of 3.35mm and not less than 40 % will pass through a sieve of 0.15mm. Between September 1986 and March 1987, 3200 tonnes of powdered limestone was applied to 640 hectares of intake land in the Esk catchment at a rate of 5 tonnes per ha (Fig.1) . This treatment was repeated between September 1987 and June 1988.

Soil pH was measured in forty fields in the summer of 1986 prior to liming, and in these and a further sixty fields in July 1987 and again in July 1988 after liming. In a separate in-stream liming exercise reported elsewhere (Crawshaw, 1987) between June 1985 and September 1987 a total of 22 tonnes of 15mm limestone chippings were added to two tributaries of the River Esk and 50 tonnes were added to a tributary of the River Duddon. It was thought that these relatively small amounts of lime would have no detectable effect on the water chemistry at Cropple How or Duddon Hall.

Supplementary Sampling

Water quality samples were taken quarterly from 15 main river sites and 39 tributary sites beginning in 1983. Benthic macroinvertebrates were kick-sampled from riffles at these sampling points mostly during winter or late spring. In addition the sites were electrofished six times between 1981 and 1988.

Results

Soil Tests

There was a small but marked general shift upwards in soil pH between 1986 and 1987 and a much larger increase between 1987 and 1988 (Fig. 2).

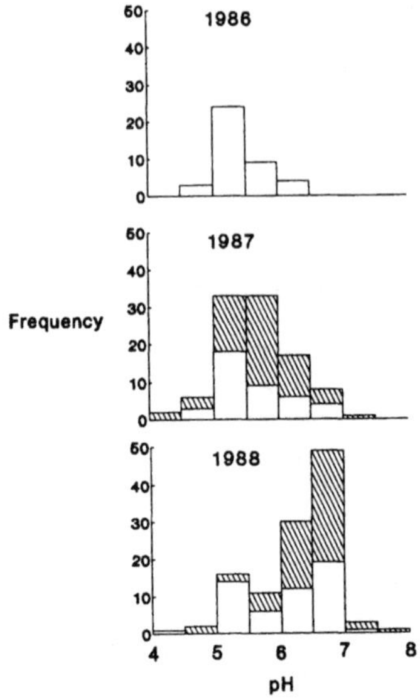

Figure 2. Frequency distribution of soil pH for fields in the Esk catchment (open bar = 40 fields sampled each year; hatched bar = 60 additional fields).

Continuous Monitor Data

Short period time series plots of continuous monitor data showed a consistent pattern with almost every rise in flow being accompanied by a decrease in pH and conductivity (Fig. 3).

These data were analysed primarily to investigate the effect of liming on the pH of the river in periods of high flow when previous conditions had, at times, been toxic to fish. For this reason these data were reduced by looking at 'flow peaks', defined as any record where the flow was greater than the ten records before and after (ie 2.5 hours before and after). These flows were then matched with the corresponding troughs in pH, which did not necessarily occur at exactly the same time. Plotting pH against flow for several periods through the seven years demonstrated that the minimum pH invariably occurred at or after the maximum flow. Therefore, the minimum pH was defined as the minimum in the period from one observation before the peak flow to ten observations after. (Slightly different definitions of both 'peak flows' and the corresponding minimum pHs were also tried with very similar results).

Plotting these minimum pHs against the natural logarithm of the flow peaks demonstrated a strong linear relationship between log(flow) and pH above approximately 1.0, whilst pH appeared not to be affected by flow when log(flow) was below approximately 0.3 (Fig. 4). The exact form of the relationship changed over time, (indicated by the increase in variation of pH with increasing flow in Figure 4) but the general form was constant for both the Esk and the Duddon. In particular, there was little variation in the position of the 'elbow' at log(flow) between 0.3 and 1.0. Therefore, the data were divided into 'low flow' (log flow < 0.3) and 'high flow' (log flow > 1.0) sets, which were analysed in different ways.

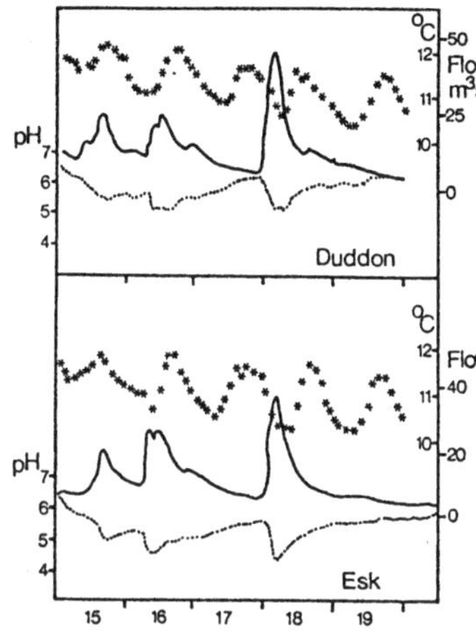

Figure 3. Time series plot of continuous monitoring showing variations in pH (....), flow (____) and temperature (****) at the Esk and Duddon catchments during the fish kills observed in 1983.

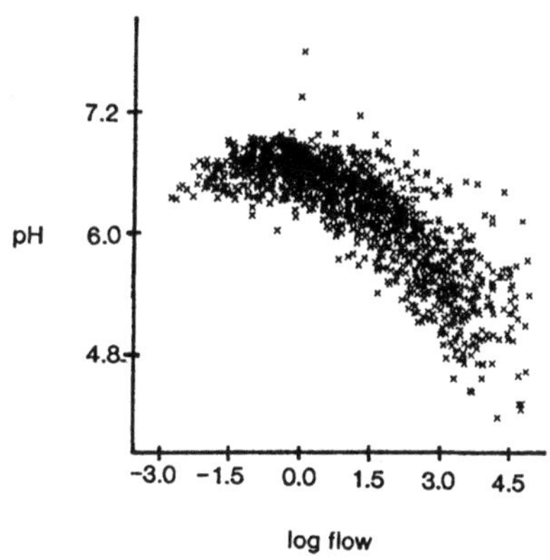

Figure 4. The relationship between log flow and pH measured at the Esk continuous monitoring station.

Low Flow Data

As the pHs in this data set were unaffected by flow, a simple mean for each month was taken. A double time series of the Esk and the Duddon showed the Duddon pH to be consistently higher than that of the Esk, by about 0.5 pH units. There was also an indication that pH in both rivers had increased slightly over the decade.

Cusum analysis of the time series for the Esk detected a step change towards the end of 1986 of approximately 0.17 pH units, and analysis for the Duddon indicated two changes of about the same magnitude. The times of these changes were not related to the liming. A time series of the difference between the two rivers showed no obvious trend, and although the cusum analysis detected two significant step changes they were of small magnitude (no more than 0.17 pH units), and the times were unrelated to liming.

High Flow Data

Data for log(flow) above 1.0 showed more interesting trends. In view of the linear relationship with log(flow), a simple linear regression was performed and the predicted pH at a log(flow) of 3.3 was obtained. This was approximately the 99 percentile flow, and so the corresponding value of pH could be interpreted as the value likely to be exceeded for 99% of the time.

The time series of predicted pH for the Esk showed an increase over time (Fig. 5), with a cusum analysis detecting a step change of 0.62 pH units at the beginning of 1987, corresponding to the first period of liming. However, the pH was known to be increasing over this period as confirmed by the time series for the Duddon (Fig. 5) where a smaller step change (0.39 pH units) occurred at the same time. The time series for the difference between the Esk and the Duddon showed a pre-liming difference of 0.5pH units (with the Duddon having the higher pH), which decreased to almost zero following the first liming (Fig. 5). The level then, surprisingly, increased to 0.35pH units at the start of the second period of liming and remained at that level. This pattern can also be seen in the double time series (Fig. 5).

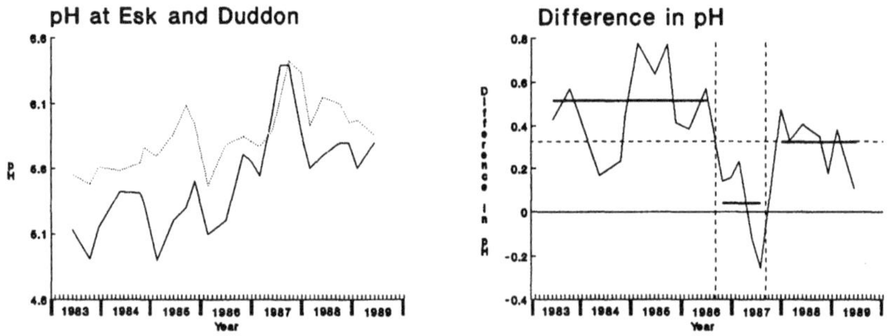

Figure 5. Time series plots of pH at the Esk (____) and Duddon (....) and the difference between the two catchments (Duddon-Esk).

Continuous Monitor Electrical Conductivity Data

A similar method of analysis was applied to the conductivity data from the continuous monitor. Conductivity showed a marked seasonality, being highest in August and lowest in February. This was allowed for by substracting the mean monthly conductivity (using all data) from each observation. Only the high flow results have been analysed. Following the first application of lime there was , surprisingly, a large and significant decrease in the electrical conductivity of the River Esk which was maintained throughout the remainder of the monitoring period (Fig. 6). The coincidence of a significant change in the difference between the conductivity of the Esk and that of the Duddon with liming indicated that the change in conductivity in the Esk was due to liming (Fig. 6). Figure 6 should be interpreted as the difference between conductivity at 99 percentile flow and the value to be expected at the mean flow at that time of year.

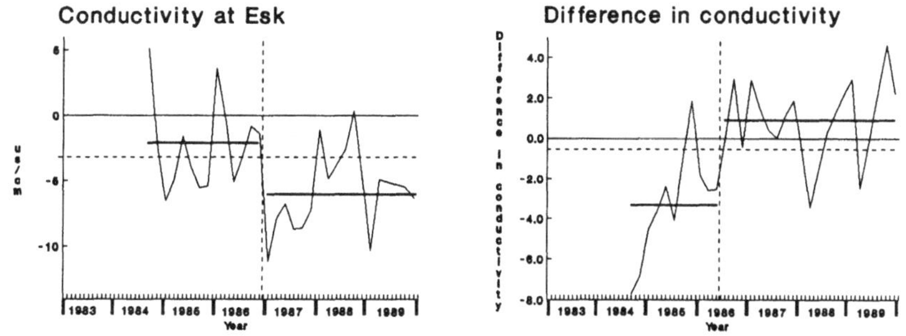

Figure 6. Time series plot of electrical conductivity measured at the Esk continuous monitoring station and the difference in conductivity between the Esk and Duddon (Duddon-Esk). Electrical conductivity was adjusted for seasonality.

Automatic Sampler Data

Due to a back-log in data filing, only auto-sampler data for the Esk upto the end of 1988 are presented. As these composite samples were made up of four successive sub-samples taken at half-hour intervals they corresponded to a range of flows rather than an instantaneous flow. However, an attempt to relate water chemistry to the flow after the start of each two hour sampling period proved fruitful.

The relationship between flow and chemistry for the auto-sampler data was not as strong as for the continuous monitor data, but it was still evident for most determinands. The raw data was biased by periods when the high flows were frequent since the number of samples collected was proportional to the length of time over which the sampler was operating. For this reason 'flow peaks' were again used, this time defined as any flow which was higher than the flow before and after it. The determinands of most immediate interest were pH, conductivity, calcium, aluminium and total humic substances. The same type of analysis was used as that described above (although only high flow data was available). In view of the uncertain relationship between the concentrations of determinands in composite samples and recorded flows and the relatively small amount of data, the results are not as reliable as those from the continuous monitors. However, it was reassuring that pH and conductivity data from the automatic samples showed similar patterns to those from the continuous monitor data (e.g. Fig 7). The step changes shown in all of the figures were identified using cusum analysis and all are significant at the 5% level.

The calcium results were surprising. The concentration of calcium fell following liming from 1.71 to 1.22 mg/l after the first liming. By the end of the second liming it had returned approximately to its pre-liming levels (Fig. 7).

As expected there was a large and significant decrease in the concentration of soluble aluminium from 118.4 to 72.2 ug/l following the first liming which was maintained during the remainder of the monitoring period (Fig.7).

Total humic substances concentration dropped sharply during the first liming from 1.3 to 0.87 mg/l and then increased to a mean level of 1.19 mg/l (Fig.7). It should be noted that this determinand is highly seasonal. This has been allowed for in the same manner as for conductivity, but the increases occurred at the times expected due to seasonality which suggested that this could be an exagerated seasonality not necessarily due to liming.

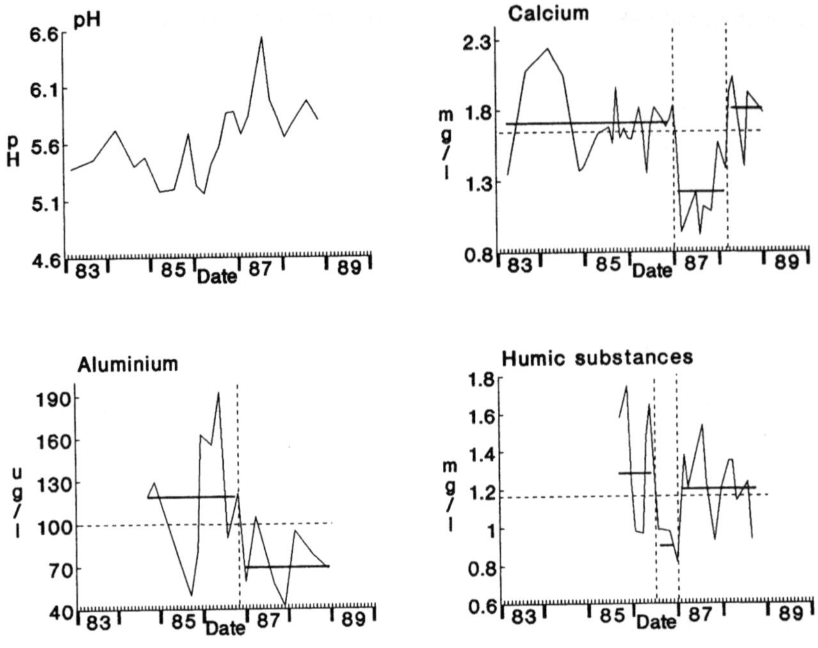

Figure 7. Time series plots of pH, Calcium, Aluminium and Humic substances measured at high flows at the River Esk. Humic substances were adjusted for seasonality.

Invertebrate Data

The invertebrate data were subjected to a simple preliminary analysis of the effect of pH on the number of species found at each sample site. The mean number of species caught per sample was calculated for the pre-liming and post-liming periods at each site, and were then plotted against the corresponding median pH. There was a positive relationship between median pH and mean number of species caught per sample at both the Esk and Duddon during the pre- and post-liming periods (Fig. 8). At a given pH, the number of species caught per sample was generally higher during the post-liming period than the pre-liming period particularly at the Duddon which suggested that water quality had improved independently of liming.

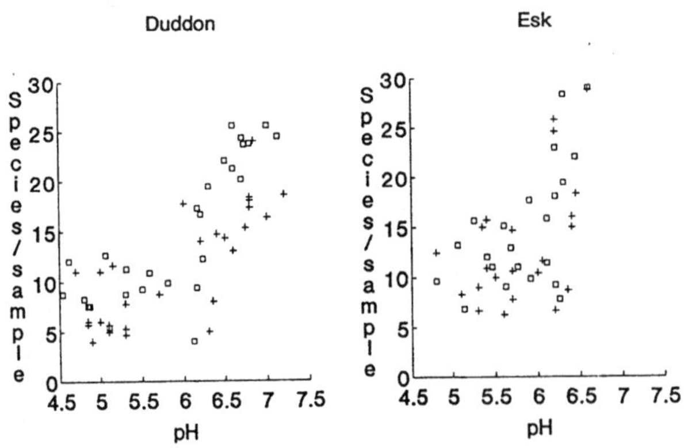

Figure 8. The relationship between pH and the total number of invertebrate species caught per sample at the Duddon and Esk during the pre-liming (+) and post liming (□) periods.

Principal component analysis of water chemistry data followed by correlation of the principal components with invertebrate data from the upland streams of the whole of North West England indicated that some groups such as Plecoptera (stoneflies) were apparently tolerant of acidity whereas other groups were very sensitive (Diamond et al., 1987). The latter 'acid-sensitive' groups included Baetidae, Heptagenidae, Gammarus, Perlodidae, Hydropsychidae, Gastropoda and Molluscs.

In order to determine the effect of catchment liming on acid-sensitive species at the Esk, the difference between the mean number of acid-sensitive species caught per sample during the post- and pre- liming periods was calculated for each site. Those sites within the limed parts of the catchment had a significantly larger increase in the number of acid-sensitive species than those within the unlimed part of the catchment ($p<0.05$) indicating that catchment liming improved water quality conditions over and above the general improvement observed (Fig. 9).

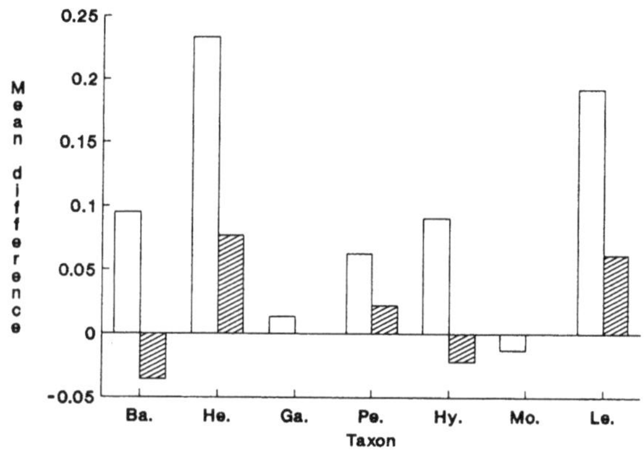

Figure 9. The mean difference in numbers of acid-sensitive species between the post and pre-liming periods (Post-Pre) within the limed (open bars) and unlimed (hatched bars) areas of the Esk catchment. Ba.=Baetidae, He.=Heptagenidae, Ga.=Gammarus, Pe.=Perlodidae, Hy.=Hydropsychidae, Mo.=Molluscs, Le.=Leeches.

Discussion

Since the early 1980's when the acid-induced mortalities of salmonids occurred, there has been a great improvement in the water quality and biological conditions in both the Esk and the Duddon. Evidence for this includes general increases in pH, calcium concentration and conductivity and an increase in the distribution of acid-sensitive invertebrates. Large increases in the densities and distribution of young salmonids in these rivers has been reported elsewhere (Crawshaw et al., 1987) and no further fish-kills have been observed. Data collected by the IFE indicates that acid deposition has declined over the last two decades. The mean weighted pH measured at a site close to the Esk and Duddon has increased from 4.4 during the 1950's and mid-1970's to 4.8 for the period 1983-1989 (T. Carrick, pers. comm.). This decrease in acid deposition may have been responsible for the general improvement in water quality observed in this study.

The validity of the use of the Duddon as a control could be questioned as, given the size of the catchments, and the known differences in the catchment geology and orientation to the prevailing (westerly) winds, some variation in the differences between the Esk and Duddon were likely to be caused by factors other than liming. However, the patterns of change of, for instance, pH in the two catchments were remarkably similar (Fig.6) except during liming. In the absence of a more suitable alternative, it was assumed that the impact of liming on the water quality of the River Esk would be indicated by a comparison with the River Duddon.

Other surface liming experiments have generally, where effective, resulted in an increase in pH, conductivity and calcium concentration and a decrease in aluminium concentration (e.g. Warfvinge and Sverdrup,1988; Hornung et al.,1990; Brown et al., 1987). Therefore the changes in the water quality of the River Esk following the first liming were surprising in that, although there was an expected increase in pH and a decrease in aluminium concentration, there was a decrease in both calcium concentration and electrical conductivity. Following the second liming calcium increased to approximately the pre-liming levels, whereas conductivity and aluminium concentration remained low.

These results are difficult to explain. The initial addition of lime may have been insufficient to completely neutralize soil acidity (as only a slight shift in soil pH was observed) and insufficient to fill all of the available cation exchange sites on the organic matter within the soil. However, as the first addition of calcium to the catchment resulted in less being released into stream water it can be deduced that adding lime stimulated the catchment to retain calcium. Investigations to determine the fate of calcium in the soil will commence in autumn 1990.

One factor which has not yet been assessed in this study is the hydrology of the catchment in relation to the areas limed which has shown to be important in other studies . For instance, in the Llyn Brianne study the application of powdered lime and NPK fertilizer to a catchment had little effect on stream water chemistry because the 20 hectares treated appeared not to be important in controlling catchment hydrology (Hornung et al., 1990). In the same study, the intensive liming of 4 hectares in a second catchment appeared to cause a great improvement in stream water chemistry because it included a hydrological source area. Warfvinge and Sverdrup (1988) stress the importance of liming hydrological source areas to effect the neutralisation of surface waters particularly during high flow events. In the Esk catchment, the importance of hydrology in relation to liming has yet to be assessed.

The increase in the distribution of acid-sensitive invertebrates at sites downstream of limed areas was further evidence for an improvement in water quality caused by liming (in excess of the general improvement). As salmonids and acid-sensitive invertebrates have similar water quality requirements (Crawshaw et al,1987) it is likely that liming has also caused an improvement in conditions for salmonids in these streams.

In conclusion, although the liming treatment was successful,in that there was a decrease in aluminium concentrations and an increase in pH during high flow periods, the other water chemistry changes are difficult to explain.

The water chemistry changes demonstrated in this study were quite large but would have been difficult to detect without the long term baseline data and the continuous monitoring of flow and pH. Therefore, it is essential, when designing monitoring programmes to relate changes in water chemistry to changes in acid deposition, and that a representative sample of sites incorporate continuous monitoring. This approach is being used by the United Kingdom's Acid Waters Monitoring Network funded by the Department of the Environment.

Monitoring of water chemistry and biology of the Esk and Duddon will continue. Future studies will include an investigation into the hydrology and soil processes of the Esk and Duddon catchments and an assessment of land-use during the monitoring period.

Acknowledgements

Thanks are due to the Department of the Environment and the European Commission who funded the early part of this study. The liming project was jointly funded by North West Water and the Department of the Environment and latterly by the National Rivers Authority.

The authors would like to thank numerous colleagues within and outside North West Water and the National Rivers Authority, North West Region,without whose continuing assistance the work would not be possible.

References

Brown D.J.A., Howells G.D. and Patterson K. (1987). The Loch Fleet Project. In Acidification and Water Pathways. Oslo, Norway: Norwegian National Commitee for Hydrology.

Crawshaw D. H. (1984). The effect of Acid Run-off on the Chemistry of Streams in Cumbria In P Ineson (Ed) ITE Symposium No 16, P25-32.

Crawshaw D. H., Diamond M., Prigg R. F., Cragg-Hine D., and Robinson J. F. (1987). Acidification of Surface Waters in Cumbria and the South Pennines. Report for the Department of the Environment, North West Water, 99p.

Crawshaw, D. H. and Diamond M. (In Press). Effects of Agricultural Liming on Surface Water.In Proceedings of Acid Rain Conference, April 1988, Cardiff, edited by H. Barth.

Diamond, M., Crawshaw D. H., Prigg R. F. and Cragg-Hine D. (1987). Stream Water Chemistry and its influence on the distribution and abundance of aquatic invertebrates and fish in upland streams in North West England. In Acid Rain: Scientific and technical advances,Selper, London, p481-488, edited by R. Perry et al.

Fraser, J. E. and Britt, D. L. (1982). Liming of acidified lakes and streams - perspectives in induced physical-chemical and biological changes. In AMS/CMOS Conference, long-range Transport of Airborne Pollutants.

Hultberg H. and Andersson I.B. (1981). Liming of acidified lakes and streams - perspectives in induced physical-chemical and biological changes. In AMS/CMOS Conference, Long-Range Transport of Airborne Pollutants.

Hornung M., Brown S.J. and Ranson A. (1990). Amelioration of surface water acidity by catchment management. In Acid Waters in Wales, p311 - 328, Kluwer Academic Publishers, Dordrecht. R.W. Edwards, A.S.Gee and J.H.Stoner (Eds).

Millward, D., Moseley, F. and Soper N. J. (1978). The Eycott and Borrowdale volcanic rocks. In The Geology of the Lake District p99 - 120 Yorkshire Geol.Soc. Occasional Publ. No 3, edited by F. Moseley.

Ormerod, S. J. and Edwards R. W. (1985). Stream acidity in some areas of Wales in relation to historical trends in afforestation and the usage of agriculture limestone. Journal of the Environmental Management 20, 189-197.

Robinson, J. F. (1984). A review of land use, farming and fisheries in the Esk and Dutton valley, North West Water Authority.

Swedish Ministry of Agriculture, Environmental Committee (1982). Acidification Today and Tommorrow. Risbergs. Tryckeri A B, Uddevalla, Sweden, 231p.

Warfvinge P. and Sverdrup H. (1984). Soil Liming and Run Off Acidification Mitigation. International Symposium on Lake and Reservoir Management, EPA 440/5-84-001, p389-393.

Warfvinge P. and Sverdrup H. (1988). Soil liming as a measure to mitigate acid run-off. Water Resources Research.

Warren S.C. et alia, (1988). United Kingdom Acid Waters Review Group Interim Report. Department of the Environment, London.

Weatherley N. S. (1988). Liming to mitigate acidification in freshwater ecosystems: a review of the biological consequences. Water, Air and Soil Pollution 39,421-437.

3

EFFECTS ON STRUCTURAL MATERIALS

BACKGROUND AND LOCAL CONTRIBUTIONS TO ACIDIC DEPOSITION AND THEIR RELATIVE IMPACT ON BUILDING STONE DECAY: A CASE STUDY OF NORTHERN IRELAND

B. J. Smith, W. B. Whalley and R. W. Magee
School of Geosciences, Queen's University of Belfast, Belfast, Northern Ireland, U.K.

ABSTRACT

Because it lies to the northwest of mainland Europe, Northern Ireland experiences little background atmospheric pollution. Urban areas such as Belfast can, however, experience high atmospheric concentrations of smoke and sulphur dioxide from local sources. In Belfast, sandstone decay (the principal building stone) can be severe and a wide range of salt weathering phenomena occur. Salts - primarily gypsum - can derive from particulate deposition, particularly of fly ash, or the reaction of mortars with atmospheric sulphur. Caution is urged in ascribing all stone decay to atmospheric pollution as similar breakdown is caused by salts from marine aerosols and spray, groundwater and road de-icing.

1. INTRODUCTION

Ever since the 1972 United Nations conference on 'The Human Environment' in Stockholm, emphasis has rightly been placed upon the role of transnational atmospheric pollution as source of acid deposition and a cause of environmental deterioration in northwest Europe. Nevertheless, it has been widely demonstrated that acidification of and damage to, for example, bio-aquatic environments is complex and can be caused or accentuated by many other factors (for a review see Park (1987, Ch. 3)). Similarly, effects can be localised in their impact. Within the field of urban stone decay the latter point has been frequently demonstrated by reference to decay in particular cities such as Venice (eg. Fassina, 1978) and exposure trials that compare rates of decay between urban and nearby rural locations (eg. Jaynes and Cooke, 1987). In these studies the importance of local pollution sources, particularly dry particulate and gaseous deposition, have been stressed. It has been difficult, however, to isolate these local contributions from those made by regional or transnational inputs. These affect not only rates of urban stone decay but also background rates obtained from supposedly clean rural locations. In an attempt to overcome this problem, Lough Navar in Northern Ireland has been chosen as a 'baseline' site within the UK materials monitoring network. The expectation is that because the site lies far to the west of mainland Europe, it is upwind of most transnational pollution and will thus provide information on background atmospheric composition and rates of stone decay. In an earlier pilot study of a small town near Lough Navar, however, (Smith et al., In Press) we noted that despite apparently low levels of background pollution, local stone decay in urban areas can be marked and that stonework in rural locations is not immune to attack - especially through salt weathering.

Northern Ireland thus provides a particular opportunity to study 'natural' patterns and causes of building stone decay and also the impact of local pollution sources. To do this we examine in this paper first, the nature of rural and urban atmospheric pollution in Northern Ireland; second, decay of sandstone buildings in Belfast together with the nature of associated particulate pollution; and finally, decay associated with atmospheric pollution is compared to that caused by other agencies in both rural and urban environments.

2. ATMOSPHERIC POLLUTION IN NORTHERN IRELAND

Regional levels of atmospheric pollution in Northern Ireland do appear to be substantially lower than for most of the rest of the British Isles. Measurements of bulk precipitation of SO_4 in 1985 by Jordan (1988) gave a rate of 32.8 kg/ha/a at Park Gate; one of three secondary stations located in Northern Ireland in addition to the primary station at Lough Navar (Figure 1). Jordan (1987) has suggested that deposition rates of sulphate sulphur might appear to be elevated because of high rainfall totals in Northern Ireland compared to England. It is significant therefore, that dry deposition at the secondary sites is particularly low, with deposition rates of typically 1-4 kgS/ha/a (Jordan 1988). This compares with background rates in most of England, outside of large connurbations, of 20-40 kgS/ha/a (Mellanby, 1988, p.44). Within Ireland as a whole it has been estimated that approximately 50% of sulphur deposition comes from long-range transport, with half of this coming on easterly and southeasterly winds from Great Britain (Fisher, 1982). The other 50% of deposited sulphur is derived from local pollutant and background sources. One background source is sea salt, high levels of which in precipitation reflect the fact that the Northern Ireland climate is dominated by weather systems originating over the Atlantic (Betts, 1982). Measurements of bulk precipitation at Parkgate by Jordan (1983) suggest that approximately 20% of sulphate in bulk precipitation was derived from sea salts. He also notes high deposition rates for both chloride (34.7 kg/ha/a) and Sodium (20.5 kg/ha/a) derived totally from sea salt (Jordan, 1988).

Levels of rainfall acidity in Northern Ireland exhibit a similar pattern to those of the rest of the UK, in that acidity is greater in the east of the Province and lower in the west (Jordan, 1983). Between 1985-1987 approximately 10% of samples collected at the three secondary sites had pH values below pH5.0. Values at Silent Valley were typically pH 4.4-4.7, whereas at Lough Navar they were pH 4.8-5.1 (Jordan, 1988). These values are consistent with those presented by Mathews et al. (1980) who, in a survey of acid rain in the Irish Republic, showed that by 1979, 48% of samples had pH values below 5.5; an

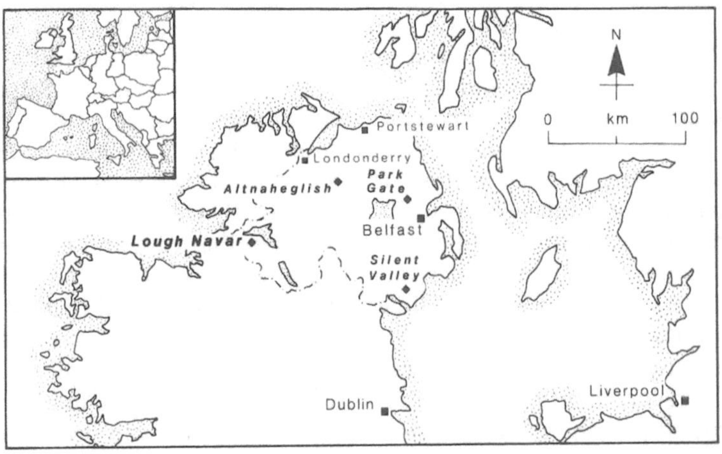

Figure One: Location Map.

increase from 27% in 1960. Thus, although rainfall can currently be considered as only slightly acidic, there is some evidence that acidity has been increasing.

These general figures mask the fact that within Ireland very high levels of pollution occur at a number of urban centres. In Dublin, for example, smogs are still a regular winter feature and between 1982-87 McGrath (1988) found that at 7 out of 17 stations smoke concentrations had increased by as much as 100% related primarily to increased domestic burning of coal. High smoke and sulphur dioxide concentrations also occur over urban areas in Northern Ireland, where effects are often accentuated by local topographic factors. Much of Belfast, for example, lies within the incised valley of the River Lagan and is particularly prone to temperature inversions during anticyclonic conditions in winter which trap pollution within the valley, particularly affecting the city centre. Long-term control of pollution has also been hampered by the delayed introduction of clean-air legislation into the Province, such that The Clean Air Act (Northern Ireland) was only passed in 1964, eight years after equivalent legislation in Great Britain. This legislation has still not been fully implemented and Belfast began in 1989 an eight year programme to enforce smoke control orders over the remaining eight districts of the City not then covered. Even within areas designated as smoke-free, compliance is often sporadic and enforcement has been particularly difficult given the history of civil unrest in the Province. It is hardly surprising therefore, that Belfast, along with Londonderry and Newry and Mourne, featured on the list of twenty eight districts submitted to the European Commission by the Department of the Environment in 1983 as areas that would not comply with emission directives by April of that year. In every case breaches of smoke limits were the problem, with the domestic burning of coal cited as the primary cause (Haigh, 1989). By 1987/8 the list of deroged areas had been reduced to seven, but this still included Belfast, and both Londonderry and Newry and Mourne had exceeded limit values on occasions during the intervening period. Despite these problems, smoke concentrations in Belfast have declined (Figure 2A) both in suburban (1) and city centre (2-4) locations. Much of this decline reflects the effects of clean air legislation, but also industrial decline and increased use of oil for domestic heating. City centre levels, nonetheless, remain approximately three times those of the suburban site and are prone to very high, inversion-related winter peak values (Figure 3A). Of the city centre sites, that in east Belfast has experienced generally higher smoke concentrations (Figure 2A, site 3). This might reflect the effects of nearby industrial complexes located around Belfast Lough but also its location downwind of any pollution carried from the west of the city by prevailing winds. The marked difference between this site and the suburban site (located only some 2 km to the east) illustrates the localised nature of much smoke pollution and, because much of the city centre lies close to sea-level, the possible significance of increased altitude in ameliorating the worst effects of pollution.

Sulphur dioxide concentrations showed a similar decline to that of smoke levels throughout the 1960's and 1970's and by the beginning of the 1980's, city centre mean annual figures were similar to those found in suburban locations (Figure 2B). During the 1980's however, there has been a consistent rise in sulphur dioxide concentrations at city centre and suburban locations. There are several possible explanations for this rise, including, perhaps, increased road traffic or increased frequencies of anticyclonic conditions. In both instances, one would have expected to see such changes reflected in increased smoke concentrations, but this was not the case. Instead, increased sulphur dioxide concentrations have coincided with the rundown and eventual closure in 1989 of the Belfast City Gas undertaking. Northern Ireland is not connected to the British natural gas network and plans to pipe natural gas from the Irish Republic were abandoned in the early 1980's. Industrial and domestic gas users were therefore given conversion grants to switch to alternative energy sources including solid fuel or oil. While no positive link can be established, the coincidence between the gas closure and increased sulphur dioxide concentrations remains intriguing. As a result of these

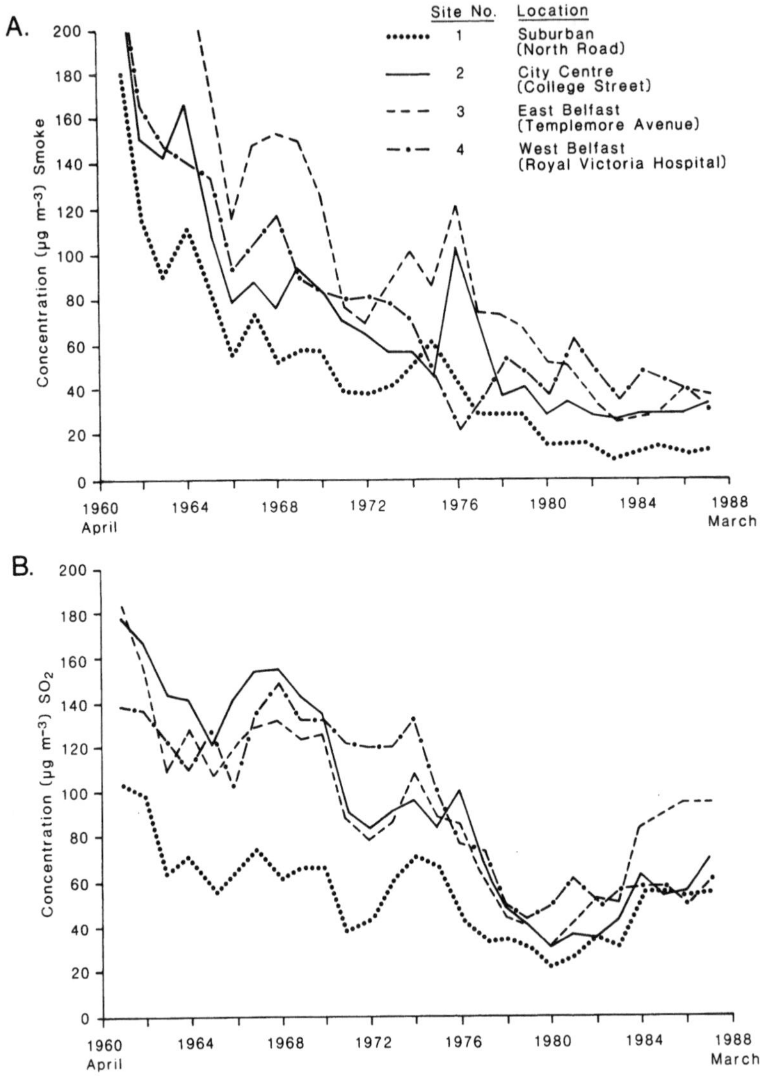

Figure Two: Changing smoke and sulphur dioxide concentraions at selected sites in Belfast. (Data from Belfast City Council)

changes, concentrations at the suburban site have returned to those experienced in the early 1960's and its current similarity with city centre values tends support to the contention that most of the sulphur dioxide derives from local domestic and vehicular sources. Inversion conditions (eg. December 1987, Figure 3B) can still accentuate the urban/suburban gradient, but generally differences in peak values are not as marked as the urban/suburban contrast in smoke concentrations (Figure 3A). Comparison with data from England suggest that mean annual sulphur dioxide concentrations in urban and suburban Belfast can exceed those found in, for example, Sheffield (of the order 50-60 $\mu g\ m^{-3}$, Mellanby (1988)) and central London (less than 30 $\mu g\ m^{-3}$ at the Guildhall during 1985, Cooke (1989)). This is despite, apparently, the absence of any significant contribution from transnational pollution.

In view of these conditions it is understandable that building stone decay, especially of the traditionally dominant sandstones, is widespread and often severe within Belfast. As a consequence, cleaning and renovation of buildings has become equally widespread in recent years. Unfortunately, this work frequently progresses in the absence of a detailed knowledge of the nature and causes of local decay and, apart from some preliminary studies (eg. McGreevy et al., 1983), there is little published information to draw upon. In the next section we present a study of sandstone decay in Belfast which, we hope, will go some way towards addressing this information gap.

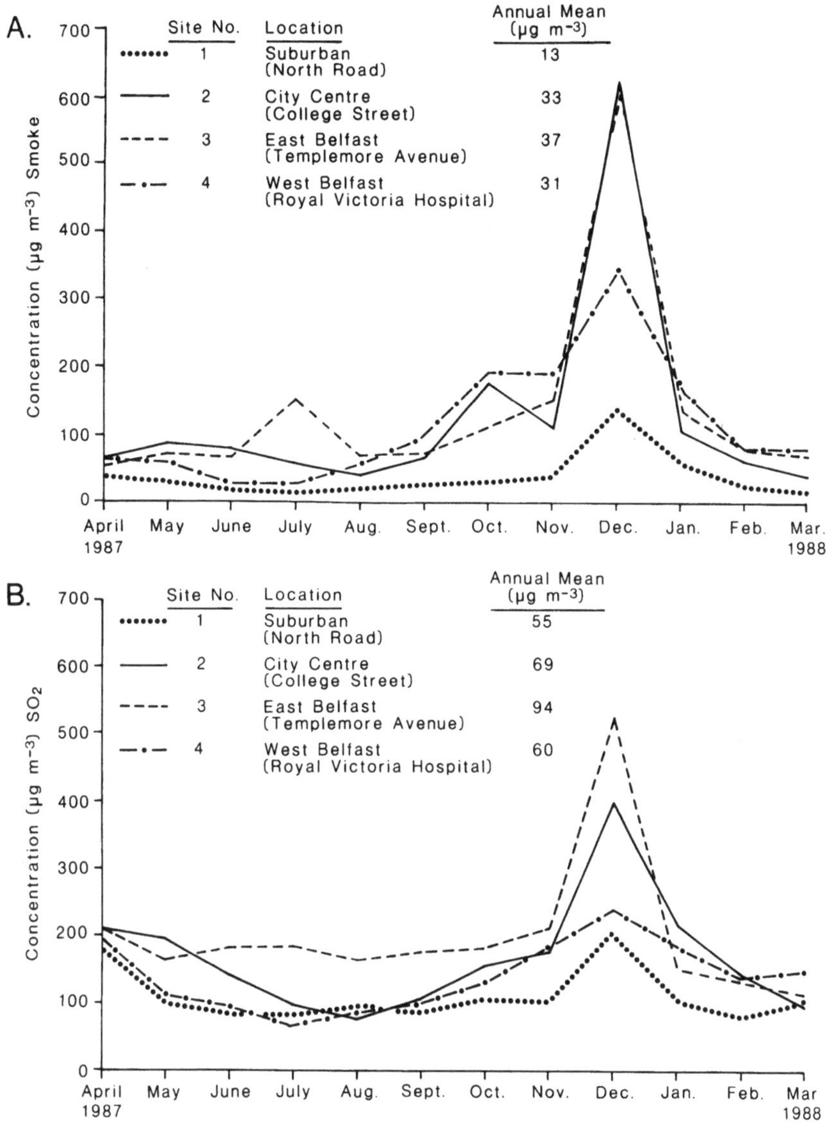

Figure Three: Peak monthly concentrations of smoke and sulphur dioxide at selected sites in Belfast for 1987/8, and annual means for the same period. (Data from Belfast City Council)

3. A SURVEY OF SANDSTONE WEATHERING WITHIN BELFAST

> 'The great majority of the numerous late Victorian churches of Belfast have knobbly stone exteriors, and exhibit a deliberate ruggedness which is ill-suited to the sooty air of a great industrial city'.

Writing in 1967, Charles Brett (Brett 1967, p.47) made this observation of buildings in a Belfast very different to the city of today. Much of the industry on which the city was founded has now gone, but the churches he makes reference to are largely still extant, many exhibiting stone decay phenomena as a result of the dual influence of background and anthropogenic pollution referred to in section two. The survey work reported here refers almost exclusively to these churches, but it should also be noted that many of the principal municipal buildings, constructed largely of Portland stone, are also suffering from the now familiar manifestations of calcareous stone decay.

Brett's comment on 'deliberate ruggedness' may well be accurate in the design sense, but it is also a likely reflection of the quality of the stone employed in many of the buildings. This is a local Triassic sandstone from Scrabo, Co. Down, which was used mainly in the period from 1850 to 1920 in the East of N. Ireland (Bell, pers comm.), and reflects the rapid industrial and population expansion of Belfast. No less than eighty-six churches were built in Victorian Belfast (Brett 1967), many using Scrabo stone for the body of the buildings. Ornamental or dressed stonework was mostly executed in different materials such as limestone or Scottish sandstones. Scrabo stone's most distinguishing characteristic is the extent of its variability (Smith and McGreevy, 1983; Bell, pers comm.), and the readily apparent inclusions of coarse and fine grained lamina containing clay minerals whose selective weathering characteristics may have a direct influence on rock breakdown (McGreevy and Smith, 1984). Similarly, these bedding features, apparent in even small blocks, have been shown to play a significant role in the development of surface flaking (ibid.). These properties could have important implications for weathering and decay phenomena related to faulty craftsmanship, in particular, face bedding, but due to the normally very apparent nature of the bedding the incorrect employment of the stone is rarely observed.

In terms of physical durability testing, the performance of Scrabo is not yet fully understood, but results to date indicate that it performs within accepted bounds for other building sandstones in the British Isles (Bell pers comm.; see Leary 1986 for details of the tests).

In dealing with a stone as extensively employed as Scrabo is in Belfast, the opportunity exists for the study of this one stone within a single urban environment, or within small, local microclimatic areas. The importance of understanding the physical and chemical properties of the stone must therefore be stressed. A preliminary survey such as this does, however, still have a role in providing observations on material response to a polluted environment. In addition, the fact that much of the stone in Belfast can be regarded as having been 'fresh' in terms of exposure to any environment outside that of the quarry is also important if any attempt is to be made to relate performance to pollution. This contrasts with the situation, for example, of many cathedral sites in Britain, where stone has often been exposed for several centuries prior to experiencing the much more aggressive atmospheric conditions resulting from industrial expansion in the nineteenth and twentieth centuries. The physical or chemical effects that such delayed exposure to pollution has had on the stone is now difficult to quantify.

3.1 The Survey

For the purposes of the survey a recording form was developed (Figure 4) one of which, was used on different elevations of a building. The aim has not been to make objective assessments of building decay states, but to record phenomena within their context on a building and, as regards the buildings themselves, in relation to their location within the city.

The descriptive terms used draw on standard terminologies from both building stone and geomorphological research. The latter is judged particularly relevant to the study of sandstone decay given the relative paucity of published material within building

BUILDING STONE DECAY — FIELD DATA COLLECTION FORM

SITE	All Soul's Church	CODE	BF1N
LOCATION	Elmwood Avenue, BT7		
NAME	R. Magee	DATE	14.7.89

DATE(S)	ASPECT	EXPOSURE
1895-6	North	Partly sheltered

ARCHITECTURAL FEATURES
Basal 'plinth' (c.6 courses) - windows of basement partly at this level. Narrow buttresses (5) along nave aisle + 4 windows. 2 string courses at c.2.5 and 5m, coping c.5.5m. Variable block size (0.12-0.2m in depth), alternating deep/shallow in elevation. Narrow jts - cement mortar. Lst for mouldings + dressings + st. courses. Sst for ashlar.

ROCK TYPE	COLOUR (unweathered + weathered)	TEXTURE	BEDDING	FINISH
Scrabo Sst	Reddish pink/+ grey → black cream-pink	Medium	Laminated	Roughly dressed
Doulting Lst	Cream + grey/green	Course 'crystalline'	Not evident	Dressed

LOCAL ENVIRONMENT EFFECTS
Busy urban location, City Hosp. c. 300m to NW. Q.U.B. adjacent. Roads (through routes) at S + W, but sheltered from direct exposure to these. Q.U.B. Geology chimney c. 100m to E. Domestic occupation surrounding site within 100-200m radius.

GENERAL DESCRIPTION OF WEATHERING (inc. location relative to other sites)
Bio decay + staining evident in lower courses - photos.
BC's in upper courses, esp. beneath upper st. course (BC's on the Sst, not really on the Lst).
Lst > prone to surface solution loss, staining, alteration skin.
Sst > prone to flaking, gran. disintegration, efflorescence + BC.
BC's generally on upper parts of blocks.

Stone decay evident at City Hosp. to NW (Sst) + Q.U.B. (Portland dressings to brick buildings + Sst) adjacent at E + churches in general vicinity.

WEATHERING		ROCK TYPE(s)	SEVERITY Severe, Moderate, Slight	FREQUENCY Abundant, Common, Rare
Flaking (<5 mm)	Fl	Sst	Mod - Sev	Common
Scaling (>5 mm)	Sc			
Gran. disintegration	Gd	Sst	Moderate	Common
Salt burst (blister)	Sb	Sst	Moderate	Rare - Common
Alveoli (honeycomb)	Ai			
Tafoni	Ti			
Solution loss	Sl	Lst	Slight	Abundant
Efflorescence	Ef	Sst	Moderate	Rare
Cryptoflorescence	Cf	Sst	Moderate	Rare
Alteration skin	As	Lst	Moderate	Abundant
Case hardening	Ch			
Stain	St	Lst + Sst	Moderate	Common
Black Crust	BC	Sst + Lst	Mod - Sev	Common
Biological	B	Lst + Sst	Sev - Mod	Common

AREAS AFFECTED	Height range		
	0-1.2m	1.2-5m	5-5.5m
Severe %	10	10	70
Moderate %	80	60	25
Slight %	10	30	5
Weathering features	B, Sl Gd, Fl	BC, B Fl, Gd Sl	BC, Fl, Gd

SAMPLES	LOCATION
BF1N1	c.2.5m, direct beneath st. course
BF1N2	c.1.5m, Sst block
BF1N3	c.2.5m, beneath st. course (photo)
BF1N4	c.1.8m, Sst block (photo)
BF1N5) BF1N6)	c.2m (photo) Sst block
BF1N7	c.1.2m Lst st. course
BF1N8	c.1.5m Sst block

NOTES, SKETCHES, etc.
This building has never been cleaned (info. from caretaker)
Visually, influence of the Lst on adjacent Sst is strong - latter exhibits more severe decay (esp. Fl + BC) in these areas.

Figure Four: Recording form as used in Belfast building survey. An example from All Saints' Church, Elmwood Avenue.

studies, but also in recognition of the established study of sandstone rock weathering in geomorphological work.

A summary of results are presented here in photographic form in Figures 5 and 6. The most common features relate to surface granular disintegration, involving individual grains, flaking (<0.5 cm) and scaling (>0.5 cm). The phenomena of 'black crusts' reported in literature on limestone decay are also prevalent. Biological degradation is generally concentrated on lower wall areas or in sheltered locations, but is not dealt with here. Features associated with, and caused by, the transport and precipitation of soluble salts are thus responsible for surface loss and crust build-up.

Figure Five:

5A:- Elevation of St Matthew's R.C. Church, Newtownards Rd, Belfast (1881-3). Note the catastrophic material loss from what was originally an extensively black-crusted wall. The roughly dressed stonework is a variety of Scrabo stone, with the window dressings and string course in a red sandstone, possibly of Scottish origin.

5B:- Elevation detail on Fitzroy Presbyterian Church, University St, Belfast (1872-4). Alveolar weathering is developing on the right, above the scale rule, possibly as a result of selective weathering of clay bands in the Scrabo stone. A large cavernous hollow is seen to the left of the scale rule, progressing inwards from a severely flaked surface.

One of the principal differences between sandstone and limestone decay, and one exhibited alarmingly on some of Belfast's buildings, is the relative amount of material loss. Solutional loss is not a characteristic of many sandstones; certainly not on non-calcareous stones, such as Scrabo. The subtle material loss and surface recession due to solution seen on limestones does not, therefore, occur. Instead, quartz sandstone decay is frequently characterised by relatively catastrophic levels of material loss. Even granular disintegration can cause the removal of surface grains at rates many times that of solutional loss and flaking and scaling can pose a genuine threat to the structural integrity of a building. Figure 5A illustrates such a threat, where this once extensively black-crusted Scrabo wall has suffered massive material loss, and where any restoration project would require replacement of most, if not all, the stonework. The church is less than one hundred and ten years old.

The bedding characteristics, and potentially selective weathering properties associated with clay mineral bands are one possible explanation of the phenomena seen

5C:- Quartz sandstone headstone, (C19th), Friars Bush graveyard, Stranmillis Rd, Belfast. Contour scaling is well developed, exposing, on the detachment of scales, a white florescence of gypsum crystals whose growth may be a direct cause of scale development.

5D:- Elevation detail on sandstone of Lyttle's Warehouse, Victoria St, Belfast (1866-7). A uniform black crust is observed, but is noticeably more severe along mortared joints. These are a potential calcium source for reaction with sulphur pollution to form gypsum - the main crystalline component of black crusts. Note, also, the blistering due to the crystallisation of soluble salts within the stonework.

in Figure 5B. Here, weathering is concentrated in bands within individual blocks to the extent that alveola (honeycomb) weathering has developed. In the block to the left of the scale rule a larger hollow can be seen. This has resulted from cavernous weathering of the sandstone through flaking and granular disintegration to produce a feature described by geomorphologists as 'tafoni'. Once formed, the interior of these hollows are protected from rainwash, salts can concentrate in them and rates of decay can be accelerated (Smith and McAlister, 1986).

Figure Six:

6A:- Severely flaked Scrabo sandstone on a buttress corner, All Souls' Church, Elmwood Ave, Belfast (1895-6). Flaking is active at different depths and has aided wind erosion in the removal of a large amount of material. A blistered remnant of black crust can be seen at the left of the photograph. Note, too, the apparent soundness of the stone immediately above and below the flaked block, illustrating the variability of Scrabo.

6B:- Contour scaling of red sandstone plinth, St Matthews R.C. Church, Newtownards Rd, Belfast (1881-3). Removal of the scale has exposed a white florescence, most probably of gypsum (see also Figure 5C).

Despite the predominant role of salt weathering in the development of flakes the flaking of Scrabo stone is aided by other processes which contribute to the physical removal of material. This can be seen in Figure 6A where flaking is evident at different levels within a single block and, due to its location on the corner of a buttress, wind erosion removes material which in turn permits further flaking. This photograph also illustrates the variability of the stone, as the blocks in the immediate vicinity of the severely flaked example appear to be relatively sound.

6C:- Characteristic blistering on a sandstone black crust, Lyttle's Warehouse, Victoria St, Belfast (1866-7). Note more severe material loss in region of mortared joints (see also Figure 6D).

6D:- Elevation detail of All Soul's Church, Elmwood Ave, Belfast (1895-6). The string course is of Doulting Limestone which has facilitated the development of a severe black crust on the Scrabo stone immediately beneath. Blocks in the lower half of the photograph exhibit early black crust development due to calcium leaching from the mortared joints (and possibly from the limestone string course). This wall is not particularly sheltered from rain washing. Crusts have nonetheless developed because of high concentrations of particulate pollutants and localised sources of calcium.

The susceptibility of Scrabo stone to flaking is such that well-defined contour scaling is rare. A contributory factor could be that contour scaling is more often observed on dressed stonework and Scrabo has almost always been used with a roughly tooled surface. Contour scaling can be seen on other sandstones in Belfast, for example, in the case of headstones (Figure 5C) and plinths (Figure 6B). The headstone example is particularly well-defined in that the lettering can still be seen where the scale has become detached. In both examples a uniformly thick layer of stone has lifted off exposing an area of white cryptoflorescence (Schaffer, 1932), originally located beneath a contour scale.

Florescences are the most direct evidence of soluble salts within stone buildings or structures; that seen in Figure 5C is, for example, composed of gypsum (McGreevy et al., 1983). This is the salt most commonly associated with limestone decay, but it is also found in sandstones where there is a potential source of calcium which can react with sulphur in the atmosphere (see section four). While it is not the primary purpose of this paper to describe the role of soluble salts in stone decay, it is, perhaps, relevant to mention that recent laboratory simulation studies have drawn attention to the role of salts in the development of granular disintegration, surface flaking and scaling of sandstones (Smith and McGreevy, 1983; 1988). In the urban environment of Belfast the presence of salts, from both natural background and anthropogenic pollution sources, must be seen to be particularly relevant to stone decay. Salts other than gypsum are presumed to be active, but these are rarely detected in analyses, possibly due to their greater solubility and preferential washing from the surface.

Gypsum is also the main component identified in 'black crusts', and an analysis is presented in Figure 7. Other characteristics of crusts are illustrated in Figure 5D, 6C and 6D. These crusts exhibit many of the characteristics observed on crusts from limestone

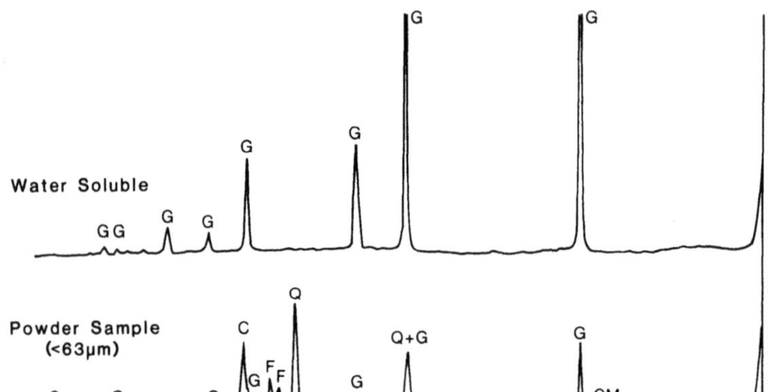

Figure Seven: XRD analyses of a black crust on Scrabo sandstone at All Souls' Church, Elmwood Ave, Belfast (1895-6). The powder sample was prepared by dispersing the ground material < 63 µm in deionised water using an ultrasonic probe. The sample was pipetted onto a glass slide and dried beneath an infra-red lamp. The water soluble sample was prepared from ground (but not sorted) material which was heated in deionised water for c. 60 min. The water soluble fraction was separated using a centrifuge, before being pipetted on to a glass slide and dried as before.
The crust is shown to be composed predominantly of gypsum, with the powder sample also containing minerals from the rock itself. [Key: G - Gypsum; C - Calcite; Q - Quartz; F - Feldspar; CM - Clay Mineral.

buildings (see Amoroso and Fassina (1983) for review), particularly their association with blistering, exfoliation and catastrophic loss of underlying material. Areas exposed when crusts break away frequently become sites of flaking and scaling (Figures 5A, 6A), which can initiate the cavernous excavation of stonework. Crusts predominantly occur on surfaces protected from direct rain washing and their distribution is patchy. Crusts do not tend to be as well developed as examples on limestone buildings, where 'cauliflower-like' accrections are often common in the most sheltered areas. Instead, crusts are invariably thin with quite a smooth surface and are more uniformly distributed over protected areas (Figure 5D). Because the substrate is non-calcareous and the main crust component is gypsum, an external source of calcium must be found. Field observations of crusts suggest that mortared joints can be a source. This has been exacerbated by the re-pointing of many buildings with a Portland cement - ironically as a first step in the alleviation of decay - as it too, provides a ready supply of calcium. Additionally, this process effectively seals the joints, so inhibiting the release of moisture from the stone, and increasing the potential for damage due to crystallisation of salts. Black crusts can thus be seen in close proximity to mortar and cement courses (Figure 5D and lower half of 6D). Leaching of mortar also causes a potential reaction surface to be distributed over much of the block below the joint leading to a more uniform crust build-up if particulates, dirt and lime are washed down the wall (Figure 5D). If a limestone is used in conjuntion with the sandstone this can also act as a calcium source, leading to crust formation on sandstones in the immediate vicinity and lower down a wall (Figure 6D). The black crust analysis in Figure 7 also reveals minor quantities of calcite in the powder sample, supporting the theory of a mortar source for the calcium in some crusts. This does not preclude other sources for calcium or gypsum, however, which could be deposited directly from the atmosphere.

In summarising the results of survey work to date, a range of features common to one particular type of stone have been identified. Flaking-related phenomena are the most common and are associated with particularly catastrophic material losses. The occurrence of black crusts highlights the potential for these to develop on non-calcareous substrates, given a ready supply of calcium and a heavily polluted atmosphere. Aspect- or height-related phenomena have not been discussed here, other than the concentration of decay in more sheltered areas. As with all building studies, such local factors depend on the location and architecture of particular buildings and only by intensive studies can they be positively identified. There is a potential, however, for such studies to be undertaken within Belfast.

4. PARTICULATE POLLUTION IN BELFAST

Particulate deposition is important on buildings because it can produce unsightly crusts as well as providing sites for sulphur oxidation. In assessing the role of particulates in stone decay, several environmental factors need to be taken into account. These include the nature of the deposition as well as the frequency with which 'deposition events' allow pollutants to interact with the material under examination. To date, nearly all exposure trials (eg. Sharp et al., 1982; Jaynes and Cooke, 1987) have provided information about the gross, averaged response of materials but say little about the actual surface response mechanisms. Additionally, although there is some agreement that particulates do contribute to building stone decay, it is not clear how important that role is; whether some particles may be more important than others or how significant local factors are. Our investigations are a start in the assessment of these problems.

4.1 Airborne particulates

Although there are studies which acknowledge airborne gaseous pollution in Northern Ireland (eg. Mellanby, 1988) there is little information on airborne particulates.

Data are collected from standard sites by, for example, Belfast City Council as smoke drawn on to filter papers, but to our knowledge there have been no studies which investigate the make-up of particulate deposition. It has been pointed out (BERG, 1989) that even though black smoke concentrations may be compared, this says nothing about the type and relative amount of different particle types and thus their potentially different reactivities. It is thus important to assess the manner and, if possible, the types of pollutant activity at different sites including:

1. The hygroscopic nature of adhering particulates
2. The hygroscopic effects of crystallised salts
3. The nature of the surface itself, eg. hydrophilic surfaces.

The third of these may be considered as a 'material constant' (Torraca, 1982) and in the present context will be held constant by considering only sandstone surfaces. Attention will, however, be given to the surface properties of adhering particulates.

Several types of spherule have been recognised from the Belfast area and elsewhere (Figure 8A). These are generally sulphur-containing particles originating from combustion of various fuels at differing temperatures (eg. Fisher et al.; 1978). It has been shown that soot particles can act as catalysts (Chang et al., 1981; Novakov et al. 1974) and that spherules will also react in a similar way (Del Monte et al., 1981; Cheng et al., 1987). Thus, particulate deposition on protected surfaces in any environment may provide a reaction site, enhanced by catalytic action as well as the ability of the particle to hold water if it has a high specific surface. Additionally, salts may also act as hygroscopic particles. In maritime environments NaCl could be an important salt adhering to buildings. If $MgCl_2$ is also precipitated then, as the latter is deliquescent, moisture will be held at the surface. If sulphuric acid is available from oxidation of SO_2 (itself produced as an oxidation product) it is possible that hydrochloric acid, as well as the salt sodium sulphate could be produced ($2NaCl + H_2SO_4 \Rightarrow 2HCl + Na_2SO_4$). Hydrochloric acid will effectively dissolve carbonates and sodium sulphate is very effective at physically disrupting stone and is used in standard durability tests. The addition of NaCl to sulphur deposition may therefore enhance decay, especially where rainwash occurs. Thus analysis of both combustion particulates and salts is important in evaluating the impact of particulate deposition.

Some studies have been carried out in projects from Queen's University and these at least provide an initial idea of the problem in Northern Ireland. Taggart (1987), for example, used a simple wet and dry (rainfall gauge type) collector and a dry collector (microscope slides with double-stick tape) to collect samples from four locations in the Belfast city centre area over a period from June to September 1987. Values for pH ranged from 4.0 to 6.0 and SO_4 concentrations ranged from 4.7 to 29.3 ppm. Particles varied from 2 mm to under 100 µm. The highest recorded levels of suspended matter were at Belfast Docks, probably due to local industrial sources. Dry deposition is difficult to estimate quantitatively over short time periods, but qualitative examination showed that fly ash

Figure Eight: Scanning Electron Micrographs showing the nature of some black crust surfaces on sandstone buildings in Belfast, together with examples of airborne particulates collected in the city.

8A:- General view of a black crust surface at low magnification showing the occurrence of particles from high temperature combustion. Scale bar = 100 µm.

8B:- Range of particle types deposited by rainfall at Belfast docks. Scale bar = 100 µm.

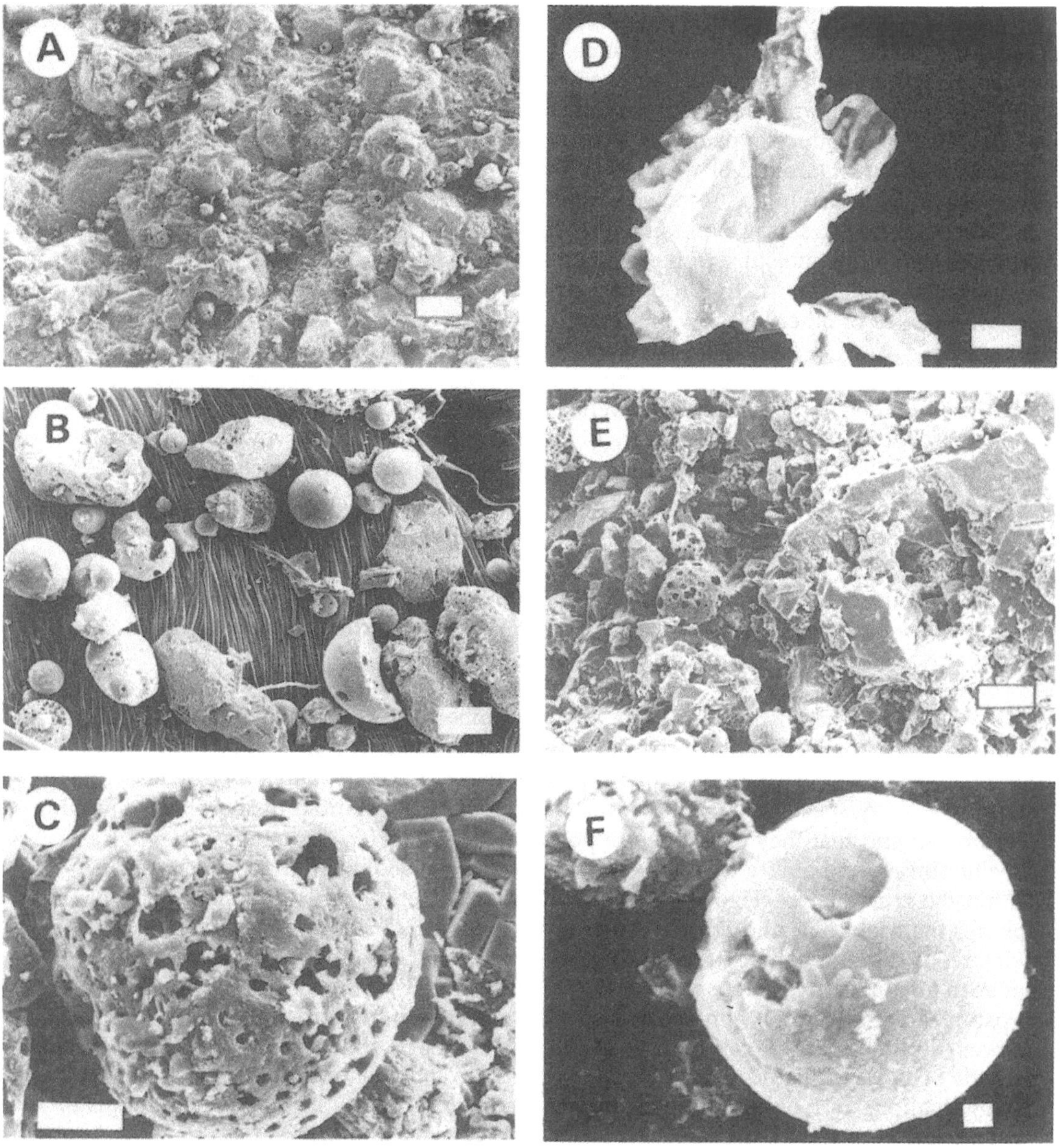

8C:- In situ view of a high temperature combustion cenosphere with gypsum crystals on its surface and in the background. A similar particle was used to gather the EDS spectrum in Figure 9A. Scale bar = 10 µm.

8D:- An exmple of a thin, high specific area carbonaecous particle collected from a car exhaust. Scale bar = 10 µm.

8E:- High temperature combustion particles on a sandstone black crust. Note, in particular, the 'solid' sphere at the bottom, middle. Scale bar = 20 µm.

8F:- High magnification view of a 'solid' particle from a disaggregated crust, similar to that used to gather the EDS spectrum in Figure 9B.

dominated, the majority of particles being < 60 µm. Cassidy (1985) also showed a variety of particle types scavenged from the atmosphere in rainfall (Figure 8B) in Belfast. When Taggart (1987) exposed a microscope slide with adhering particles to wetting and drying cycles a gypsum deposit was seen after several cycles. This suggested that deposited sulphur was combining with calcium ions, perhaps under the influence of fly ash acting as a catalyst, to produce gypsum. It is also possible that fly ash particles themselves contain sulphur. Duration of wetness is likely to be an important variable in allowing such reactions to take place. One concommitant factor, therefore, could be the nature of hygroscopic effects. It is also possible that the high specific surface of fly ash and related particles will allow moisture to be retained at sites where the particles reside and where evaporation should normally produce a dry surface. This may allow some chemical activity at sites of otherwise low-susceptibility.

The nature of dry deposition may thus be particularly important in short-term response of stonework to pollution events. Because of this, both existing pollution sources in Belfast as well as the nature of black crusts are being surveyed in order to obtain a possible historical perspective. Preliminary findings are presented here.

4.2 Particles on existing sandstone buildings

The surfaces of existing crusts from a number of buildings across Belfast have been examined by scanning electron microscopy (SEM) and electron microprobe (EPMA) to characterise the nature of particles.

Crusts characteristically contain a high number of fly ash particles and, although domestic pollution is thought to be most significant, fly ash from higher temperature combustion shows clearly (Figure 8A). Most of these are cenospheres (Figure 8C). Energy dispersive X-ray spectra (EDS) show that these contain substantial proportions of sulphur (Figure 9A) and that they are therefore possible sites of surface reactions. The high specific surface and ability to hold water are also obvious. Gypsum crystals are seen within the crust matrix (Figure 8C) which may have come from component parts of the spheres themselves (with high Ca content: Figure 9A). On limestone, gypsum (with its deletereous disaggregation effect) is to be expected as a consequence of the reaction between a sulphur rich atmosphere and the substrate. It is thus of interest to see that gypsum can also form on non-calcareous rocks. If fly ash, perhaps with other forms of carbon-sulphur particles, can bind to chemically rather inert sandstone surfaces they will provide salts which could ultimately be responsible for phenomena such as granular disaggregation. In looking for such sources we have sampled the exhaust emissions of vehicles and found thin, high specific surface carbonaceous particles (Figure 8D). Almost certainly these contribute to the black colour of the crusts, but it is not yet known what role they play in catalysing surface reactions.

As well as the 'perforated' cenospheres, smaller 'solid' spheres are found (Figures 8E and 8F) which EDS analyses show to be less sulphur-rich (Figure 9B). They do, however, contain high amounts of silica plus trace metals such as titanium, and a significant calcium component. These are similar to examples described by Fisher et al. (1976) as products of coal combustion. They may act as secondary sources of gypsum if sulphur is available either directly from the atmosphere or from sulphur-rich cenospheres produced from oil combustion.

5. BACKGROUND PATTERNS OF STONE DECAY

In section three we have described a range of decay phenomena found on sandstone buildings in Belfast attributable to the effects of atmospheric pollution. It is

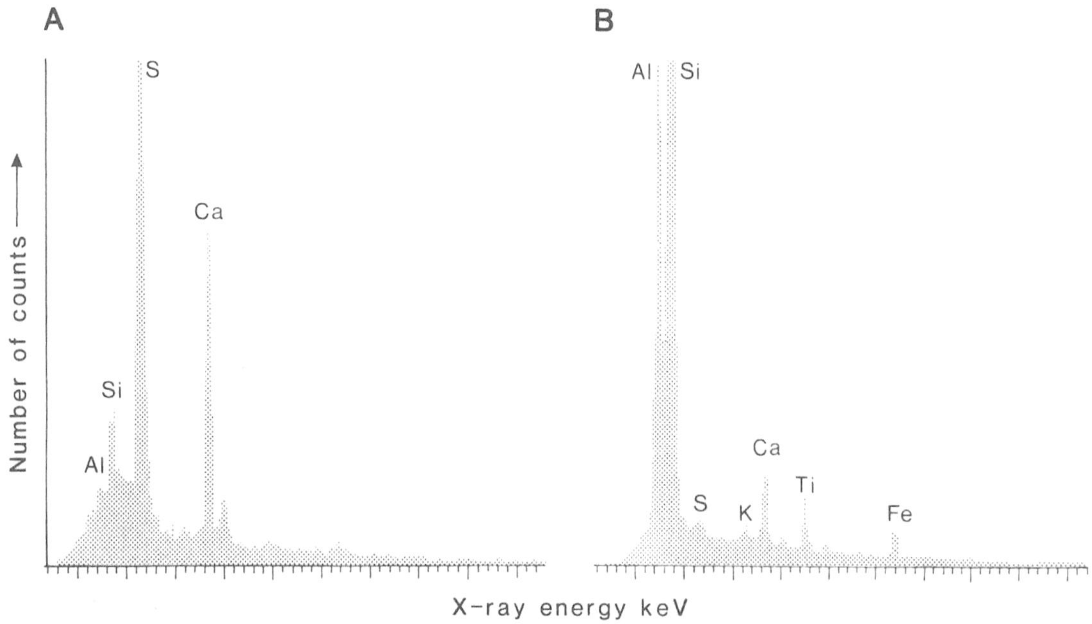

Figure Nine: Energy dispersive X-ray spectra of particulate pollutants in disaggregated black crusts from Belfast. A = large cenosphere similar to Figure 8C, B = small, solid sphere similar to Figure 8F. Crusts disaggregated using ultrasonic water bath.

also possible to identify within the city, and in rural areas of Northern Ireland, other examples of salt-induced decay wherein salts do not derive principally from local atmospheric pollution. What follows are a series of case studies which seek to exemplify these background causes of decay on to which pollution related effects must be superimposed.

5.1 Stone decay in a maritime environment

The two examples used come from the town of Portstewart on the north coast of Northern Ireland (figure 1). The first is the St Mary's, Star of the Sea Church which is located on the sea front some 100 m inland and was built in 1916. The second is the Dr Adam Clarke Memorial Chapel, located one street behind the sea front but in a prominent, exposed location on a hill 10-15 m above sea-level. Both buildings are constructed of local, roughly cast basalt with yellow sandstone doorways, window frames and sills. The basalt appears little altered, but in places the sandstone is severely decayed. Normally this commences with the contour scaling of the outer, dressed surface (Figure 10A) but is followed, appparently quite rapidly, by granular disintegration of the newly exposed surfaces which are hollowed out to form a mixture of tafoni and honeycombs (Figure 10B). Visual inspection suggests that weathering is more advanced on the Star of the Sea Church on the sea front. The features observed are frequently ascribed to the process of salt weathering and to identify any salts that are present samples of weathered sandstone were collected for examination by X-ray diffraction. Where possible, samples were collected from sites protected from rainfall but open to the atmosphere. Results from these analyses are presented in Figure 11 and consist of two traces for each of the samples shown; one is from a powder sample and the other from a water soluble extract.

The traces for powder samples from both buildings (Figure 11A and B) indicate a quartz sandstone containing substantial quantities of kaolinite and traces of muscovite.

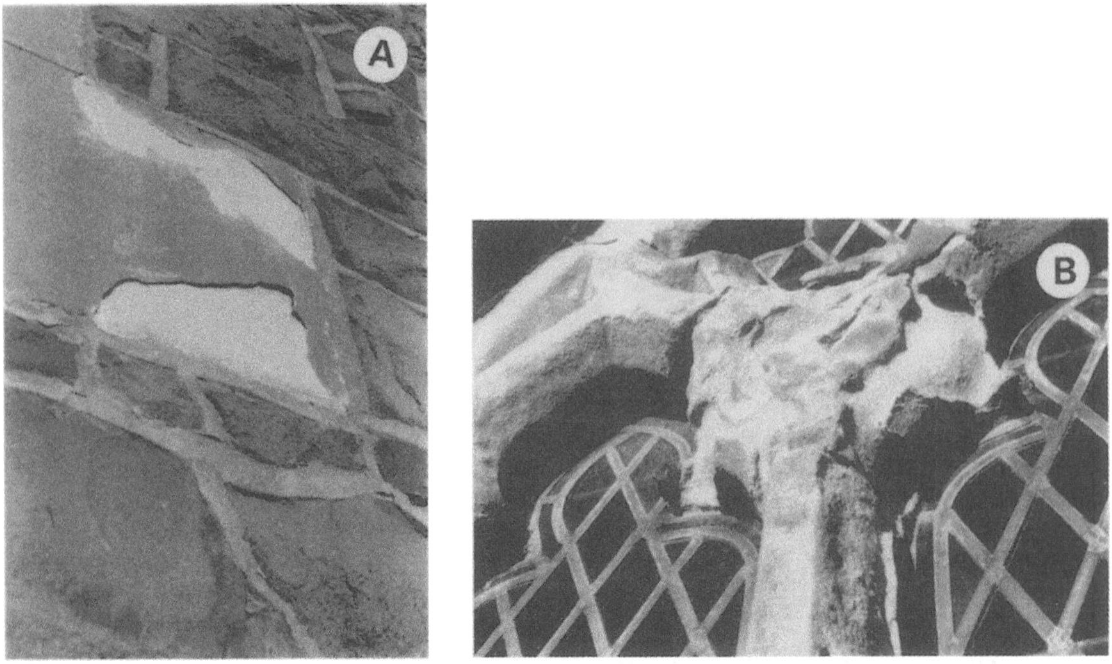

Figure Ten: Photographs of contour scaling (A) and subsequent cavernous weathering (B) of yellow sandstone on the Adam Clarke Memorial Chapel, Portstewart (see Figure 1).

The similarity of the two traces suggest that both sandstones are from the same source. On the powder trace from the Star of the Sea Church there are also minor peaks for gypsum but, if present, this is swamped by quartz and kaolinite on the trace for the Adam Clarke Chapel. Analysis of the 'water soluble' trace for the latter shows that gypsum is present, and reveals major peaks for this salt. Interestingly, there is a minor peak that could indicate halite. No such confusion exists on the 'water soluble' trace for the Star of the Sea Church, where multiple, major peaks occur for gypsum together with a major peak for halite. To complement these analyses samples of rock meal from both buildings were analysed for chlorine content by a titration technique using silver nitrate with a potassium chromate indicator. Results from this are given in Table 1. This shows surprisingly high concentrations of chlorine in the samples from the Chapel, but

Table One: Chlorine content of weathered sandstone samples from Portstewart and Belfast.

Location	Sample	Chlorine Content (ppm of dry weight)
Adam Clarke Mem. Chapel (Portstewart)	1	1065
	2	639
Star of the Sea Church (Portstewart)	1	35
	2	106
	3	1384
	4	3124
Scrabo Sandstone Wall (Bedford Street, Belfast)	1	4792

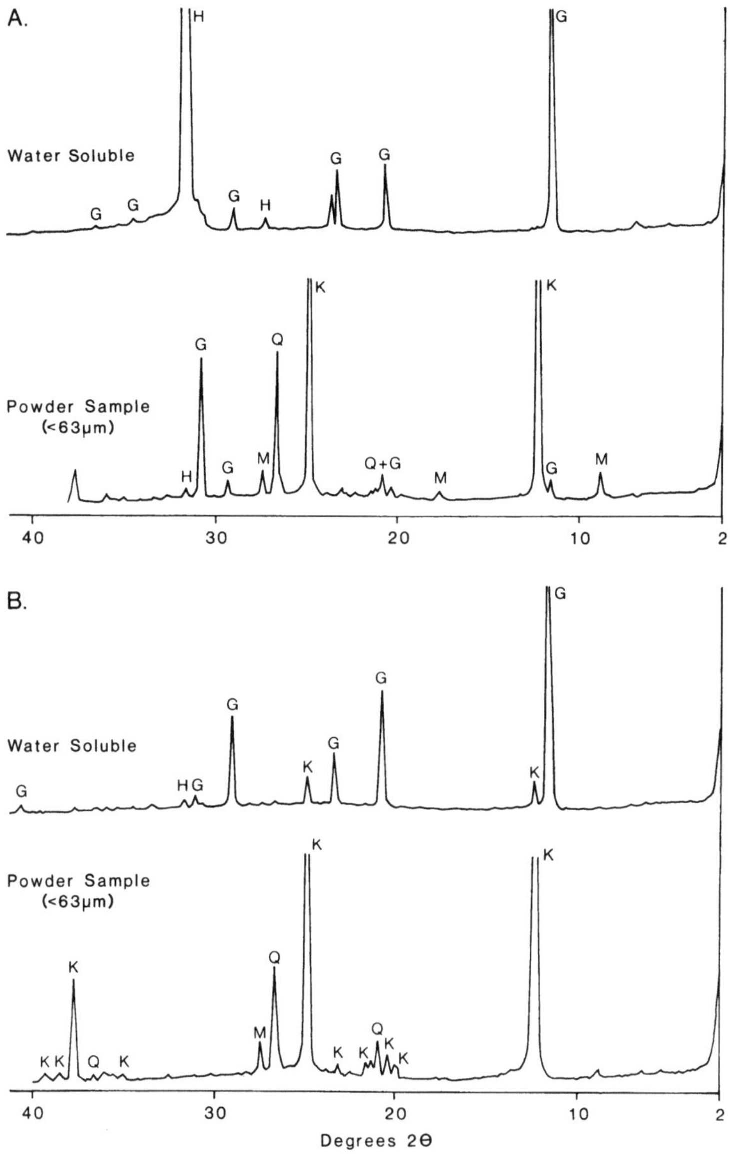

Figure Eleven: X-ray diffraction traces of surface scales from St Mary's Star of the Sea Church (A) and the Adam Clarke Memorial Chapel, both Portstewart (B) (See Figure 1).
[Key: H = Halite, G = Gypsum, Q = Quartz, K = Kaolinite, M = Muscovite.

confirm that much higher concentrations can be found on the seafront church. The variability in measurements from the church also confirm that retention of soluble salts such as halite in stonework is particularly dependent upon protection from removal in rainwash.

The clear evidence for halite is very interesting in that previous studies of a similar Carboniferous quartz sandstone at nearby Ballycastle by McGreevy (1985) found only near-surface accumulations of gypsum in outcrops within the spray zone. In his discussion he noted that of the salts that can be derived from evaporation of sea water, gypsum will crystallize out before salts such as halite and that, if present, halite and other soluble salts will be the first to be removed by rainwater. Other studies of coastal weathering (eg. Mothershead, 1982 and 1989) have inferred the effectiveness of halite in causing disaggregation through crystallization or differential thermal expansion in pores and fractures. Both Mothershead (1982) and Dibb et al. (1983) also demonstrated that halite precipitated from seawater under laboratory conditions can induce rock breakdown. In neither study, however, do they present evidence of halite being found precipitated in coastal rocks. The role of 'transient' salts such as halite in causing natural stone breakdown thus remains conjectural, but this may reflect the tendency for previous studies to concentrate on exposed, frequently wetted rocks near the shoreline. In the case of the Star of the Sea Church, it is far enough from the shoreline to be wetted by spray only infrequently and it would seem that if areas are protected from rainfall or rainwash then halite can accumulate. Once there, it can complement the action of other salts such as gypsum (which can also exert stresses through hydration) in causing stone breakdown. The apparently lower levels of halite at the Chapel suggests that its persistence is enhanced in protected locations which are occasionally saturated in sea water. Elsewhere, it will be supplied to stonework within rainfall and is therefore equally prone to removal by rainwash. Only less soluble salts will therefore be retained when absorbed moisture evaporates from stone surfaces. Absorption of, and washing by saline water could also increase rates of granular disintegration in quartz sandstones through enhanced silica dissolution. Evidence of this solution in the form of surface etching was presented by McGreevy (1985) and has been proposed elsewhere (eg. Young, 1987) as a possible mechanism of cavernous weathering.

It is difficult to assess what contribution local, regional or transnational pollution makes to stone decay in Portstewart, but the location and limited size of the town, together with its aspect which exposes it to Atlantic winds, argue against any major contribution. Instead, it would seem that the sometimes marked decay of sandstones, achieved over a relatively short time-span, is largely attributable to sea salts.

5.2 Stone decay and road deicing

The roads of Belfast, perhaps because of its location which makes it prone to frosts, or perhaps because of a ready supply of rock salt in County Antrim, are frequently 'gritted' during the winter months. Coincident with the areas splashed by the resulting dissolved 'road salt', many of Belfast's sandstone buildings exhibit accelerated flaking and granular disintegration immediately above pavement height. It is also common for salt efflorescences to develop near ground level during dry periods and for rock meal to accumulate on the ground beneath buildings. Such salt accumulations and concentrated weathering could equally be explained through salts derived from below in rising groundwater (see next section), by a concentration of vehicle emissions or by the accumulation of atmospheric-derived salts washed down the side of a building. To investigate these possibilities samples of severely weathered Scrabo Sandstone were analysed from a stone wall in Bedford Street, central Belfast. Samples were taken of surface flakes and rock meal at approximately 0.5-0.75 m above pavement level and approximately 3 m away from the road. Samples were analysed by X-ray diffraction and were prepared for analysis in the same way as the coastal samples. The traces produced are shown in Figure 12A. The powder sample again reflects the bulk composition of the sandstone, with major peaks for quartz and feldspars and only minor peaks for gypsum. The trace for water soluble material, however, shows major peaks for gypsum and, in particular, halite. In addition to the major peaks for halite, further analysis of the rock meal revealed a very high chlorine concentration of 4792 ppm (Table 1). The presence of halite in such concentrations would seem to implicate road salting in the ground-level decay of stonework, but the presence of gypsum would suggest that decay may represent the combination of road salt- and pollution- or groundwater-derived effects.

5.3 Stone decay and groundwater

In addition to the widespread concentration of weathering near to ground level described in the previous section, occasionally salt efflorescences and associated decay are seen to develop for several metres up the sides of buildings in Belfast. Generally this is attributable to the absence or failure of damp coursing in the building and the upward migration of groundwater-derived salts or washing out of salts contained within the stonework or mortar. Within central Belfast this is a particular problem as much of the city is built upon reclaimed marsh land. The possibility remains, however, that these salts could be partly derived from other sources, including atmospheric pollution or road salt. To investigate decay associated with rising groundwater, a site was chosen in the quadrangle of Queen's University which is detached from both road salting and vehicular emissions. The main building at Queen's was built in 1848 and is primarily brick with sandstone ornamentation. In places both brickwork and sandstone is

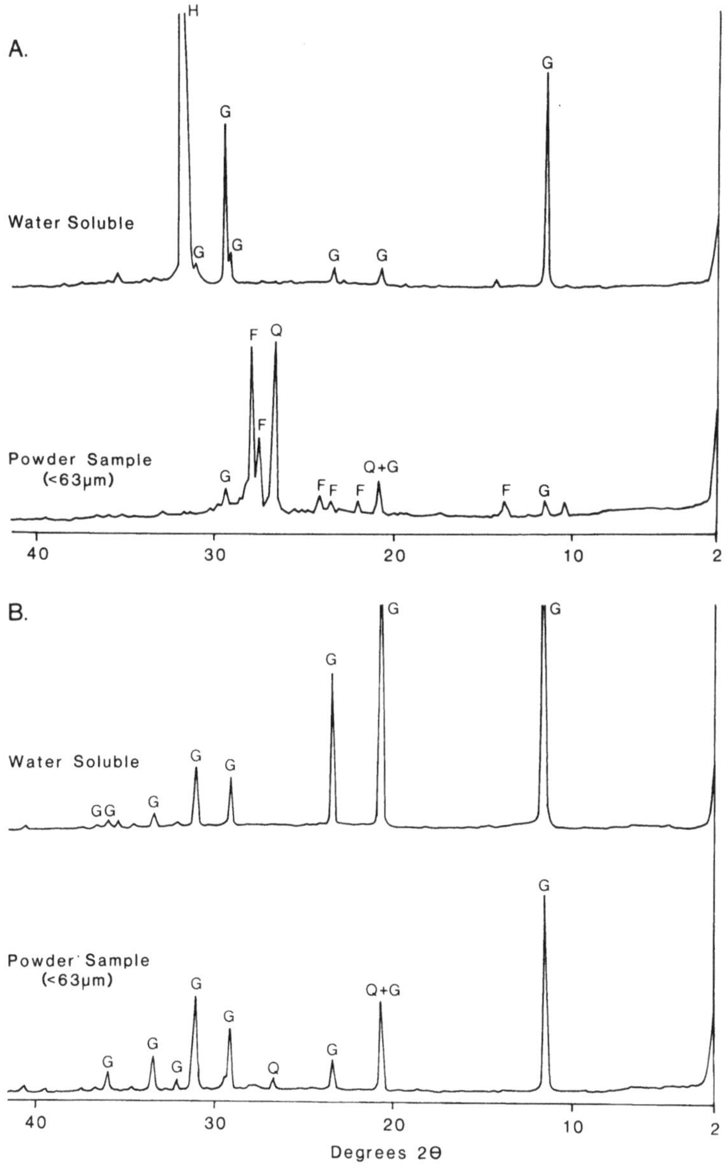

Figure Twelve: X-ray diffraction traces of surface scales and rock meal from severely weathered Scrabo Sandstone, Bedford Street, Belfast (A) and Quartz sandstone of unknown provenance from the main building, The Queen's University of Belfast (B).
[Key: G = Gypsum, Q = Quartz, F = Feldspar, H = Halite]

episodically covered in white salt efflorescence and both building materials can be severely decayed to a height of 2-3 m. On the sandstone this takes the form of blistering followed by contour scaling and granular disintegration. Samples of the sandstone flakes and rock meal were taken for X-ray diffraction analysis and the results from characteristc 'powder' and 'water soluble' samples are shown in Figure 12B. The powder sample shows a mixture of quartz and multiple peaks for gypsum. The water soluble sample shows multiple, major peaks for gypsum only. While these results are not representative of the possible complexity of salts that can be derived from groundwater, they are of significance for the identification and interpretation of decay attributable to atmospheric pollution. The location of the weathering and the patterns and seasonal variations in efflorescences on the buildings all indicate a groundwater source for the large quantities of salts observed. Chemical analysis, however, reveals only gypsum - a salt that could equally be derived directly from atmospheric pollutants (section four) or the reaction between atmospheric sulphur and mortar (section three). Furthermore, as previously outlined, when such pollution-derived salts are mobilised by rainwash they could accumulate towards the base of a building and thus be further confused with salts derived from upward migration in groundwater. Whatever the source of the gypsum, its association with well-developed scaling and granular disintegration clearly demonstrates its effectiveness as an agent of stone decay. The fact that it can be derived from a variety of sources illustrates the dangers of ascribing all gypsum-related decay the effects of atmospheric pollution.

6. SUMMARY

Because it lies to the northwest of mainland Europe, Northern Ireland is generally considered to experience low levels of transnational atmospheric pollution. This therefore provides the opportunity to examine both 'natural' patterns and causes of building stone decay and also the impact of local pollution sources. The effects of the latter are further enhanced by incomplete enactment of clean-air legislation and only sporadic compliance in cities such as Belfast.

In this paper a review of air pollution and acidic deposition across Northern Ireland has been presented, together with observations on the nature of particulate deposition within Belfast. This information is compared with observations on the nature, extent, severity and causes of building stone decay, particularly of sandstones in Belfast. Results from this and previous studies (eg. Smith et al., In Press) suggest that background rates of decay in non-urban locations can be significant. Salt-induced disaggregation and scaling is especially prevalent and is thought to reflect high concentrations of sea salts carried by prevailing westerly air masses. Within the urban environment of Belfast, rates of solution loss from limestones do not appear to be high. However, black crust formation and salt induced breakdown of sandstone in response to locally high levels of particulate deposition is particularly widespread and severe. Deposition and subsequent mobilisation of particulates is further enhanced by frequent temperature inversions over the City which both 'trap' pollutants and encourage occult deposition.

Although few data exist on the decay of sandstones within urban environments of the United Kingdom, results from Northern Ireland suggest that long-term rates of breakdown, despite the absence of solution processes, can be high. Sandstones exhibit a particular susceptibility to salt weathering mechanisms, which are in turn predominantly functions of salts derived from local particulate pollutants or from reactions between mortars and acid rain. Where salt weathering predominates in this way, however, care must be taken in isolating other effects such as the liberal use of road salt and upward migration of saline groundwater - both of which are common in Belfast.

Once this is done it is apparent that the effects of acidic deposition have, to date, been largely aesthetic and principally manifested in a loss of architectural detail. Nonetheless, this has generated extensive programmes of stone cleaning and replacement and, when combined with spatially concentrated attack by groundwater-derived salts, is beginning to threaten the structural integrity of certain buildings.

ACKNOWLEDGEMENTS

We would like to thank the Department of the Environment for funding this project (contract number PECD7/12/08) and Belfast City Council for providing data on atmospheric pollution. Preparation of the manuscript would not have been possible without the assistance of the Electron Microscope Unit and Photographic Unit at Queen's University and the technical, cartographic and secretarial staff of the School of Geosciences.

REFERENCES

Amoroso, G.G. and Fassina, V. (1983). *Stone decay and conservation.* Elsevier, Amsterdam, 453p.

Betts, N.L.B. (1982). Climate. In Cruickshank, J.G. and Wilcox, D.N. (ed.), *Northern Ireland - Environment and Natural Resources.* Queen's University of Belfast and the New University of Ulster, Belfast: 9-42.

Brett, C.E.B. (1967). *Buildings of Belfast.* Friars Bush Press, Belfast, 86p.

Buildings Effects Review Group (BERG) (1989). *The effects of acid deposition on buildings and building materials in the United Kingdom.* London, H.M.S.O., 106p.

Cassidy, J.G. (1985). *An investigation into the chemical variability of rainfall over Belfast, February-August 1984.* B.Sc. dissertation, Department of Geography, The Queen's University of Belfast, 97p.

Chang, S.G., Toosi, R. and Novakov, T. (1981). The importance of soot particles and nitrous acid in oxidising SO_2 in atmospheric aqueous droplets. *Atmos. Env.*, 15: 1287-1292.

Cheng, R.J., Jih, R.H., Jung, T.K. and Show-Mei, L. (1987). Deterioration of marble structures. The role of acid rain. *Analytical Chemistry*, 59: 104A-106A.

Cooke, R.U. (1989). Geomorphological contribution to acid rain research: studies of stone weathering. *Geogr. J.*, 155: 361-366.

Del Monte, M. and Sabbioni, C. (1984). Morphology and mineralogy of fly ash from a coal-fueled power plant. *Arch. Met. Geoph. Biocl.*, Ser. B 35: 93-104.

Del Monte, M., Sabbioni, C. and Vittori, O. (1981). Airborne carbon particles and marble deterioration. *Atmos. Env.*, 15: 645-652.

Dibb, T.E., Hughes, D.W. and Poole, A.B. (1983). The identification of critical factors affecting rock durability in marine environments. *Q. J. Eng. Geol.*, 16: 149-161.

Fassina, V. (1978). A survey of air pollution and deterioration of stonework in Venice. *Atmos. Env.*, 12: 2205-2211.

Fisher, B.E.A. (1982). Deposition of sulphur and the acidity of precipitation over Ireland. *Atmos. Env.*, 16: 2725-2734.

Fisher, G.L., Chang, D.P.Y. and Brummer, M. (1976). Fly ash collected from electrostatic precipitators: microcrystalline structures and the mystery of the spheres. *Science*, 192: 553-555.

Fisher, G.L., Prentice, B.A., Silberman, D., Ondov, J.M., Biermann, A.H., Ragaini, R.C. and McFarland, A.R. (1978). Physical and morphological studies of size-classified coal fly ash. *Env. Sci. and Tech.*, 12: 447-451.

Haigh, N. (1989). *EEC Environmental Policy and Britain* (2nd revised edition). Longmann, Harlow, 382p.

Jaynes, S.M. and Cooke, R.U. (1987). Stone weathering in southeast England. *Atmos. Env.*, 21: 1601-1622.

Jordan, C. (1983). The precipitation chemistry of a rural site in Co. Antrim, Northern Ireland. *Record Agric. Res. (N.I.)*, 31: 89-98.

Jordan, C. (1987). The precipitation chemistry at rural sites in Northern Ireland. *Record. Agric. Res. (N.I.)* , 35: 53-66.

Jordan, C. (1988). Acid Rain in Northern Ireland. In Montgomery, W.I., McAdam, J.H. and Smith, B.J. (ed.), *The High Country: Landuse and landuse change in northern Irish uplands.* Institute of Biology (N.I.), Belfast: 47-55.

McGrath, R. (1988). Analysis and prediction of air pollution in the Dublin area. *Ir. Geogr.*, 21: 88-100.

McGreevy, J.P. (1985). A preliminary SEM investigation of honeycomb weathering of a sandstone in a coastal environment. *Earth Surf. Proc. Landforms*, 10: 509-518.

McGreevy, J.P. and Smith, B.J. (1984). The possible role of clay minerals in salt weathering. *Catena*, 11: 169-175.

McGreevy, J.P., Smith, B.J. and McAlister, J.J. (1983). Stone decay in an urban environment: observations from south Belfast. *Ulst. J. Arch.*, 46: 167-171.

Mathews, R.O., McCaffrey, F. and Hart, E. (1980). Acid rain in Ireland. *Ir. J. Env. Sci.*, 1: 47-50.

Mellanby, K. (ed.) (1988). Air pollution, acid rain and the environment. Elsevier, London, 129p.

Mottershead, D.N. (1982). Coastal spray weathering of bedrock in the supratidal zone at East Prawl, South Devon. *Field Stud.*, 5: 663-684.

Mottershead, D.N. (1989). Rates and patterns of bedrock denudation by coastal salt spray weathering: a seven year record. *Earth Surf. Proc. Landforms*, 14: 383-398.

Novakov, T., Chang, S.G. and Harker, A.B. (1974). Sulfates as pollution particulates: Catalytic formation on carbon (soot) particles. *Science*, 186: 259-261.

Park, C.C. (1987). *Acid Rain: Rhetoric and Reality.* Methuen, London, 272p.

Schaffer, R.J. (1932). *The weathering of natural building stones.* Dept. of Scientific and Industrial Research (U.K.), Building Res. Spec. Rep., 18, 149p.

Sharp, A.D., Trudgill, S.T., Cooke, R.U., Price, C.A., Crabtree, R.W., Pickles, A.M. and Smith, D.I. (1982). Weathering of the balustrade on St Paul's Cathedral, London. *Earth Surf. Proc. Landforms,* 7: 387-390.

Smith, B.J. and McAlister, J.J. (1986). Observations on the occurrence and origins of salt weathering phenomena near Lake Magadi, southern Kenya. *Zeit. für Geomorph.,* 30: 445-460.

Smith, B.J. and McGreevy, J.P. (1983). A simulation study of salt weathering in hot deserts. *Geogr. Annlr.,* 65A: 127-133.

Smith, B.J. and McGreevy, J.P. (1988). Contour scaling of a sandstone by salt weathering under simulated hot desert conditions. *Earth Surf. Proc. Landforms,* 13: 697-706.

Smith, B.J., Whalley, W.B. and Magee, R.W. (In Press). Building stone decay in a clean environment. In: Sors, A. (ed.) *Science Technology and the European Cultural Heritage,* Proc. Bologna Conf. June 1989.

Taggart, J.D. (1987). *The effects of acid rain on limestone structures in Belfast.* B.Sc. dissertation, Department of Geography, The Queen's University of Belfast, 91p.

Torraca, G. (1982). *Porous building materials* (2nd edition). ICCROM, Rome, 145p.

Young, A. (1987). Salt as an agent in the development of cavernous weathering. *Geology,* 15: 962-966.

SOME ASPECTS OF THE SURFACE CHEMISTRY INVOLVED IN THE WEATHERING OF BUILDING LIMESTONE

R. H. Bradley
Institute of Surface Science and Technology,
Loughborough University, Loughborough, Leics, U.K.

Abstract. The surfaces of samples of two types of building limestone in pristine condition and weathered either in or out of direct rainfall have been studied using SEM, LAMMS and XPS. The surfaces of Portland and Monks Park limestones in pristine condition are shown to be composed essentially of Ca, C and O, as would be expected. Examination of weathered samples reveals significant levels of sulphur containing species due to atmospheric corrosion. The surfaces of sheltered samples appear to be composed entirely of $CaSO_4$ whilst exposed samples appear to have lower levels suggesting that SO_4 is leached from the sample surface by rainwater. Solubility data are presented to support this theory. SEM, LAMMS and XPS are shown to give useful informatiom about the surfaces of samples studied.

1. Introduction

The weathering of limestones used for building purposes appears to be attributable to two major factors: chemical attack e.g. by atmospheric pollutants in either gaseous form or dissolved in rain water; and the effects of frost upon the stone structure when an appreciable volume of its porosity is filled with water (Honeyborne and Harris 1959). In the present environmental conditions chemical weathering is the predominant process leading to deterioration of historic buildings. This process occurs by reaction between atmospheric pollutants - such as Cl, SO_2 and NO_2, either in gaseous form or in solution - and the building stone. These reactions are, therefore, essentially surface reactions of the gas-solid, liquid-solid type which lead to degradation/erosion of the building surface via the formation of environmentally unstable reaction products.

Atmospheric levels of pollutant species have shown a marked increase over the last century due to greater industrialisation and the use of fossil fuels. It has been shown that pollutants tend to return to the ground at or near to their source, according to prevailing weather conditions (Fenn et.al. 1963). Therefore in urban and industrial areas deposition of these corrosive species can result in marked deterioration of the surfaces of stone buildings. Excellent reviews covering atmospheric pollutants and their behaviour have been published (Winkler 1973) (Amoroso and Fassina 1983).

Although the effects of weathering are often obvious e.g. the loss of fine detail from stone statues, the precise nature of the chemistry involved is not fully understood and there is little published information concerning possible mechanisms of decay. The work reported results from a broader study the objective of which was to investigate in detail the changes in building limestone resulting from environmental attack. A range of surface sensitive techniques have been used in order to assess which, if any, give the most useful elemental and chemical information about these surfaces. Samples of stone in pristine condition and weathered at sites which were either sheltered from or exposed to direct rainfall have been investigated using scanning electron microscopy (SEM), secondary ion mass spectrometry (SIMS), X-ray photoelectron spectroscopy (XPS), laser ablation microprobe mass spectrometry (LAMMS) and Fourier transform infrared spectroscopy (FTIR). Results are reported here for three of these techniques: SEM, LAMMS and XPS.

2. Experimental

SEM work has been carried out using a Cambridge Stereoscan 360 instrument which was opperated at a residual vacuum of 10^{-6} Torr. Samples were sputter coated with Au prior to investigation.

LAMMS is a laser based solid sampling mass spectrometric technique. The spectra presented have been measured using a LIMA 2A instrument manufactured by Cambridge Mass Spectrometry Ltd. The system uses a pulse from a 266 nm. wavelength, frequency quadrupled neodynium-YAG laser to ablate a volume of material (typically 3µm wide x 0.5µm deep) from the sample surface. The ionised fraction of this volume is then analysed by a time of flight mass spectrometer. Positive and negative ion spectra can be measured for all ions. The instrument is fitted with a 1µ resolution optical microscope and a He-Ne spotting laser which allows specific sample areas to be targeted. A residual vacuum of 10^{-7} torr was maintained within the system during experiments. Full details have been published (Southen et.al. 1984).

X-ray photoelectron spectroscopy was carried out on a Vacuum Generators ESCALAB MK1 spectrometer. Spectra were measured at a residual vacuum of 10^{-8} torr using Al K_α X-rays (energy 1486.4ev) and analyser pass energies of 80 eV and 20eV. The instrument was set up to sample an area of 0.4 cm^2. Binding energies have been referenced to the lowest binding energy C 1s peak at a value of 284.6ev. Data were recorded and processed on an IBM AT microcomputer. Compositions were determined by measurement of peak areas, following subtraction of a Shirley type background, and the use of the relevant sensitivity factors. The sample was assumed to be homogeneous over the sampling depth and corrections were made for photoelectron cross-section, angular asymmetry in photoemission, the transmission of the energy analyser and the energy dependence of the inelastic mean free path.

3. Samples

Results are presented here for two types of stone, namely; Portland limestone and Monks Park limestone. The structure of these materials has been characterised in the past (Honeyborne and Harris 1959). Both of these rock types are composed of small spherical $CaCO_3$ particles, known as oolites, in a calcium carbonate rich cement. The oolites themselves have been formed by nucleation in a $CaCO_3$ rich oceanic environment. Both of the stones are carboniferous in origin. Tiles 50x50x8 mm were supplied by the Building Research Station (BRS) in the following condition:

Pristine. - Cut and stored unweathered.

Sheltered. - Exposed to environmental weathering but not to direct rainfall.

Exposed. - Exposed to environmental weathering including direct rainfall.

For analysis the tiles were cut into small blocks approximately 15x15x8 mm and analysis performed on the 15x15 mm faces. The as supplied tiles each had a central hole, used for mounting during weathering, the areas around these holes have been avoided during analysis as have the sample edges.

4. Results and Discussion

4.1 Scanning Electron Microscopy

The structure of these materials, as revealed by SEM, is shown in figure 1. The oolites appear to have diameters of approximately 200-300 µm. No striking morphological differences are evident between the pristine, sheltered and weathered samples studied by SEM however, in some instances the oolites appear to be more distinct in the exposed sample of Portland limestone than in either the pristine or sheltered samples. This may be due to the removal, during the weathering process, of the cement material which surrounds these oolites.

4.2 Laser Ablation Microprobe Mass Spectrometry (LAMMS)

Generally the positive ion spectra (relative ion current against m/z) for all of the samples show peaks for calcium and calcium oxide which are associated with the general $CaCO_3$ matrix (both oolites and cement) e.g. figure 2 for the Portland exposed sample. Peaks due to sodium, magnesium and aluminium, which are intrinsic at low levels in stones of this type, are also evident in the spectra (m/z 23, 24 and 27 respectively). The negative ion spectra given in figures 3, 4 and 5 show much more structure. Peaks for Carbon and C_n are evident and carbon related peaks e.g. CN (m/z=26) and CNO (m/z=42) probably arise from the presence of nitrate at the surface. Prominant peaks are also present for S, Cl, SO, SO_2, and SO_3 suggesting the presence of chloride, sulphite and/or sulphate in the sheltered sample. The peak observed at m/z=96 may be due to SO_4 but, because of peak overlap with C_8 at this point in the spectrum, this cannot be proven conclusively using this technique, evidence for the presence of SO_4^{2-} is presented in section 4.3. Examination of the exposed sample indicated much lower levels of Cl and little evidence of SO_3^{2-} or SO_4^{2-}.

The Monks Park samples give similar spectra (not shown) to those from the Portland. Spectra measured for the sheltered samples contain significant peaks due to the presence of S, Cl, SO_2 and SO_3 compared to the pristine. The peaks attributable to sulphur species are present at lower intensities in the spectra for the exposed sample.

The spectra shown are not quantitative but have been measured at similar laser intensities and therefore allow relative comparisons to be made between species present. Used in this way the technique is useful for fingerprinting the corrosion surface in terms of the elements present whilst also giving some chemical/molecular information about the surface. The fragmentation patterns of minerals such as $CaCO_3$, $CaSO_4$, $CaNO_3$ and $CaCl_2$ need to be studied in greater detail before this technique will allow a quantitative approach.

4.3 X-ray Photoelectron Spectroscopy (XPS)

The surface composition of the limestone samples has been investigated using XPS which probes to a depth of about 3 nm. Figure 6 shows broad scan spectra for the Portland samples. Whilst the pristine sample shows only calcium, carbon and oxygen peaks, associated with the matrix calcium carbonate, the sheltered and exposed samples also give prominent sulphur peaks. These are most striking in the spectrum for the sheltered sample. Integration of the peak areas for the elements present in the spectra recorded for both the Portland and monks Park

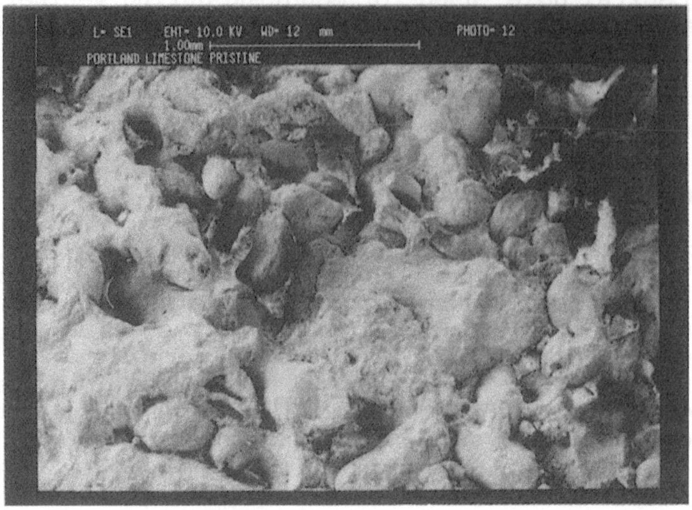

a) Pristine surface.

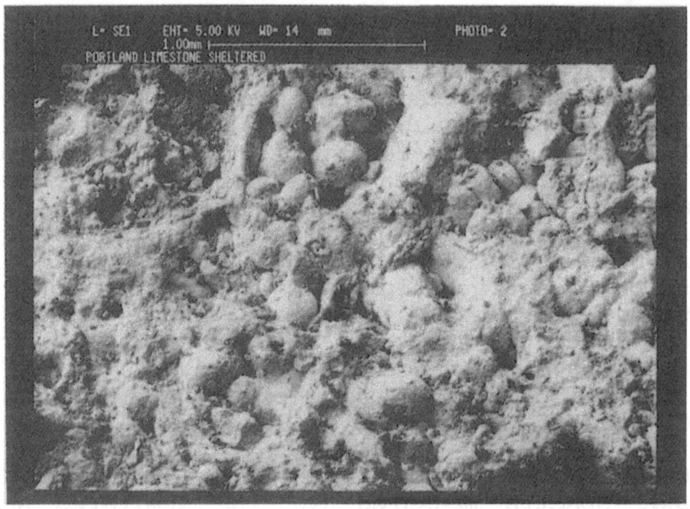

b) Weathered but sheltered from direct rainfall.

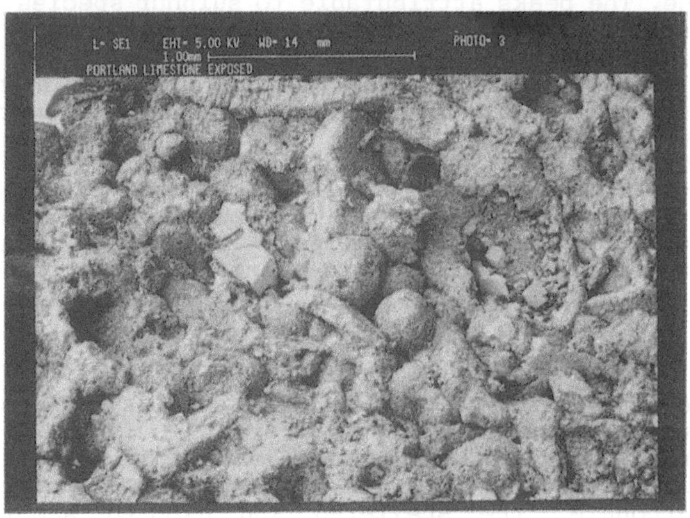

c) Weathered in direct rainfall.

Figure 1. Scanning electron micrographs of Portland Oolitic limestone samples

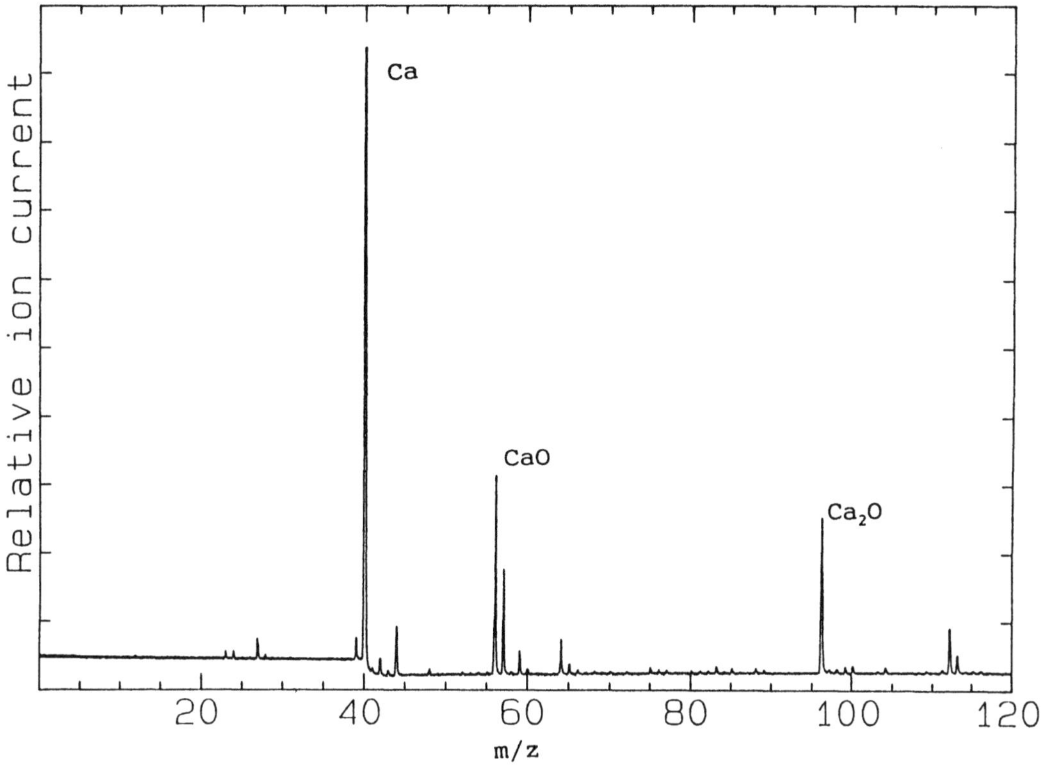

Figure 2. LAMMS positive ion spectrum from Portland exposed sample showing peaks due to calcium and calcium oxide..

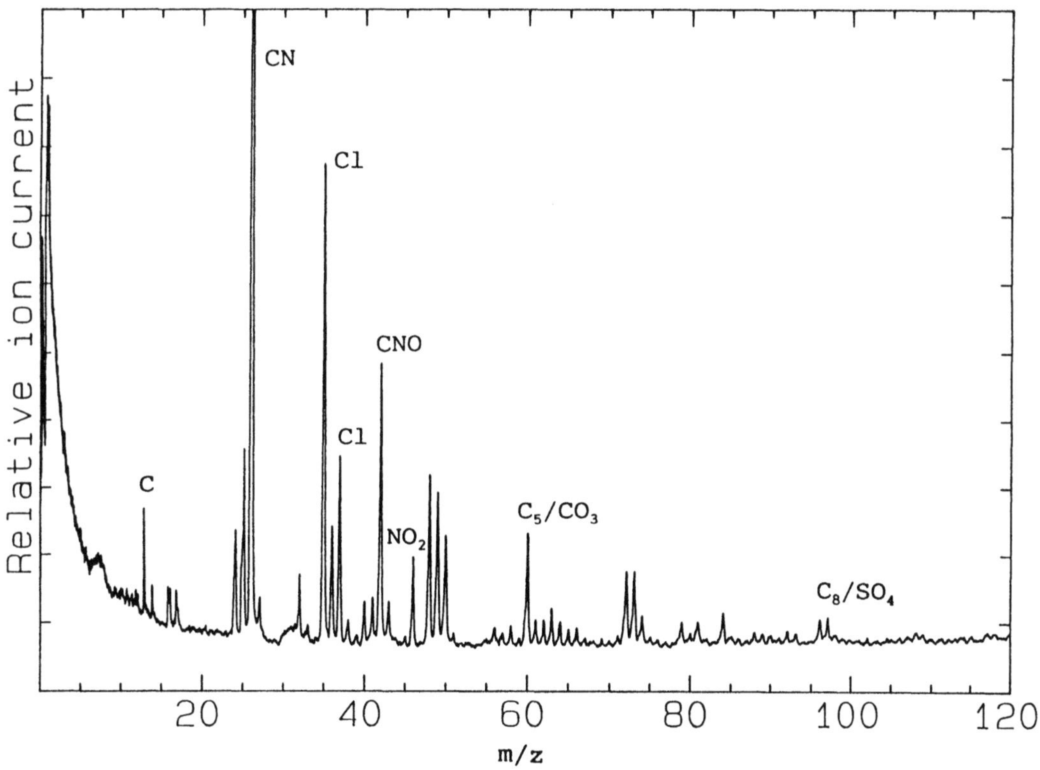

Figure 3. LAMMS negative ion spectrum from Portland pristine sample.

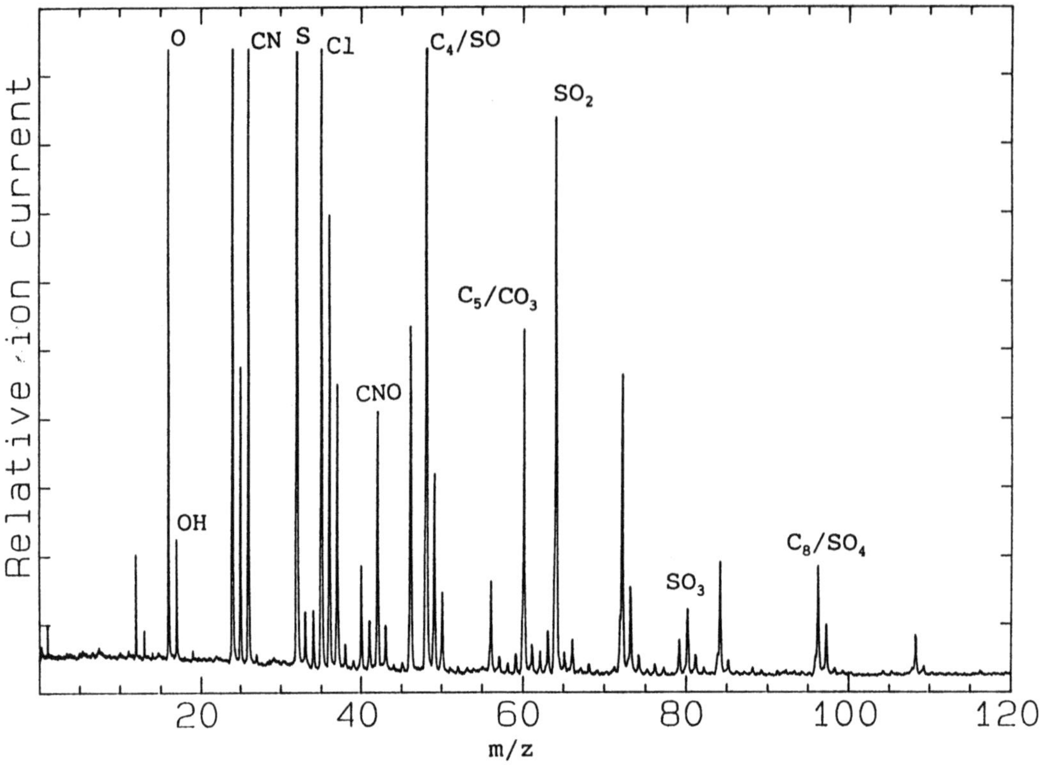

Figure 4. LAMMS negative ion spectrum from Portland sheltered sample.

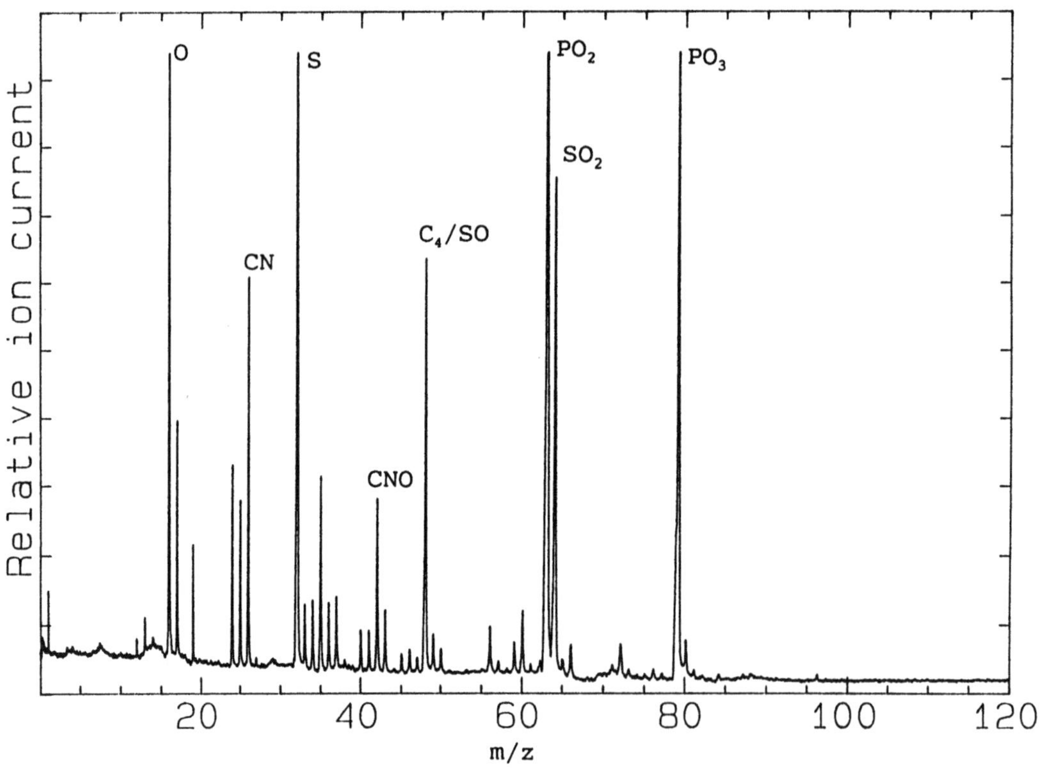

Figure 5. LAMMS negative ion spectrum from Portland exposed sample.

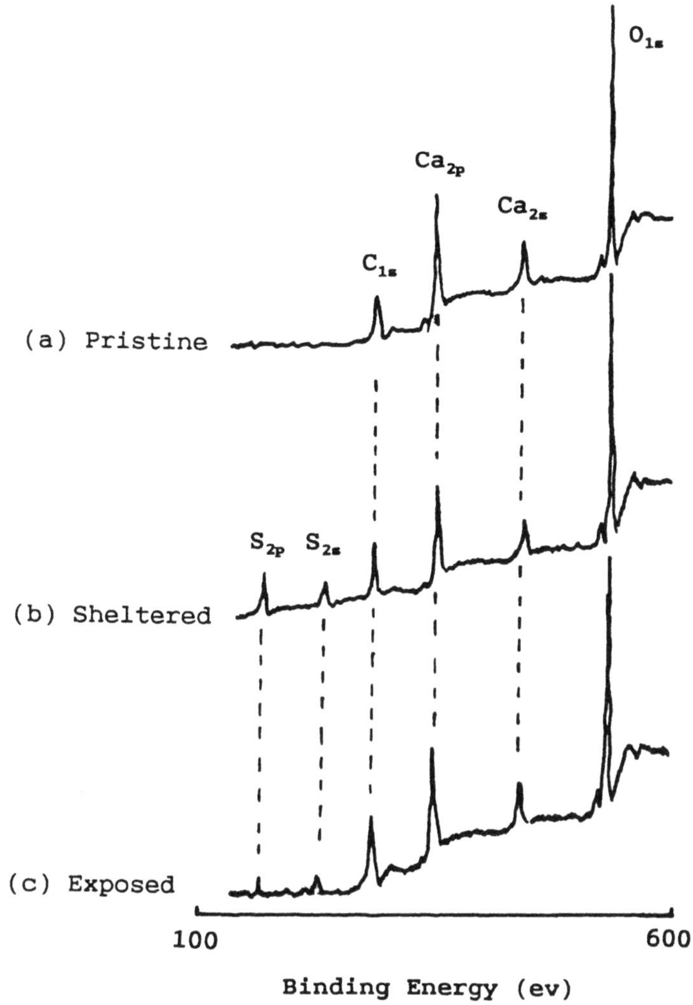

Figure 6. XPS broad scan spectra from a) pristine Portland surface. b) Portland sheltered surface (showing marked S_{2p} peaks). c) Portland exposed sample.

samples leads to the surface composition values (atom %) given in table 1. Absolute values must be regarded as approximate however, accurate comparisons can be made between the respective samples. Considerable levels of sulphur are evident at the surfaces of the sheltered samples. For the exposed samples the measured level decreases for the Portland sample and non is detected for the Monks Park sample. The metal species (sodium, magnesium and aluminium) which give peaks in the LAMMS positive ion spectra shown in figure 2, are not evident in the XPS spectra and are therefore probably present in the samples at levels below the detection limits of the XPS technique (0.5 atom %). Similarly, no peaks are observed for N or Cl in the XPS spectra.

High resolution spectra allow precise determination of an elements binding energy and can therefore be used to obtain detailed chemical information about surface chemistry. Figure 7 shows the high resolution carbon 1s spectra obtained from the Pristine and sheltered Portland samples. The spectrum for the pristine sample (figure 7a) shows two peaks, one at a binding energy of 284.6ev attributable to adsorbed hydrocarbon and a second peak at 289ev due to carbon present as

Table 1. **Surface composition (atom %) of limestone samples from XPS experiments.**

Portland.

	Ca	C	O	S	Si	Mg
pristine	12.2	45.5	42.3	0.0	0.0	0.0
sheltered	12.1	30.2	47.0	10.7	0.0	0.0
exposed	14.9	36.1	47.6	1.4	0.0	0.0

Monks Park.

	Ca	C	O	S	Si	Mg
pristine	14.0	32.4	50.3	0.0	3.2	0.0
sheltered	11.8	27.2	47.5	10.7	2.5	0.3
exposed	9.9	47.5	30.0	0.0	3.7	0.0

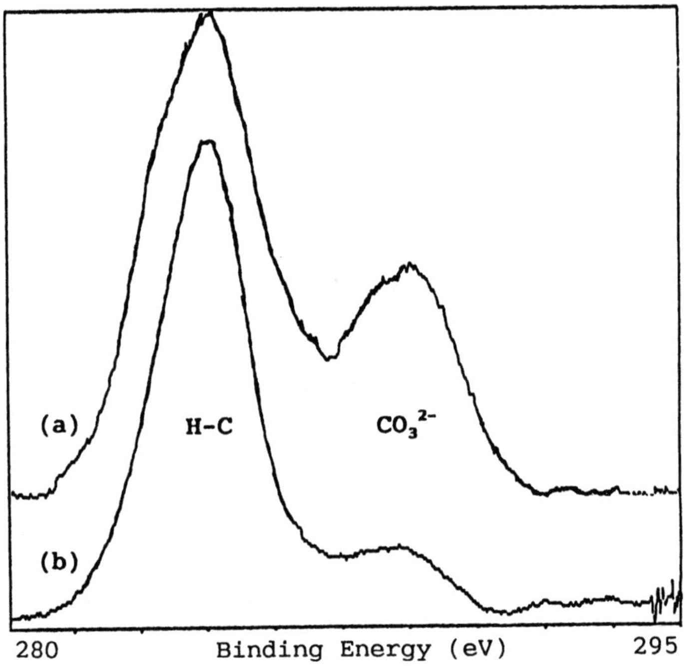

High Resolution C 1s Peaks
(a) Pristine (b) Sheltered

Figure 7. Carbon 1s peaks from Portland a) pristine sample (showing peaks due to adsorbed hydro-carbon at 284.6_{ev} and carbonate at 289_{ev}. b) sheltered sample showing low intensity of carbonate peak.

carbonate (CO_3^{2-}). The spectrum for the sheltered sample is shown in figure 7b over the same range of binding energies. The most striking difference between the two is the marked reduction in the size of the carbonate peak for the sheltered sample. This can be explained by reference to figure 8 which shows the prominant sulphur (S2p) peak for the sheltered sample. The measured binding energy for this peak (168.5ev) is consistent with that previously recorded for S^{6+} in the SO_4 anion (Christie et.al. 1983). The relative intensities of these peaks indicate that whilst the surface of the pristine sample is essentially composed of calcium carbonate that of the sheltered sample is predominantly calcium sulphate. The exposed Portland sample gives spectra which show a marked CO_3^{2-} peak but also indicate some SO_4^{2-} to be present. The data are consistent with the LAMMS spectra presented in section 4.2 which show sulphur and sulphur related peaks for the sheltered samples which are absent or much less prominent for the exposed samples. The XPS data suggest that one of the major reactions occuring in the weathering process is the conversion of the limestone ($CaCO_3$) surface to $CaSO_4$ by reaction with atmospheric SO_2. In an exposed environment the sulphate is then leached from the surface by rainwater leaving behind a surface nearer in composition to that of the pristine samples which is then capable of further reaction. This type of reaction process has also been observed by other workers (Amoroso and Fassina 1983) (Leysen et.al 1987). Because no peaks attributable to nitrate or chloride are present in the XPS spectra it appears that the formation of these anions is a secondary process for the samples studied.

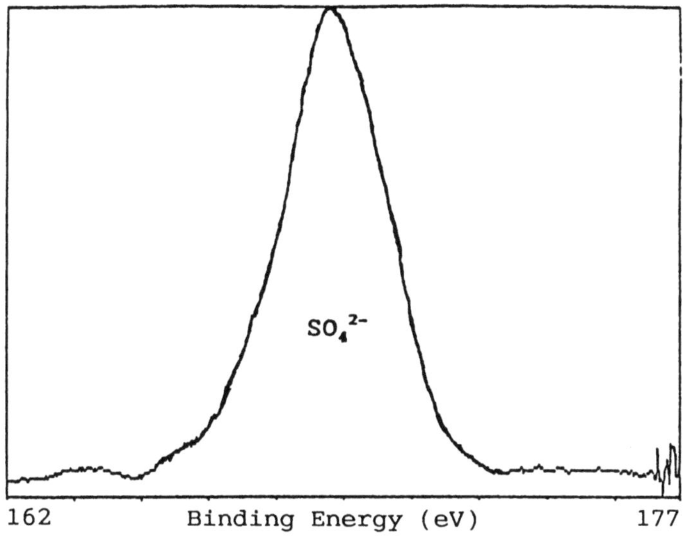

Figure 8. Sulphur 2p peak for Portland sheltered sample. Binding energy of 168.4_{ev} indicates S^{6+} as SO_4^{2-}.

5. Weathering Product Solubility

In order to study the ease with which the sulphur species can be removed from a limestone surface a sample of Portland sheltered stone the surface of which gave a LAMMS spectrum with prominant S, SO_2, SO_3 and possibly SO_4 peaks (figures 4) was washed overnight in deionised water which had been shown to contain no detectable SO_4^{2-} ions using barium chloride. After overnight stirring the rock sample was removed. The washing water was again tested with barium chloride but this time a white precipitate of barium sulphate was obvious indicating that SO_4^{2-} had been leached from the sample. The sample was dried for 3 hours at 333K and then overnight at 378K and its surface was then re-analysed using LAMMS in order to utilize the high sensitivity of the mass spectrometer. The resulting negative ion spectrum is shown in figure 9. Peaks for sulphur at m/z 32 are greatly diminished, compared to those from the unwashed sheltered sample shown in figure 4, whilst those for SO_2, SO_3 and SO_4 (m/z 64, 80 and 96 respectively) are absent. Peaks for chlorine (m/z 35 and 37) and NO_2 (m/z 46) also appear at lower intensities.

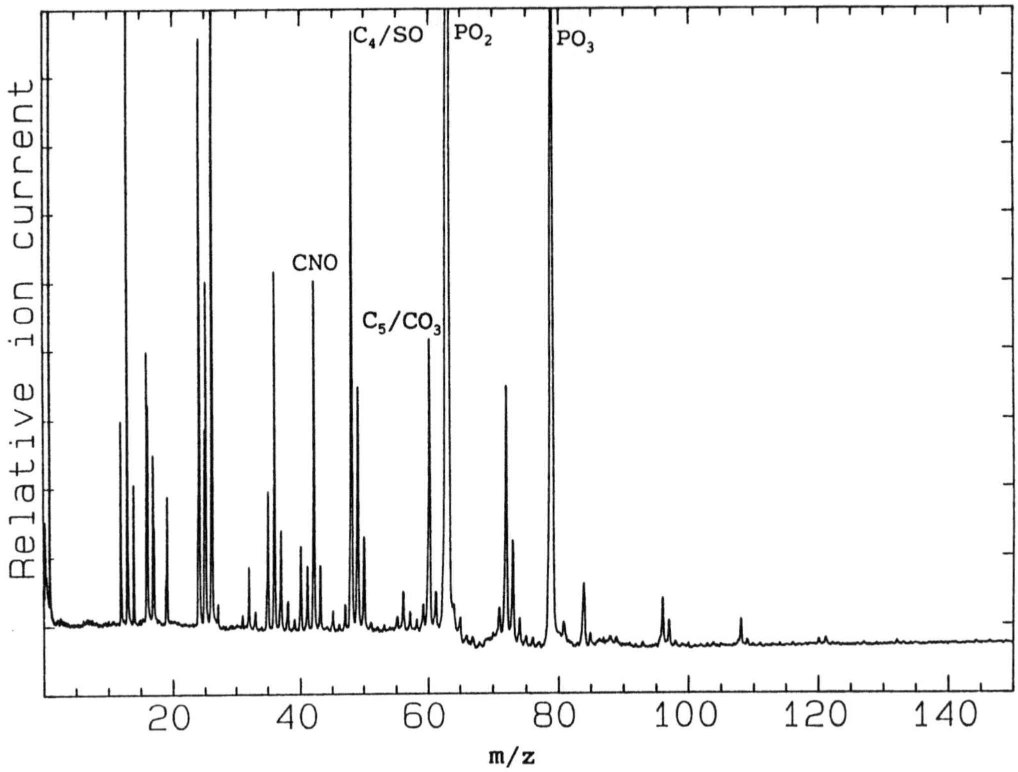

Figure 9. LAMMS negative ion spectrum for the washed sheltered sample. Peaks for S(32) and other S species have lower intensities compared to those in the spectrum given in figure 4.

The experiment clearly demonstrates the effectivness of water washing in removing the species mentioned from stone surface of the type under investigation. Table 2 contains solubility data for some of the minerals which are likely to be present at the surfaces of samples of this type. The values for the sulphates of calcium and magnesium are significantly higher than those of their carbonates or sulphites. If the conversion of $CaCO_3$ to $CaSO_4$ by SO_2 is one of the major reactions occurring at the stone surface then rapid subsequent dissolution of the sulphate by rainwater seems highly likely.

Table 2. Mineral solubilities in cold water ($g.cm^{-3}$.)
(Handbook of Chemistry and Physics, Ed. Weast, Pub. CRC Press, Cleveland, Ohio (1974).

Mineral.	**Solubility in Cold Water. $g.\ 100\ cm^{-3}$.**
$CaCO_3$	0.0015
calcite	0.0014
$CaSO_4$	0.209
$CaSO_4.2H_2O$	0.241
$CaSO_3.2H_2O$	0.0043
$MgCO_3$	0.016
$MgSO_4$	26.0
$MgSO_3.6H_2O$	1.25

6. Concluding Remarks

The objectives of this study were to investigate the suitability of a range of surface sensitive analytical techniques for identifying morphological and chemical changes which occur at the surface of limestones during the weathering process. The SEM allows resolution of the limestone structure but no conclusive differences are apparent between the pristine and weathered samples of either one rock type. Because of the type of reactions likely to occur during the weathering process information about the molecular chemistry of these surfaces is of interest. In particular the anions present and the oxidation state of elemental species is important. For this reason LAMMS and XPS appear to offer the most suitable combination of techniques.

LAMMS allows rapid analysis, either at several sites on one sample or of several samples, it also gives some molecular information but at present it is non-quantitative and may cause fragmentation of some surface species due to the energy of the laser. The LAMMS spectra indicate the presence of sulphur and sulphur species as well as chlorine and nitrogen containing species on the surface of the sheltered samples. In the corresponding exposed samples these species are either absent or present at low levels.

XPS is quantitative and also fulfills the requirement to determine anions and oxidation state. The surface of the pristine samples have been shown to be comprised of calcium carbonate whilst the corresponding sheltered samples appear to be predominantly calcium sulphate. Exposed samples of these rock types show both carbonate and sulphate.

The data presented suggest that the primary chemical weathering process occurs by the reaction of SO_2, either in gaseous form or dissolved in rain-water, with the limestone surface. This leads to a build up of sulphur rich layer at the surface which is shown to be predominantly SO_4^{2-}. A more detailed study of this reaction mechanism should allow identification of reaction intermediates such as SO_3. The formation of chlorine and nitrogen compounds appears to be a secondary process for the samples studied. The data presented and the results of the experiment described in section 5 suggest that subsequent dissolution of sulphate, by rain-water, is a significant part of the weathering mechanism.

Acknowledgements. The author wishes to thank Dr. R. Butlin and Mr. T. Coote of BRE and also acknowledges the help of colleagues at Loughborough university who have contributed to this study.

References

G.G.Amoroso and V. Fassina, *Materials Science Monographs*, 11, Stone Decay and Conservation, Elsevier (1983).

A.B. Christie, J. Lee, I. Sutherland and J.M. Walls, *Appl. Surface Sci.* 15, 224 (1983).

D.B. Honeyborne and P.B. Harris, in '*Proceedings of the Tenth Symposium of the Colston Research Society*', p343, pub. Butterworth, London, (1958).

L. Leysen, E. Roekens, Z. Komy and R. Van Gieken, *Analytica Chimica Acta.*, 195 247, (1987).

M.J. Southen, M.C. Witt, A. Harris, E.R. Wallach and J. Myatt, *Vacuum* 34, 903 (1984).

E. M. Winkler, *Stone: Properties, Durability in Mans Environment*, p.92, Springer-Verlag, New York (1973).

CRITICAL LOADS AND CRITICAL LEVELS FOR THE EFFECTS OF SULPHUR AND NITROGEN COMPOUNDS

M. J. Chadwick and J. C. I. Kuylenstierna
Stockholm Environment Institute at York, University of York, U.K.

1. INTRODUCTION

Estimates of the man-induced global annual sulphur flux for the mid-1980s are put at about 93 (±14) million tonnes S, of which some 80 million tonnes S were emitted in the form of SO_2 (Brimblecombe *et al.*, 1989). Of this, 75 per cent results from the combustion of fuel, mainly fossil fuel. Although the man-induced sulphur represents only about one-third of the total global sulphur flux, because over 90 per cent occurs in the northern hemisphere, there are areas here (particularly in Central Europe) where man-induced sulphur deposition far exceeds the deposition from natural sources.

The direct effects of sulphur dioxide tend to be local and spatially related to the emission source. However, sulphur dioxide may be transformed chemically in the atmosphere and transported hundreds of kilometres before being deposited where harmful effects may result, far distant from the source. Thus, solely national efforts to reduce the harmful effects of such deposition by abating emissions are unlikely to be very effective. Transboundary pollution phenomena must be taken into account and it has proved to be the impetus for evolving regional co-ordinated abatement strategies for SO_2 and NO_x emissions.

The United Nations Economic Commission for Europe (UN-ECE) Convention on Long Range Transboundary Air Pollution (LRTAP) came into existence in 1979 and into force in 1983. An "SO_2 Protocol" stipulated specific agreements on the reduction of national annual sulphur emissions, or their transboundary fluxes, by 1993. Nineteen countries signed the Protocol in 1985 in Helsinki, agreeing to at least a 30 per cent reduction over 1980 emissions. In November, 1988, the "NO_x Protocol" was agreed in Sofia. In addition, EC countries have agreed a directive on emissions from large combustion plants that should result in an overall reduction of 60 per cent in SO_2 emissions by 2003 and reductions of around 30 per cent of NO_x by 1998, within the Community.

However, as understanding of the basic ecological effects of acid depositions improves, it is becoming apparent that emission abatement to achieve the necessary deposition reductions, in some areas, will be called for that are considerably in excess of those envisaged by the existing international agreements. To take one example, it has been estimated that a reduction of approximately 75 per cent in S deposition (from approximately 80 keq km^{-2} yr^{-1} to approximately 20 keq km^{-2} yr^{-1}) would be required to return certain Scottish lochs to their non-acidifed status

(Batterbee, 1989). Such a situation also applies to many areas in Europe. In fact, the approach that is increasingly being adopted is one in which the acid buffering capacity and rate, of a site, are being determined. This is employed, along with other site features, such as the sensitivity of living systems to acid conditions, to set a **critical load** value for acidic depositions below which damage to the ecosystem structure or function is unlikely to occur. A similar approach has been adopted to the direct effects of air pollutant concentrations, such as SO_2 or NO_x, by setting **critical levels**.

This critical load/level approach has come into prominence as more stringent abatement levels are sought. Where there is uneven access to emission control technologies and ability to meet the costs of overall abatement, concentration on a cost-effective strategy involving the attainment of deposition reductions to meet targets, based on scientific criteria, becomes an attractive and practical method of progress. Deposition reduction targets can be set to allow for the variation in the level of deposition that has been determined will not cause damage to ecosystems. This may be set against using an overall percentage reduction, irrespective of spatial variations in sensitivity.

Critical load estimates are not simple to obtain. A critical load is "a quantitative estimate of an exposure to one or more pollutants below which significant harmful effects on specified sensitive elements of the environment do not occur according to present knowledge" (Nilsson and Grennfelt, 1988). Essentially the concept is a dose-response or exposure-effect one. Hence, all kinds of questions arise:

i) What constitutes a realistic measure of 'dose' over a period of time and should 'chronic' and 'acute' doses be represented differently?

ii) How are doses in multiple-pollutant systems dealt with?

iii) What represents a satisfactory measure of response in complicated systems and how is the 'harmful effects' threshold recognized particularly where variability masks clear effects?

iv) Should alterations of both **structure** and **function** be taken into account and, if so, how would this allow for new equilibria situations?

Questions of a different nature also arise from a critical loads approach. These chiefly concern the logic of not polluting 'up to' a critical load if this threshold can be accurately determined; alternatively, there is the question of why it should not be desirable to use the best available technology (BAT) to abate emissions below critical load values where such technology exists. There are answers, of a sort, to both such sets of questions but, as critical loads and critical levels estimates increasingly indicate the need for more stringent levels of abatement, the debate becomes less and less relevant to practice. Not only is it evident that, in many areas, BAT will need to be applied to attain critical load values but also, even with all BAT applied, in certain areas, reaching critical loads will not be a feasible goal. Nevertheless, the critical load and critical level concept does provide a relatively firmly-based scientific yardstick by which to judge progress towards pollution minimization. Their use as a basis for target load establishment also provides a practical means of effecting progress in pollution abatement agreements that incorporate a cost-effective approach.

2. THE CRITICAL LOAD-CRITICAL LEVEL CONCEPT

The critical load concept applied to ecosystems assumes a threshold response to acid depositions in terms of the onset of harmful effects. Below the threshold it is postulated that acidic depositions will not give rise to deleterious effects on animal, plant and other life and that these will only begin to occur once the critical load is exceeded. Critical loads for certain elements, such as nitrogen, will have to take account of the possibility of the existence of limitation of supply and also nutrient imbalances induced by accumulation as deposition proceeds. The critical load is thus a site specific, "pollutant" and ecosystem specific value. It will depend upon a combination of site factors which influence the sensitivity of the site to depositions and hence the response of the biota.

Gaseous pollutant concentration standards for direct effects on plants and vegetation have also been suggested and are known as critical levels (UN-ECE, 1988). These are also based on a threshold response assumption, the critical level being the concentration of such substances as SO_2, O_3 and NO_2 respectively above which damage is caused.

The critical load concept used for acidic depositions owes its origin to the "target loading" concept proposed by Canadian scientists, in relation to the deposition of sulphates to aquatic systems, during the transboundary pollution control negotiations with the USA (MOI, 1983; Brydges and Neary, 1984). An upper limit of deposition was proposed to safeguard all but the most sensitive of lake ecosystems. The concept was further developed in Scandinavia and the term "critical load" was used (Nilsson, 1986). Critical load values were suggested for a range of aquatic and terrestrial sites in Scandinavia that would take account of both sulphur and nitrogen depositions. A critical load was defined as (Nilsson, 1986), "the highest load that will not cause chemical changes leading to long-term harmful effects on the most sensitive ecological systems". The phrase "long-term" was taken to mean 50 years or more (Nilsson, 1986a).

3. CRITICAL LOAD CRITERIA

Critical loads applied to acidic depositions rely on the assumption that there are rate dependent processes in ecosystems which act as sinks for acids and are able to fully neutralize the acidic deposition. It is assumed that there is enough capacity to neutralize acidic depositions in the long term. These rate dependent processes include the rate of weathering of soil minerals in the catchment and microbial reduction of sulphate and nitrate. It is postulated that if the total input of acidic substances to the ecosystem is less than the rate at which the acids are neutralized then there will be no net acidification of the system and hence no harmful effects will occur.

It has not yet been possible to directly relate all the major facets of biotic **response** to acidic deposition (**dose**). The basis of the critical load concept is generally the response of biologically relevant chemical characteristics. One approach is to require that no decrease in base saturation occurs so that all acidic inputs are consumed by processes such as weathering and since this will guard against changes in soil chemistry there will be no deleterious response by the biota. Other threshold values of soil chemical parameters also may be set. Among the most commonly used for terrestrial systems are pH and alkalinity, calcium:aluminium ratio and the change in concentration of NO_3^- in soil solution. Sverdrup and

Warfvinge (1988) set critical loads to safeguard coniferous forest ecosystems by assuming a pH > 4.0-4.3 should be maintained in the O, A and upper B horizons and a soil solution alkalinity of >0.03 meq l^{-1} at the base of the B horizon. For lakes and streams runoff from the soil should have an alkalinity of at least 0.05 meq l^{-1} which equates with pH 6.0. In aquatic systems the water pH affecting fish response is a commonly used variable but base cation concentration and the toxicity of biologically active aluminium species have also been included.

Minns, Elder and Moore (1988) have attempted to take the criteria closer to actual biological responses. The basis for surface waters' critical loads assessment is the loss of one fish species from amongst the twelve most commonly occurring species in the region. It is suggested that, although a single species loss is not a viable threshold to choose when dealing with other taxa such as invertebrates, it is possible to select a loss of a proportion of the various taxa. Inclusion of these (such as molluscs or leeches) will lead to more stringent critical loads being set.

4. LIMITATIONS OF THE CRITICAL LOAD CONCEPT

The critical load concept provides a way of setting a deposition standard based on environmental effects and criteria. There are, however, limitations to this approach and these relate to dose considerations, considerations relating to response and identification of the target units, and also to operational disadvantages of the concept.

Dose-related limitations include difficulties of integrating chronic dosages over time to produce single value estimates and decisions related to using criteria of that sort rather than 'peak' or acute, episodic dose criteria. A second difficulty arises due to the non-comparability, in dose terms, between a system that is already undergoing acidification and one that is in a relatively pristine condition but may be at risk; a similar critical load value might be applied to both. Furthermore, the use of critical loads, so far, assumes that the trajectory of recovery, in relation to dose, is the same as degradation under initial increasing acidification. This assumption ignores hysteresis effects which may occur. For example, buffering mechanisms involving adsorption and desorption are known to show hysteresis as are biological systems when density changes occur during colonization and perturbation at different rates. A third difficulty relates to the complicating effect, on straightforward threshold dose recognition, of changes in other pollutants or interacting factors. Indeed, the most satisfactory way of representing the dose-response relationship of a mixture of pollutants may be in terms of a linear no threhold model (Butler, 1978).

From the response point of view, it may be difficult or impossible to detect with certainty a threshold of damage. It is to be expected that where relatively simple 'pollutant' substances contain essential elements, a threshold model is, *a priori*, the most likely relationship. However, uncertainty boundaries in the dose-response relationship might equally well accommodate both a threshold or a linear no threshold model response.

In addition, the many elements of an ecosystem may react differently to variation in pollutant level or rate of deposition. Fish are often taken as the target organism in aquatic ecosystems (Minns, Elder and Moore, 1988); however, there are more sensitive biotic elements that will decline before fish are affected (Eilers, Lien and Berg, 1984). The integration of ecosystem response in relation to particular critical loads or levels is extremely difficult, particularly as both structural (species diversity, species number, food web features, population density, standing crop) and functional (rate of photosynthesis, rate of respiration, maintenance efficiency, nutrient turnover) attributes, at least theoretically, should be taken into account. There is also a further response consideration that is fundamental to the critical load concept. The concept is a steady state, static variable attempting to describe a response of a dynamic system. Ecosystems may acidify naturally; depositions of acids are accelerating this process. Therefore, the critical load referring to the existing equilibrium state is also taken to refer to the ecosystem at a changed equilibrium state, sometime in the future. The problem of whether critical load values should allow for ecosystem adjustment to some future equilibrium condition is partially dealt with by suggesting a time frame within which the critical load estimate should guard against adverse changes (Nilsson, 1986). This is important as the time lag between the delivery of a dose and the certain detection of an effect may be a matter of years. This is one reason why the early warning systems being developed (Wolfenden *et al.*, 1988) are of such interest.

The operational disadvantage to the critical load approach is that logically there can be no reason not allow deposition (or pollutant concentration) to rise to the critical load/level set. Even if 'best available technology' could keep values well below the critical load/level, it could be argued that there would be no point in doing so as, by definition, no damage will occur up to that value. In practice, it is doubtful whether such an unwise attitude would gain acceptance for at least two reasons. Firstly, critical load values are only estimates and have uncertainties associated with the values. Secondly, sensitive areas tend to be relatively evenly spread, over Europe at least, and rather widely dispersed (Chadwick and Cooke, 1987). By meeting more stringent critical loads it will be likely that other less-sensitive areas would be kept well below the higher critical load value assigned to them.

5. TARGET LOADS

A critical load is considered an inherent property of an ecosystem. But it is possible to use critical load estimates as a basis to set a target load (Streets, Hanson and Carter, 1984; Chadwick, 1986; Nilsson and Grennfelt, 1988). The target load value can form the basis of emission reduction strategies to reach area deposition goals. A target load value may be set above the critical load (if a certain degree of damage is regarded as acceptable, in order to set a realistically achievable deposition goal or to meet some socio-political requirement). Target loads might be set below a critical load value if it is thought this would allow more rapid ecosystem recovery to occur.

Target loads are useful in another way: at present it is only possible to estimate deposition rates over a wide area from meteorological atmospheric transfer models such as those developed by the Co-operative Programme for Monitoring and Evaluation of the Long Range Transmission of Air Pollutants in Europe (EMEP). The estimates of atmospheric transfer from emitting sources to receptor areas (for sulphur and nitrogen) is based on a grid of 150 x 150 km squares (the "EMEP grid square"). The relation between emission abatement and reductions in deposition to meet a critical or target load is thus initially most easily managed in abatement strategy terms of allotting an aggregated critical load value for a grid square. It may be a value based upon the mode for that grid square, a mean value (or weighted mean) or simply be the most stringent value represented in the grid square. Thus, a value not inherently representative of the total area may be identified that is more correctly described as a target load

(Minns, Elder and Moore, 1988) than a critical load.

6. CRITICAL LOAD ESTIMATIONS

Four main approaches to the derivation of critical load values have been adopted:

i) estimations based on empirical relationships;

ii) estimations based on proton deposition/production and consumption considerations (budget studies);

iii) estimations based on 'process models' of eco-systems;

iv) estimations based on palaeolimnological investigations.

6.1. Empirical Methods

In some of the earliest work that led to critical load estimations, Elder and Brydges (1983) and Brydges and Neary (1984) derived an interim "target loading" which considered that the lakes had assumed a steady state with respect to acidification. They used observations of the distribution of damage to fish stocks along a gradient of sulphur deposition in Canada in lakes with similar catchment characteristics. They recorded the deposition at which damage to fish stocks became evident. Effects were seen as deposition reached 25 kg SO_4 ha^{-1} yr^{-1} but were not measurable in areas where the wet sulphate deposition was less than 10-15 kg SO_4 ha^{-1} yr^{-1}. The interim "target loading" at which all but the most sensitive lakes were protected was set at 20 kg SO_4 ha^{-1} yr^{-1}. In areas where surface water alkalinity exceeded 200 ueq l^{-1} higher loading rates were considered acceptable.

Henriksen (1980) developed an empirical model based on the hypothesis that the acidification of freshwaters can be envisaged as a titration of a bicarbonate solution with strong mineral acids and showed how it was possible to estimate the pre-acidification alkalinity. pH in the lake is considered to be a function of the original alkalinity of the lake water and the acidity of the rain falling on the watershed and it was therefore possible to predict the pH of a lake of known Ca + Mg concentration. Henriksen, Dickson and Brakke (1986) defined the critical load as depositions which will maintain a lake water at pH 5.3 since above this value lakes are still within the bicarbonate buffering system, there are no strong acids present and aluminium concentrations are below toxic levels. Using the empirical model, which has become known as the Steady State Water Chemistry Method for critical load evaluation (Sverdrup, de Vries and Henriksen, 1989), it is possible to predict what deposition value will keep the lake water above pH 5.3. It is suggested that this is 20-40 keq km^{-2} yr^{-1} for the most sensitive Norwegian lakes.

Acidification will lead to an increase in the Ca levels in lakes which will result in an overestimate of the original alkalinity. It will therefore take a higher deposition to depress lake pH by the same amount as if there hd been no increase in Ca concentrations. The so-called F-factor (base cation change factor) was initially set at 0.2 (Henriksen, Dickson and Brakke, 1986). The F-factor is defined as:

$$F = \frac{(Ca^* + Mg^*)}{SO_4}$$

where * indicates 'non-marine'. F will vary between 0 and 1 (Henriksen *et al.*, 1988). The F-factor is assumed to vary as a function of the base cation concentration and increases with the base cation concentration (Henriksen *et al.*, 1988). It is used to adjust the critical load estimate.

It follows that lakes of different original alkalinities will follow different titration curves of acidification and Dickson (1986) used observed data from Swedish lakes which had different non-marine Ca + Mg concentrations and received different amounts of sulphate deposition to predict the pH of lakes under a given deposition (Figure 1). Dickson (1986) set the threshold at pH 6.0 assuming that there is damage to aquatic ecosystems below this pH. For the most sensitive lakes this resulted in a critical load of between 10 and 20 keq km^{-2} yr^{-1}.

Another empirical approach to setting critical loads is to set a target sulphate concentration in runoff or leaching rate to lakes (Henriksen, Dickson and Brakke, 1986). Dickson (1986) found that sensitive lakes will be acidified if sulphur leaching exceeds 0.3-0.5 g S m^{-2} yr^{-1}. The critical load would be the deposition causing this leaching rate.

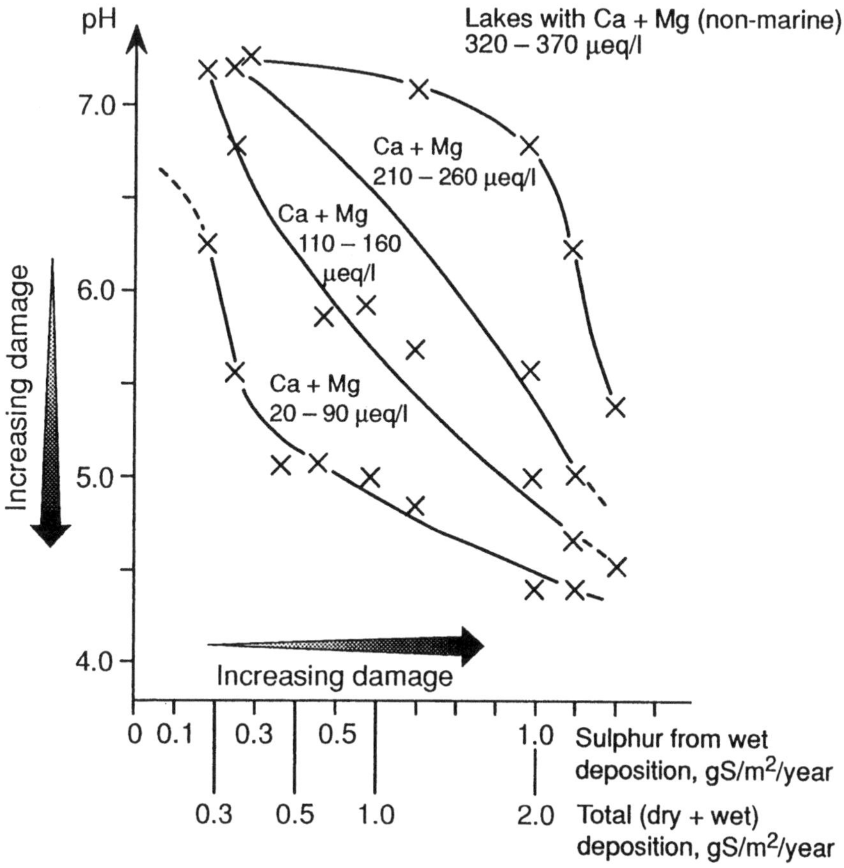

FIGURE 1. The pH value in lakes (x) in Sweden of varying base production potential at different sulphur loadings. Total deposition was calculated from concentrations in runoff water (Dickson, 1986).

Minns, Elder and Moore (1988) carried out critical load determinations based on the response of fish to changing lake pH. They used an empirical model developed by Jones, Browning and Hutchinson

(1987) to describe the response of lake pH to acidic deposition. This model calculates the steady state alkalinity of a lake from the original alkalinity and the change in alkalinity due to acidic depositions. The alkalinity is then related to pH using an empirical relationship derived from a surface water database for Eastern Canada. Regionalization of the model was carried out by dividing the study area into "sub-regions" based on site characteristics such as land use and hydrology. Using data from within the sub-regions it was possible to generate pH profiles which could then be related to the biological impacts through pH relationships in a biotic response model developed from observational data. The biological impacts model was based on pH minima curves for freshwater fishes (the minimum pH at which the fish species had been observed in lakes) in particular regions. Using a weighted random sampling technique for regional species-pH minima data, Minns, Elder and Moore (1988) were able to relate pH decrease to species loss, in terms of the original pH level of the lake and the rate of deposition through an empirical lake chemistry model. Critical load estimates of 20 keq km^{-2} yr^{-1} were derived for lakes in eastern Canada that would give a high chance of protecting most lakes.

Meijer (1986) set critical loads based on Dutch experience of nitrogen deposition on various ecosystems. The critical loads were based on the NH_4:K, NH_4:Mg or Al:Ca ratios in the soils of forest ecosystems. Observations of forest health in response to changes in these parameters were made. These ratios should not exceed 5, 5 and 1 respectively which corresponds to a nitrogen deposition of about 20 kg N ha^{-1} yr^{-1} for most forest ecosystems in The Netherlands. However, in order to minimize nitrate leaching to the drinking water standard (50 mg l^{-1}) the deposition was adjusted to 10 kg N ha^{-1} yr^{-1}. The critical value for NH_4:Mg ratio is revised to 10 (Houdijk, 1988) which gives rise to a critical load of about 15 kg N ha^{-1} yr^{-1} for sensitive sites. Van Dijk and Roelofs (1988) reviewed experiments to determine the deposition at which nitrogen caused vegetation changes. In small softwater ecosystems the critical load to prevent vegetational change was 20 kg N ha^{-1} yr^{-1}.

Liljelund and Torstenson (1988) reviewed the N deposition rates in various studies which were implicated in changes in plant community composition. Sites with deposition rates of 20 kg N ha^{-1} yr^{-1} were found to have had changes in floristic composition and the critical load therefore would have to be set at a lower value (15 kg N ha^{-1} yr^{-1}) for most ecosystems but at 10 kg N ha^{-1} yr^{-1} for very sensitive ecosystems.

Gundersen and Rasmussen (1988) have suggested that the deposition of nitrogen should not exceed the uptake by plants (5-15 kg N ha^{-1} yr^{-1} in Nordic coniferous forest ecosystems) in order to prevent acidification of the soil. Rosén (1988) calculated critical loads on the basis of the capacity of woody biomass to accumulate nitrogen and also the increased depletion of base cations from the soil in response to increased growth rates. The critical load set in this way varies with site quality and for Scandinavian forests will probably vary between 5-15 kg N ha^{-1} yr^{-1}.

Nilsson and Grennfelt (1988) give critical loads for nitrogen (CL_N) using:

$$CL_N = N_{upt} + IM_{acc} + L_{acc}$$

where N_{upt} is the uptake in biomass, IM_{acc} is the acceptable accumulation in the soil and L_{acc} is the acceptable leaching from the system. Using this method they derive critical load ranges for nitrogen, dependent on the productivity of the site, which vary from 3 to 20 kg N ha^{-1} yr^{-1}.

6.2. Ion Budget Studies

Nilsson (1986b) calculated critical loads based on a "proton budget" approach. The critical load is calculated by ensuring that the total input rate of hydrogen ions (H^+_{in}) is less than or equal to the weathering rate (WR) of base cations:

$$WR \leq H^+_{in}$$

The input of hydrogen ions consists of the following components: wet deposition (H^+_{wet}), dry deposition (H^+_{dry}), internal processes (H^+_{int}), excepting the biological transformations of nitrogen, and the net effect of the deposition of nitrogen (assuming NH_4^+ leaching is negligible):

$$H^+_{in} = H^+_{wet} + H^+_{dry} + H^+_{int} + NH^+_{4\,in} + (NO^-_{3\,out} - NO^-_{3\,in})$$

The weathering rate may be calculated from:

$$WR = BC_{out} + BC_{upt} - BC_{dep}$$

where BC_{out} is the output of base cations in drainage water, BC_{upt} is net uptake in perennial biomass and BC_{dep} is the base cation deposition. These factors must be measured or estimated in order to derive an estimate of the critical load by making certain assumptions.

Nilsson (1986b) simplified the calculation and derived the critical load or "permissible deposition" (H^+_{perm}) from:

$$H^+_{perm} = WR - H^+_{in}$$

Critical loads were calculated for catchments in Scandinavia and North America which varied from 0 to 75 keq km^{-2} yr^{-1}, with critical loads for most of sensitive catchments of between 10-20 keq km^{-2} yr^{-1}.

Sverdrup and Warfvinge (1988) also used budget study data to derive critical loads for specific catchments using:

$$CL = WR - BC_{upt} - \text{nitrogen acidity} - BC_{out}$$

where WR is weathering rate and BC base cation uptake and outflux. The critical load estimates using this method vary between 0 and 120 keq km^{-2} yr^{-1} for terrestrial ecosystems with most Scandinavian systems having critical loads around 30 keq km^{-2} yr^{-1}.

Weathering rates may be calculated by budget studies or measuring the depletion of cations from the upper part of the soil profile. Tamm (1940) calculated the annual depletion rate of a podsol to be 7.5 keq km^{-2} yr^{-1} over 9000 years.

Eriksson (1986) calculated the critical load for groundwater from budget studies by assuming that the critical load is equal to the alkalinity production down to any given depth of the soil. The critical load was set by reference to the suitability of the water for human consumption from dug wells. Critical loads for groundwater vary between 3-24 keq km^{-2} yr^{-1} at a depth of 3m and between 10 and 80 keq km^{-2} yr^{-1} at a depth of 10m, dependent on the type of soil.

6.3. Process Models

Various process models have been used to estimate critical loads. These include empirical models using process-oriented mechanisms in their structure, such as the MAGIC model (Cosby, Hornberger and Galloway, 1984), the RAINS Lake Model (RLM) of Kämäri and Posch (1987) and the Model to Assess a Critical Acid Load (MACAL) of de Vries (1988). More complex models such as the Steady State Soil Chemistry Model (PROFILE) developed by Sverdrup and Warfvinge (1989) and RESAM, the Regional Soil Acidification Model (de Vries and Kros, 1989) have been developed.

MAGIC (Cosby, Hornberger and Galloway, 1984; Cosby, Wright, Hornberger and Galloway, 1985; Cosby, Hornberger, Wright and Galloway, 1986) simulates the fluxes to and from the pool of exchangeable cations in the soil which will be affected by acidic deposition and give rise to changes in surface water chemistry. MAGIC is a whole catchment lumped parameter model attempting to describe catchment response without incorporating the detail of less easily measured mechanisms; it concentrates only on a small number of important soil processes. The model consists of two main sections:

i) a soil-soil solution chemical equilibrium section taking into account sulphate adsorption, cation exchange, aluminium chemistry and dissolution of inorganic carbon;

ii) a mass-balance or budget section explaining the flux of major ions in terms of deposition, the weathering rate, base cation uptake and loss to runoff.

The equilibrium equations assume an instantaneous equilibrium but since the mass-balance section includes rates of weathering, uptake and deposition, this is a dynamic model which may be used to investigate the time aspects of surface water response to deposition. The output of the model is the streamwater chemistry including pH, alkalinity and aluminium concentrations. MAGIC can be used to derive critical loads by setting a critical value for either pH or alkalinity and running the model with different deposition loads.

The RAINS Lake Model RLM (Kämäri and Posch, 1987) assumes that only a few reactions between the soil and soil solution need to be considered to describe surface water chemistry. These include relationships between soil base saturation and soil solution pH and between the solid and liquid phases of aluminium. The change in base cations and therefore base saturation is determined by the movements of acids and bases into and out of the soil. This change is defined as:

$$\frac{dBC}{dt} = WR + BC_{dep} - SO_{4\,dep}$$

where $SO_{4\,dep}$ is the total sulphate deposition.

The model considers two soil layers and assumes that all runoff goes through the A layer and the routing of waters through the two soil layers is calculated from the maximum possible flow rate through the lower B layer. The output of the model is lake pH and alkalinity and again critical loads have been estimated by setting up a critical pH or alkalinity. Wright, Kämäri and Forsius (1988) used a critical value of pH 5.5 to calculate the critical loads for Finnish catchments.

MACAL is a model which can be used to assess the critical load of forest soils (de Vries, 1988). Here, critical values for aluminium concentrations and Ca:Al ratios are set at 0.2 $mmol_c\ l^{-1}$ and 1.0 respectively and the model calculates the yearly averaged calcium and aluminium concentrations in soil compartments of 10 cm up to 80 cm depth. The model predicts the element flux and water flux in given soil compartments at given levels of acidic deposition:

$$EL_{flux}(X) = EL_d + EL_e + EL_m + EL_w(X) - EL_u(X)$$

where $EL_{flux}(X)$ is the element flux at depth X, El_d is the deposition of the element, El_e i the foliar exudation of the element, El_m is the element mineralization, $El_w(X)$ is the accumulated element weathering to depth X and $El_u(X)$ is the accumulated element uptake to depth X.

MACAL simplifies soil processes by numerous assumptions. The model is run using different deposition scenarios and the Al concentrations and Ca:Al ratios at different soil depths are compared to the critical values mentioned.

The model developed by Sverdrup and Warfvinge (1988) attempts to find the steady state equilibrium soil chemistry at any particular deposition rate and therefore ignores the time development of acidifcation or recovery. The soil is divided into any number of horizons and the Acid Neutralizing Capacity (ANC) production by various processes is calculated for each soil horizon. The processes considered are cation exchange, the weathering rate and biochemical processes including the transformation of nitrogen. In each soil horizon there is a mass-balance equation for ANC which calculates the change in ANC for that horizon from the ANC of the solution entering it, the ANC of the solution leaving it and the production of ANC within the soil horizon. The weathering rate is calculated in a separate sub-model which uses experimentally defined chemical reactions for most of the common soil minerals. The model is run until there is no change in ANC in any horizon and the new steady state is achieved. The model requires many input prameters for each soil horizon including the mineralogy, density and exposed surface area of the soil, the carbon dioxide partial pressure and base cation uptake rate by vegetation. The output is the soil solution composition including estimates of pH, calcium and aluminium concentrations and base saturation. Critical loads are estimated by running the model until the deposition produces a soil solution chemistry conforming to one of the three target values mentioned previously.

6.4. Palaeolimnological Reconstructions

Flower amd Battarbee (1983), Battarbee *et al.* (1985) and Renberg and Helberg (1982) reconstructed the historical pH changes from the changes in the composition of diatom assemblages preserved in the sediments of lakes in Scotland and Scandinavia. From these pH reconstructions it has proved possible to suggest a critical load by either finding or estimating the deposition at which the rate of decrease in lake pH becomes significant or when it reaches a given value. Batterbee (1989) plotted sulphur deposition against calcium concentration (as an indication of the lake sensitivity) of different British lakes and on this graph indicated which lakes had been acidified according to diatom analyses. There would seem to be an approximate sulphur deposition to calcium concentration ratio which separates the acidified lakes from the non-acidified which can then be used to indicate the critical load for any given lake.

7. CRITICAL LOAD ESTIMATES AND UNCERTAINTY

Critical load estimates are subject to uncertainties that arise from their method of calculation and from the accuracy of the input data. For example, empirical determinations have the virtue of simplicity and of emphasizing dose-response relationships but ignoring time related aspects; dynamic models take account of time in the response in making some of the calculations. In both methods different criteria may be adopted: the critical value for lake water in relation to adverse threshold effects on biota varies between pH 5.3 and 6.0. The outcome of such uncertainties can be seen in comparisons of critical loads arrived at for catchments using different methods of estimation (Table 1).

TABLE 1. *Comparison of critical load estimates for a range of catchments.*

Catchment	Aquatic (Aq)/ terrestrial (Te)	Critical load estimate (keq km^{-2} yr^{-1})	Method	Reference
Gårdsjön (S.W. Sweden)	Aq	0-5	Ion budget	Sverdrup and Warfvinge (1988)
	Aq	20	Process model	Wright, Kämäri and Forsius (1988)
	Aq	8-38	Process model	Sverdrup, de Vries and Henriksen (1989)
	Aq	8	Empirical model	Sverdrup, de Vries and Henriksen (1989)
	Te	0-50	Ion budget	Sverdrup and Warfvinge (1988)
	Te	10-33	Ion budget	Nilsson (1986)
	Te	20	Process model	Sverdrup, de Vries and Henriksen (1989)
	Te	15-40	Process model	Sverdrup and WSarfvinge (1988)
Sogndal (Norway)	Aq	0	Ion budget	Sverdrup and Warfvinge (1988)
	Aq	21	Process model	Wright, Kämäri and Forsius (1988)
	Te	5-55	Ion budget	Sverdrup and Warfvinge (1988)
Birkenes (S. Norway)	Aq	25	Process model	Wright, Kämäri and Forsius (1988)
	Aq	4	Process model	Sverdrup, de Vries and Henriksen (1989)
	Te	25-44	Ion budget	Nilsson (1986b)
Woods Lake (USA)	Aq	20	Process model	Kämäri (1986)
	Aq	0-70	Ion budget	Sverdrup and Warfvinge (1988)
	Te	15-46	Ion budget	Nilsson (1986b)
	Te	0-45	Ion budget	Sverdrup and Warfvinge (1988)
	Te	25-35	Process model	Sverdrup and Warfvinge (1988)
Kloten (C. Sweden)	Aq	0	Ion budget	Sverdrup and Warfvinge (1988)
	Te	2-28	Ion budget	Nilsson (1986b)
	Te	0-20	Ion budget	Sverdrup and Warfvinge (1988)
Kullarna (C. Sweden)	Aq	0-55	Ion budget	Sverdrup and Warfvinge (1988)
	Te	15-85	Ion budget	Sverdrup and Warfvinge (1988)
	Te	4-6	Ion budget	Nilsson (1986b)

8. MAPPING CRITICAL LOADS

Once critical load estimates are available it may be of interest to produce critical load maps. It is possible that such maps will be particularly useful in comparison with rates of acid deposition, in order to see the balance between the two and also where most effort is required to limit and control depositions so that ecosystems (and eventually buildings and structures, and human health) can be protected. In fact, as already indicated, such activities may form the basis of linking fuel use and other acid emitting processes, S and N emissions, atmospheric transport and deposition to cost-effective abatement procedures, in a cost-optimized way, to produce co-ordinated abatement strategies.

Critical load maps of acid deposition may be generated by calculation methods or by surrogate methods that attempt to assess the sensitivity of ecosystems to acidic depositions from readily available environmental information.

8.1. Calculation Methods

Three different approaches to the calculation of critical loads, for the purpose of mapping, have recently been recommended (Sverdrup, de Vries and Henriksen, 1989). They are essentially the methods already described and comprise:

i) the Steady State Water Chemistry Method;

ii) the Steady State Mass-Balance Method;

iii) the Dynamic Modelling Method.

The three methods differ significantly in the data requirements.

8.1.1. The steady state water chemistry method

This applies to lakes and water courses. The pre-acidification base cation concentration is calculated using the present base cation concentration, site-adjusted employing an empirical F-factor, and an estimate of the background sulphate concentration is combined with this to obtain the original lake alkalinity. The deposition of strong acid anions that cause the original alkalinity to fall to a chosen alkalinity value is the site critical load.

8.1.2. The steady state mass-balance method

This method may be used for soils, and for groundwater and surface runoff water from catchments. It identifies the mean, long-term sources of acidity and alkalinity in the system and then determines the maximum acid input that will bring about a balance that is biologically 'safe'. Weathering rates, biomass acidity input, acid inputs from nitrogen transformations and alkalinity outflux are estimted. Models like MACAL referred to above can be used or more simple mass balance calculation performed.

8.1.3. Dynamic modelling

Methods usirg dynamic modelling incorporate time (rate factors) into the estimation of critical loads. They are applicable to soils, ground and surface waters, and to episodic events. There are quite a large number of models. Data on up to 30 parameters may be needed to run the model. This data will need to be obtained from field measurements or from site characteristics using site functions. These models attempt to make use of an understanding of processes in the system and should, when employed in a satisfactory manner, give the most realistic and reliable critical load estimates.

8.2. Other Methods

Progress has been made in mapping calculated critical loads: in Norway (Henriksen, 1988) and in Sweden (Sverdrup and Warfvinge, 1988). However, abatement strategy models being developed as a tool for investigating optimum strategies for emission abatement require information for a whole region so that target loads can be identified.

As an interim measure, an assessment of the relative sensitivity of ecosystems to acidic depositions might prove useful so that tentative critical load values could be applied to the relative sensitivity classes and form the basis of target loads that could be used in abatement strategy models. One such attempt at mapping the relative sensitivity of ecosystems to the indirect effects of acidic depositions is now described.

9. MAPPING THE RELATIVE SENSITIVITY OF ECOSYSTEMS

An assessment of the relative sensitivity of ecosystems to acidic depositions can be obtained from environmental site factors which are most likely to influence the response of the ecosystems to those depositions. Use of a limited number of factors, which are readily accessible, give a relatively simple and widely applicable assessment amenable to mapping in reasonable detail over large areas. For example, Europe can be mapped using bedrock lithology, soil type, land use and mean annual rainfall statistics. When these factors are combined using weights, the resulting sensitivity refers to the likely effect of acidic depositions on ecosystem function and structure.

9.1. Environmental Factors

Within each factor a limited number of categories are distinguished according to the way in which they might affect sensitivity. Rock type is assigned to one of two categories depending on the weathering rate, reflecting the importance of mineral weathering in the buffering of acidic inputs to the system. Soil type is assigned to one of two categories dependent on the physical and chemical properties which determine the likelihood of the soil chemistry entering the aluminium buffer range where most of the acid buffering is carried out by aluminium compounds in the soil and where aluminium concentrations in the soil begin to rise. Land use is assigned to one of four categories based on the effect that the vegetation has on soil formation, the interaction between the canopy and deposition and on the effect that the vegetation has on site hydrology. The amount of rainfall determines the site hydrology and also the amount of ions leaching from the soil and is assigned to one of two categories.

9.1.1. Rock type

The relative rates of weathering of different rock types has been reviewed and consideration given to the mineralogical and chemical composition of the different rock types. The resulting classification is shown in Table 2. The slow weathering rate rock types in Group A (Table 2) are designated Category I. Unfortunately, the map of Europe-wide geology shows sedimentary rocks by age and not type. This has been overcome by assigning to Group A rocks of Pre-Cambrian and Lower Palaeozoic age, which are generally slow-weathering and low in carbonates (Kinniburgh and Edmunds, 1986). Other, faster weathering rock types (Groups B, C and D) are assigned to Category II. Sites with Category I rock types have higher sensitivity than those with rock types of Category II.

TABLE 2. *The acid neutralizing ability of rock types.*

Group	Acid neutralizing ability	Rock type
A	None - low	Granite, syenite, granite-gneisses, quartz sandstones (and their metamorphic equivalents) and other siliceous (acidic) rocks, grits, orthoquartz, decalcified sandstones, some quaternary sands/drifts
B	Low - medium	Sandstones, shales, conglomerates, high grade metamorphic felsic to intermediate igneous, calcsilicate gneisses with no free carbonates, metasediments free of carbonates, coal measures
C	Medium - high	Slightly calcareous rocks, low-grade intermediate to volcanic ultramafic, glassy volcanic, basic and ultrabasic rocks, calcareous sanstones, most drift and beach deposits, mudstones, marlstones
D	"Infinite"	Highly fossiliferous sediment (or metamorphic equivalent), limestones, dolostones

Source: Norton (1980); Kinniburgh and Edmunds (1986); Lucas and Cowell (1984).

9.1.2. Soil type

The physical and chemical soil attributes which are assumed to give an indication of whether soils are in danger of entering the aluminium buffer range are shown in Table 3. The data for typical soil profiles for all the soil types in the FAO classification are given in the book accompanying the FAO Soil Map of the World (FAO-UNESCO, 1981) and the Soil Map of the European Communities (EEC, 1985). On the basis of this information the soil types with low pH, base saturation (V), Ca content and high proportion of sand were assigned to Category I, and other soil types to Category II. Table 3 shows summary data for the two soil categories. Sites with Category I soil types are more sensitive than those with Category II soil types.

TABLE 3. *A summary of the soil data for the two soil categories.*

Soil category		pH	CEC (meq $100g^{-1}$)	V %	Sand %	Ca content (meq $100g^{-1}$)
I	Mean	4.2	23	8	61	1.52
	s.d.	0.27	9.5	2.5	21	1.7
	Range	3.8-4.5	14-33	6.13	30-94	0.1-4
II	Mean	6.7	33	57	30	18
	s.d.	1.01	39	31	26	20
	Range	4.9-8.4	2-182	7-100	5-97	0.2-100

Source: FAO-UNESCO (1981); EEC (1985).

9.1.3. Land use

Four categories of land use are taken to be the minimum number which have to be considered in the assessment of sensitivity to acidic deposition (coniferous forest; deciduous forest; intensively managedagricultural land; other, relatively unmanaged land such as heath or rough grazing).

Coniferous forest is considered to increase site sensitivity to the greatest extent. This is due to the way in which conifers confer certain hydrological characteristics to sites (Miller, 1985), and because of the characteristics of the typical acid mor organic layer formed under coniferous forest stands (Mikola, 1985). The organic layers are an important part of the plant rooting zone and substantial amounts of water reaching lakes may travel through the organic layer, particularly in high rainfall areas (Kinniburgh and Edmunds, 1986; Cresser *et al.*, 1986). Trees act as funnels for acidic rain water and acidic stem flow may cause a decrease in pH at the base of trees. The filter effect by which pollutants are efficiently scavenged from the air, increases the deposition of pollutants (Hultberg and Grennfelt, 1986; Ulrich, 1985). Coniferous trees are particularly efficient at filtering acidic pollutants from the air and as coniferous forest cover increases within catchments, so does the deposition of sulphur compounds (Hultberg, 1985).

Rough grazing and heathland vegetation produce mor humus and so this type of vegetation is also considered to increase site sensitivity, though to a lesser extent than coniferous forest as the filter effect is not so large (Hultberg, 1985) and hydrological modification not so great (Munn *et al.*, 1973).

Deciduous forest vegetation produces less acid, mull humus which has a higher decomposition rate and lower organic acid production rate than a mor humus (Mikola, 1985). Hardwoods often have deep roots which bring up nutrients from lower soil horizons leading to a certain amount of surface soil layer enrichment (Black, 1968). Sites with deciduous forest vegetation, therefore, have a lower relative sensitivity than sites with vegetation causing the production of mor humus.

It is assumed that practices such as fertilizer and lime application will artificially maintain high pH and base saturation levels on intensively managed land and reduce sensitivity accordingly (Bache, 1983).

9.1.4. Amount of rainfall

As the amount of rainfall increases, the base cation, aluminium and other acid ion leaching rates increase due to the increased flow of water in the soil. Build-up of the organic layer occurs in high rainfall sites, reducing the ability to buffer acidic inputs (Kinniburgh and Edmunds, 1986). Regions with an annual average rainfall greater than 1200 mm are categorized as having a higher sensitivity than those receiving less.

9.2. **Combining Factors**

The sensitivity of an ecosystem is the result of the combined influence of the site factors determining ecosystem response to acidic depositions. Therefore, in order to derive a relative sensitivity ranking, the factors described have been combined by addition, since this is the clearest and simplest form of combination. The relative importance of the different factors in determining sensitivity is reflected by the weights shown in Table 4 which are used in the combination.

Land use is weighted most heavily (Table 4) due to the large difference between the effect that coniferous forest and arable land has on the soil. The high weighting given to the rock type reflects the relevance of mineral weathering to the neutralization of acids. Soil is weighted less heavily since part of the effect of soil in neutralizing acids, namely the weathering of minerals, is assumed to be reflected by the bedrock lithology. The effect of rainfall is not considered to affect the sensitivity to the same degree as either mineral weathering or land use.

The combination using the weights produces eight classes of relative sensitivity to acidic depositions (0-7). In view of the restricted amount of critical load information in Europe and also in consideration of the use of

TABLE 4. *Division of site factors into categories and associated weights for use in combination.*

Factor	Weight	Category	Weighting
Rock type	2	I siliceous, slow weathering rocks	1
		II faster weathering rocks	0
Soil type	1	I major acid buffering <pH 4.5	1
		II major acid buffering >pH 4.5	0
Land use	3	I coniferous forest	1
		II rough grazing	2/3
		III deciduous forest	1/3
		IV arable land	0
Rainfall	1	I >1200 mm (annual average)	1
		II <1200 mm (annual average)	0

the map as a basis for applying these critical loads, the seven sensitivity classes have been reduced to five by merging some of the sensitivity classes. The higher sensitivity rankings were combined in pairs (2+3; 4+5; 6+7) and with Classes 0 and 1 gives sensitivity classes 1 to 5 which are shown on the map (Figure 2).

The map is designed to show that for a unit of acidic deposition aquatic and terrestrial ecosystems in the more sensitive areas will be more affected than those in less sensitive areas. It is assumed here that the sensitivity of aquatic and terrestrial ecosystems is controlled by the same factors and that the relative sensitivity will be the same in the same catchment. However, this does not mean that the absolute sensitivity will be equal for the aquatic and terrestrial ecosystems and, in many cases, the critical loads suggested for the same catchment are different and often higher for terrestrial ecosystems. The validity of the relative sensitivity map (Figure 2) will depend upon a number of assumptions:

i) that the four factors chosen account for the major sources of the variation in sensitivity;

ii) that the division of the factors is correct;

iii) that the weights applied to the factors reflect the real importance of the different factors in controlling site sensitivity;

iv) that the method of combination is the optimal way to derive a measure of site sensitivity from the controlling factors;

v) that the same methodology can be applied to the whole region.

9.3. Use of Critical Loads with the Sensitivity Map

The map produced shows relative sensitivity. Target load values are assigned to the classes of relative sensitivity. The target loads are based on critical load estimates derived using the methods outlined above. Here the targets have been set at the same level as the critical loads for the ecosystems.

The target values have been obtained by comparing the critical load values for specific sites to the sensitivity class of the region in which the site lies. Many of the terrestrial and aquatic ecosystems in Scandinavia, within regions having the highest sensitivity (class 5) according to Figure 2, have been allocated critical load values of 20 keq km^{-2} yr^{-1} (Nilsson, 1986a; Nilsson and Grennfelt, 1988). This level of deposition has been designated the target for this sensitivity class. By similar comparisons targets in Table 5 have been applied to the five classes of relative sensitivity.

TABLE 5. *Target deposition levels applied to the relative sensitivity classes.*

Relative sensitivity class	Targets (keq H^+ km^{-2} yr^{-1})
1	>160
2	160
3	80
4	40
5	20

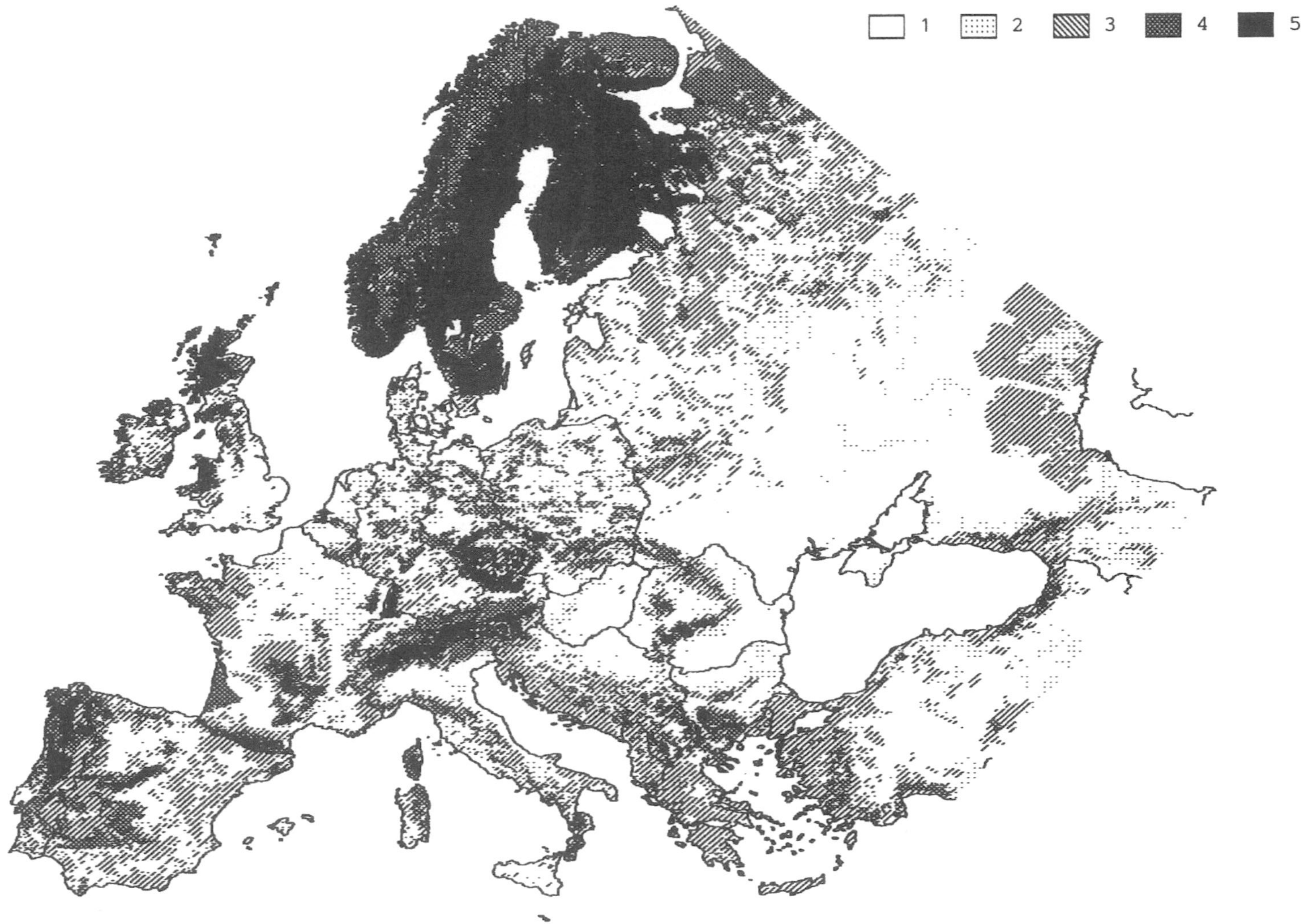

FIGURE 2. The relative sensitivity of ecosystems in Europe to acidic depositions (1, least sensitive; 5, most sensitive).

The values in Table 5 compare well with the range of critical loads suggested by Nilsson and Grennfelt (1988) for soils with different bedrock types and with minerals of different weathering rates.

10. THE CONTRIBUTION OF NITROGEN COMPOUNDS TO ACIDIFICATION

The critical load for acidification refers to the total anthropogenic deposition of acidifying substances (from sulphur sources and proportion of nitrogen leached). Sulphur deposition usually accounts for much of the acidic input to ecosystems but nitrogen may also be a major contributor to the acidification of soils and waters. Sulphur and nitrogen deposition will only acidify the soil when deposited sulphate and nitrate anions are leached. Plant requirements for sulphur are low, and the sulphate adsorption capacity of most soils in Europe is also relatively low. This means that most of the deposited sulphur will be free to leach from the soil and cause acidification. Nitrogen, however, is a major plant nutrient and is often a factor limiting growth in European ecosystems. Not all nitrogen deposited therefore leaches from sites and so only a proportion of deposited nitrogen will acidify ecosystems. It is necessary to obtain estimates of the proportion of deposited nitrogen that will be subject to leaching if the 'nitrogen component' of total acidity is to be included in the estimates.

In order to predict the degree of regional nitrate leaching it is necessary to:

i) determine the regional uptake/removal of nitrogen;

ii) determine the degree of immobilization.

The degree of acidity production (N_{acid}) due to nitrogen depositions (i.e. the amount of nitrogen leaching) may be estimated from:

$$N_{acid} = (N_{deposition} - N_{uptake}) \times (1 - N_{immobilized})$$

where $N_{immobilized}$ is the proportion of excess N immobilized in the soil and ($1-N_{immobilized}$) is the proportion of the excess of deposition over uptake that is leached.

10.1. Nitrogen Uptake

Nitrogen uptake rates depend upon biomass productivity and the optimal internal nitrogen concentration of plant species. The biomass productivity depends on climatic factors, the plant species making up the vegetation and site fertility (Tinker, 1979; Tamm, 1988; Rosén, 1988). The optimal range of internal nitrogen concentration is an inherent feature of species.

Table 6 summarizes nitrogen uptake/removal rates for four main categories of land use. The estimates are approximate due to variation from site to site within a land use category. Regional uptake rates are illustrated in Figure 3.

10.2. Immobilization of Nitrogen

Leaching of soil nitrogen not subject to uptake by plants and other organisms is controlled by factors affecting the rate of release of nitrate from organic matter (mineralization and nitrification), denitrification and solute leaching. Mineralization would seem to be affected to a large degree by the A horizon C:N ratio (Kriebitszch, 1978; Gundersen and Rasmussen,

1988). Soils with A horizon C:N ratios of >25 tend to exhibit negligible mineralization and nitrate leaching rates, with C:N ratios of 15-25 intermediate leaching rates and C:N ratios <15, high leaching rates. The pH, often quoted as an important variable affecting nitrification rates, correlates well with C:N ratio, at least on a broad scale (Figure 4).

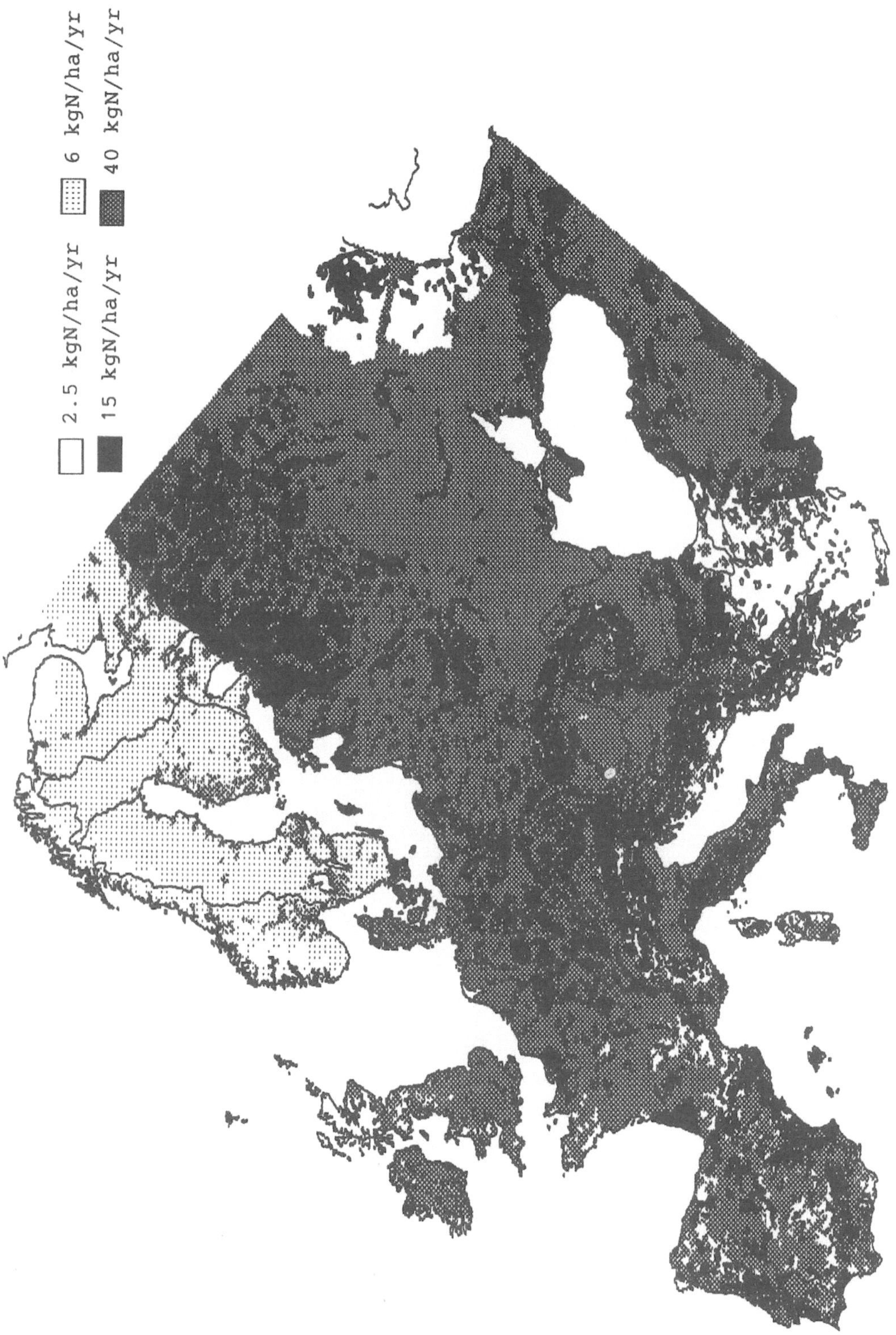

FIGURE 3. Regional nitrogen uptake rates (kg N ha^{-1} yr^{-1}).

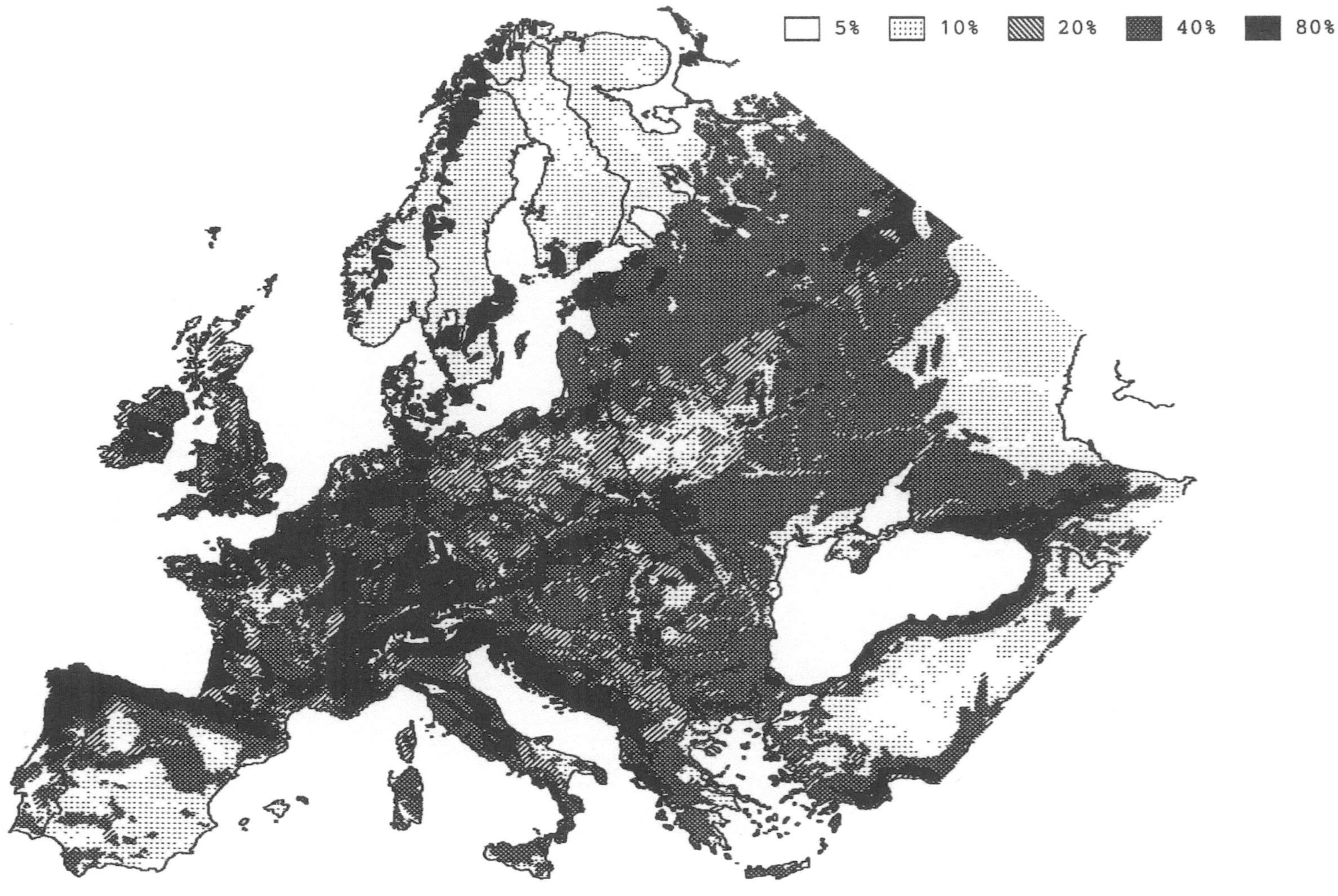

FIGURE 4. Percentage of excess nitrate leaching regionally from soil.

Soil drainage, dependent on soil texture, is a factor which affects the rate and degree of solute (e.g., nitrate) leaching. Soil moisture affects nitrate leaching rates. Dry soils leach little nitrate compared to wetter soils. The precipitation to potential evapotranspiration ratio (P:E ratio) is used as an indicator of soil moisture conditions:

TABLE 6. *Approximate nitrogen uptake removal rates (N uptake) for different land use types.*

Land use category	N removal rate ($kg\ N\ ha^{-1}\ yr^{-1}$)
Heath/rough grazing	2.5
Forest - Northern Boreal	6
- Rest of Europe	15
Arable/rich pasture	40

Source: Rosén (1988); Haynes *et al.* (1986); Gundersen and Rasmussen (1988); Tinker (1979).

i) P/E = <0.5: very dry soils;

ii) P/E = 0.5 - 1.0: moderately moist soils;

iii) P/E = >1.0: moist soils.

Steep topography increases the rate of leaching. Altitude may be taken as a crude substitute for topography. Regions with an altitude >1000 m are considered to have steep topography compared with regions of an altitude <1000 m.

Table 7 shows the categories of the main factors, together with estimates of the degree to which the different categories affect nitrate leaching. The combined effect of the factors on nitrate leaching is considered to be a multiplicative. For example, sites with a C:N ratio of >25 are considered to only leach 10 per cent of the available nitrate in the soil, the rest being immobilized. If this site also is poorly drained, then only 50 per cent of the 10 per cent already leaching will be leached (i.e., only 5 per cent of the available nitrate will leach). Single values have been allocated to the relevant ranges shown in Table 8. These values are used in calculations of the percentage of excess N ($N_{deposition}-N_{uptake}$) leached. The distribution of these percentages are illustrated in Figure 4.

Once N deposition is known it is possible to obtain an estimate of the regional N leaching, that constitutes the N contribution to total acidity, from the maps for N uptake (Figure 3) and the proportion (or percentage) of the excess nitrogen that leaches ($1 - N_{immobilization}$) as shown in Figure 4. The amount of acidity produced by nitrogen depositions can then be summed with that due to sulphur depositions to give a value of total deposition of acidifying substances from anthropogenic sources. The total acid load should include ammonium depositions, estimates of which can now be derived (Eliassen *et al.*, 1988; Saltbones, Sandnes and Eliassen, 1989).

TABLE 7. *The degree to which site factors affect the leaching of available nitrate. Numbers refer to the proportion of available nitrate that will be leached.*

Factor	Category	Proportion of Nitrate Leached
C:N Ratio	<15	1.0
	15-25	0.5
	>25	0.1
Drainage	Well drained	1.0
	Poorly drained	0.5
P/E Ratio	>1.0	1.0
	0.5-1.0	0.5
	<0.5	0.1
Relief	Sharp	1.0
	Shallow	0.75

TABLE 8. *Values for the proportion of excess nitrate immobilized.*

Calculated values of N immobilized	Value used in calculations	Percentage excess nitrate leached
0.00 - 0.075	0.05	5
0.075 - 0.15	0.10	10
0.15 - 0.30	0.20	20
0.30 - 0.60	0.40	40
0.60 - 1.00	0.80	80

10.3. Estimates of the Contribution of N-Depositions to Total Acidity

Data from European sources (Popovic, 1985; Buldgen and Remack, 1984) suggest that observed leaching rates range from 0.1 to 20 kg N ha^{-1} yr^{-1}. Nitrogen deposition rates of 2 to 50 kg N ha^{-1} yr^{-1} have been measured. The leaching rates estimated by the method outlined above are in the range of 0-38 kg N ha^{-1} yr^{-1}.

Popovic (1985) measured deposition rates in Central Sweden of approximately 10 kg N ha^{-1} yr^{-1} and observed leaching of about 0.1 - 0.4 kg N ha^{-1} yr^{-1}. The leaching rate calculated from the method suggested here for this area is 0.4 kg N ha^{-1} yr^{-1}. In the Haute Ardenne Buldgen and Remack (1984) measured leaching of 13 kg N ha^{-1} yr^{-1} under an approximate deposition of 35 kg N ha^{-1} yr^{-1}. The calculation made here is of leaching at 8 kg N ha^{-1} yr^{-1} so there is relatively good agreement although much more checking against measured rates is required.

11. CRITICAL LEVELS FOR DIRECT EFFECTS OF POLLUTANT GASES

11.1. The Sensitivity of Vegetation to SO_2

Various site factors alter the response of plants to SO_2, and the assessment of sensitivity to direct effects of sulphur dioxide concentrations is designed to show where damage to plants by SO_2 are most likely to occur.

SO_2 may have chronic effects on plants at low concentrations during long exposures or in high concentrations may cause acute damage after short exposure times. In certain respects these two types of damage are quite different and have to be treated in a separate manner. It is important to determine which is most relevant to effects on plants in order to predict where plants will be most sensitive.

11.1.1. Methodology

The approach used to estimate relative sensitivity to direct effects of pollutant gas concentration is to assess the environmental factors which influence the action of SO_2 on plants and from knowledge of their effect on plant response deduce the broad regional distribution of ecosystem relative sensitivity in Europe. The method used identifies a small number of the most important environmental site factors influencing plant sensitivity and divides these into a small number of categories. Weights are used to combine the controlling variables which broadly represent the degree to which the site factors influence plant sensitivity.

Site factors influencing the response of plants to SO_2. Site factors were selected on the basis of the importance of their effect on sensitivity to SO_2, the unambiguous way in which they influence sensitivity, the general availability of information on them in a Europe-wide context and susceptibility to mapping procedures, their overall integrative nature, and the relative permanence of applicability to an area.

The factors used to assess the sensitivity of plants to SO_2, which fulfil these criteria are:

i) the sensitivity of different plant species;

ii) factors affecting stomatal movement (soil moisture);

iii) soil nutrient status;

iv) Winter conditions (mean January temperature).

The presence of other gaseous pollutants is an important factor influencing the response to SO_2 but is not directly included in this sensitivity assessment.

Species sensitivity to SO_2 - Plant species vary in their sensitivity to chronic and acute effects of SO_2 uptake (Last, 1982; Guderian, 1977). The relative sensitivity of plant species has been investigated using experimental determinations and observation of impacted sites. Much of the information is based on experiments using acute doses of SO_2 and the correlation between sensitivity to acute and chronic effects is poor (Guderian, 1977).

It would appear in general that evergreen vegetation (especially coniferous trees) are more sensitive to SO_2 than deciduous trees (Guderian, 1977; Tamm and Aronsson, 1972; Jäger and Schulze, 1988; Last, 1982;

Bell, 1986). "Natural vegetation" has also been considered to be very sensitive to SO_2 and changes in the species abundance within the community may occur at low concentrations. Within each of these broad classes there is wide variation in the sensitivity of different species and this seems particularly true of agricultural species which exhibit a broad spectrum of sensitivity. It would seem, however, that many grain crops and important grass species have similar sensitivity to deciduous trees (Guderian, 1977; Last, 1982; Jäger and Schulze, 1988). A simple separation of sensitivity of different land use types to SO_2 has been derived (Table 9).

TABLE 9. *The sensitivity of land use types to SO_2.*

Sensitivity	Land use
Medium	Deciduous forest; arable or improved land
High	Coniferous forest/evergreen vegetation and natural vegetation

Factors affecting stomatal opening (soil moisture) - The clearest relationship between pollutant uptake, subsequent injury, and an environmental factor is with relative humidity and soil moisture content. Leaf cells of plants with a limited water supply will have a low turgor pressure and stomata will tend to close, reducing the dose of SO_2 to the plant. Generally, more damage has been found to occur to plants under well-watered than under dry conditions (Katz, 1937; Applegate and Durrant, 1969; Setterström and Zimmerman, 1939) and at a high rather than a low relative humidity (Norby and Kozlowski, 1982). It is assumed that soil moisture status and relative humidity are the factors which most consistently affect the uptake rate of SO_2 and that, on a broad scale, relative humidity will tend to follow soil moisture content. The annual precipitation to potential evapotranspiration ratio (P:E ratio) is assumed to be an approximate indicator of soil moisture (Wilsie, 1962). Three ranges of P:E ratio have been chosen to represent soil moisture contents in Europe: <0.5 (very dry); 0.5-1.0 (medium moisture); >1.0 (wetter soils).

Soil nutrient status - There are many experiments and field observations which indicate that SO_2 damage to plants increases under conditions of low soil nutrient status (Kisser, 1966; Zahn, 1963; Enderlein and Kastner, 1967; Cowling and Koziol, 1982). The nutrient poor, acid soils, which are assumed to increase sensitivity of regions to indirect effects of acidic depositions (Section 9.1.2), are used to show areas which will tend to be more sensitive to SO_2 depositions due to low nutrient status.

Winter conditions - The severity of conditions in Winter affects the injury caused to plants by SO_2 due to interactions between the action of SO_2 and the climatic stresses on plants and effects of shorter day length and low irradiance (Materna, 1985; Huttunen, 1981, 1985; Guderian, 1977; Freer-Smith, 1984; Bell, 1982; Davies, 1980; Davison and Bailey, 1982). In order to account for the effects of Winter conditions on sensitivity to SO_2 described, the mean January temperature is used. Areas where the mean January temperatures are less than -2.5°C are considered to be exposed to significant frost episodes and other Winter stresses, causing greater effects of SO_2 on plants.

Weighting procedures. Weights employed should be such that when the factors are combined the resultant distribution of sensitivity should relate to the observations and results from field sites. The species sensitivity and soil moisture status are considered the most important factors influencing sensitivity to SO_2 due to the large difference in the sensitivity of species and the importance of the uptake of SO_2 to subsequent injury. The weights applied to the factors are shown in Table 10.

TABLE 10. *Weights used to combine the factors influencing sensitivity to direct effects of sulphur depositions.*

Factor	Weigh	Category	Category Weighting
Soil moisture (P:E ratio)	2	<0.5	0
		0.5-1.0	0.5
		>1.0	1
Species (Land use)	2	Deciduous forest and arable land	0.5
		Evergreen forest and natural vegetation	1
Soil nutrient status	1	High status	0
		Low status	1
Winter conditions (January temperature)	1	> -2.5°C	0
		< -2.5°C	1

11.1.2. Mapping relative sensitivity to direct effects

The factors used to determine the relative sensitivity to direct effects have been digitized so that they can be combined. The land use types from the *Land Use Map of Europe* (FAO-Cartographia, 1980) and the *Types of Agriculture Map of Europe* (Kostrowicki, 1984); annual rainfall and January temperatures from the *Climatic Atlas of Europe I* (WMO-UNESCO-Cartographia 1970); soil types from the *Soil Map of the World*, Volume **V**, Europe (FAO-UNESCO, 1981) and potential evapotranspiration rates from the *Agro-ecological Atlas of Cereal Growing in Europe: agro-climatic Atlas of Europe*, Volume 1 (Thran and Broekhuisen, 1965).

Six relative sensitivity classes result from the combination of the factors using the weights given in Table 11. The use of the map as a basis for assigning target concentrations, based on critical levels, is only feasible with a reduced number of classes due to the inability to achieve many different targets in view of the broad dispersion of SO_2 and errors associated with critical level estimates. Some of the sensitivity classes have very low cover in Europe and so are not justifiable as separate entities. The allocation of the six possible classes to three is shown in Table 11 and the resulting map is shown in Figure 5.

TABLE 11. *Combination of the sensitivity classes to the three classes to which targets are applied.*

Original sensitivity class	Combined sensitivity class
1-2	1
3-4	2
5-6	3

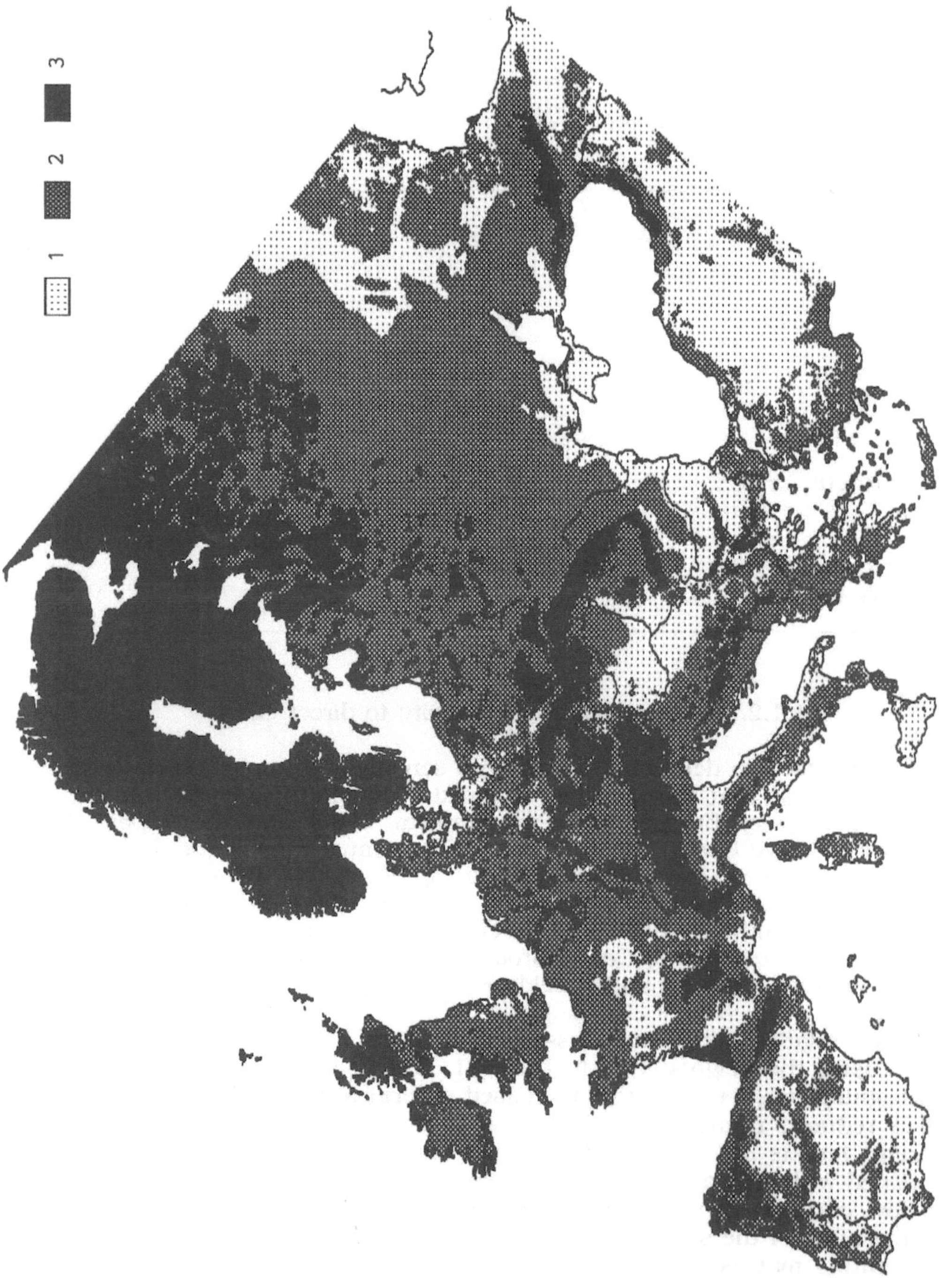

FIGURE 5. Relative sensitivity of ecosystems to SO_2 in Europe (1, least sensitive; 3, most sensitive).

11.2. **Critical levels**

As with critical loads, critical levels assume that there is a threshold concentration above which effects start to occur. The threshold concentration may change dramatically in the presence of other gaseous pollutants since interactions between gases may be synergistic. It is therefore necessary to set critical levels that take some account of the existence of other pollutant concentrations.

The experimental results on which assessments of critical levels are based tend only to specify the dose and plant species used in the experiment and do not describe the edaphic or climatic variables and in fact, are generally fumigated under optimal conditions. It is therefore difficult to collect relevant data in order to set target concentrations to relative sensitivity classes which are partly based on such factors. Many fumigation experiments use high gas concentrations over relatively short time periods. This may lead to unrealistic results since Garsed and Rutter (1984) found injury to *Pinus sylvestris* only after the second year of fumigation. Due to the form of the experiments carried out, the difficulty in reproducing experiments, and the lack of relevant information, setting critical levels can be difficult.

From experimental and observational data the critical levels for SO_2 in Table 12 have been suggested (Jäger and Schulze, 1988). These refer to SO_2 acting alone but assume that the other gaseous pollutants are kept below their respective critical levels (Table 13).

TABLE 12. *Critical levels for SO_2 acting alone.*

Species	Duration of dose	Critical level ($\mu g\ m^{-3}$)
Sensitive plants	annual	20
Agricultural and horticultural	annual	30
All	24 hour (mean)	70

11.3. **Target Concentrations**

In order to set target concentrations on the basis of critical levels it is necessary to compare the conditions typical of the various relative sensitivity classes with the conditions of the experiments or field observations used to produce critical level estimates. It is also necessary to decide how to deal with the presence of other phytotoxic gases. It is considered that targets may be set as annual means. The potential importance of fluctuating levels is stressed by Jacobson and McManus (1985) but Garsed and Rutter (1984) conclude that for regions with an annual mean of about 100 $\mu g\ m^{-3}$, the main effect on plants under fluctuating regimes is attributable to the mean rather than the peak concentrations.

TABLE 13. *Critical levels for NO_2 and O_3.*

	NO_2 ($\mu g\ m^{-3}$)	O_3
1. Acting alone		
vegetation period	60	
Winter 1/2 year	40	
24 h mean		50
8 h mean		60
4 h mean		80
2 h mean		110
1 h mean	800	150
0.5 h mean		300
2. In combination with SO_2 and O_3:		
long term	20	

Source: UN-ECE (1988).

The targets applied to the relative sensitivity classes are set in the expectation that other interacting gaseous pollutants will be present (NO_2 and O_3) and this will lower the target levels of SO_2 concentration. The target concentrations applied to the sensitivity classes are shown in Table 14.

TABLE 14. *Target SO_2 concentrations applied to the relative sensitivity classes.*

Relative sensitivity class	Target SO_2 concentration ($\mu g\ m^{-3}$ - annual average)
1	50
2	25
3	12

12. REFERENCES

Applegate, H.G. and Durrant, L.C. (1969). Synergistic action of ozone: sulphur dioxide on peanut plants. *Environmental Science and Technology*, **3**, 759-60.

Bache, B.W. (1983). The implications of rock weathering for acid neutralization. In *Ecological Effects of Acid Deposition.* Report PM 1636, pp. 175-87. National Swedish Environment Protection Board, Solna.

Battarbee, R.W. (1989). *The Acidification of Scottish Lochs and the Derivation of Critical Sulphur Loads from Palaeolimnological Data.* Research Paper No. 36. Palaeoecology Research Unit, University College London.

Battarbee, R.W., Anderson, N.J., Appleby, P.G., Flower, R.J., Fritz, S.C., Haworth, E.Y., Higgitt, S., Jones, V.J., Kreiser, A., Munro, M.A.R., Natkanski, J., Oldfield, F., Patrick, S.T., Richardson, N.G., Rippey, B. and Stevenson, A.C. (1988). *Lake Acidification in the United Kingdom 1800-1986 - Evidence from Analysis of Lake Sediments.* Palaeoecology Research Unit, University College London. HMSO, London.

Bell, J.N.B. (1982). Sulphur dioxide and the growth of grasses. In *Effects of Gaseous Air Pollution in Agriculture and Horticulture* (ed. by M.H. Unsworth and D.P. Ormrod), pp. 225-46. Butterworth, London.

Bell, J.N.B. (1986). Effects of acid deposition on crops and forests. *Experientia*, **42**, 363-71.

Black, C.A. (1968). *Soil-Plant Relationships* (2nd edition). Wiley, New York.

Brimblecombe, P., Hammer, C., Rohde, H., Ryoboshapko, A. and Boutron, C.F. (1989). Human influence on the sulphur cycle. In *Evolution of the Global Biogeochemical Sulphur Cycle* (ed. by P. Brimblecombe and A.Y. Lien), SCOPE 39. Wiley, Chichester.

Brydges, T.G. and Neary, B.P. (1984). Target loadings to protect surface waters. Ministry of the Environment, Ontario.

Buldgen, P. and Remack, J. (1984). Sulphur budget in forested watersheds in the Haute Ardenne region (Belgium). In *Belgian Research on Acid Deposition and the Sulphur Cycle* (ed. by O.L.J. Vanderborght), 163-72. SCK/CEN Mol, Belgium.

Butler, G.C. (1978). Approaches for protection standards for ionizing radiation and combustion pollutants. *Environmental Health Perspectives*, **22**, 13-15.

Chadwick, M.J. (1986). Co-ordinating economic and ecological goals. In *The Assessment of Environmental Problems* (ed. by G.R. Conway). ICCET, London.

Chadwick, M.J. and Cooke, J.G. (1987). Discussions. Towards a targetted emission reduction in Europe. *Atmospheric Environment*, **21**, 1675-9.

Cosby, B.J., Hornberger, G.M. and Galloway, J.N. (1984). Modeling the effects of acid deposition: estimation of long-term water quality response in a small forested catchment. *Water Resources Research*, **21**, 1591-601.

Cosby, B.J., Hornberger, G.M., Wright, R.F. and Galloway, J.N. (1986). Modeling the effects of acid deposition: control of long-term sulphate dynamics by soil sulphate adsorption. *Water Resources Research*, **22**, 1283-91.

Cosby, B.J., Wright, R.F., Hornberger, G.M. and Galloway, J.N. (1985). Modeling the effects of acid deposition: assessment of a lumped parameter model for soil water and stream water chemistry. *Water Resources Research*, **21**, 51-63.

Cowling, D.W. and Koziol, M.J. (1982). Mineral nutrition and plant response to air pollutants. In *Effects of Gaseous Air Pollution in Agriculture and Horticulture* (ed. by M.H. Unsworth and D.P. Ormrod), pp. 349-75. Butterworth, London.

Cresser, M.S., Edwards, A.C., Ingram, S., Skiba, U. and Peirson-Smith, T. (1986). Soil-acid deposition interactions and their possible effects on geochemical weathering rates in British uplands. *Journal of the Geological Society*, **143**, 649-58.

Davies, T. (1980). Grasses more sensitive to SO_2 pollution in conditions of low irradiance and short days. *Nature*, **284**, 483-5.

Davison, A.W. and Bailey, I. (1982). SO_2 pollution reduces the freezing resistance of ryegrass. *Nature*, **297**, 400-2.

Dickson, W. (1986). Critical loads for nitrogen on surface waters. In *Critical Loads for Nitrogen and Sulphur* (ed. by J. Nilsson), Report 1986:11, pp. 199-210. Nordic Council of Ministers, Copenhagen.

van Dijk, H. and Roelofs, J. (1988). Critical loads for nitrogen on natural vegetation. In *Critical Loads for Sulphur and Nitrogen* (ed. by J. Nilsson and P. Grennfelt), Miljørapport 1988:15, pp. 313-8. Nordic Council of Ministers, Copenhagen.

EEC (1985). *Soil Map of the European Communities* (1:1,000,000). Directorate-General for Agriculture, Commission of the European Communities, Luxembourg.

Eilers, J.M., Lien, G.J. and Berg, R.G. (1984). *Aquatic Organisms in Acidic Environments: A Literature Review*. Technical Bulletin No. 150, Department of Natural Resources, Wisconsin. Madison.

Eliassen, A., Hov, Ø., Iversen, T., Saltbones, J. and Simpson, D. (1988). *Estimates of Airborne Transboundary Transport of Sulphur and Nitrogen*, EMEP/MSC-W Report 1/88. Norwegian Meteorological Institute, Oslo.

Enderlein, H. and Kastner, W. (1967). Welchen Einfluss hat der Mangel eines Nahrstoffes auf die SO_2 Resistenz 1 jahriger Kierfen. *Arch. Forstwes*, **16**, 431-5.

Eriksson, E. (1986). Critical loads for acid deposition on groundwater. In *Critical Loads for Nitrogen and Sulphur* (ed. by J. Nilsson), Report 1986:11, pp. 71-86. Nordic Council of Ministers, Copenhagen.

FAO-CARTOGRAPHIA (1980). *Land Use Map of Europe* (1:2,500,000). Cartographia. Budapest.

FAO-UNESCO (1981). *Soil Map of the World, Volume V, Europe* (1:5,000,000). UNESCO, Paris.

Freer-Smith, P.H. (1984). The responses of six broad leaved trees during long-term exposure to SO_2 and NO_2. *New Phytologist*, **97**, 49-61.

Garsed, S.G. and Rutter, A.J. (1984). The effects of fluctuating concentrations of sulphur dioxide on the growth of *Pinus sylvestris* L. and *Picea sitchensis* (Bong.) Carr. *New Phytologist*, **97**, 175-95.

Guderian, R. (1977). Air pollution phytoxicity of acidic gases and its significance in air pollution control. *Ecological Studies*, **22**. Springer-Verlag, Heidelberg.

Gundersen, P. and Rasmussen, L. (1988). Nitrification, acidification and aluminium release in forest soils. In *Critical Loads for Sulphur and Nitrogen* (ed. by J. Nilsson and P. Grennfelt), Miljørapport 1988:15, pp. 225-68. Nordic Council of Ministers, Copenhagen.

Haynes, R.J., Cameron, K.C., Goh, K.M. and Sherlock, R.R. (1986). *Mineral Nitrogen in the Plant-Soil System.* Academic Press, London.

Henriksen, A. (1980). Acidification of freshwaters - a large scale titration. In *Ecological Impact of Acid Precipitation* (ed. by D. Drabløs and A. Tollan), pp. 68-74. SNSF, Oslo.

Henriksen, A., Dickson, W. and Brakke, D.F. (1986). Estimates of critical loads for sulphur to surface waters. In *Critical Loads for Nitrogen and Sulphur* (ed. by J. Nilsson), Report 1986:11, pp. 87-120. Nordic Council of Ministers, Copenhagen.

Henriksen, A., Lien, L., Traaen, T.S., Sevaldrud, I.S. and Brakke, D.F. (1988). Lake acidification in Norway - present and predicted chemical status. *Ambio*, **17**, 259-66.

Houdijk, A. (1988). Nitrogen deposition and disturbed nutrient balances in forest ecosystems. In *Critical Loads for Sulphur and Nitrogen* (ed. by J. Nilsson and P. Grennfelt), Miljørapport 1988:15, pp. 297-312. Nordic Council of Ministers, Copenhagen.

Hultberg, H. (1985). Budgets of base cations, chloride, nitrogen and sulphur in the acid Lake Gårdsjön catchment, SW Sweden. *Ecological Bulletins*, **37**, 133-57.

Hultberg, H. and Grennfelt, P. (1986). Gårdsjön project: lake acidification, chemistry in catchment runoff, lake liming and microcatchment manipulations. *Water, Air and Soil Pollution*, **30**, 31-46.

Huttunen, S. (1981). Seasonal variation of air pollution stresses in conifers. *Mitteilungen der Forstlichen Versuchanstalt*, **137**, 103-13.

Huttunen, S. (1985). Ecophysical effects of air pollution on conifers. In *Symposium on the Effects of Air Pollution on Forest and Water Ecosystems*, pp. 23-4. Foundation for Research of Natural Resources in Finland, Helsinki.

Jacobson, J.S. and McManus, J.M. (1985). Pattern of atmospheric sulfur dioxide occurrence: an important criterion in vegetation effects assessment. *Atmospheric Environment*, **19**, 501-4.

Jäger, H.-J. and Schulze, E. (1988). *Critical Levels for Effects of* SO_2. Working Paper for UN-ECE Critical Levels Workshop, pp. 15-50, Final Draft Report. Umweltbundesamt, Berlin.

Jones, M.D., Browning, H.R. and Hutchinson, J.G. (1986). The influence of mycorrhizal associations on Paper Birch and Jack Pine seedlings when exposed to elevated copper, nickel or aluminium. *Water, Air and Soil Pollution*, **30**, 441-8.

Kämäri, J. (1986). Critical deposition limits for surface waters assessed by a process-oriented model. In *Critical Loads for Nitrogen and Sulphur* (ed. by J. Nilsson), Report 1986:11, pp. 121-42. Nordic Council of Ministers, Copenhagen.

Kämäri, J. and Posch, M. (1987). Regional application of a simple lake acidification model in northern Europe. In *Systems Analysis in Water Quality Management* (ed. by M.B. Beck), pp. 73-84. Pergamon, Oxford.

Katz, M. (1937). Report on the effect of dilute sulphur dioxide on alfalfa. In *Trail Smelter Question*, Document Series DD, Appendix DD3. National Research Council of Canada, Ottawa.

Kinniburgh, D.G. and Edmunds, W.M. (1986). *The Susceptibility of UK Groundwaters to Acid Deposition.* Hydrogeological Report, British Geological Survey No. 86/3. British Geological Survey, London.

Kisser, J. (1966). Forstliche Rauchschaden aus der Sicht des Biologen. *Wien: Osterr. Agrarverlag. Mitt. Forstl. Bundesversuchanst. Mariabrunn,* **73**, 7-46.

Kostrowicki, J. (ed.). (1984). *Types of Agriculture Map of Europe* (1:2,500,000). Polish Academy of Sciences, Wydawnictwa Geologisczne, Warsaw.

Kriebitzsch, W.U. (1978). Stickstoffnachlieferung in sauren Waldböden Nordwestdeutschlands. *Scripta Geobotanica,* **14**, 1-66.

Last, F.T. (1982). Effects of atmospheric sulphur compounds on natural and man-made terrestrial and aquatic ecosystems. *Agriculture and the Environment,* **7**, 299-387.

Liljelund, L.-E. and Torstensson, P. (1988). Critical load of nitrogen with regard to effects on plant composition. In *Critical Loads for Sulphur and Nitrogen* (ed. by J. Nilsson and P. Grennfelt), Miljørapport 1988:15, pp. 363-74. Nordic Council of Ministers, Copenhagen.

Lucas, A.E. and Cowell, D.W. (1984). Regional assessment of sensitivity to acidic deposition for eastern Canada. In *Geological Aspects of Acid Deposition* (ed. by O.P. Bricker), *Acid Precipitation Series* **7**, *Ann Arbor.* Butterworth, Boston.

Materna, J. (1985). Results of the research into air pollutant impact on forests in Czechoslovakia. In *Symposium on the Effects of Air Pollution on Forest and Water Ecosystems,* pp. 127-38. The Foundation for Research of Natural Resources in Finland, Helsinki.

Meijer, K. (1986). Critical loads for sulphur and nitrogen deposition in The Netherlands. In *Critical Loads for Nitrogen and Sulphur* (ed. by J. Nilsson) Report 1986:11, pp. 223-32. Nordic Council of Ministers, Copenhagen.

Mikola, P. (1985). *The Effects of Tree Species on the Biological Properties of Forest Soil.* Report 3017. National Swedish Environmental Protection Board, Solna.

Miller, H.G. (1985). The possible role of forests in stream-water acidification. *Soil Use and Management,* **1**, 28-9.

Minns, C.K., Elder, F.C. and Moore, J.E. (1988). *Using Biological Measures to Estimate Critical Sulphate Loadings to Lake Ecosystems.* Department of Environment, Ontario.

MOI (1983). *United States-Canada Memorandum of Intent on Transboundary Air Pollution.* Impact Assessment Report. Ottawa.

Munn, D.A., McLean, E.O., Ramirez, A. and Logan, T.J. (1973). Effect of soil cover, slope and rainfall factors on soil and phosphorus movement under simulated rainfall conditions. *Soil Science Society of America Proceedings*, **37**, 428-31.

Nilsson, J. (ed.). (1986a). *Critical Loads for Nitrogen and Sulphur.* Miljørapport 1986:11. Nordic Council of Ministers, Copenhagen.

Nilsson, S.I. (1986b). Limits for nitrogen deposition to forest soils. In *Critical Loads for Nitrogen and sulphur* (ed. by J. Nilsson) Report 1986:11, pp. 37-69. Nordic Council of Ministers, Copenhagen.

Nilsson, J. and Grennfelt, P. (eds.). (1988). *Critical Loads for Sulphur and Nitrogen.* Miljørapport 1988:15. Nordic Council of Ministers, Copenhagen.

Norby, R.J. and Kozlowski, T.T. (1982). The role of stomata in sensitivity of *Betula papyrifera* seedlings to SO_2 at different humidities. *Oecologia*, **51**, 33-6.

Norton, S.A. (1980). Geologic factors controlling the sensitivity of aquatic ecosystems to acidic precipitation. In *Atmospheric Sulphur Deposition: Environmental Impact and Health Effects*, pp. 539-53. Ann Arbor, Michigan.

Popovic, B. (1985). Leaching of nitrate from forest soil types with different air deposition levels with and without fertilization. Nitrogen saturation. *Abstracts from a Workshop at Uppsala.* University of Uppsala.

Renberg, I. and Hellberg, I. (1982). The pH history of lakes in southwestern Sweden as calculated from the subfossil diatom flora of the sediments. *Ambio*, **11**, 30-3.

Rosén, K. (1988). Effects of biomass accumulation and forestry on nitrogen in forest ecosystems. In *Critical Loads for Sulphur and Nitrogen* (ed. by J. Nilsson and P. Grennfelt). Miljørapport 1988:15, pp. 269-93. Nordic Council of Ministers, Copenhagen.

Saltbones, J., Sandnes, H. and Eliassen, A. (1989). *Estimated Reductions of Deposition of Sulphur and Oxides of Nitrogen in Europe due to Planned Emission Reductions.* EMEP/MSC-W Note 3/89. Norwegian Meteorological Service, Oslo.

Setterström, C. and Zimmermann, P.W. (1939). Factors influencing susceptibility of plants to sulphur dioxide injury. *Contrib. Boyce Thompson Inst.*, **10**, 155-81.

Streets, D.G., Hanson, D.A. and Carter, L.D. (1984). Targetted strategies for control of acidic deposition. *Journal of the Air Pollution Control Association*, **34**, 1187-97.

Sverdrup, H.U. and Warfvinge, P.G. (1988). Assessment of critical loads of acid deposition on forest soils. In *Critical Loads for Sulphur and Nitrogen* (ed. by J. Nilsson and P. Grennfelt). Miljørapport 1988:15, pp. 269-93. Nordic Council of Ministers, Copenhagen.

Sverdrup, H., de Vries, W. and Henriksen, A. (1989). *Mapping Critical Loads: a Guide to the Criteria, Calculations, Data Collection and Mapping of Critical Loads.* UN-ECE and Nordic Council of Ministers.

Tamm, O. (1940). *Den Nordsvenska Skogsmarken.* Norrlands Skogsvårdsforbund, Stockholm.

Tamm, C.O. (1988). *Nitrogen in Terrestrial Ecosystems. Questions of Productivity, Vegetational Changes and Ecosystem Stability.* Department of Ecology and Environmental Research. The Swedish University of Agricultural Sciences, Uppsala.

Tamm, C.O. and Aronsson, A. (1972). *Plant Growth as Affected by Sulphur Compounds in Polluted Atmosphere: A Literature Survey.* Research Notes No. 12. Royal College of Forestry, Stockholm.

Thran, P. and Broekhuizen, S. (1965). *Agro-ecological Atlas of Cereal Growing in Europe. Volume I Agro-climatic Atlas of Europe.* Centre for Agricultural Publications and Documentation (PUDOC). Elsevier, Amsterdam.

Tinker, P.B.H. (1979). Uptake and consumption of soil nitrogen in relation to agronomic practice. In *Nitrogen Assimilation of Plants* (ed. by E.J. Hewitt and C.V. Cutting), pp. 101-22. Academic Press, London.

Ulrich, B. (1985). Interaction of indirect and direct effects of air pollutants in forests. In *Air Pollution and Plants* (ed. by C. Troyanowski), pp. 149-81. VCH Verlagsgesellschaft. Weinheim.

UN-ECE (1988). *ECE Critical Levels Workshop.* Bad Harzburg Workshop. Final Draft Report. Umweltbundesamt, Berlin.

de Vries, W. (1988). Critical deposition levels for ammonia, nitrogen and sulphur on forest ecosystems in The Netherlands. Background document, ECE meeting on *Critical Loads for Sulphur and Nitrogen*, Skokloster, Sweden.

de Vries, W. and Kros, H. (1989). The long-term impact of acid deposition on the aluminium chemistry of an acid forest soil. In *Regional Acidification Models* (ed. by J. Kämäri, D.F. Brakke, A. Jenkins, S.A. Norton and R.F. Wright), pp. 113-28. Springer-Verlag, Heidelberg.

Wilsie, C.P. (1962). *Crop Adaptation and Distribution.* Freeman, San Francisco.

WMO-UNESCO-CARTOGRAPHIA (1970). *Climatic Atlas of Europe* I. *Maps of Mean Temperature and Precipitation.* Cartographia, Budapest.

Wolfenden, J., Robinson, D.C., Cape, J.N., Paterson, I.S., Francis, B.J., Mehlhorn, H. and Wellburn, A.R. (1988). Use of caretenoid ratios, ethylene emissions and buffer capacities for the early diagnosis of forest decline. *New Phytologist*, **109**, 85-95.

Wright, R.F., Kämäri, J. and Forsius, M. (1988). Critical loads for sulphur: modelling time of response of water chemistry to changes in loading. In *Critical Loads for Sulphur and Nitrogen* (ed. by J. Nilsson and P Grennfelt). Miljørapport 1988:15, pp. 201-4. Nordic Council of Ministers, Copenhagen.

Zahn, R. (1963). Uber den Einfluss verschiedener Umweltfactoren auf die Pflanzenempfindlichkeit gegenüber Schwefeldioxid. *Zeitschrift Pflanzenkrankheit Pflanzenshcütz*, **70**, 81-95.

4
MITIGATION, CONTROL AND MANAGEMENT

CONTROL OF NITROGEN OXIDES FROM COAL COMBUSTION

Anna-Karin Hjalmarsson
IEA Coal Research, Gemini House, 10-18 Putney Hill, London, U.K.

ABSTRACT

NO_x emission from coal combustion can be reduced through measures during combustion, cleaning of the flue gases or a combination of both. There is currently a total installed capacity of over 150 GWe equipped with combustion measures, mainly low NO_x burners and air staging in furnace. The rates of NO_x reduction in, for example, the FRG have been 25-40%. Higher reduction rates have been obtained with advanced low NO_x systems. To achieve further NO_x reduction, flue gas treatment is necessary. The most common process presently in use is selective catalytic reductions (SCR). A capacity of 35 GWe was equipped with SCR by the end of 1989. Most SCR plants in operation are designed to achieve about 80% NO_x reduction to meet emission standards of 200 mg/m^3.

INTRODUCTION

The importance of the role of nitrogen oxides in acidification, forest damage and impacts on human health has received increased attention in recent years in many countries. This has led to the introduction of regulations on emissions. NO_x ($NO + NO_2$) emissions from coal combustion are now subject to or planned to have regulatory controls in several countries, through international, national and local agreements. Two international agreements were adopted in 1988; the EC directive covering twelve countries and the UNECE agreement covering twenty-five countries. Thirteen countries have national NO_x control regulations currently in force or agreed for implementation. The driving force behind the introduction of NO_x control measures is the regulation of NO_x emissions.

NO_x formed during combustion is mainly NO with a small part, on average less than 10%, of NO_2. The amount of NO_x formed during combustion depends on operational conditions including coal characteristics as well as combustion conditions. Some of the important parameters are the nitrogen and volatile content of the coal, combustion temperature, and the level of excess air. NO_x emission cannot easily be predicted from the nitrogen content of the coal in the way that the sulphur emissions can be predicted from the sulphur content. Different combustion systems give different levels of NO_x emissions.

Where the concentration of NO_x in flue gases after the boiler exceeds the actual emission standards, it is necessary to reduce the NO_x. Different values of uncontrolled NO_x emissions and variations in limits for NO_x emissions, mean that different measures for NO_x reduction are required depending on boiler type and fuel characteristics. These control measures may be either measures during combustion, or cleaning of the flue gases or a combination of both.

REGULATIONS LIMITING NO_x EMISSIONS

Recently, there have been significant developments in NO_x control legislation, both nationally and internationally.

Limits on NO_x emissions may vary but most countries have chosen to set standards for the concentration of NO_x emitted in the flue gases. Emission standards are based mainly on two different principles. One principle is to set the standards in order to achieve a certain reduction of emissions, assuming that a technology of the required efficiency will be developed. The other philosophy is to choose a suitable available technology, and to set emission standards at a level which that technology can achieve.

An alternative to these two major philosophies is to set a target for reduction of total NO_x emissions from a group of sources (eg utilities) and to allow the polluters, at least in theory, to choose where to take action to reach this target. This "bubble" approach has been adopted in Denmark, although constraints on emissions from individual plants may also be applied.

The way in which emission standards for NO_x are set varies in many different ways between the countries. The differences lie mainly in the categories of plants subject to limits, in the units used for emission standards, the time base over which emissions are measured and the levels of the standards.

One of the two main international agreements on controlling NO_x emissions is the EC Directive on the limitation of emissions of certain pollutants into the air from large combustion plants, which was agreed in 1988, and applies to twelve European states (Official Journal of the European Communities, 1988). The directive specifies country by country target reductions in aggregate emissions from existing combustion plants compared with 1980 emission levels. It covers combustion plants having a thermal input of more than 50 MWt, and existing combustion plants are defined as those in place before July 1987.

The reduction targets for existing plants will give an overall 10% reduction in EC total emissions by 1993 and a 30% reduction by 1998 compared to 1980 baseline figures. To achieve this total reduction for the EC, different targets have been set for different countries (see Table 1). The EC Directive also includes emission standards for new plants which must be implemented in each member country by 1990. The limits are 650 mg/m^3 (6% O_2, dry) for plants over 50 MWt or 1300 mg/m^3 for plants firing coal with volatile content less than 10%.

The other international agreement, a legally binding protocol to the United Nations Economic Commission for Europe (UNECE) Convention on Long-Range Transboundary Air Pollution, was signed in November 1988 freezing NO_x emissions at 1987 levels (or levels of any earlier year) by 1994. This covers all NO_x emissions both from stationary and mobile sources. The signatory countries are also required to use the best available technology to control emissions from both

Table 1 NO_x emissions reduction targets for existing plants over 50 MWt in the EC (Official Journal of the European Communities, 1988)

Country	1980 baseline (10^3 t NO_x)	1993 target % change	1998 target % change
Belgium	110	- 20	-40
Denmark	124	- 3	-35
FRG	870	- 20	-40
Greece	36	+ 94	+94
Spain	366	+ 1	-24
France	400	- 20	-40
Ireland	28	+ 79	+79
Italy	580	- 2	-26
Luxembourg	3	- 20	-40
Netherlands	122	- 20	-40
Portugal	23	+157	+178
UK	1016	- 15	-30
Total	3678	- 10	-30

stationary and mobile new sources. Twenty-seven member countries signed the protocol, which by the end of 1989 has been ratified by eight countries.

More than ten countries have national limits on NO_x emissions from coal combustion plants currently in force or agreed for future implementation. The main countries are Austria, Belgium, Denmark, the Federal Republic of Germany (FRG), Italy, Japan, the Netherlands, Sweden, Switzerland and the United Kingdom (UK) as well as the United States (USA). Other countries, such as Finland, have drafts or proposals for legislation. As a summary, Table 2 shows the size of plants covered by NO_x emission standards for new combustion plants as well as for existing plants and Table 3 shows the level of standards applied in g NO_x/m^3, at 5-7% O_2.

Table 2 Size of plants (MWt) covered by NO_x control regulations, currently in force or agreed for implementation (Hjalmarsson, 1990)

Countries	new	retrofit
Austria	>3	>3
Belgium	>50	none
Denmark	>50	-
EC	>50	>50
FRG	>1	>1
Italy	>100	>400
Japan	all	all
The Netherlands	all	all
Sweden	all	>30
UK	>700 >50(1)	none
USA	>73(2) >29(3)	none

Note: (1) EC requirements apply
(2) utility boilers
(3) industrial boilers

Table 3 National NO_x emission standards, currently in force or planned for implementation (Hjalmarsson, 1990)

	new plants mg/m^3	existing plants mg/m^3
Austria	200-400	200-400
Belgium	200-800	-
Denmark	650(1)(2)	(1)
EC	650-1300	-
FRG	200-500	200-1300
Italy	650	1200
Japan	410-510	620-720
The Netherlands	400-800	1100
Sweden	140	140-560
Switzerland	200-500	200-500
UK	650(2)	-
USA	615-980(3) 740-860(4)	-

Note: (1) in addition to "bubble principle" for utilities
(2) EC Standards apply
(3) industrial boilers
(4) utility boilers

MEASURES TO CONTROL NO_X EMISSIONS

Where the concentration of NO_X in flue gases after the boiler exceeds the applicable emission standards, measures to reduce NO_X emissions are required. Different levels of uncontrolled NO_X and variations in limits for NO_X emissions mean that different measures for NO_X reduction are required depending on boiler type and fuel characteristics.

The amount of NO_X formed during combustion depends on several factors such as fuel composition and combustion conditions. The first step to reduce NO_X emissions is usually to take measures at the combustion stage, to reduce NO_X formation during the combustion. Where the NO_X emission standards cannot be met by combustion control alone, flue gas treatment must be installed. In this case the NO_X already formed during combustion is removed from the flue gases. It is usually most economical to reduce NO_X as far as possible through combustion modification and thus keep the required reduction rate for flue gas treatment down.

Combustion measures

Combustion measures were introduced mainly in Japan, the USA and the FRG in the 1970s. The technologies which have been developed subsequently have led to a new generation of low NO_X systems. Combustion measures have since been extended to coal fired plants in several countries including Austria, Canada, Denmark, Finland, Italy, the Netherlands, Sweden and the UK. There is now a total installed capacity of over 150 GWe, with different types of combustion measures used to control NO_X emissions. (Hjalmarsson, 1990)

The formation of NO_X depends mainly on oxygen partial pressure, temperature and coal properties such as nitrogen content and volatile content. Measures can be taken to modify the combustion conditions so that they are less favourable for NO_X formation. There are a number of options for combustion modification measures such as reduction of combustion temperature, reduction of residence time in high temperature zones and reduction of excess air.
Different approaches can be used to reduce the NO_X emissions from the boiler such as:

* operational measures, for example reduced excess air;
* air staging
  - in burner, for example low NO_X burners
  - in furnace, for example over fire air;
* fuel staging (reburning);
* flue gas recirculation.

The implementation of low NO_X techniques in combustion is quite different for an existing boiler compared to a new boiler. Low NO_X measures on existing boilers can affect the combustion as well as the equipment. There can also be difficulties in retrofit due to limitations in the furnace such as available space, volume and so on. Combustion measures especially on existing boilers are very specific to each boiler, consequently it is difficult to transfer experience directly. It is also important to realise that even after combustion modifications the NO_X emissions do not stay at a constant level.

The most widely-used measures for controlling NO_X are reduced excess air and low NO_X burners. Air staging in a furnace is often called a two stage combustion (TSC) process, in which air to the combustion zone is reduced and additional air is introduced after the combustion zone to complete combustion. Air can be introduced at different sites in the furnace through separate openings to stage the combustion and also to prevent corrosion and slagging in the furnace. Generally, 10-20% of the total combustion air is used for staging, but the possibility of using 20-30% is now being investigated for further NO_X reduction. The most commonly used air staging method in the furnace is the addition of over-fire air. Installing low NO_X burners on an existing boiler usually involves more than just changing burners. As the new burners often have different dimensions, the installation may involve changing whole tube panels of the furnace wall. The same requirement often applies to installation of new air ports for air staging .

In fuel staging or reburning the aim is to reduce the nitrogen oxides already formed during combustion back to nitrogen. The technique involves injecting fuel into a second substoichiometric combustion zone in order to let hydrocarbon radicals from the secondary fuel reduce the NO_x produced in the primary combustion zone. This technique is now being tested using natural gas or pulverised coal as reburning fuel, however, no commercial installations are in operation yet.

The purpose of flue gas recirculation is to reduce the level of available air through dilution and to reduce the flame temperature. Flue gas recirculation can also be used to improve mixing. The success of reducing NO_x through flue gas recirculation depends on combustion conditions. In most coal fired boilers the result is minimal, especially compared to results in oil and gas fired boilers. There are exceptions, though, in special cases such as wet bottom (or slag tap) boilers.

The results of different combustion measures varies with the specific conditions. In general 25-40% NO_x reduction has been reached by combustion modification measures in the FRG, reaching 600-800 mg NO_x/m^3 at 6% O_2 on dry bottom boilers and 900-1200 mg NO_2/m^3 at 6% O_2 on wet bottom boilers, using low NO_x burners in combination with air staging in the furnace. Advanced low NO_x burners have recently been installed at some plants, where about 50% NO_x reduction has been achieved. NO_x emissions in the range of 300-600 mg NO_2/m^3 are reported from Japan with a combination of combustion measures.

New combustion technologies

There are a number of new combustion technologies where emission control is integrated into the process, the so-called advanced combustion technologies. Atmospheric fluidised bed combustion (FBC) is a technology used in an increasing number of plants and unit boiler sizes are also increasing. The combustion conditions of FBC, such as low combustion temperature, favour low NO_x formation. Similar low temperatures are found in pressurised fluidised bed combustion (PFBC). Three demonstration plants using this technology are currently under either construction or start up in Sweden, Spain and the USA; two of the plants have guaranteed NO_x emission levels of 135 and 250 mg/m^3. Measures such as ammonia addition may be required to meet the lower emission level.

Another new technology which is at demonstration scale is coal gasification/combined cycle (IGCC). The reducing atmosphere which exists during gasification means that most fuel-bound nitrogen is converted to nitrogen gas although a small fraction may form ammonia. The most important source of NO_x in IGCC is the combustion chamber of the gas turbine. The conditions here, such as intensive combustion, high excess air and high temperature, favour NO_x formation. Emissions of NO_x can be reduced by up to 80% using water or steam injection, though this may result in increased turbine water, lower combustion efficiency and higher CO emissions.

FLUE GAS TREATMENT

Where the limits on NO_x emissions cannot be met by combustion control, NO_x has to be removed from the flue gases through installation of flue gas treatment equipment. The processes presently in use for reduction of NO_x in flue gases are mainly selective catalytic reduction, selective non catalytic reduction and combined processes for sulphur and nitrogen oxide reduction. By the end of 1989, 35 GWe of coal fired plants worldwide were equipped with flue gas treatment equipment.

Selective catalytic reduction (SCR)

Up to now selective catalytic reduction has been the completely dominant method of flue gas NO_x treatment. In this method ammonia is injected in the flue gas in the presence of a catalyst, commonly titanium oxide based, to reduce NO and NO_2 to nitrogen and water. The catalyst can be placed in different positions in the flue gas flow, the important factor being that conditions such as the flue gas temperature are optimal (usually 300-400°C). The positions used for catalyst are high dust, where the catalyst is placed between the economiser and the air pre-heater; low dust, with the catalyst situated after a hot gas precipitator and before the air pre-heater; and tail end, with the catalyst situated after the desulphurisation plant (see Figure 1). The most widely used position worldwide is high dust, where untreated flue gas containing sulphur dioxide and particulates passes through the catalyst.

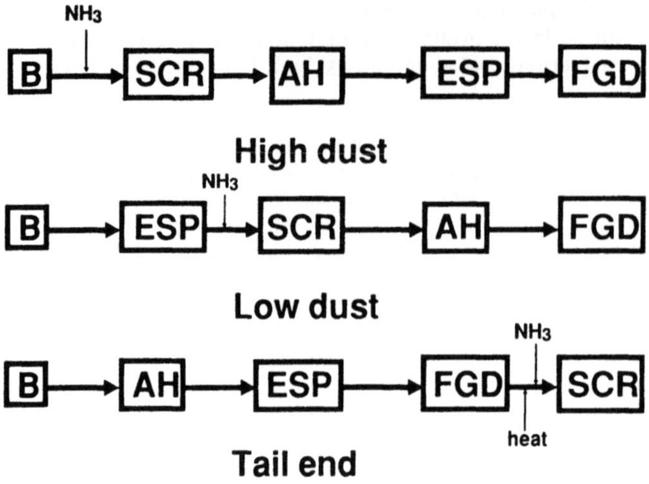

B = boiler
AH = air preheater
ESP = electrostatic precipitator (or other dust collector)
SCR = selective catalytic reduction
FGD = flue gas desulphurisation

Figure 1 Position of catalyst

The SCR technology was developed in Japan, where the first plant on a coal fired unit started to operate at the end of 1980. The first system in Europe began operating in the FRG at the end of 1985. The next country to implement this technology was Austria. A full scale demonstration plant has been operating in the Netherlands since 1987 and pilot plants have been operating in Denmark and Sweden. Tests at pilot scale are planned in the USA. At the end of 1989 there was a total installed capacity of 35 GWe equipped with selective catalytic reduction, with the majority located in the FRG (see Figure 2). Most plants are designed for a 70-80% NO_x reduction to meet emission levels of 200 mg NO_x/m^3 at 6% O_2. A few plants in the FRG are designed for reduction of emissions by over 90% where the NO_x concentration after the boiler has been particularly high.

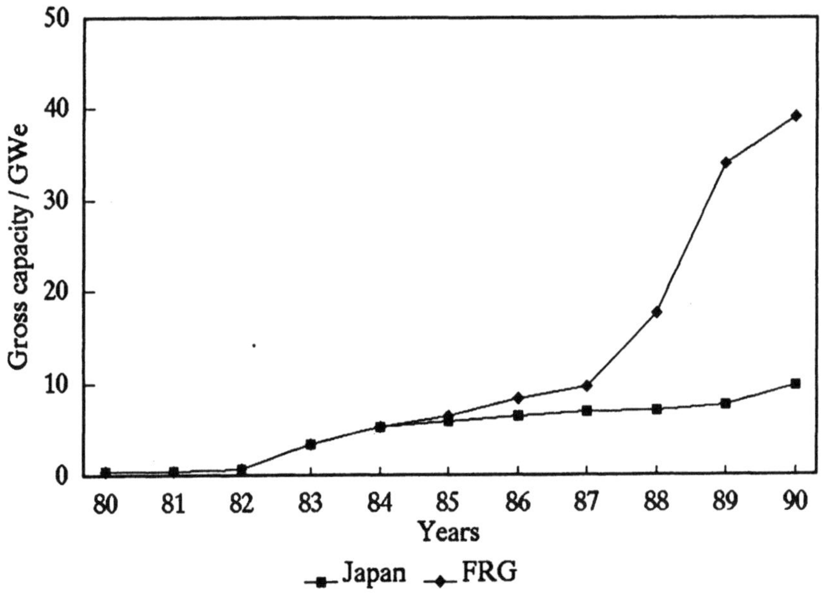

Figure 2 Total installed capacity of SCR in Japan and the FRG (IEA Coal Research NOx Installations Data Base)

Selective non catalytic reduction (SNCR)

Selective non catalytic reduction is an attractive method from the point of view that no costly catalyst is required. Nitrogen oxide can be controlled through thermal reactions by using appropriate reducing chemicals. The reaction usually occurs at temperatures of 900-1100°C. Ammonia and urea are generally used as the reducing chemicals. Methods also exist using urea plus additives where the additive enables the reactions to occur at lower temperatures. Large scale tests have been conducted using urea with or without additives, and using ammonia in concentrated form or in solution.

The method results in less NO_x reduction than selective catalytic reduction, although higher consumption of chemicals is required. There are now eight commercial installations (1700 MWe) in operation in the FRG and Austria. The first circulating fluidised bed boiler to be equipped with selective non-catalytic reduction began operating in the USA in 1988. This has been followed by other similar installations in the USA. Selective non catalytic reduction is expected to reduce NO_x by 30-50%. Higher reduction, up to 70-80%, is supposed to be attainable under favourable conditions.

Combined SO_x/NO_x processes

Numerous processes have been developed for combined desulphurisation and denitrification of the flue gases. Most processes are still at the laboratory scale and there are very few plants operating at full commercial scale. Most combined processes are considered to be complex and expensive. That opinion may change, however, as more stringent standards are introduced for both sulphur and nitrogen oxide emissions together with restrictions on residue disposal. Many combined processes produce a refined by-product, (eg sulphur, sulphuric acid).

The activated coke process is one of the processes that has reached furthest on the commercial scale. There are currently two commercial plants in operation in the Federal Republic of Germany and one in Japan. Another combined approach is catalytic conversion of both SO_2 and NO_x. Two such processes are in use, SNO_x and $DESONO_X$. Pilot and demonstration plants have been operating in both Denmark and the Federal Republic of Germany and one is planned for the USA. The first full scale SNO_x plant has been ordered in Denmark and it will begin operation in 1991 (Richter and others, 1989).

Additional processes currently planned for pilot scale operation include irradiation processes. Work using the corona effect is proceeding in Denmark and Italy, and using E-beam in Germany and Japan; Pilot plants of 5MW are planned in the USA for the copper oxide process, the NOXSO process (Halsbeck and others 1988) and the $SO_xNO_xRO_xBO_x$ process (Kitto, 1988).

REFERENCES

Haslbeck G L, Neal L G, Ma W T (1988) NOXSO demonstration gives encouraging results. Modern Power Systems, 8(7); 39-43, (July 1988)

Hjalmarsson A-K (1990) NO_x control technologies for coal combustion, London, UK, IEA Coal Research, pp 150 (in press)

IEA Coal Research (1990) IEA Coal Research FGD and NO_x control installations database. London, UK, IEA Coal Research (1990)

Kitto J B (1988) $SO_xNO_xRO_xBO_x$ uses hot catalytic scrubber. Modern Power Systems, 9(1); 24-25 (Jan > 1989)

Official Journal of the European Communities (1988) Council Directive of 24 November 1988 on the limitation of emissions of certain pollutants into the air from large combustion plants. (88/609/EEC). Official Journal L 336; pp 1-13; (7 December 1988)

Richter E, Knoblauch K, Wunnenberg W (1989) Simultaneous SO_x and NO_x removal process gain acceptance. Modern Power Systems: 9 (4); 21-25 (April 1989)

Selective non catalytic reduction (SNCR)

Selective non catalytic reduction is an attractive method from the point of view that no costly catalyst is required. Nitrogen oxide can be controlled through thermal reactions by using appropriate reducing chemicals. The reaction usually occurs at temperatures of 800-1100°C. Ammonia and urea are generally used as the reducing chemicals. Methods are based on additives where the addition enables the reactions to proceed at lower temperatures, and tests have been conducted using urea with or without additives injected in atomised form or in solution.

The method results in [illegible] reduction than selective catalytic reduction [illegible]

[illegible]

USING SCIENCE TO DEVELOP AND ASSESS STRATEGIES TO REDUCE ACID DEPOSITION IN EUROPE

R.W. Shaw
International Institute for Applied Systems Analysis
Laxenburg, Austria

ABSTRACT

The Regional Acidification Information and Simulation (RAINS) model that has been developed at IIASA can be used to assess the environmental effects of a given pattern of emissions in Europe or, given an environmental target, develop cost-effective international emission reduction strategies. In this paper, the RAINS model has been used to assess the effect of emissions from the United Kingdom at four receptor points in the United Kingdom, southwestern Norway, southern Sweden and the German Democratic Republic. The United Kingdom is, of course, the dominant contributor to sulfur deposition in the UK and is the largest national contributor (outside of the substantial contribution of background sulfur) to deposition in southwestern Norway. It is not important for deposition in southern Sweden or the German Democratic Republic.

1. Introduction

Science helps us to deal with many problems, including environmental ones, by increasing our understanding of cause-effect relationships. The problem of regional acidification from transboundary air pollution can be expressed as a chain of events, each linked by cause-effect relationships, some of which are better known than others. A cause-effect relationship can be used in two ways: in a "forward" sense, one can assess the effects of a certain input action on policy; or, in a "reverse" sense one can, after deciding what the desired effect is, calculate what input, action or policy is needed to achieve it.

This paper will deal with the Regional Acidification Information and Simulation (RAINS) model. This is a tool developed at the International Institute for Applied Systems Analysis (IIASA) that embodies scientific knowledge about cause-effect relationships for acidic deposition in what is termed an "integrated assessment model". The RAINS model can:

(i) Help one to assess the environmental effects (in terms of acidic deposition and acidification of ecosystems) of a given pattern of energy use and emissions in Europe.

(ii) Given an environmental target expressed in terms of maximum concentration or deposition of air pollutants, develop the geographical distribution of emission reductions that will accomplish the target in the most cost-effective manner.

A brief description of the model will be given, followed by examples of its use.

2. Cause-Effect Relationships in the RAINS Model

The chain of events leading to regional acidification may be expressed as follows:

(i) The burning of fossil fuels for energy leads to emissions of sulfur oxides and nitrogen oxides. The sulfur oxides are produced from the sulfur contained in the fuel; the nitrogen oxides mainly from the nitrogen in the combustion air. In addition, agriculture produces emissions of ammonia, mainly from the decomposition of animal wastes.

(ii) The above pollutants are diluted by turbulence in the atmosphere, and transported by the winds, usually within a layer whose thickness is no more than 1.5 km above the earth's surface. They are also deposited to the surface by "dry" processes (diffusion, impaction) and by "wet" processes (snow, rain and fog). However, the deposition processes are slow enough that the atmospheric lifetime of these pollutants extends to several days, allowing them to be transported across national boundaries in both Europe and in North America. In addition, during their lifetimes, the sulfur and nitrogen oxides will be transformed into sulfuric and nitric acid. In combination with other pollutants such as volatile organic compounds (VOCs), nitrogen emissions will take part in the formation of photochemical oxidants.

Upon coming into contact with sensitive receptors through impaction or deposition, the pollutants can cause serious damage through acidification of freshwater lakes (thereby harming aquatic life) or acidification of forest soils (causing leaching of nutrients from the soil, mobilization of toxic metals and a general lowering of the resistance of trees to other stresses such as drought or insects). Another effect is the direct damage of forest vegetation by gaseous pollutants such as sulfur dioxide and ozone; these direct effects are exacerbated by the weakening of the trees from soil acidification. Space here does not permit a detailed description of the processes involved, or the extent to which they have been observed but the reader is referred to, for example Nilsson and Duinker (1987) and Rosseland *et al.* (1986).

Following the above three-link chain of events, Figure 1 shows a simplified flow chart for the RAINS model. A complete description of the model, including a more detailed flow chart may be found in Alcamo *et al.* (1987 and 1990).

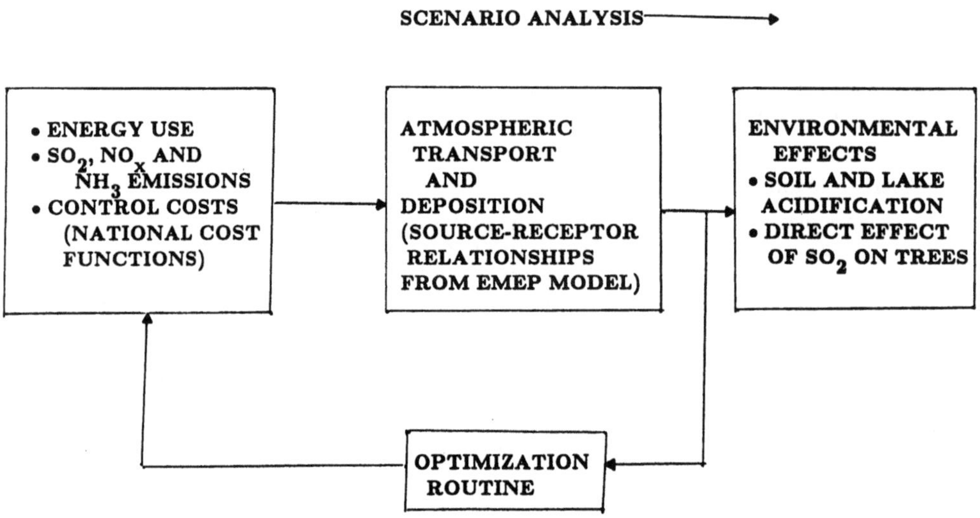

Figure 1: Simplified flow chart for the RAINS model.

(a) Energy, Emissions and Costs of Control

The left-hand box in Figure 1 represents the Energy/Emission/Costs (ENEM) submodel of RAINS (Amann, 1990). In it are data bases for energy use, and sulfur and nitrogen dioxide emissions for the 27 countries of Europe, broken down by economic sector (power plants, industrial emissions, transportation, etc.) and by fuel type (coal, oil, etc.). Sulfur dioxide emissions are calculated on a mass balance basis; nitrogen dioxide emissions on the basis of a regression analysis between fuel use and emissions (Lübkert, 1987). Ammonia emissions are taken from Buijsman *et al.* (1987).

The ENEM submodel also contains various control steps for reducing emissions of sulfur and nitrogen dioxides, and the costs of doing so. Such control steps include flue gas desulfurization and fuel switching for reducing SO_2 emissions, and catalytic converters and low NO_x burners for reducing NO_x emissions. There are as yet no control steps for ammonia emissions. In each country, the control steps can be arranged in order of increasing marginal cost (cost per tonne of pollutant removed) to produce a National Cost Curve (Amann and Kornai, 1987). An example, for SO_2 control in the year 2000 in the United

Kingdom, is shown in Figure 2. The solid line in Figure 2 represents the marginal cost, which increases in a discrete, stepwise fashion from the right hand point on the curve, located at the unabated emissions of 4200 kilotonnes SO_2. The marginal costs for the first control steps are about DM 700 per tonne SO_2, for the use of low sulfur fuel in power plants. The marginal costs increase until they are greater than DM 30,000 per tonne SO_2, for regenerative flue gas desulfurization in refineries and industrial boilers. The dashed curve represents the accumulated control cost (in DM per year, capital and operating costs) starting of course, at zero for the unabated emissions and increasing to 10 billion DM per year at the upper left hand point of the curve which represents the application of Best Available Technology to reduce the emissions in the United Kingdom to 600 kilotonnes SO_2 per year.

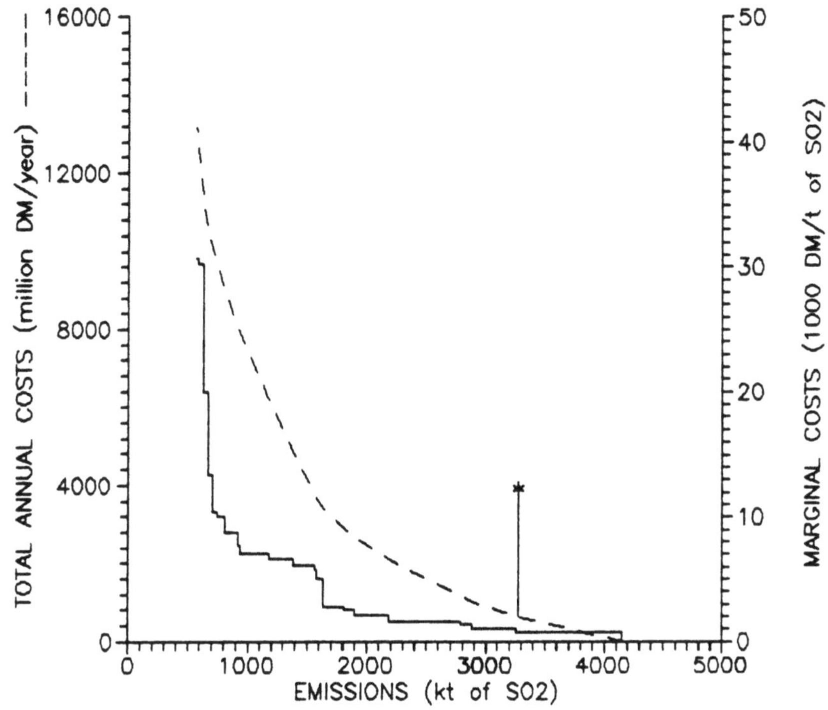

Figure 2: National Cost Curve for SO_2 emission control in the United Kingdom in the year 2000. Solid curve represents marginal cost; dashed curve the total accumulated cost. Right-hand end points of curves represent no controls, the left-hand end points represent the application of best available control technology.

As will be seen later, the National Cost Curves, in conjunction with the atmospheric source-receptor relationships to be discussed next, are key elements in the development of cost-effective strategies to reduce acidic deposition in Europe.

(b) Atmospheric Transport and Deposition

The emissions (abated or unabated) that are calculated by the ENEM submodel must be transformed into concentration and deposition fields over Europe, as depicted by the middle box of Figure 1. On a case-by-case basis, it would be possible to do this by running a full-blown long range atmospheric transport and deposition model. However, these models are time-consuming and costly to run and, because RAINS is meant for practical use with a rapid turn-around time, it was decided instead to store pre-processed source-receptor relations as follows: the deposition D_j at a receptor point j is expressed as

$$D_j = \sum_{i=1}^{27} t_{ij} Q_i. \tag{1}$$

Equation (1) says that the deposition at j is a sum of contributions from the 27 countries of Europe; each contribution is proportional to the emission rate Q_i, in country i. The constant of proportionality t_{ij} is determined solely by the strength of the atmospheric linkage between source country i and receptor point j. The use of this scheme implies that the processes of atmospheric chemistry and deposition are linear, i.e. they are proportional to atmospheric concentrations. This assumption appears to be valid for relatively large areas and long averaging times.

In RAINS, the constants t_{ij} are calculated by the long range atmospheric transport model of the Norwegian Meteorological Institute (Eliassen and Saltbones, 1983), known as the "EMEP" model, using a 150 × 150 km receptor grid in Europe (for a total of 546 land-based receptor points), and meteorology averaged over eight years. This means that the quantities t_{ij} form the elements of a 27 × 546 matrix; calculating deposition at each of the 546 receptor points is a simple matter of multiplying the emission vector of 27 values through the matrix. It follows that assessing the effect upon deposition of an altered emission scenario is equally simple.

The source receptor matrix t_{ij} is also useful in determining cost-effective emission reduction strategies in Europe. If one wants to reduce deposition at receptor point j, with the least reduction of total emissions, one should reduce emissions first in those countries for which there is a large value for t_{ij}, i.e. which have strong atmospheric links to receptor j. If one wants to also reduce *costs* to a minimum in reducing deposition at a given point, one should combine the use of the source-receptor matrix with the National Cost Curves described above. This is done in the Optimization Submodel of RAINS which is shown at the bottom of Figure 1. This submodel (which is described in Batterman *et al.* 1988 and Batterman 1988) will calculate optimized country-by-country emission reductions to meet specified targets, while minimizing either total sulfur removed, or costs.

(c) Ecological Effects

The RAINS model contains three submodels to calculate the ecological effects of a given deposition pattern: the Forest Soil Acidification submodel; the Lake Acidification submodel and the Direct SO_2 Forest Impact submodel. The Forest Soil Acidification submodel calculates, over a 1° longitude by 0.5° latitude grid over Europe, the change in pH or base saturation in forest soils with time due to a given deposition scenario. Several buffering reactions in the top 0.5 m of the soil are considered (Kauppi *et al.* 1990).

The lake acidification submodel will calculate the change in the distribution of lake pH in each of several regions in Norway, Sweden and Finland (Kämäri *et al.* 1990). It is based upon a single lake acidification model incorporating meteorological, hydrological, soil chemistry and lake chemistry processes; this model is regionalized using a Monte Carlo procedure.

The submodel predicting the risk of damage to forest vegetation from SO_2 gas in the ambient air is based upon a dose-response relationship observed in the Erzegebirge region of Czechoslovakia (Mäkelä and Schöpp, 1990). An accumulated dose of mean annual SO_2 concentration (calculated by the RAINS atmospheric transport submodel) is compared to a critical dose which in turn depends upon factors such as growing degree days.

Space in this paper does not permit a thorough discussion of the results of the effects submodels; one example will be given of the output of the soil acidification submodel. Instead, emphasis will be placed upon deposition fields since the environmental targets governing control strategies are usually based upon deposition. At present the United Nations Economic Commission for Europe, under its Convention on Long Range Transboundary Air Pollution, is developing critical loads for sulfur and nitrogen which Nilsson and Grennfelt (1988) define as: "a quantitative exposure to one or more pollutants below which significant harmful effects on specified sensitive elements of the environment do not occur according to present knowledge". These critical loads will be used in conjunction with the RAINS model to develop new sulfur and nitrogen emission protocols under the Convention.

In this paper, emphasis will be placed on sulfur deposition (although nitrogen deposition is acknowledged as an important cause of acidification), because the parts of the RAINS model dealing with nitrogen emission control strategies are still under development. These are expected to be completed at the end of 1990.

3. Using the RAINS Model to Develop and Assess Strategies

(a) Scenario Analysis

Figure 3 shows the expected sulfur deposition pattern in 1995 over a large part of

Europe resulting from emissions in all European countries, as calculated by the RAINS model. (The model can produce similar maps for all of Europe or for individual countries.) For consistency with calculations to be discussed later, the map in Figure 3 assumes that none of the presently committed reductions have taken place.

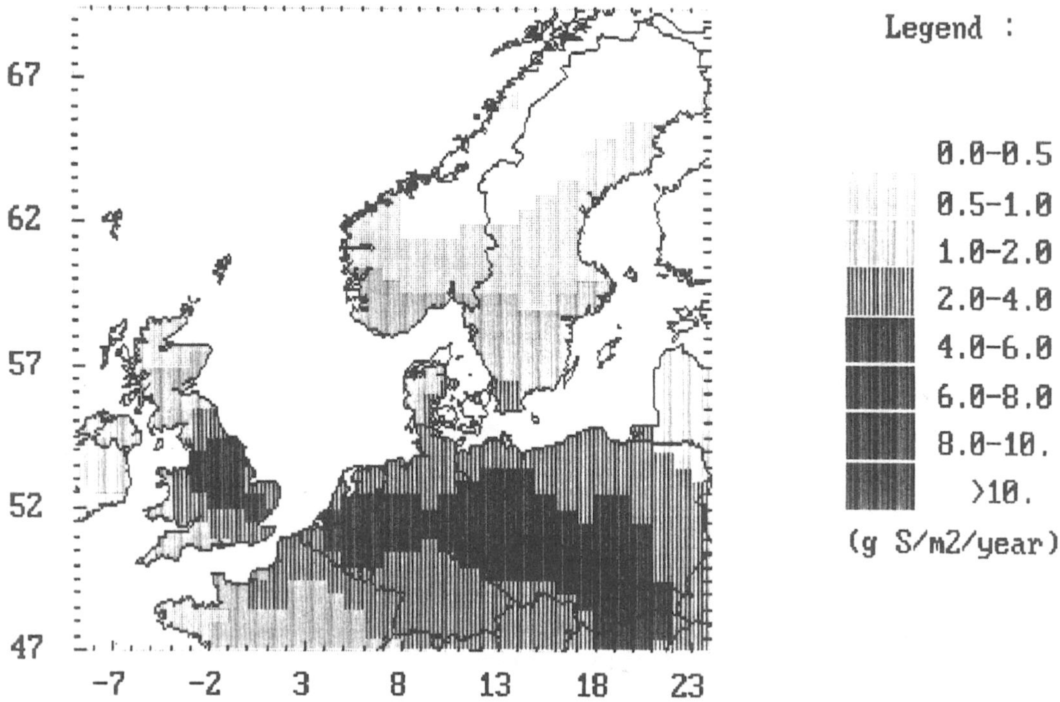

Figure 3: Sulfur deposition (g S m^{-2} a^{-1}) in northwestern Europe in the year 1995, resulting from emissions in all countries and assuming that none of the presently committed emission reductions are implemented.

There is a band of relatively high deposition (greater than 4 g S m^{-2} a^{-1} extending from the United Kingdom through to Central Europe. The region of deposition greater than 0.5 g S m^{-2} a^{-1} extends over most of the map, and even over the southern half of the Scandinavian peninsula, where there are sensitive bodies of fresh water which may be harmed by deposition rates greater than 0.5 to 1.0 g S m^{-2} a^{-1}.

What part does the United Kingdom play in this deposition? One advantage of the RAINS model is that one can examine the contributions to acidic deposition by individual countries. First, Figure 4 shows a map of the emission flux in the United Kingdom, with the greatest emission density centered on the Midlands. The total emission from the United Kingdom (in the absence of the committed reductions) is expected to be about 2100 kt S, about 7.7% of the European total of about 27,100 kt S a^{-1}. (Even with the currently committed reductions, the United Kingdom will still contribute 7.6% of the total European emissions.)

Figure 5 shows how the emissions are transported by the atmosphere and deposited both within the United Kingdom and beyond its borders, resulting in contributions of greater than 0.5 g S m^{-2} a^{-1} to the total deposition fluxes as far away as the Federal Republic of Germany, Denmark and southwestern Norway.

Table 1 shows the expected percentage contribution by each European country to deposition at four locations: central United Kingdom, the southwest coast of Norway, southern Sweden (Skåne) and the southern German Democratic Republic.

As expected, in the central United Kingdom, emissions from the UK contribute about 91% of the deposition, according to the RAINS model. Other sources are unimportant. Other RAINS calculations indicate that, of the 598 kt S that would be deposited in the United Kingdom in 1995 in the absence of controls, about 78% will be due to domestic sources. (Under the current reduction plans, that percentage would remain unchanged.) That means that control strategies to reduce deposition in the United Kingdom must begin at home.

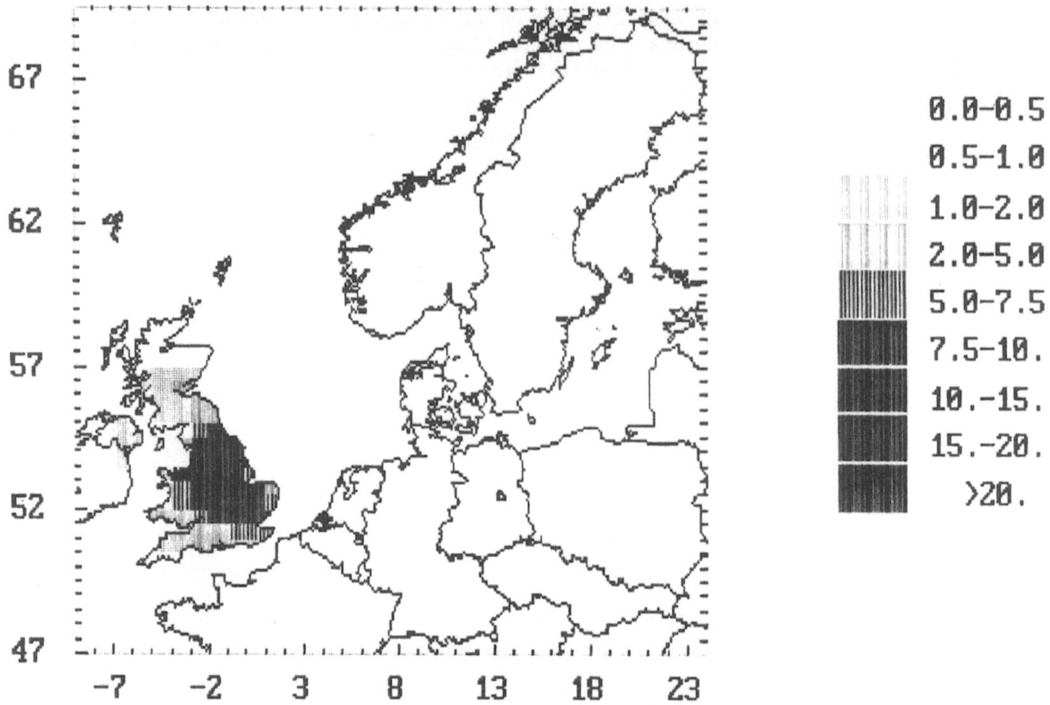

Figure 4: Sulfur emission flux (g S m^{-2} a^{-1}) in the United Kingdom in 1995, assuming no emission reductions are implemented.

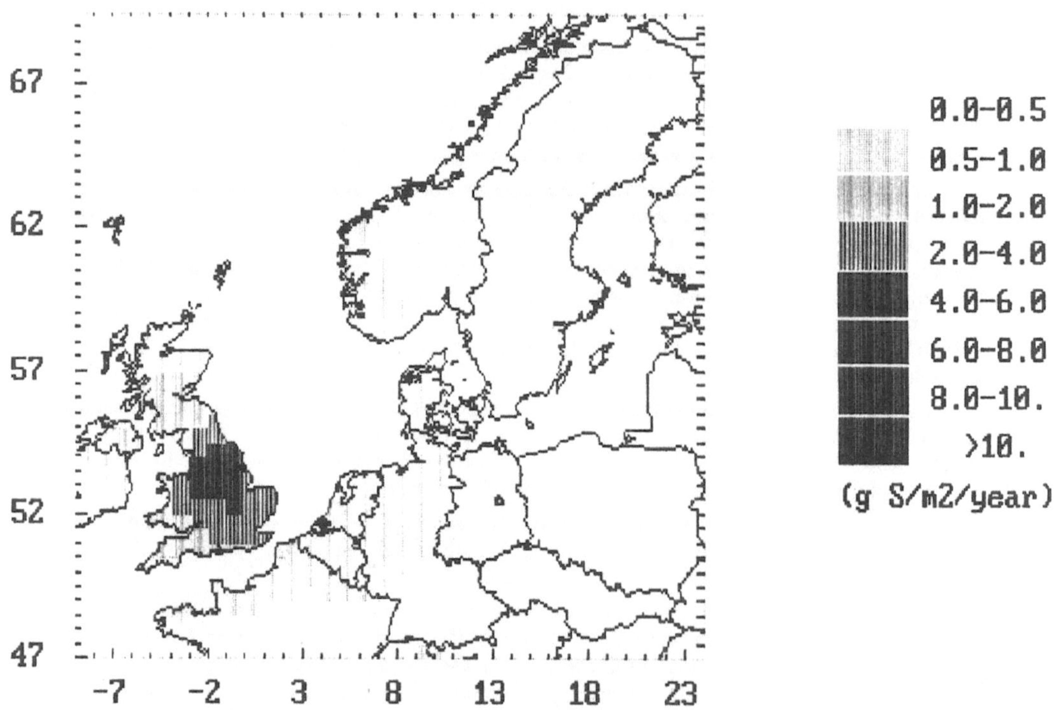

Figure 5: As in Figure 2, but for emissions from the United Kingdom only.

Table 1: Percentage contribution of each European country to deposition at four receptor points in 1995, as calculated by the RAINS model assuming none of the committed controls are implemented.

	Receptor Point			
Source	United Kingdom[1]	S.W. Norway[2]	S. Sweden[3]	S. GDR[4]
Albania	0	0	0	0
Austria	0	0	0	0
Belgium	0	1	1	1
Bulgaria	0	0	0	0
Czechoslovakia	0	3	5	25
Denmark	0	4	16	0
Finland	0	0	0	0
France	1	2	2	1
Fed. Rep. Germany	1	6	11	4
German Dem. Rep.	1	12	22	60
Greece	0	0	0	0
Hungary	0	1	1	1
Ireland	1	1	0	0
Italy	0	0	1	0
Luxembourg	0	0	0	0
Netherlands	0	1	1	0
Norway	0	3	0	0
Poland	0	4	9	5
Portugal	0	0	0	0
Romania	0	0	0	0
Spain	0	0	0	0
Sweden	0	1	9	0
Switzerland	0	0	0	0
Turkey	0	0	0	1
United Kingdom	91	17	7	0
USSR	0	1	1	0
Yugoslavia	0	1	1	0
Background	5	42	13	2
Total Deposition ($g\ S\ m^{-2}\ a^{-1}$)	5.3	1.2	2.0	12.0

1. Latitude 54.0°N, Longitude 1.0°W
2. Latitude 59.0°N, Longitude 7.0°E
3. Latitude 56.0°N, Longitude 14.0°E
4. Latitude 51.0°N, Longitude 14.0°E

On the southwest coast of Norway, Table 1 shows that the United Kingdom is the largest national contributor to sulfur deposition there at 17%. However, the background deposition (from non-European sources and European emissions, unattributable to a particular country) account for over 40%. Background deposition is a result of emissions which have resided in the upper troposphere for a considerable time and have re-entered the surface mixed layer. The German Democratic Republic is the second largest national contributor at 12%. If one considers all of Norway, the United Kingdom would contribute about 9% of the 194 kt S that would be deposited in 1995 in the absence of controls. (With the committed controls, the percentage is expected to be 10%.) In summary, to reduce deposition in Norway, emission reductions in the United Kingdom are an important, but not sufficient step. Fortunately, as will be seen the deposition in Norway does not need to be reduced very much.

In Skåne in southern Sweden, the United Kingdom is expected to contribute only about 7% of the total deposition of 2 g S m^{-2} a^{-1} from all sources (Table 1). Denmark, the Federal Republic of Germany, the German Democratic Republic contributed more; their total contribution is about one-half of the total. The RAINS model calculates that, with respect to deposition in Sweden as a whole, the United Kingdom would contribute about 6% of the total (7% with the currently committed reductions).

At the receptor point in the southern German Democratic Republic, the deposition is about 12 g S m^{-2} a^{-1}, of which the United Kingdom contributes only 1%. The GDR itself contributes about 60%; and Czechoslovakia 25%. In the GDR as a whole, the GDR emissions would contribute 62% of the 804 kt S annual deposition; the next largest contributors are Czechoslovakia and the Federal Republic of Germany at 12 and 11%, respectively. As is the case in the United Kingdom, measures to reduce deposition in the GDR must begin with domestic sources.

(b) Optimization - Development of Control Strategies

As described in Section 2, the calculation of optimized, cost effective emission reductions to meet specific targets makes use of the source-receptor relations, some examples of which were given in Section 3(a), and the national cost curves which describe the costs of emission controls in each country. Table 2 shows, for each country, the percentage emission reductions from uncontrolled 1995 emissions, the total European cost, and each country's share of it, that are needed to reduce emissions, in a cost-effective manner, to the specified deposition target at each of the four receptor points discussed in Section 3a.

To reduce deposition at the receptor point in the United Kingdom to a target value of 1.5 g S m^{-2} a^{-1}, emission reductions in only the United Kingdom are required, costing about 4 billion DM per year. However, as the deposition target is lowered to 1.0 g S m^{-2} a^{-1}, expenditures for emission reductions are also needed in other countries, especially the Federal Republic of Germany and the German Democratic Republic. In the last two countries, the cost of emission reductions is 30% of the total European cost; not surprisingly, those in the United Kingdom comprise 45% of the total. The reason for emission reductions being required outside the United Kingdom is that domestic emissions in the U.K. are lowered to the minimum value, i.e. that specified by the application of the best available abatement technology; emission reductions are then sought in other countries with the strongest meteorological linkages to the United Kingdom, and/or the lowest marginal costs.

The effect upon forest soil acidification in the United Kingdom of no controls, the currently committed reductions and of reducing deposition in the United Kingdom is shown in Figure 6. Each of the four curves shows the change with time of percentage area of acidified forest soil, a threshold of pH = 4.0 was chosen below which the soil is assumed to be acidified. (The curves are jagged because the minimum resolution of the soil model is 1° longitude by 0.5° latitude, resulting in the counting of a relatively small number of grid squares in the U.K.).

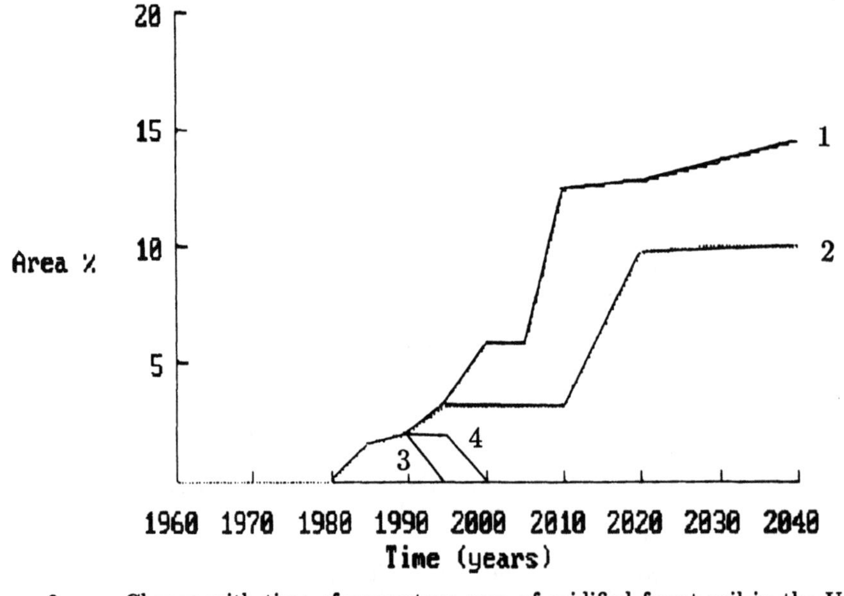

Figure 6: Change with time of percentage area of acidified forest soil in the United Kingdom (with pH < 4.0) for several emission scenarios: Curve 1, no controls; Curve 2, presently committed emission reductions; Curve 3, reduction of deposition in the central United Kingdom to 1.5 g S m^{-2} a^{-1} by the year 1995; Curve 4 as in Curve 3 but with a five-year delay.

Curve 1 shows the effect of no emission controls; the percentage area of acidified forest soil will increase to about 15% by the year 2040 according to the RAINS model. The effect of the currently committed emission reductions, in the United Kingdom about 30% from 1980 values, will reduce this percentage to about 10% (Curve 2). Curve 3 shows the effect of reducing deposition at the U.K. receptor point to 1.5 g S m^{-2} a^{-1}. The model calculates that the forest soil in the United Kingdom will recover very quickly from acidification, not surprisingly considering the 80% emission reduction there. Because an 80% emission reduction in the U.K. by 1995 is certainly impossible, curve 4 shows the effect of a five-year delay.

It would appear from Figure 6 that, according to the RAINS model, it is not necessary to reduce deposition in the U.K. to 1.0 g S m^{-2} a^{-1} (at a total European cost of four times that to reduce it to 1.0 g S m^{-2} a^{-1}) to reverse soil acidification there.

Table 2 shows that no emission reductions are required anywhere to reduce deposition on the southwest coast of Norway to 1.5 g S m^{-2} a^{-1}. To reduce the deposition to 1.0 g S m^{-2} a^{-1}, emission reductions are required in the German Democratic Republic, the United Kingdom, Czechoslovakia, Norway, Denmark and Poland, in order of national percentage of the total European cost of 3.8 billion DM per year. Substantial percentage emission reductions (from unabated 1995 levels) are required in the German Democratic Republic, Denmark, and Norway; the last two because of their relative proximity to the receptor site, the first because of the relatively low marginal costs for sulfur emission control there. Note that emissions in Norway are not lowered to the minimum possible value but only by 45%; after that point it is more cost-effective to reduce emissions elsewhere.

Table 2. Columns (1) show the percentage emission reductions from 1995 value in each country that are needed to reduce deposition to the target value at the same four receptor points as in Table 1. Columns (2) show the percentage in each country of the total European cost.

	Receptor Point													
	United Kingdom				Norway coast				Southern Sweden				German Dem. Rep.	
= Target (g S m^{-2} a^{-1})	1.5		1.0		1.5		1.0		1.5		1.0		3.0	
Emittor Country	(1)	(2)	(1)	(2)	(1)	(2)	(1)	(2)	(1)	(2)	(1)	(2)	(1)	(2)
Albania	0	0	0	0	0	0	0	0	0	0	0	0	0	0
Austria	0	0	0	0	0	0	0	0	0	0	0	0	0	0
Belgium	0	0	61	4	0	0	0	0	0	0	0	0	34	2
Bulgaria	0	0	0	0	0	0	0	0	0	0	0	0	0	0
Czechoslovakia	0	0	28	3	0	0	0	13	0	0	0	0	69	15
Denmark	0	0	30	1	0	0	72	3	0	0	83	17	0	0
Finland	0	0	0	0	0	0	0	0	0	0	0	0	0	0
France	0	0	32	3	0	0	0	0	0	0	0	0	23	3
Fed. Rep. Germany	0	0	56	16	0	0	0	0	0	0	0	0	74	39
German Dem. Rep.	0	0	73	14	0	0	73	62	0	0	73	63	82	38
Greece	0	0	75	3	0	0	0	0	0	0	0	0	0	0
Hungary	0	0	51	2	0	0	0	0	0	0	0	0	0	0
Ireland	0	0	81	2	0	0	0	0	0	0	0	0	0	0
Italy	0	0	0	0	0	0	0	0	0	0	0	0	0	0
Luxembourg	0	0	0	0	0	0	0	0	0	0	0	0	0	0
Netherlands	0	0	67	3	0	0	0	0	0	0	0	0	27	1
Norway	0	0	0	0	0	0	45	4	0	0	0	0	0	0
Poland	0	0	24	3	0	0	6	3	0	0	24	11	12	2
Portugal	0	0	0	0	0	0	0	0	0	0	0	0	0	0
Romania	0	0	0	0	0	0	0	0	0	0	0	0	0	0
Spain	0	0	23	2	0	0	0	0	0	0	0	0	0	0
Sweden	0	0	0	0	0	0	0	0	0	0	45	8	0	0
Switzerland	0	0	0	0	0	0	0	0	0	0	0	0	0	0
Turkey	0	0	0	0	0	0	0	0	0	0	0	0	0	0
United Kingdom	80	100	87	45	0	0	23	18	0	0	0	0	0	0
USSR	0	0	0	0	0	0	0	0	0	0	0	0	0	0
Yugoslavia	0	0	0	0	0	0	0	0	0	0	0	0	0	0
Total European Cost (10^6 DM a^{-1})	4000		16600		0		3800		0		3800		10500	

Because of relatively high marginal costs, only moderate emission reductions are required in the United Kingdom despite the fact that it is the largest national contributor to deposition in southwestern Norway (Table 1).

A similar situation exists for the receptor point in southern Sweden. No emission reductions are required to reduce deposition to 1.5 g S m^{-2} a^{-1}; however, substantial percentage reductions are required in Czechoslovakia, the German Democratic Republic, Poland and Sweden to reduce the deposition to 1.0 g S m^{-2} a^{-1}. In the latter case, almost two-thirds of the total European cost of 3.8 billion DM per year (the same expenditure as that required for the same target deposition at the receptor point on the Norwegian coast) must be spent in the German Democratic Republic because of the low marginal costs and the large potential for emission reductions there. No emission reductions are required in the United Kingdom, because of its weak meteorological linkages to southern Sweden.

Because of the very high unabated deposition flux at the receptor point in the southern German Democratic Republic, it was not possible to reduce deposition there to 1.0 or 1.5 g S m^{-2} a^{-1}, even with the application of the best available abatement technology everywhere in Europe. Instead, a deposition target of 3.0 g S m^{-2} a^{-1} was chosen. Although this target deposition was relatively high, emission reductions of greater than 25% are still required in six countries including of course the GDR itself. Emission reductions are required even in Belgium and the Netherlands despite the fact that they are not closely linked via the atmosphere to the GDR (Table 1). In fact, the RAINS model calculates that the minimum attainable deposition at that receptor point, even with the application of best available technology in all countries is about 2.3 g S m^{-2} a^{-1}.

It should be pointed out that the preceding optimizations were carried out at individual receptor points to illustrate the factors affecting the calculations. In actual use, it is more effective to develop strategies for an entire country or group of countries, because emission reductions in one country simultaneously reduce deposition in any country which is linked meteorologically to the source country. The RAINS model has been extensively used in this manner (Shaw, 1988; Shaw *et al.* 1988a and b; Amann, 1989; Shaw, 1989; Shaw *et al.* 1990) and is a key tool in developing new sulfur and nitrogen emission protocols under the United Nations Economic Commission for Europe Convention on Long-Range Transboundary Air Pollution.

4. Concluding Remarks

The IIASA Regional Acidification Information and Simulation (RAINS) model is a science-based tool capable of assessing and developing strategies for the control of acidifying pollutants in Europe. The model contains submodels for creating sulfur and nitrogen emission scenarios (including control strategies and costs), calculating atmospheric transport and deposition, and predicting environmental effects, namely acidification of forest soils and freshwater lakes, and risk of damage to forest vegetation from sulfur dioxide gas. The RAINS model also contains a routine for calculating the country-by-country emission reductions that would be needed to reduce deposition at specified locations in Europe to given target values.

The contribution of a given source country to deposition at a given receptor point depends upon its rate of emissions, and show strongly the source is linked via the atmosphere to that receptor. These source-receptor relations, in the form of a matrix, are calculated for meteorology averaged over several years by the long range atmospheric transport model of the Norwegian Meteorological Institute and are incorporated into the RAINS model. In the examples given in this paper, emissions from the United Kingdom are the largest contributor to deposition in the central U.K. and, apart from background deposition, the largest national contributor to deposition in southwestern Norway. They are not important for deposition in southern Sweden or the German Democratic Republic.

To achieve the most cost-effective emission reductions to reduce emissions at a given receptor point, the RAINS model uses not only the source-receptor relationships but information on the marginal costs of emission control in each European country, and the maximum allowable emission reduction. This information is based upon a thorough technical evaluation in each country. The choice of target receptor point, and the desired deposition target, have a significant influence upon the resulting strategy.

5. References

Alcamo, J., Amann, M., Hettelingh, J.-P., Holmberg, M., Hordijk, L., Kämäri, J., Kauppi, L., Kauppi, P., Kornai, G. and Mäkelä, A. (1987). Acidification in Europe: A Simulation Model for Evaluating Control Strategies. *Ambio*, 16(5):232–245.

Alcamo, J., Shaw, R.W., and Hordijk, L. (eds.) (1990). *The RAINS Model of Acidification: Science and Strategies in Europe*. Kluwer Publishers, Dordrecht (in press).

Amann, M. (1989). Using Critical Loads as the Basis for Abatement Strategies in Europe. Working paper submitted to the UN-ECE Task Force Meeting on Integrated Assessment Modelling, 27–29 November 1989, Geneva.

Amann, M. (1990). Energy Use, Emissions, and Abatement Costs. In Alcamo, J., Shaw, R.W., and Hordijk, L. (eds.) *The RAINS Model of Acidification: Science and Strategies in Europe*. Kluwer Publishers, Dordrecht (in press).

Amann, M., and Kornai, G. (1987). Cost Functions for controlling SO_2 Emissions in Europe. Working Paper WP-87-65. International Institute for Applied Systems Analysis, A-2361 Laxenburg, Austria.

Batterman, S.A. (1988). Influential Receptors in Targetted Emission Control Strategies. In: *Systems Analysis Modelling Simulation*. Journal of Mathematical Modelling and Simulation in Systems Analysis, 5(6):519–532.

Batterman, S., Amann, M., Hettelingh, J.-P., Hordijk, L., and Kornai, G. (1988). Optimal SO_2 Abatement Policies in Europe: Some Examples. In: *Systems Analysis Modelling Simulation*. Journal of Mathematical Modelling and Simulation in Systems Analysis, 5(6):533–559.

Buijsman, E., Maas, J.M., and Asman, W.H. (1987). Anthropogenic NH_3 Emissions in Europe. *Atmospheric Environment 21*:1009–1022.

Eliassen, A., and Saltbones, J. (1983). Modeling of Long-Range Transport of Sulphur over Europe: A Two-Year Model Run and some Model Experiments. *Atmospheric Environment 17* (8): 1457–1473.

Kämäri, J., Hettelingh, J.-P., Posch, M., and Holmberg, M. (1990). Regional Freshwater Acidification: Sensitivity and Long-Term Dynamics. In: Alcamo, J., Shaw, R.W., and Hordijk, L. (eds.) *The RAINS Model of Acidification: Science and Strategies in Europe*. Kluwer Publishers, Dordrecht (in press).

Kauppi, P., Posch, M., Kauppi, L., and Kämäri, J. (1990). Modeling Soil Acidification in Europe. In: Alcamo, J., Shaw, R.W., and Hordijk, L. (eds.) *The RAINS Model of Acidification: Science and Strategies in Europe*. Kluwer Publishers, Dordrecht (in press).

Lübkert, B. (1987). A Model for estimating Nitrogen Oxide Emissions in Europe. Working Paper WP-87-122. International Institute for Applied Systems Analysis, A-2361 Laxenburg, Austria.

Mäkelä, A., and Schöpp, W. (1990). Regional-Scale SO_2 Direct Forest Impact Calculations. In: Alcamo, J., Shaw, R.W., and Hordijk, L. (eds.) *The RAINS Model of Acidification: Science and Strategies in Europe*. Kluwer Publishers, Dordrecht (in press).

Nilsson, J., and Grennfelt, P. (eds.) (1988). *Critical Loads for Sulphur and Nitrogen*. Nordic Council Report NORD 1988:15.

Nilsson, S., and Duinker, P. (1987). The Extent of Forest Decline in Europe. *Environment 29*(9): 4–31.

Rosseland, B.O., Skogheim, O.K., and Sevalrud, I.H. (1986). Acid Deposition and Effects in Nordic Europe: Damage to Fish Populations in Scandinavia continue to Apace. *Water, Air and Soil Pollution 30*:65–74.

Shaw, R.W. (1988). Transboundary Acidification in Europe and the Benefits of International Cooperation. Presented at the Conference "Pollution knows no Frontiers: Priorities for Pan-European Cooperation" held in Varna, Bulgaria on 16–20 October 1988, organized by Professors World Peace Academy (PWPA), Athens, Greece.

Shaw, R. W. (1989). Using an Integrated Assessment Model for Decision-Making in Transboundary Air Pollution in Europe. In Bresser, L.J., and Mulder W.C. (eds.) *Man and His Ecosystem*, Vol. 2:177–182. Elsevier Science Publishers, Amsterdam.

Shaw, R.W., Amann, M. and Schöpp, W. (1988a). The Regional Acidification INformation and Simulation (RAINS) Model - A Tool to Develop Emission Strategies and to Assess their Effects. Presented at the European Conference "Combustion-Pollution-Reduction: New Techniques in Europe", 15–16 May 1988, Hamburg, organized by Verein Deutscher Ingenieure, Dusseldorf, FRG.

Shaw, R.W., Alcamo. J., Chadwick, M., and Hordijk, L. (1988b). Integrated Assessment Models for Developing Strategies for Reducing Regional Acidification. Proceedings of Air Pollution in Europe: Environmental Effects, Control Strategies and Policy Options. Stockholm, 26–30 September.

Shaw, R.W., Alcamo, J., and Hordijk, L. (1990). Strategy Development and Assessment Using RAINS. In: Alcamo, J., Shaw, R.W., and Hordijk, L. (eds.) *The RAINS Model of Acidification: Science and Strategies in Europe*. Kluwer Publishers, Dordrecht (in press).

"FORGET THE ENVIRONMENT — THE REAL BATTLE'S ABOUT JOBS, COAL AND POLITICS AS USUAL". CLEAN AIR LEGISLATION AND FLUE GAS DESULPHURISATION IN THE USA

David Gibbs, Department of Environmental and Geographical Studies, Manchester Polytechnic, Manchester, U.K.

Abstract

This chapter examines the context of clean air legislation in the United States, with particular reference to flue gas desulphurisation as a method of compliance. It provides detail of recently proposed changes to the 1970 Clean Air Act and the subsequent responses of environmentalists, utilities, legislators and industry. As the title of the paper suggests, the environmental issues behind the proposed legislation quickly became subsumed in arguments over its economic impact. In conclusion, parallels are drawn with the potential impact of the European Commission's Directive on Large Combustion Plant in the United Kingdom.

1. Introduction

This chapter arises out of a research programme being undertaken at Manchester Polytechnic into the economic and environmental impact of flue gas desulphurisation (FGD) in the United Kingdom. As part of this research a comparative study of the USA was undertaken[2], which was particularly concerned with those aspects of the US which may have relevance to the future development of emissions control in the UK. While the UK's FGD programme will not have any impact on emissions until 1993 at the earliest (and even this will only involve the retrofit of one power station), the US has a much longer history of FGD utilisation and the major utilities are closely involved in the development and implementation of a variety of systems to reduce emissions, from more advanced flue gas desulphurisation (or "scrubber") technology to clean coal technologies. The problems encountered by the US utilities may offer guidelines for the future development of both FGD technology and emissions control policy in the UK.

Moreover recent proposed amendments to the US Clean Air Act of 1970 have led to a number of possible compliance scenarios for the US utilities industry. As the title of this chapter suggests, the beneficial impact upon the environment has largely been lost in the arguments over the economic impacts. Similar debates are likely to arise in the UK in response to the European Commission's Large Plant Combustion Directive which will involve comparable discussions over waste disposal, the impact on the domestic coal industry and the choice of emission control technologies.

2. Emissions control in the USA

The 1970 Clean Air Act

In 1970, the US Congress passed a Clean Air Act which established a national programme for reductions of major by-products of fuel combustion. The law directed the Environmental Protection Agency (EPA) to establish National Ambient Air Quality Standards (NAAQS) to be met through state-set, federally-approved emission limits. In 1971, the EPA also set strict federal limits on emissions from new power plants. In 1977, Congress revised the law and required new coal-burning power plants to be equipped with stack gas scrubbers or other technological control systems to provide for even greater control of sulphur dioxide (SO_2) emissions (Edison Electric Institute, 1989).

The 1970 Act established two categories of NAAQS based on scientific criteria. There are both primary and secondary standards:

- primary standards are designed to protect even the most sensitive members of the population against any adverse health effect caused by air pollution. The primary standards adopted to date establish maximum allowable concentrations for seven air pollutants: sulphur dioxide, nitrogen oxides, particulate matter (soot and dust), carbon monoxide, hydrocarbons, lead and ozone. The standards must be strict enough to provide a margin of safety to protect the elderly and infirm.

- secondary standards are to protect non-health or "public welfare" concerns, such as soil, crops, forests, animals, buildings, statues, man-made material and aquatic systems from any "known or anticipated " adverse effects.

States are required to tell the EPA how they plan to achieve compliance with the federal air quality standards (NAAQS) through a State Implementation Plan (SIP). The plans outline the strategy a state will use in regulating polluting sources. Each plan establishes enforceable schedules for installing needed pollution control equipment and instituting other control measures. These plans must be approved by the EPA (CAWG, 1989; Vernon, 1989).

New Source Performance Standards (NSPS) set the limits for facilities built after 1971. They require that new coal-fired plants meet more stringent federal emission standards than plants that have already been constructed. In areas of the US that already meet national standards, the Prevention of Significant Deterioration provisions require that the quality of the air does not get appreciably more polluted. These standards allow increases in pollution concentrations only to a certain level. Once this level has been reached, any further emissions must be offset by reductions from existing sources. This programme also monitors compliance with the visibility section of the Act.

Under the Act, coal-fired power plants fall into one of three categories:

- new plants under construction since 1978. Under the NSPS, these plants are required to reduce overall SO_2 emissions by 70 to 90 per cent and meet a stringent sulphur dioxide emissions limit in the range of 0.6 to 1.2 pounds per million Btu (740 - 1480 mg/m^3) depending on the type of fuel burned. These plants are required by law to use stack gas scrubbers or other pollution control devices.

- plants under construction between 1972 and 1978. These plants are allowed a maximum emission rate of 1.2 pounds per million Btu (1480 mg/m^3) and can meet emission standards through various methods such as the use of low sulphur coal, coal washing or by adding control equipment.

- plants built before 1972. These older plants are not subject to the NSPS, but they come under the jurisdiction of the SIPs.

Impact of the 1970 Act

The immediate impact of the 1970 Act was to enforce FGD in a very short time scale (Kyte, 1989). Costs for the utilities have risen accordingly. The overall cost of all air pollution control in the UShas been estimated to exceed $35 billion each year (Edison Electric Institute, 1989), and air pollution costs can account for more than one-third of a new coal-fired power plant's cost. A single FGD system or scrubber may cost $100 million or more, and capital costs for adding scrubbers on older plants can equal or exceed the original plant investment. All this investment has obviously had an impact upon emissions. The EPA calculate that from the peak year of 1973, total SO_2 emissions were down nearly 21 per cent by 1987, while utility coal use rose by 88 per cent. In addition, from 1975 to 1987, the average sulphur content of coal purchased and then burned by the electric utility industry has decreased by approximately 37 per cent (Edison Electric Institute, 1989). The average sulphur content of coal used in coal-fired utility boilers in the US is around 1.5 per cent (Melia et al., 1988), although in certain regions the average sulphur content of coal utilised can be considerably higher. In Ohio, for example, average sulphur content of coal burnt in utility boilers is 3.5 per cent (OCDO, 1989).

The 1970 Act was introduced at a time when little attention was being directed to acid rain. The Act focused on SO_2 as a health hazard, whose effects were mostly visible in the immediate neighbourhood of the worst SO_2 offenders; plants burning high sulphur coal. The result of the Act was not altogether what environmentalists expected. Until the EPA ended the practice, the utility industry adopted a "tall stacks" policy, building tall chimneys to reduce local concentrations of sulphur and nitrogen oxides to meet EPA standards. This had the effect of "wafting the crud toward the Northeastern forests" (McLoughlin, 1989, 51). Indeed, despite the prohibition of a tall stacks policy by Congress and federal courts, 108 were erected after the Clean Air Act was passed in 1970. An anti-utilities view of this is that:

> "By building the stacks - typically 400 to 600 feet high - utilities have been able to continue burning cheap, high-sulphur coal, avoid the expense of sulphur removal devices and still met limits on sulphur dioxide near the plant. The power companies were able to do this because they pushed the Environmental Protection Agency until they got what they wanted, blanketing the government with lobbyists, exploiting ambiguities in the law, ignoring court rulings and manipulating public opinion" (Cincinnati Enquirer, 25 June 1989).

In time the introduction of the NSPS led to the adoption of the simplest throwaway lime FGD system, and the resultant sludge was disposed of in ponds, disused mines and behind dams (Kyte, 1989). Despite the impact of the 1970 Act, the electric utility industry remains the largest source of SO_2 emissions in the USA, and coal-fired units continue to be responsible for about 90 per cent of emissions by the industry (Keener and Keener, 1987). Acid-forming emissions are not evenly spread over the USA. Ten states in the central and upper Midwest - Missouri, Illinois, Indiana, Tennessee, Kentucky, Michigan, Ohio, Pennsylvania, New York and West

Virginia - produce 53 per cent of total US SO_2 and 30 per cent of total US NO_x (EPA, 1986).

3. FGD in the USA

FGD processes

The Clean Air Act relies heavily upon the use of technological standards and technology-implementing regulations to accomplish the objectives of protecting and enhancing US air quality (Radian Corporation, 1980). In general FGD or scrubber systems can be subdivided into the types of process, or in terms of the utilisation of the process end- and by-products. According to the type of process involved, FGD systems can be classified as wet or dry systems,. A wet FGD system uses an aqueous scrubbing solution to remove SO_2 from the gas stream. A dry FGD system uses either a dry solid sorbent to absorb SO_2 or transforms the spent material into a solid as an end product of the process. A throwaway or non-regenerable process disposes of the spent chemicals and by-products of the process without any regeneration. These waste materials usually require further treatment and/or proper disposal techniques. Conversely, regenerable processes are designed so that spent chemicals can be regenerated and reused and/or the by-products can be recovered as a saleable product (Keener and Keener, 1987).

FGD processes in use in the US

Table 1 summarises the status of FGD systems in the US in 1987. The operating FGD capacity in the US has grown significantly each year since 1972, although since 1977 the capacity under construction has been fairly stable. The planned capacity reported by the utilities increased each year until 1980, when it reached its peak and then dropped. Projections for future capacity estimate that the total power generating capacity of the US electric utility industry will be 751 GW by the end of 1997 (US Department of Energy, 1987). Approximately 335 GW or 44 per cent of the 1997 total, is estimated to be produced by coal-fired units. Of this 25.7 per cent of coal fired generating capacity in 1997 is expected to be controlled by FGD, compared to 20.9 per cent in 1987. Equivalent figures for total generating capacity are 9.1 per cent and 11.4 per cent (Melia et al., 1988).

Table 1

Number and Total Capacity of FGD Systems

Status	No. of Units	Total Controlled Capacity (MW)	Equivalent Scrubbed Capacity (MW)
Operational	151	65,520	60,643
Under construction	7	4,135	3,990
Planned	50	41,068	24,612
Total	**208**	**94,823**	**89,245**

Source: Melia et al. , 1988.

US FGD is dominated by throwaway-product process systems at present and it is estimated that these will continue to be the most important at the end of the century (see Table 2). In throwaway systems, sorbent is introduced after coal combustion into the flue gas. The by-product is captured along with the fly ash in the particulate removal equipment. Throwaway lime and limestone scrubbers have been installed on over 150 coal-fired utility boilers throughout the US, and these systems dominate the utility market. (Dalton and Syrett, 1989; McIlvaine, 1986). Indeed, the NSPS introduced by the Clean Air Act mandated their use on utility boilers constructed since 1978. The dominance of limestone systems is largely due to the widespread availability of limestone within the USA and its relative cheapness, at around $15 per ton delivered (1989 prices). Lime is more expensive, at around $60 per ton delivered, but tends to be used in areas where it has a high magnesium content, such as the Ohio Valley, which gives it good scrubbing qualities. These lime/limestone scrubbers have been developed to reliably remove over 90 per cent

Table 2.

Summary of US FGD Systems by Process

(Per cent of total MW)

Process	By product	1987	2000
Throwaway-product process			
Wet systems			
Lime		19.5	16.9
Limestone		48.4	48.0
Lime/alkaline fly ash		7.2	5.4
Limestone/alkaline fly ash		2.5	1.8
LIMB/dry injection		0.5	0.4
Dual alkali		3.4	2.6
Sodium carbonate		3.3	3.7
Process undecided		-	0.6
Dry systems			
Lime		8.5	11.1
Sodium carbonate		0.7	0.5
Process undecided		-	2.3
Dry injection			
Sodium bicarbonate		-	0.4
Trona/dry injection		-	0.6
Saleable-product process			
Lime	Metals/fly ash	<1	<1
Limestone	Gypsum	1.3	2.3
Magnesium oxide	Sulphuric acid	1.5	1.1
Wellman Lord	Sulphuric acid	3.2	2.3
Total		**100.0**	**100.0**

Source: Melia et al., 1988.

of the SO_2 emitted from high sulphur coal-burning utility units. Most systems operate along similar lines; flue gas containing sulphur dioxide passes through a lime or limestone slurry in a large absorber vessel, which causes the sulphur dioxide to react with the calcium to form calcium sulphite or calcium sulphate (OCDO, 1989).

Such FGD systems are expensive:

> "For utility application, the FGD system is the second most expensive capital cost exceeded only by the boiler itself while the maintenance cost for the FGD systems may range as high as 20 times the maintenance cost for the rest of the plant" (Keener and Keener, 1987, 1).

These costs may be compounded by the regulatory requirement to have two scrubbers if a utility needs or wants to bypass the FGD unit. This adds considerably to utilities' costs and the legislated spare capacity was reckoned by utility representatives interviewed in Ohio to be a significant cost burden.

Waste disposal from FGD

Costs are also increased by the necessity of waste disposal. Waste by-products from conventional scrubbers can be substantial, thus for every tonne of SO_2 removed these conventional systems generate approximately twice as much waste by-product. Moreover, the material is difficult to handle; most systems produce a calcium sulphite ($CaSO_3$) by-product that, when mixed with fly ash, has "a consistency and texture much like watered-down toothpaste" (OCDO, 1989, 18). This material may allow the leaching of metals when placed in landfill disposal sites. In consequence, some installations process the calcium sulphite sludge with forced oxidation to produce a drier, more stable, calcium sulphate ($CaSO_4$) by-product, which is then mixed with dry fly ash prior to disposal. Most disposal is to landfill which necessitates large areas for disposal. Clay lining of landfill sites may also be necessary.

Summary

In terms of market share the majority of systems in the USA at present are limestone natural oxidation, followed by natural oxidation lime scrubbing. There has recently been a revival of interest in forced oxidation (FO) lime/limestone scrubbing in the US because of the longer-term liability and costs associated with land-fill. For instance, one utility visited in Ohio with a 40 acre landfill site had no plans to reclaim a 40 acre landscaped landfill site for future recreational use due to the possibility of future litigation. In addition there have been problems at some sites with leaching into ground water and it has proved difficult to fix the sludge (Kyte, 1987, 1989). Forced oxidation systems may also help to avoid the technical problems of scaling that occur with natural oxidation methods. The growth of interest in FO and concern over disposal is also related to the changing nature of the utilities industry in the USA. A representative of EPRI, when interviewed in June 1989, could only think of one plant in Ohio (a nuclear to coal-fired conversion) and two low sulphur units in Texas under construction. The major market for scrubbing is therefore not for new plant, but for retrofit at older plants in the Northeastern states of the USA. These are usually older units in more built-up areas, where space constraints on landfill are important.

However, to date FGD retrofit has only been undertaken at a handful of US installations. Interviews with a leading US environmental consultancy indicated that there have been major headaches associated with retrofit in the USA. These problems may be compounded by recently proposed legislation to update the 1970 Clean Air Act which will impose stricter emission controls upon utilities and may lead to FGD introduction at a larger number of locations.

4. The Bush proposals for clean air legislation

The proposals

After more than a decade of legislative stalemate, President Bush, the self-proclaimed "First Environmentalist", introduced proposals on June 12 1989 to improve US air quality. Under the Bush proposals, sulphur dioxide emissions were to be reduced by 10 million tons by 2000, representing a 50 per cent reduction on 1989 emissions. The proposals also planned to reduce nitrous oxide emissions by 2 million tons. Half of the reductions need to be accomplished by 1995, and the rest by 2000 under the original Bush proposals. Companies would be allowed to decide how to meet compliance levels; they could install FGD equipment, burn coal with a low sulphur content, encourage conservation or adopt new "clean coal" technologies. Those opting for the latter option were allowed until 2003 to comply with the regulations (New York Times, 13 June 1989). These proposals were accepted by the US Senate by 89-11 on April 3 1990, and are virtually certain to become law in some form.

The proposals were intended to allow companies to buy and sell the "right to pollute" and to let the market decide the cheapest way to contain emissions. Plants within the same state were to be allowed to trade pollution rights. Thus if they exceeded the required reductions, they could sell rights to emit extra pollution to other companies or transfer rights to other plants within the same company. The first phase of required cuts in pollution would have affected 107 power plants in 18 states which emit more than 2.5 pounds of sulphur for every million Btus they produce. The proposals failed to meet the demands of environmentalists who had called for a reduction of 12 million tons of sulphur dioxide, while the utility industry and its supporters contended that mandated controls are unnecessary because sulphur pollution has been declining in recent years and that clean coal technologies would assure future gains (New York Times, 10 June 1989).

Costs and benefits

The costs of compliance with the legislation were estimated at $30 million by Robert Crandall of the Brookings Institute (Wall Street Journal, 16 June 1989) and at $14-18 billion per annum by William Reilly of the EPA (New York Times, June 13 1989). The costs of compliance and the associated economic impact quickly became one of the major debates over the legislation. For example, Senator Robert Byrd (Democrat, West Virginia) was reported as saying that:

> "the President's proposal would "decimate" the companies that mine high-sulphur coal, which are largely in Eastern states like his, because it would make a switch to low-sulphur coal, which comes mainly from mines in the West, the easiest way to achieve the reductions" (New York Times, June 13, 1990).

Thus, while as already noted sulphur dioxide emissions have a specific geography so does acid rain politics. Thus the costs and benefits of control will not be internal to polluting states. Those states currently relying upon high sulphur coal for power generation could experience a two- or three-fold increase in electricity rates compared to states in the west. In Ohio, estimates of electricity rate increases from the initial Bush proposals were estimated to be between 15-25 per cent by American Electric Power, one of the largest utilities in that state (Cincinnati Enquirer, 10 June 1989). The highest increases under any scenario are envisaged to arise in West Virginia, Ohio, Wisconsin, Kentucky and Indiana (Wang, Ham and Byrne, 1988). Conversely, the Environmental Protection Agency (1989) argues that increases in electricity rates will be relatively small. They cite figures to suggest that their estimated extra national cost of $3-5 billion only represents some 2-3 per cent of the total national electric bill of $160 billion per annum. They go on to say that "for a typical family this represents about one dollar per month. The states hardest hit generally are those with the lowest current rates" (EPA, 1989, 2).

Similarly there are also regional differentials in the benefits from control. Massachusetts, Michigan, New Hampshire, New York, Wisconsin and Minnesota have already introduced state-level sulphur dioxide control legislation. In some cases the motivation can be as much economic as environmental. In Minnesota, for example, acid rain legislation in 1982 was intended to protect the state's lakes which are an important source of revenue from sport fishing and tourism (Wang, Ham and Byrne, 1988).

Following acceptance of the Bush proposals by Senate, Midwestern and Appalachian states were unhappy at having to pay for reductions in acid rain in Northeastern and Pacific states. "They fear that the price will be fewer jobs in power utilities, mines, chemicals plants, steel works and car factories" (Financial Times, 6 April 1990). Measures approved by the House of Representatives in May 1990 attempted to deal with this by approving the provision of $250m in additional unemployment and training assistance for workers who lose their jobs as a result of industry's efforts to reduce emissions (Financial Times, 25 May 1990). These provisions were narrowly rejected by Senate and may be the subject of Presidential veto. Moreover, proposals from the House Energy and Commerce Committee approved on April 6 1990 required smaller cuts by the most polluting plants than the Senate approved plans and would set twice as stringent standards for some coal burning Department of Energy plants in the Midwest. A lengthy Senate/House conference is expected to resolve the differences in the summer of 1990, with the approval of legislation in some form by the autumn.

Prior to the announcement of the proposed legislation, internal debate between Presidential advisors was concerned over whether the bulk of the sulphur dioxide reduction requirements should be imposed on the 20 most heavily polluting plants, all situated in the Midwest, or spread around all US plants (New York Times, 10 June 1989). Original proposals, and earlier ones from Senators George Mitchell (Democrat, Maine) and Robert Byrd, would have forced the retrofit of FGD equipment or scrubbers. The eventual Bush proposals were directed at 107 plants in 18 states and allowed them to take several routes to compliance.

5. The Debate over the Impact of the Legislation

The impact on the US economy

The proposed environmental impact of the legislation soon became secondary to the debate over the economic impacts of the policy. The proposed new legislation has come under heavy attack from the radical right (as well as from some environmentalists who feel that the measures do not go far enough). At a macro-economic scale, there was concern that the increased cost of electricity could increase costs for US manufacturers and thus "diminish the ability of some companies to compete globally" (New York Times, 14 June 1989). Conversely, there are those who argue that by introducing such measures the US is imposing a major burden upon utilities and, indirectly, upon industry generally. One estimate is that the passage of the bill would raise US pollution control costs to around 2.3 per cent of GNP, compared to less than 0.8 per cent for Western Europe (Brookes, 1990). The arguments against the introduction of such measures propose that an unacceptable burden is being placed on utilities and industry, together with the lack of hard scientific evidence to support environmentalists' claims that such measures are necessary, or will be effective. One estimate is that reducing sulphur dioxide levels by 50 per cent will cost $5-7 billion per annum (Brookes, 1990). Brookes goes on to cite the results of the report to Congress directed by James Mahoney which concluded that there is no widespread forest or crop damage from current levels of acid rain in the US. The utility industry itself has argued, citing the President's Council on Environmental Quality, that the SO_2 problem, from a public health impact standpoint, has largely been eliminated (Edison Electric Institute, 1989). The Clean Air Working Group, a consortium of industry representatives and trade associations, argues that the proposed clean air legislation "amounts to an emergency programme that would cause job dislocation, economic disruption and higher electricity rates for millions of Americans. Such a programme is unwarranted because current controls are working, despite dramatic increases in coal use" (CAWG, 1989).

A specific attempt to analyse the potential employment impact of legislation has been undertaken by Hahn and Steger (1990) for the Clean Air Working Group. This report concentrates upon the "jobs at risk" element of any economic impact. As well as the impact upon the utility industry, Hahn and Steger also expect the legislation to affect energy-intensive sectors of industry, such as plastics, steel, motor vehicles, chemicals and foundries. In total they estimate that 1,719 plants in 96 counties in 20 states will be affected by acid rain provisions. These plants employ 232,000 workers, representing around 4 per cent of the total employment in these states. Particular impacts they feel will be in states such as Ohio, Missouri, Illinois and Pennsylvania. However, Hahn and Steger's analysis is over-simplistic. It merely adds up the number of plants in industries termed energy-intensive, calculates total employment in these plants, and assumes that these jobs will somehow be "at risk" if legislation is introduced. Obviously, this is naive. Not all the jobs would vanish, but they do not place any figures on what proportion of the jobs at risk they expect to be displaced. In addition the potential job gains in pollution control sectors are explicitly excluded from the study.

Indeed, the legislation could have potential advantages for some sections of manufacturing industry, particularly those manufacturing pollution control equipment. Moreover, other substantial economic benefits have been proposed as a consequence of the legislation. For example, reduced health care costs were estimated by the American Lung Association to be around $16-40 billion annually. Even for the utilities themselves, investment in modern equipment could, it was proposed, have advantages of increasing efficiency and competitiveness. The fact that the legislation looks set to become law indicates the strong movement towards environmental issues in the USA. While the perceived impact of the measures differ depending upon the protagonists, it can be argued that the legislation will place the USA in the forefront of the environmental movement. Where one of the major world industrial powers leads, other nations may be forced to follow. It has also been proposed that the US needs to take a lead on this increasingly important global issue if it is to continue to lead the western alliance of nations, and such control measures would facilitate working with the USSR on environmental issues (Wall

Street Journal, 19 June 1989). In addition, the fact the the USA is willing to introduce such measures exposes the claims of industrialists elsewhere that such measures are impossible to contemplate. It has been claimed that "companies with experience of the new measures in the US will be in a position to respond quickly and flexibly as similar provisions are introduced elsewhere" (Financial Times, 6 April 1990).

The impact on pollution control technologies

The freedom of choice contained within the legislation also gave rise to debate over the impact upon different sections of the economy. While the legislation does give some incentive for the introduction of clean coal technologies, there was also concern amongst utilities that the marginally extended timescale for compliance (to 2003) would kill off developments in these technologies (Shepard, 1988). CAWG (1989) argue that "requirements in some of the proposed clean air legislation would force states to opt for higher-cost, outdated emission control technologies that could delay development of new technology". In addition, utilities complained that they had to make a decision on which route they would follow on compliance by 1995, and it was pointed out during the research that in order to comply with this 1995 deadline technologies would need to enter the marketplace by 1991-92.

It has been proposed that:

> "mandating technology cuts off the competition that selects the most efficient solutions. What appears to some legislative staffer as the most "cost-effective" technology may turn out to harm the environment it's supposed to protect. Mandating technology doesn't allow for failure, a flaw that may price the investment risk on a mandated technology out of the market. And mandating technology invites legislation that satisfies the special interests of constituents at the expense of rational engineering" (Wall Street Journal, June 19 1989).

However, initial reactions by utilities to the proposed legislation were not all negative. Most were keen on the flexibility in the proposals (Wall Street Journal, June 14, 1989), although the Edison Electric Institute insisted that the options for utilities must be as flexible as possible to encourage the adoption and continued development of clean coal technology (Edison Electric Institute, 1989). The argument would be countered by the Environmental Protection Agency which argues that the regulations do not mandate particular technologies, but that waiting for clean coal technology (CCT) is not the answer:

> "Since most cost-effective reductions will come about through fuel switching waiting for CCT would cause an unnecessary delay in achieving emission reductions. But, by having a market-based approach with incentives for technology and extensions for CCT, we are providing maximum encouragement for expeditious demonstration and deployment of these new technologies" (Environmental Protection Agency, 1989, 12).

Suppliers of pollution control technologies interviewed in the US expected that clean coal technologies would become the dominant means of emissions control in the future, but not for ten years or so. These are likely to be centred around some form of fluidised bed combustion. The general consensus amongst those interviewed during the course of the research study in the US was that the US utilities are conservative organisations and that they will comply with new regulations by using proven technologies. The prospect of taking clean coal technology from the bench scale and scaling it up to operational size would not usually be considered an option by most utilities it was thought, as the problems are too great. It was therefore believed that regulation is not likely to result in the immediate introduction of new technologies, even though the proposed legislation allows for this in its timescale. Thus unless legislation and regulation are written to force the utilities to use the most cost-effective technologies, then most utilities are likely to opt for wet scrubbing. This is an expensive option for utilities, but the cost will be passed on to the consumer. Indeed the utilities interviewed in Ohio planned to adopt wet throwaway systems as their response to new legislation. Different systems are only likely to be introduced if the resultant legislation insists upon site specific controls. Utilities will need to look at waste disposal options, the age of the plant (they are unlikely to retrofit on plant over 40 years old), size and location. One consultant to the industry pointed out that "most plants listed for clean-up are large base-load units. There's no way utilities are going to opt for new technologies and have a 1000MW hole blasted out their generating system if it doesn't work".

A study by DRI/McGraw Hill Electricity Service, an economic analysis firm, estimated that of the 1,300 coal-fired utility plants in the USA, about 800 would be affected. Approximately 450 would probably switch from high to low sulphur coal, while 133 would opt to retrofit scrubbers. Of the remaining 217 plants, DRI expected that most would utilise embryonic clean coal technologies (Wall Street Journal, June 14, 1989). The interviews conducted in the US, however, indicate that this may be a substantial overestimate of the market for clean coal technologies.

The impact of new legislation will also have an impact upon manufacturers of pollution control systems. This impact will obviously vary according to the methods of compliance by the utilities. A major shift to low sulphur coal would probably have the most impact as not only would high sulphur coal mining jobs be lost, but also the job gains from manufacturing scrubber equipment would not materialise. Conversely, a shift to retrofit technology would benefit manufacturers of this equipment. One study estimates that major growth in FGD equipment sales will occur if acid rain legislation is passed (McIlvaine, 1988). The study estimated that acid rain legislation would lead to a peak of sales in the US two years after legislation is passed, with a total of 40,000MW retrofitted. This may not necessarily benefit the US economy, as European and Japanese manufacturers of such equipment, with considerable expertise, could be expected to launch major sales campaigns in the USA. The dearth of new orders for generating capacity in the US in recent years may mean that prospective US FGD equipment suppliers are in short supply or non-existent. The lack of development in FGD systems in the US could mean that any large scale retrofit programme may benefit foreign firms if this has led to attrition in FGD process design and capability. Construction jobs may be generated in the USA, but the design and implementation of such systems (the value-added component) will be found abroad.

In the case of saleable by-product systems, distance from markets for saleable by-products may be important in the decision over the adoption of such systems. Despite an interest by utilities in the production of saleable by-products, there have been problems in following this strategy (Ellison and Luckevich, 1984; Rosenburg, 1986). Thus utilities have had problems in producing gypsum, for example, to the quality demanded by the US wallboard industry and the wallboard industry itself has, for various reasons, been less than enthusiastic about using by-product gypsum. Studies by the Battelle Corporation for Ohio Coal Development Office also indicate that the entire US market could be supplied from a small number of plants (Battelle, 1989a; 1989b). Few of those interviewed in the US saw these types of control system or regenerative systems such as the Wellman Lord process becoming important in the USA in the immediate future.

Overall therefore the response of different utilities will vary. The decision is made more difficult by the fact that utilities may not have to introduce controls at all units. The important question for the utilities is whether they decide to take one large unit and reduce emissions by, say, 90 per cent, or whether they reduce emissions at lower rates at more units, but at a lower capital cost. The important factor is obviously which of these options is cheaper? The preferred strategy of one Ohio utility would be "to retrofit as few locations as possible with the highest efficiency system available" (Forney, pers. com). Such choices may become more complicated with the introduction of "pollution rights" able to be bought and sold. Under the free market proposals:

> "A company would have the right to keep a particular plant above the legal limit for emissions if it agreed to offset those emissions by a requisite reduction in emissions from a plant already meeting the standard. These "pollution rights" could be bought and sold among companies, thus letting the marketplace decide the cheapest way to contain smokestack emissions." (New York Times, 14 June 1989).

It could be profitable for a utility to reduce its own emissions to a high degree, and then sell "pollution credits" to other utilities.

The impact on coal producers

In part utility strategies will depend on their coal sources. If at all possible utilities are likely to switch to low sulphur coal as an easy method of compliance. The decision on compliance by the utilities will thus have an impact upon US coal producers. If utilities decide to meet the new targets by burning low sulphur coal (less than 2 per cent sulphur content) , then this will boost sales for companies mining low sulphur coal and have a dramatic impact upon high sulphur coal mining companies. Again, there is a spatial dimension to this. Low sulphur coal reserves are predominantly located in the Western US, while high sulphur coal is mined mainly in Eastern states like West Virginia, Ohio and Kentucky. In Ohio, for example, the predominantly high sulphur coal industry has an annual output worth in excess of $1.0 billion and employs over 7,500 workers (OCDO, 1989). Conversely, if the utilities discover that installing scrubbers is a cheaper option, then the impact on the latter producers will be less severe. Obviously coal companies and their lobbyists will press utilities to install scrubber equipment. Much of the political bargaining still remains to be resolved. In Phase I of the Bush proposals to 1995, state governors will be allowed to override utility executives in a state by imposing a mechanism for curbing air pollution. It is thought that this is especially likely in states such as Ohio and West Virginia where large amounts of coal are both mined and burned. The direct conflict in such states is between preserving mining jobs and higher electricity rates.

6. Conclusion

This chapter has examined the form of emissions control in the USA to date and focused particularly upon the potential impact of proposed new legislation to counter acid emissions by utilities. The reasons for the emphasis on the latter is the increasing concern being shown by governments, environmentalists, industry and the general public towards the need for environmental protection. This chapter has tried to show that although the objectives of environmental policy may be relatively easy to formulate, the implementation of such policies are extremely problematic. In the US example, it is interesting to note that the proposed environmental and health benefits soon became lost in a debate over the potential economic impact upon both industry and the consumer. The debate is not only important for the wider issues of employment and competitiveness, but is also vital for the future development of the utility industry in the US. Thus if the legislation does effectively mandate FGD technology, then the future environmental benefits from clean coal technologies may be set back several years in their development. Conversely, legislation may encourage the development of new technology as firms seek to reduce costs and take advantage of the pollution trading rights.

To return to the parallells drawn with the UK in the introduction in this chapter, the EC Directive on Large Combustion Plants adopted in 1988 calls for a 20 per cent reduction in UK SO_2 emissions relative to 1980 levels by 1993, followed by a 40 per cent and 60 per cent reductions by 1998 and 2000 respectively. How the UK achieves these targets is likely to be the subject of lively debate. The same mix of factors as in the USA are present. These include the debates over low and high sulphur coal, which in the UK case is given added piquancy by the fact that low sulphur coal will need to be imported (possibly from South Africa). In addition there will also be concern over potential job losses in UK coal mining areas. The adoption of FGD systems may assist the domestic coal industry, but in turn could lead to the production of large amounts of gypsum waste which necessitates landfill disposal (Moyes, 1989).

Overall the instance of acid rain legislation and its impacts provides an interesting example of the problems and issues that surround environmental policy. Formulation of policy on environmental grounds may prove to be the easy part of the equation in future years. The more difficult problems are to ensure not only compliance with the legislation, but also to achieve the initial long-term environmental gains that are the driving force behind the legislative process.

Notes:

[1] The quote for the title of this chapter was a sub-head in an article entitled "Our Dirty Air" in US News, June 12 1989.

[2] The research programme is funded by the National Advisory Body on Higher Education and also involves Jim Longhurst and Bridget Heath of the Acid Rain Information Centre. I am extremely grateful to the following individuals who assisted with the research programme in the USA; Mary Ann Miller; Howard Stafford,Tim Keener (University of Cincinnati); John Funke, Mike Vandebuegger and Bernie Huff (Cincinnati Gas & Electric); Howard Johnson (Ohio Department of Development); Bill Spires (Ohio Environmental Protection Agency); Bernie Laseke, Mike Szabo and Yatendra Shah (PEI Associates Inc., Cincinnati); Chuck Dene and Stu Dalton (Electric Power Research Institute, Palo Alto); Paul Nolan (Babcock & Wilcox); Bill Megonnell and Robert Beck (Edison Electric Institute, Washington DC); Suzanne Clark (Clean Air Working Group, Washington DC) and David Forney (American Electric Power, Columbus). All errors and omissions remain the responsibility of the author.

7. References

Battelle Corporation (1989a) Baseline of Knowledge Concerning By-product Characteristics. Final Report on Technical Support for the Ohio Clean Coal Technology Program. Vol. I, Battelle, Columbus, Oh.

Battelle Corporation (1989b) Baseline of Knowledge Concerning Process Modification Opportunities, Research Needs, By-product Market Potential, and Regulatory Requirements, Final Report on Technical Support for the Ohio Clean Coal Technology Program, Vol. II, Battelle, Columbus, Oh.

Brookes, W. (1990) "Fresh look at the Clean Air Act", ***Financial Times***, 17 May, 23.

Clean Air Working Group (1989) Clean Air Working Group Briefing Papers, CAWG, Washington DC.

Dalton, S. M. and Syrett, B.C. (1989) "Flue gas desulphurisation progress in the USA" in Desulphurisation in Coal Combustion Systems, Institution of Chemical Engineers Symposium Series No. 106, Hemisphere, London, 161-173.

Edison Electric Institute (1989a) Why Clean Coal Technology?, Edison Electric Institute, Washington DC.

Edison Electric Institute (1989b) Acid Rain and the Clean Air Act, Edison Electric Institute, Washington DC.

Edison Electric Institute (1989c) The Clean Air Act and the Electric Utility Plants, Edison Electric Institute, Washington DC.

Ellison W. and Luckevich, L.M. (1984) "FGD waste: long-term liability or short-term asset?", ***Power***, June, 79-83.

Environmental Protection Agency (1986) Acid Rain , OPA-86-009, EPA, Washington DC.

Environmental Protection Agency (1989) The Acid Rain Amendment, EPA, Washington DC.

Hahn, R. and Steger, W.A. (1990) An Analysis of Jobs-at-Risk and Job Losses Resulting From the Proposed Clean Air Act Amendments, CONSAD Research Corporation, Pittsburgh, Pa.

Keener, T.C. and Keener, S.U. (1987) Current Status of Flue Gas Desulphurisation in the United States, Paper presented to National Conference on Environmental Engineering, July 8-10, Cincinnati, Oh.

Kyte, W.S. (1987) Technical possibilities for achieving target reductions in air pollution levels using clean technologies, Paper presented at the International Pollution Abatement Conference, 6-8 April, Birmingham.

Kyte, W.S. (1989) "Technologies for the removal of sulphur dioxide from coal combustion" in Desulphurisation in Coal Combustion Systems, Institution of Chemical Engineers Symposium Series No. 106, Hemisphere, London, 15-28.

Mcllvaine, R.W. (1986) "The present and future status of FGD in the United States", EPRI Report, Proceedings: Ninth Symposium on FGD, Session 2: Commercial Status of FGD, CS-4390, 1-16.

Mcllvaine, R.W. (1988) "The air pollution control market in the 1990s", ***JAPCA***, 38(3), 248-251.

McLoughlin, M. (1989) "Our dirty air", ***US News***, June 12, 48-54.

Melia,M.T., McKibben, R.S., Hance, S.L. and Jones, F.M. (1988) Project Summary: Utility FGD Survey, January-December 1987, US Department of Energy, Washington DC.

Moyes, A. (1989) "Flue gas desulphurisation: an environmental backhander", ***Geography***, 74(2), 169-170.

Ohio Coal Development Office (1989) Ohio Coal Development Agenda, Ohio Department of Development, Columbus, Oh.

Radian Corporation (1980) National Commission on Air Quality Study of Air Pollution Control Technology, Research and Development: Public and Private Roles in Undertaking and Stimulating Innovation: Survey of Eight Air Pollution Control Innovations, NCQ 15-AQ-7421, Radian Corporation, Austin, Tx.

Rosenburg, H.S. (1986) "Byproduct gypsum from flue gas desulphurisation processes", ***Industrial and Engineering Chemistry Product Research and Development***, 25(2), 348-355.

Shepard, M. (1988) "Clean coal technologies for a new age", ***EPRI Journal***, Jan/Feb, 6-17.

United States Department of Energy (1987) Inventory of Power Plants in the United States, 1986. Energy Information Administration, Office of Coal, Nuclear, Electric, and Alternative Fuels. DOE/EIA-0095 (86), Washington DC.

Vernon, J. "Regulatory control of SO_2 emissions: current status and future trends" in Desulphurisation in Coal Combustion Systems, Institution of Chemical Engineers Symposium Series No. 106, Hemisphere, London, 141-152.

Wang, Y., Ham, K. and Byrne, J. (1988) "The political geography of acid rain: the US case", ***Regional Studies Association Newsletter***, 157, 3-6

MIX
Papier aus verantwortungsvollen Quellen
Paper from responsible sources
FSC® C105338

If you have any concerns about our products,
you can contact us on
ProductSafety@springernature.com

In case Publisher is established outside the EU,
the EU authorized representative is:
Springer Nature Customer Service Center GmbH
Europaplatz 3, 69115 Heidelberg, Germany

Printed by Libri Plureos GmbH
in Hamburg, Germany